Agamemnon Despopoulos · Stefan Silbernagl

Color Atlas of Physiology

4th edition, revised and enlarged

156 color plates
by Wolf-Rüdiger Gay and Astried Rothenburger

Translations by Joy Wieser

D0360734

1991
Georg Thieme Verlag Stuttgart · New York
Thieme Medical Publishers, Inc., New York

Professor Dr. Agamemnon Despopoulos
formerly: Ciba Geigy, CH-4004 Basel

Professor Dr. Stefan Silbernagl
Head, Department of Physiology,
University of Würzburg
Röntgenring 9, D-8700 Würzburg, Germany

Wolf-Rüdiger Gay and
Astried Rothenburger
Stuttgart, Germany

Joy Wieser, Rum/Tirol, Austria

CIP-Titelaufnahme der Deutschen Bibliothek

Despopoulos, Agamemnon:
Color atlas of physiology / Agamemnon Despopoulos ; Stefan Silbernagl. 154 color plates by Wolf-Rüdiger Gay and Astried Rothenburger. Transl. by Joy Wieser. – 4th rev. and enl. ed. – Stuttgart ; New York : Thieme ; New York : Thieme Medical Publ., 1991
Dt. Ausg. u. d. T.: Silbernagl, Stefan: Taschenatlas der Physiologie
NE: Silbernagl, Stefan:

1st German edition 1979
2nd German edition 1983
3rd German edition 1988
4th German edition 1991

1st English edition 1981
2nd English edition 1984
3rd English edition 1986

1st Czechoslovakian edition 1985
1st Dutch edition 1981
1st French edition 1985
1st Greek edition 1989
1st Italian edition 1981
1st Japanese edition 1982
1st Spanish edition 1982
2nd Spanish edition 1985
1st Turkish edition 1985

This book is an authorized translation from the 4th German edition, published and copyrighted 1991 by Georg Thieme Verlag, Stuttgart, Germany. Title of the German edition: Taschenatlas der Physiologie.

Some of the product names, patents and registered designs referred to in this book are in fact registered trademarks or proprietary names even though specific reference to this fact is not always made in the text. Therefore, the appearance of a name without designation as proprietary is not to be construed as a representation by the publisher that it is in the public domain.

This book, including all parts thereof, is legally protected by copyright. Any use, exploitation or commercialization outside the narrow limits set by copyright legislation, without the publisher's consent, is illegal and liable to prosecution. This applies in particular to photostat reproduction, copying, mimeographing or duplication of any kind, translating, preparation of microfilms, and electronic data processing and storage.

© 1981, 1991 Georg Thieme Verlag, Rüdigerstraße 14, D-7000 Stuttgart 30, Germany
Thieme Medical Publishers, Inc., 381 Park Avenue South, New York, N.Y. 10016

Typesetting by Tutte, Druckerei GmbH, Salzweg-Passau, Germany
Printed in Germany by R. Oldenbourg, München

ISBN 3-13-545004-X (GTV, Stuttgart)
ISBN 0-86577-382-3 (TMP, New York)

1 2 3 4 5 6

Preface to the Fourth Edition

The overwhelming response this book is still receiving has kept motivating me to revise it again. To its attentive readers and to my colleagues, especially Prof. Dr. *Theodor Heinrich Schiebler*, Prof. Dr. *Joachim Lutz* and Mr. cand. med. *Bruno Wagner*, I am particularly indebted for their considerate and helpful criticism. The highest praise is also due to Mr. *Rüdiger Gay* and Ms. *Astried Rothenburger* for their extraordinarily reliable cooperation on the occasion of designing several new color tables, and to Mrs. *Joy Wieser* for her highly competent and most careful translation of the revised text. I should like to extend my sincere thanks to the publishers for their continued generosity and confidence. I am indebted to Ms. *Angelika Reuß* and Ms. *Barbara Zahn* for their help with the typescript.

It is my hope that this fourth edition of the *Color Atlas of Physiology*, too, will help students of the subject gain a thorough understanding of the functions of the body, and that it will provide physicians and scientists with new information as well as a handy means of reference.

Würzburg, December 1990 *Stefan Silbernagl*

Preface to the First Edition

In the modern world, visual pathways have outdistanced other avenues for informational input. This book takes advantage of the economy of visual representation to indicate the simultaneity and multiplicity of physiological phenomena. Although some subjects lend themselves more readily than others to this treatment, inclusive rather than selective coverage of the key elements of physiology has been attempted.

Clearly, this book of little more than 300 pages, only half of which are textual, cannot be considered as a primary source for the serious student of physiology. Nevertheless, it does contain most of the basic principles and facts taught in a medical school introductory course. Each unit of text and illustration can serve initially as an overview for introduction to the subject and subsequently as a concise review of the material. The contents are as current as the publishing art permits and include both classical information for the beginning students as well as recent details and trends for the advanced student.

A book of this nature is inevitably derivative, but many of the representations are new and, we hope, innovative. A number of people have contributed directly and indirectly to the completion of this volume, but none more than *Sarah Jones*, who gave much more then editorial assistance. Acknowledgement of helpful criticism and advice is due also to Drs. *R. Greger, A. Ratner, J. Weiss*, and *S. Wood*, and Prof. *H. Seller*. We are grateful to *Joy Wieser* for her help in checking the proofs. *Wolf-Rüdiger* and *Barbara Gay* are especially recognized, not only for their art work, but for their conceptual contributions as well. The publishers, Georg Thieme Verlag and Deutscher Taschenbuch Verlag, contributed valuable assistance based on extensive experience; an author could wish for no better relationship. Finally, special recognition to Dr. *Walter Kumpmann* for inspiring the project and for his unquestioning confidence in the authors.

Basel and Innsbruck, Summer 1979

Agamemnon Despopoulos
Stefan Silbernagl

From the Preface to the Third Edition

The first German edition of this book was already in press when, on November 2nd, 1979, *Agamemnon Despopoulos* and his wife, *Sarah Jones-Despopoulos* put to sea from Bizerta, Tunisia. Their intention was to cross the Atlantic in their sailing boat. This was the last that was ever heard of them and we have had to abandon all hope of seeing them again.

Dr. Agamemnon Despopoulos

Born 1924 in New York; Professor of Physiology at the University of New Mexico. Albuquerque, USA, until 1971; thereafter scientific adviser to CIBA-GEIGY, Basel.

Without the creative enthusiasm of Agamemnon Despopoulos, it is doubtful whether this book would have been possible; without his personal support it has not been easy to continue with the project. Whilst keeping in mind our original aims, I have completely revised the book, incorporating the latest advances in the field of physiology as well as the welcome suggestions provided by readers of the earlier editions, to whom I extend my thanks for their active interest.

Würzburg, Fall 1985 *Stefan Silbernagl*

Table of Contents

"... If we break up a living organism by isolating its different parts it is only for the sake of ease in analysis and by no means in order to conceive them separately. Indeed when we wish to ascribe to a physiological quality its value and true significance we must always refer it to this whole and draw our final conclusions only in relation to its effects in the whole."
Claude Bernard (1865)

The Body: an Open System with an Internal Milieu

Life in its simplest form is best illustrated by the example of a unicellular organism. In order to survive, even such a simple organism has to meet two seemingly conflicting requirements. On the one hand, it has to protect itself from the disorder of its inanimate surroundings; on the other hand, as an **open system** (→ p. 19) it is dependent on the exchange of heat, oxygen, nutrients, waste substances, and information with its environment.

Protection is mainly the job of the **cell membrane**, whose hydrophobic properties prevent the lethal mixing of the hydrophilic constituents of the watery solutions inside and those outside the cell. The required permeability of the membrane barrier to certain substances is achieved by protein molecules in the cell membrane, either in the form of **pores** or of other transport proteins, the so-called **carriers** (→ p. 10f.). For gases, however, the permeability of the membrane is relatively good. Although this is advantageous for the essential exchange of O_2 and CO_2, it also means that the cell is at the mercy of toxic gases such as carbon monoxide. The presence in the external medium of higher concentrations of such gases or other lipophilic agents (e.g. organic solvents), represents a threat to the survival of the cell.

For the reception of signals from the environment the cell membrane contains certain proteins known as **receptors**, which convey the information into the cell. Only lipophilic signal substances can pass through the membrane without this process of mediation, and they meet up with their receptor proteins inside the cell.

A unicellular organism, if thought of in the environment constituted by the primeval ocean (→ **A**) exists in a more or less **constant milieu**. The extraction of nutrients and the deposition of useless waste matter cause no appreciable alteration in the composition of the cell's environment. Nevertheless, even this unicellular organism is capable of a motor reaction to signals from its environment (e.g. to changes in concentration of nutrients). It moves about with the help of pseudopodia or flagella.

Evolution from the unicellular to the multicellular animal, the specialization of cell groups to form organs, the development of heterosexuality and of a social way of life, as well as the transition from water to land, brought a vast increase in efficiency, chances of survival, radius of activity, and independence of the organism. This could only be achieved by the simultaneous development of a complex infrastructure within the organism. The individual cells of the body still require the milieu of the primeval ocean in order to survive and to function. The constant milieu is now provided by the **extracellular fluid** (→ **A**), but its volume is no longer infinitely great. Due to their metabolic activity, the cells would soon exhaust the O_2 and nutrient contents of this fluid; the extracellular space would be flooded with waste products if the organism had not developed organs that are specialized, among other things, for the uptake, processing, metabolism, and storage of nutrients, for the uptake of O_2, and for the release of metabolic products. Teeth, salivary glands, esophagus, stomach, intestines, liver, lungs, kidneys, and the urinary bladder are all involved.

The specialization of cells and organs for particular functions obviously requires **integration**. This is achieved by convective transport over longer distances, by humoral transfer of information via the circulatory system, and by the relay of electrical signals in the nervous system, among other means. In addition to carrying "provisions" and

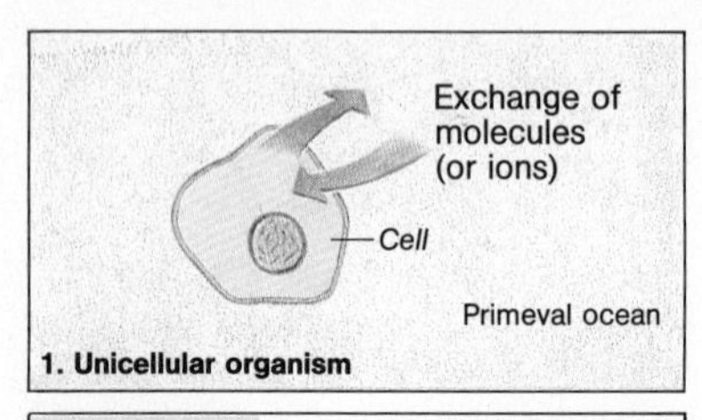

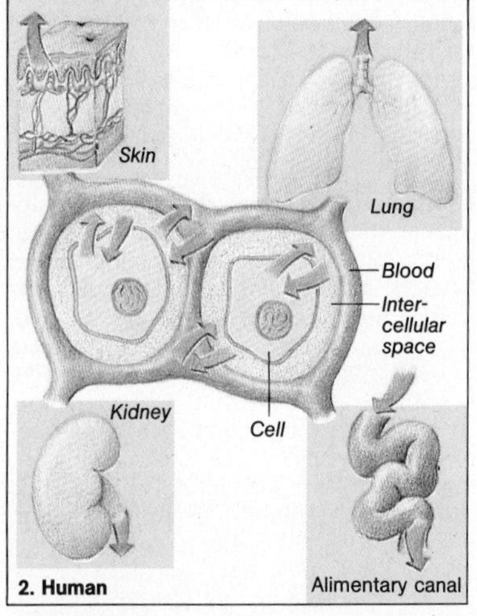

A. The milieu in which the cell lives
(**1**) The first cell arose in the primeval ocean. The unicellular organism exchanged material with the infinitely large ocean, without causing any appreciable change in its composition (**2**) The cells of the human body are bathed in the extracellular fluid (ECF), whose volume is smaller than the cellular volume (→ p. 138). This "internal milieu" would therefore very quickly alter if the intercellular space were not connected via the blood stream to organs and systems that take up fresh nutrients, electrolytes, and water, and excrete end products in stool and urine. The regulation of the "internal milieu" is chiefly carried out by the kidneys (H_2O and electrolytes) and by respiration (O_2 and CO_2). Important constituents of the ECF are continually being lost via the lungs (H_2O) and the skin (H_2O and electrolytes)

waste products, and thus contributing to the maintenance of a constant **internal milieu** even under extreme stress, the same mechanisms control and regulate functions that serve the purpose of survival in a broader sense, that is, **species survival**. This includes not only the timely development of the reproductive organs and the availability of mature germ cells at sexual maturity, but also the control of erection, ejaculation, fertilization and nidation, coordination of the functions of the maternal and fetal organism during pregnancy, and the regulation of the process of parturition and of the period of lactation.

The **central nervous system** (CNS) — which on the one hand processes signals from peripheral sensors, sensory cells, and sensory organs, and on the other hand activates effectors such as the skeletal muscles, as well as influencing the endocrine organs — assumes a focal position when human **behavior** is brought into the discussion. It is involved not only in the search for food and water, in seeking protection against cold and heat, in the choice of a partner, the care of offspring until long after their birth, and integration in social systems, but also in the origin, expression, and assimilation of what we associate with the terms desire or lack of desire, curiosity, happiness, anger, fear, and envy, as well as in creativity, self-discovery, and responsibility. However, this far exceeds the limits of physiology in the strictest sense, as the science of the functions of the body, which is the subject of this book. Ethology, sociology, and psychology are some of the disciplines bordering on physiology, although convincing links between these fields and physiology have only been established in exceptional cases.

The Cell

Cell Theory:

1. All free-living organisms are composed of cells and their products.
2. All cells are basically similar in their chemical construction.
3. New cells are formed from preexisting cells by cell division.

4. The activity of an organism is the sum of activities and interactions of its cells.

The cell is the smallest structural unit of living organisms. It is surrounded by a **cell membrane** within which are found the cell **cytoplasm**, or matrix, and the subcellular structures, the cell **organelles**, which are also surrounded by membranes.

Cells may be described as **prokaryotic** or **eukaryotic**. Prokaryotic cells, such as bacteria, have a minimum of internal organization and no membrane-bound intracellular organelles.

The organelles of the eukaryotic cell are highly specialized: the genetic material of the cell is concentrated in the nucleus, the "digestive" enzymes in the lysosomes; the oxidative adenosine triphosphate (ATP) production takes place in the mitochondria and the protein synthesis in the ribosomes.

Although many cells of the organism are specialized for different functions, their contents, the **cell organelles**, have much in common.

The **nucleus** contains a fluid known as the *karyolymph*, as well as the *chromatin network* and the *nucleolus*. Chromatin contains the *desoxy ribonucleic acids* (DNA) that are the carriers of the genetic information. Two strands of DNA (*double helix*, up to 7 cm in length) are twisted and folded to form the *chromosomes*, which are about 10 μm in length.

The normal human cell nucleus contains 46 chromosomes: two sets of 22 *autosomes* and 2 *X-chromosomes* in the female or 1 *X*- and 1 *Y-chromosome* in the male.

DNA is a long-chain molecule composed of four different nucleotides: adenosine, thymidine, guanosine, and cytidine. Its "backbone" is composed of the pentose sugar deoxyribose as well as phosphoric acid residues, with side chains consisting of the bases adenine, thymine, guanine, and cytosine, respectively. The sequence in which the bases are arranged constitutes the **genetic code**. The two DNA strands forming the double helix are held together by hydrogen bonds between base pairs, whereby adenine binds only with thymine, and guanine only with cytosine. The nucleotide composition of the two strands is therefore complementary, the structure of one determining that of the other, so that one strand can serve as template for the synthesis of a complementary strand containing identical information. RNA is a single strand containing ribose in place of deoxyribose and uracil in place of thymine.

The process of *protein synthesis* is fundamentally a transfer of information, initially encoded in the gene (DNA) as a polynucleotide, to form the final protein, a polyamino acid. It has been estimated that a typical cell synthesizes roughly 100000 different proteins during its life cycle.

The **nucleolus** contains *ribonucleic acid* (**RNA**), the so-called messenger RNA (**mRNA**). The mRNA conveys the genetic information that it has received from the DNA molecules (*transcription*) to the *ribosomes*, where the information is used for the process of protein synthesis (*translation*). mRNA and other large molecules traverse the two layers of the *nuclear membrane* through the *nuclear pores*. Transfer RNA (tRNA) is involved in the transfer of individual amino acids in protein synthesis, in which ribosomal RNA (rRNA) plays an essential role.

The first step in protein synthesis requires formation of RNA in the nucleus (**transcription**) according to the informational template on the gene (DNA). Every amino acid (e.g. lysine) of the protein is coded by three bases (in this example –C–T–T–). This is the *codogen*. During transcription, the complementary base triplet (–G–A–A–), the *codon*, is built into the *mRNA*. Formation of RNA is controlled by a **polymerase**, which is normally prevented from acting by a repressor protein on the DNA and is activated when the repressor is removed (*derepression*). This precursor of mRNA undergoes excision and rejoining of selected segments and modification of its terminals in the second intranuclear step, **posttranscriptional modification**. The mRNA next binds to polyribosomes in the cytoplasm and assembles amino acids (polymerization) carried to it by tRNA. The base triplets (**anticodons**) of this tRNA match corresponding bases in the codons of the mRNA (–C–U–U– in the above example). The rate of polymerization is approximately four to eight amino acids per second. This step, **translation**, ends in the formation of a polypeptide chain. The last step, **posttranslational modification**, involves cleavage of bonds within the new protein, modification of selected amino acids in the chain, for example γ-carboxylation of glutamate residues in certain clotting proteins (→ p. 74), and folding of the protein into its characteristic configuration. The final protein is then delivered to its site of action, for example, to the nucleus, to cytoplasmic organelles, or out of the cell into the blood.

The **rough endoplasmic reticulum RER** (→ **B** and **C**) consists of flat vesicles which are connected to form a *network of channels* throughout the cell. Proteins synthesized by the **ribosomes** on the surface of the endoplasmic reticulum (ER) are transported inside vesicles that have been split off from the RER (see below). The **ribosomes** are attached to the outside of the RER (thus *rough* ER; → **B** and **C**, RER) or are found free in the protoplasm. They contain transcripts (RNA) of the nuclear DNA. ER without ribosomes is called *smooth ER* and is chiefly engaged in the synthesis of lipids (e.g. for lipoproteins, → pp. 220–223). The **Golgi apparatus** (→ **B** and **C**) consists of stacks of flat vesicles, from which smaller vesicles can be constricted off. Its main functions are connected with secretion. For example, it takes over the proteins from the RER to synthesize glycoproteins and it produces polysaccharides; the substances are condensed and surrounded with a membrane. The **secretory granules** thus formed migrate to the cell boundary, fuse with it (→ **B**) and are extruded to the extracellular space (ECS): **exocytosis** (e.g. hormone secretion; → e.g. p. 240), which is an energy-dependent process. **Endocytosis** is the reverse process by which bulk material, either solid or in solution (pinocytosis), can enter the cell (→ p. 12).

The **mitochondria** (→ **B** and **C**) are the power station of the cell. They contain enzymes of the citric acid (Krebs) cycle and of the respiratory pathway. They are the principal site for oxidative reactions that generate energy. The energy thus produced is stored primarily in chemical form in the adenosine triphosphate (ATP) molecule. Synthesis of ATP provides almost all of the immediately accessible energy stores of the body; breakdown of ATP by various enzymes (phosphatases, ATPases) liberates energy for utilization in cellular reactions. The mitochondria also contain ribosomes and can synthesize certain proteins.

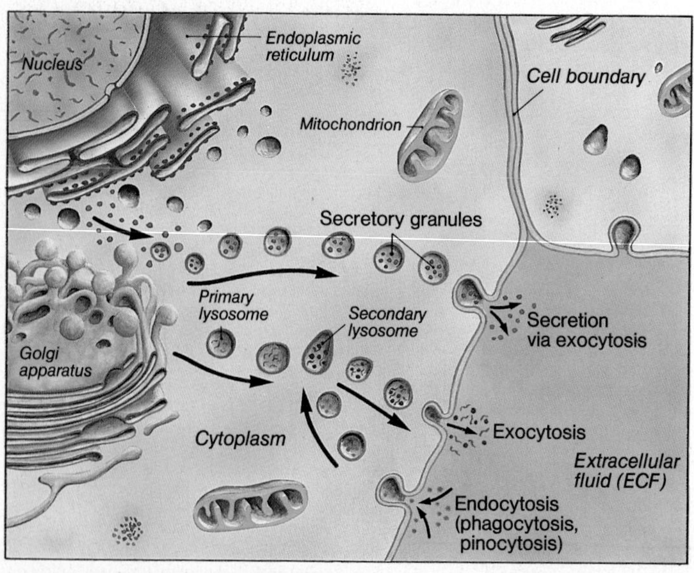

B. The cell. Endocytosis and exocytosis

Actively metabolizing cells, such as liver cells, or epithelial cells specialized for *transport* (→ **C**), are rich in mitochondria. The mitochondria are surrounded by a smooth outer membrane, inside which is another membrane deeply folded (*cristae*) to increase its surface area. The inner membrane, unlike the outer, is highly impermeable,

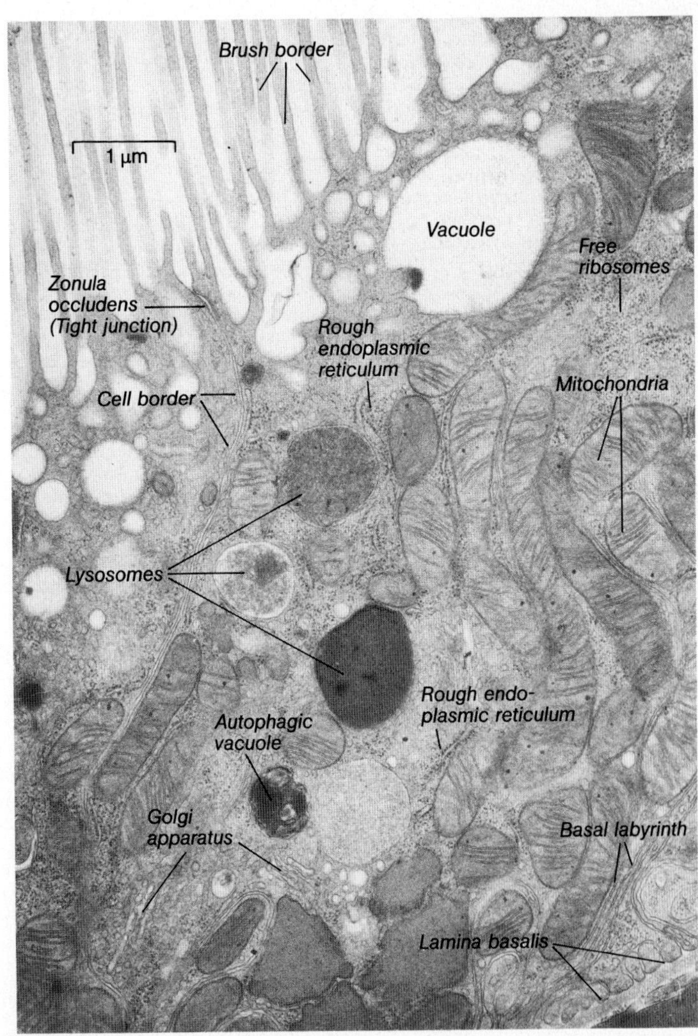

C. Cell Structure. Electron micrograph (enlargement 13000 ×), by courtesy of Prof. W. Pfaller, Innsbruck University

letting through only substances for which an active transport mechanism (→ p. 11) is present (malate, pyruvate, citrate, Ca^{2+}, phosphate, Mg^{2+}, etc.: → **H**).

Lyosomes are enzyme-containing vesicles, arising in most cases from the ER and the Golgi apparatus (*primary lysosomes*). They are also involved in protein transport and the *breakdown* of substances taken up in the cell by *phagocytosis* (→ p. 12) or by *pinocytosis* (→ e. g. p. 129, D) (*phagolysosomes and secondary lysosomes*). The breakdown of the cell's own organelles takes place in the lysosomes, too (*cytolysosome* or *autophagic vacuole*). The indigestible remnants are transported to the periphery of the cell, which they leave by *exocytosis* (→ **B**).

The *centrioles* are involved in cell division; *microtubules* are thought to play a role in cell rigidity but may also be important for other cellular functions.

The **cell membrane** may be either smooth or folded (e. g. *brush border; basal labyrinth* → **C**) and consists of phospholipids, cholesterol, and other lipids. The hydrophobic portions of the lipid molecules are arranged in a *double layer* facing one another, the hydrophilic portions jutting out into the watery surroundings. Proteins, many of them mobile (→ e. g. p. 242), are incorporated into the membrane, some extending through its entire thickness and serving as carriers or pores through which polar (hydrophilic) substances can pass (→ e. g. **F**). Other transient "holes" in the membrane itself are bounded by hydrophilic lipid components arranged face to face and are unlikely to be of any significance in the passage of polar substances.

Some of the functions of the cell membrane are the *protection* of the interior of the cell from its surroundings, *transport* (see below), the *recognition* of hormones (→ pp. 234 ff.), and the *adhesion* of cells to one another.

The electron micrograph in **C** shows cells of the proximal tubule of the rat kidney. The membrane surface of these specialized transport cells (→ pp. 126 ff.) is increased about 30- to 60-fold by invagination on the side towards the tubule lumen (*brush border*) and on the blood side (*basal labyrinth*). A number of mitochondria can be recognized (site of production of ATP, in this case mainly supplying the energy required for active transport processes); several *lysosomes, a cytolysosome, Golgi apparatus, RER* with ribosomes attached, *free ribosomes*, and the boundary between two cells. The cells are in relatively close contact at the *tight junctions*. The nuclei are outside the picture.

Transport: a Fundamental Process in Living Systems

As already described above, the interior of the cell is protected from the very different composition of the extracellular fluid (ECF) by the lipophilic cell membrane. Hence, with the help of metabolic energy it is possible for the kind of milieu essential for life and survival of the cell to be achieved and maintained. Pores, carriers, ion pumps, and the process of cytosis make possible the transmembranal transport of specific substances only, whether it be the import of substrates for cellular metabolism, the export of intermediary or end products of metabolism, or the vectorial transport of ions with which the cell potential is established and which also provides the basis for the excitability of nerve and muscle cells. The consequences of the influx and efflux of substances against which the cell membrane is not an effective barrier (e. g. for H_2O and CO_2), can be counteracted or at least mitigated by vectorial transport of other substances. Undesirable changes in cell volume and intracellular pH value, for example, can be offset by this kind of regulation.

Since the cell is divided up into distinct compartments (with respect to contents) by the diverse membranes of the cell organelles, there also exists a wealth of specific **intracellular transport systems**. Examples are the RNA export and hormone import of the cell nucleus, protein transport from the RER to the Golgi apparatus, active uptake and signal-dependent release of Ca^{2+} into and out of the RER, specific transport processes in the mitochondria (→ **H**), and axonal transport in the nerve fibers (→ p. 22) over distances of up to 1 m.

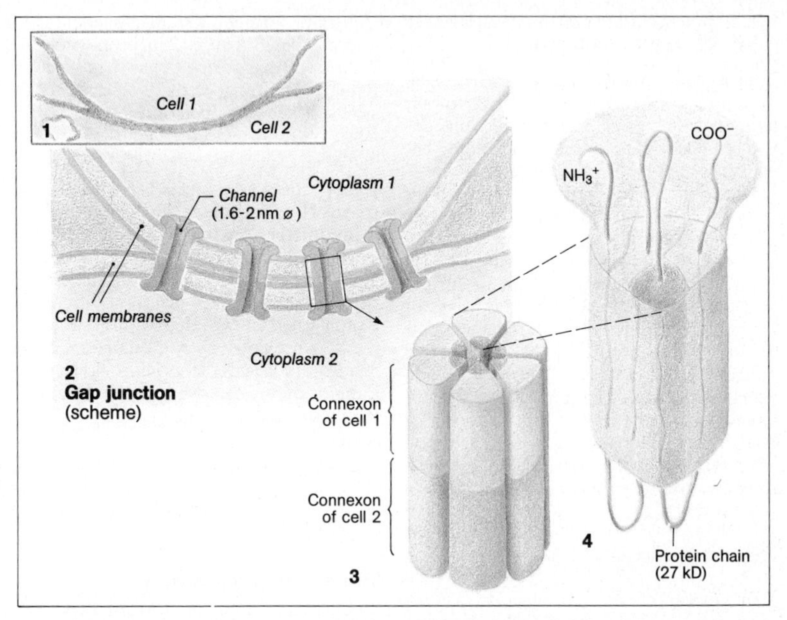

D. A **gap junction** contains communication channels between adjacent cells (e.g. in smooth muscle, epithelia, glia, liver) (**1**) Electron micrograph of two liver cells whose cell membranes are in contact inside a gap junction (**2**) Schema showing the channels (**3**) A channel protein complex (*connexon*) of cell 1 is connected end-to-end with a similar complex of cell 2 by a channel in such a way that the cytoplasmic spaces of the two cells are connected (**4**) A connexon consists of 6 subunits, each with a molecular mass of 27 kD. Both ends of the peptide chains face their own cytoplasm, while the two loops of the chain connect the connexons with one another at the other end. The part of the chain colored blue in (4) forms the channel wall. (After Evans W.H. BioEssays 1988; 8:3–6.)

In the multicellular organism, **transport** also takes place **between neighboring cells**, either by diffusion through the extracellular space (e.g. paracrine hormone action) or through channels – the **gap junctions** – directly connecting two adjoining cells (→ **D**). Gap junctions permit the passage of substances with a molecular mass of up to several hundred Dalton. Since ions can also use this route, the cells form an intimate functional association (**syncytium**), examples of which are epithelia (see below), smooth muscle, and the glia of the CNS. The electrical (ionic) coupling, for example, allows excitation of smooth muscle cells to spread to adjacent cells, thus setting off a wave of excitation throughout an entire organ (stomach, intestines, bile duct, uterus, ureter, and so on; see also p. 44). The presence of the gap junctions makes it possible for the work involved in the transport and barrier functions (see below) of glia and epithelia to be shared by the entire cell association. If, however, the Ca^{2+} concentration rises in one of the constituent cells, as for example in the extreme case of a "leaky" membrane, the gap junctions close down. In other words, in the interests of overall function, the individual cell is left to cope with its own problems.

Transcellular Transport. In multicellular organisms the function of the cell membrane as a barrier between "inside" and "outside" is often taken over by **cell associations**. The *epithelia* (e.g. skin, gastrointestinal tract, genitourinary tract, respiratory tract), the *endothelium* of the blood vessels, and the *glia* of the CNS constitute extensive barriers of this kind. They separate the general ECS from other spaces whose contents have a much different composition, such as air (skin and bronchial epithelia), the contents of the gastrointestinal tract, the spaces containing bile or urine (gall bladder and urinary bladder, kidney tubules), the aqueous humor in the eye, the cerebrospinal fluid (CSF) (blood-CSF barrier), the blood space (endothelia), and the ECS of the CNS (blood-brain barrier). Nevertheless, certain substances must be transported through these cell associations; this is effected by vectorial **transcellular transport**, in which import on one side of the cell is combined with export on the opposite side. Unlike other cells (e.g. blood cells) whose plasma membrane is endowed with uniform properties over their entire surface, epithelial and endothelial cells, as far as structure (e.g. **C**) and transport function are concerned, are *polar cells*.

Transport can take place not only transcellularly (across cells, as described above), but also between the cells: **paracellular transport**. Some epithelia (e.g. in the small intestine and in the proximal renal tubules) are, in this respect, relatively permeable or leaky, whereas others are less leaky (e.g. distal nephron, colon). This depends upon the state of the **tight junctions** (→ **C**) by means of which the cells adhere to one another. The paracellular pathway and its degree of permeability (it can, for example, also be cation-specific), are important functional elements of an epithelium. Large molecules can cross the endothelial barrier of the blood vessel walls by a combination of endocytosis on the one side and exocytosis on the other, a process called **transcytosis** (→ p. 13). However, in the case of transendothelial transport, paracellular transport through the intercellular space appears to play a quantitatively greater role. Anionic macromolecules (such as albumin) that are required in the bloodstream for the sake of their oncotic action (→ p. 158), are held back by electric charges on the epithelial cell walls.

Lastly, there is a need for **long-distance transport** from one organ to another and from different organs to the outside world. The main transport principle involved in such cases is **convection**. "Communication" between organs takes place via blood- and lymph-streams; exchange with the environment is provided by the airstream in the respiratory tract, the urinary flow in the kidneys and urinary tract, and by transport through the mouth-stomach-intestinal tract.

The following sections briefly describe the types, phenomena, and laws of transport mechanisms, and their functional significance for the organism.

Passive transport processes

One of the fundamental transport processes is **diffusion**. *Net* diffusion can take place if the substance concerned is present in a higher concentration at the *starting point* than at the *destination*, that is, if there is a *concentration difference*.

As an example, think of the way in which the scent of flowers on a table fills the entire room, even if the air is still. The substance responsible for the scent *diffuses* from the site where its concentration is higher (the flowers) to the surroundings where the concentration is lower.

Unidirectional diffusion can take place in the absence of a concentration difference. In this case the diffusion rate is the same in each direction, and net diffusion is therefore zero.

In the air (and in gases in general), the process of diffusion is relatively fast. It is slower in liquids and is particularly slow in the body tissues. How long a substance takes to diffuse depends on the *distance* involved as well as on the *surface area* available for diffusion (the larger the area, the faster the diffusion). The speed of diffusion is also dependent upon the *nature of the diffusing substance*. Oxygen, for example, diffuses through the alveolar membrane (alveoli, → p. 92) 20 times more slowly than CO_2. The reason for the difference, in this case, is the low solubility of O_2, as compared to CO_2, in the "watery" environment. Other factors influencing diffusion are described below.

Fick's First Law of Diffusion defines these relationships quantitatively as follows: the quantity, Q, of a substance diffusing per unit of time (t) at any moment (dQ/dt), is proportional to the *diffusion coefficient*, D, the area (e. g. of the cell membrane) available for exchange, A, and the concentration difference, ΔC, and is inversely proportional to the distance over which diffusion takes place, l:

$$\frac{dQ}{dt} = \frac{D \cdot A}{l} \cdot \Delta C [mol/s]. \quad (1)$$

Thus, the larger A, ΔC, and D, and the smaller l, the greater is the quantity of a substance diffusing/time. In nonideal solutions, *activities* instead of concentrations have to be inserted.

For diffusion through a biological **membrane**, D contains the gas constant R (→ p. 14), the absolute temperature (T), the radius of the diffusing molecule (r), the viscosity of the membrane (η), and the oil-water distribution coefficient (k), which is a measure of the lipid solubility of the diffusing molecule in the phospholipid membrane:

$$D = R \cdot T \cdot k/(6\pi \cdot \eta \cdot r) [m^2 \cdot s^{-1}]. \quad (2)$$

Since the thickness of the cell membrane can usually be regarded as constant, the **permeability coefficient** (P) is often used instead of D/l. Inserting into equ. (1) the net transport rate, dQ/dt, per area A can be calculated from

$$\frac{dQ}{A \cdot dt} = P \cdot \Delta C [mol \cdot m^{-2} \cdot s^{-1}]. \quad (3)$$

The quantity of substance diffusing at any moment, per area and time ($mol \cdot m^{-2} \cdot s^{-1}$), is thus proportional to the concentration difference (ΔC, here in mol/m^3 solvent) and the permeability coefficient (P, in m/s).

For the **diffusion of gases**, ΔC in equ. (1) is replaced by $\alpha \cdot \Delta P$ (where P is the partial pressure of the gas; → p. 98). Thereafter, the product of the two constants, $\alpha \cdot D$, is renamed **K** ($m^2 \cdot s^{-1} \cdot Pa^{-1}$). K is called **Krogh's diffusion coefficient**, or **diffusion conductivity**. Finally, volume V (m^3) is used instead of quantity Q. In this form, Fick's law of diffusion reads:

$$\frac{dV}{A \cdot dt} = K \cdot \Delta P/l \; [m \cdot s^{-1}].$$

Since **ions** carry an electric charge, membrane permeability of the ion "x" can be converted to the **electrical conductance** *per membrane area* (**g**) of this ion x (→ pp. 10 and 14):

$$g_x = P_x \cdot z_x^2 \cdot F^2 \cdot R^{-1} \cdot T^{-1} \cdot \bar{c}_x [S \cdot m^{-2}],$$

whereby z, F, R, and T have their usual meanings (→ above and p. 14) and $\bar{c}_x$ is the *"mean"* ion activity across the membrane (i = inside, o = outside).

$$\bar{c} = (c_o - c_i)/(\ln c_o - \ln c_i).$$

Note that, in contrast to P, g varies with varying concentrations of the ion, leading to a change of $\bar{c}$. If, for instance, the extracellular K^+ activity rises from 4 to 8 mmol/kg H_2O (intracellular activity of 160 mmol/kg H_2O may be unchanged), $\bar{c}_K$ increases from 42 to 51 mmol/kg H_2O, resulting in a 20% increase of g.

In **nonionic diffusion**, the uncharged form of a weak acid (e. g. uric acid) or base (NH_3) passes through a membrane more readily than the charged form. The cell membrane is, for example, much more permeable to NH_3 than to NH_4^+ (→ pp. 146f.). Since the pH of the solution determines whether or not such substances are charged (pK, → p. 334), the diffusion of weak acids and bases is clearly dependent upon the pH value.

For **transport over large distances** diffusion is too slow a process. For a unicellular organism like an ameba, which is very close to the outer environment, diffusion is adequate as a means of transporting, for instance, oxygen. A multicellular organism, on the other hand, requires additional transport mechanisms; the diffusion of oxygen from the body surface of a human being to the internal organs would take a number of months. In other words, at the normal O_2 consumption of about 0.3 l/min, a sufficiently high O_2 pressure would only be achieved to a depth of less than 0.3 mm from the skin surface. More deeply situated tissues would suffer from anoxia.

In order to cover the larger distances, the fluid or the gas that contains the substance to be transported moves at the same time (transport by **convection**). The circulation of the blood (→ p. 154 ff.), the ventilation of the lungs (→ p. 78 ff.), and the excretion of urine and feces all represent examples of this type of transport mechanism.

For example, in the transport of *carbon dioxide* in the body (→ p. 96), the processes of diffusion (D) and convection (C) alternate: D, from tissues into the blood; C, in the blood, from tissues to the lungs; D, from the blood into the alveolar space; C, from the alveolar space to the outside air.

Convection is also involved in the **transport of heat** in the blood and in the loss of heat in the form of warm air (→ p. 192 f.).

Filtration is yet another form of transport taking place through various membrane barriers in the organism, provided that the *membrane is permeable to water*. If a **pressure difference** builds up between the two sides of the barrier (e.g. a relatively higher pressure in the blood capillaries than in the interstitial space, → p. 158), the fluid is *pressed* through the membrane. Substances too large to pass through the pores of the membrane (e.g. proteins in the blood capillaries) are retained, while low molecular substances such as Na^+ and Cl^- are filtered through the membrane with their solute (also a form of convection; see above). Some low molecular substances which, on their own, would be filtered, bind to plasma proteins (**plasma-** or **protein-binding**) and thus, to a greater or lesser extent, avoid filtration at the blood capillaries, for example, at the glomerular filter of the kidney (→ p. 120 ff.).

This is illustrated in the following example: roughly 20 % of the plasma fluid flowing through the kidney, and consequently 20 % of a freely filtrable substance, are filtered in the renal glomerulus. If 9/10 of this substance is bound to plasma proteins, then only 1/10 is freely filtrable. This means that only 2 % instead of 20 % are filtered per passage through the kidney.

Functions of protein binding: (a) it protects many substances (e.g. heme) from excretion; (b) it provides a method of transport for certain substances such as iron ions; (c) it makes possible the "storage" of important plasma ions (such as Ca^{2+} and Mg^{2+}) in a reservoir from which they can immediately be mobilized when required. Plasma binding plays an important role in the administration of therapeutic agents, since the part that is protein-bound is neither pharmacologically active nor is it freely filtrable (delayed renal excretion), but it may have an allergenic effect (→ p. 72).

In its *passage across epithelia* (e.g. intestinal wall, renal tubule), *water* can drag dissolved substances along with it. This is termed the **solvent drag**. How much of the dissolved substance can be transported in this manner depends on its concentration, on the magnitude of the water flow, and on the ease with which the particles succeed in passing through the epithelium, or, rather, the proportion of particles that do not succeed in crossing the call wall and are "reflected." The latter value can be expressed as the **reflection coefficient σ**. For larger molecules that are totally reflected (i.e. are not transported by solvent drag), $\sigma = 1$; for smaller molecules, $\sigma < 1$. For urea in the proximal renal tubule, for instance, $\sigma = 0.68$.

In the case of *electrically charged particles (ions)*, a **potential difference** across a cell membrane can constitute a driving force (→ p. 14). The positively charged ions (cations) migrate to the negatively charged side of the membrane and the negatively charged ions (anions) to the positive side. This kind of transport can only take place if the membrane is permeable to the ion, which is indicated by the *permeability coefficient P* (→ p. 9).

At a given driving potential, the quantity of ions transported per unit of time and area (*ionic current*) is determined not only by P, but also by the charge (z) on the ion, and the "mean" ion activity ($\bar{c}$) on the two sides of the membrane (→ p. 9). Taking these factors into account, P can be converted in electrical terminology into *conductance* g (→ p. 9). If g is substituted for resistance in Ohm's Law (→ p. 14) then:

ionic current $= g_{ion} \cdot$ driving potential; (4)
(see also p. 14).

Facilitated diffusion is a passive form of transport across a membrane with the help of a *carrier* molecule. Since most biologically important substances are highly polar, their simple diffusion (→ p. 8) across the membrane would take too long. In the case of glucose, Na^+, and many other substances, this is overcome by their binding to carrier proteins built into the membrane itself, from which they then separate on the other side of the membrane. Whether the carrier itself diffuses through the membrane, rotates, or otherwise changes its conformation in the process is largely unknown. Transport of this type is *saturable* and *specific* for classes of substances closely related in structure which may compete for the carrier (**competitive inhibition**). Facilitated diffusion differs from active transport (see below) in that it can only take place *downhill*, that is, along an electrochemical gradient (→ p. 14).

Active Transport

At many locations in the organism, substances have to be transported against the sum of a concentration gradient, and an electrical gradient (potential), i. e. *against an electro-chemical gradient*.

This *cannot* be achieved by means of passive transport processes (see above), which proceed in the opposite direction (downhill), but requires the participation of **active transport mechanisms**. The latter require **energy** since they have to transport *uphill*. Much of the chemical energy supplied to the organism in the form of food is converted into energy-rich compounds such as ATP (→ p. 19f.) that can be utilized in a wide variety of energy-requiring cellular reactions, and which also provide the energy for active transport.

In **primary active transport**, ATP hydrolysis *directly* provides energy for the transport or "pump" mechanism. These pumps, therefore, are also termed ATPases. The ubiquitous **Na^+-K^+(activated)-ATPase**, the sarcoplasmic **Ca^{2+}-ATPase**, and the **H^+-ATPase** of the renal collecting duct are examples of such pumps, transporting Na^+ and K^+, Ca^{2+}, or H^+, respectively.

Secondary active transport means that the active transport of a compound (e. g. glucose) is coupled by a carrier to the passive transport of an ion (e. g. Na^+). In this case, the Na^+-gradient is the driving force that has to be maintained indirectly by primary active transport of Na^+ at a different site on the cell membrane. It is called **cotransport** (or **symport**) if the substance involved is actively transported in the same direction as the driving ion (e. g. Na^+ with glucose), or **countertransport** (**antiport**) if the Na^+-gradient, for instance, drives H^+ in the opposite direction.

Examples of primary and secondary active transport processes are the Na^+, glucose, and amino acid transport out of the renal tubule (→ p. 128); the uptake of these substances from the intestine (→ p. 224); the secretion of gastric acid (→ p. 208); Na^+ transport across the nerve membrane (→ p. 24); and so on.

Active transport mechanisms are characterized by the following properties:

- They are *saturable*, that is, above a maximum rate of transport they are unable to move additional loads (e. g. glucose reabsorption in the kidney, → p. 128).
- They are more or less *specific*, that is, only certain substances, as a rule chemically very similar, can be transported by a particular system, and may compete for it (e. g. excretory function of the liver, → p. 214 ff.).
- Although chemically similar, such substances are often transported to differing extents, that is, they have *different affinities* ($\sim 1/K_m$; see below) for the transport system.
- They are inhibited when the *energy supply* of the cell is disrupted.

The transport rate J_{sat} of a saturable transport process can in most cases be calculated by applying the Michaelis-Menten equation:

$$J_{sat} = J_{max} \cdot C/(K_m + C) [mol \cdot m^{-2} \cdot s^{-1}],$$

where C is the actual concentration of the substance to be transported, J_{max} is its maximum rate of transport, and K_m its concentration at half saturation, that is, at $0.5\ J_{max}$ (→ p. 333).

Cytosis

Cytosis is a completely different type of active transport. It comprises the formation, with the expenditure of ATP, of membrane-enclosed **vesicles**, roughly 50–400 nm in diameter, which are constricted off from the plasma membrane or from the membranes of the cell organelles (RER, Golgi apparatus, → p. 4). By specific cytosis, mainly macromolecules (proteins, polynucleotides, and polysaccharides) are taken up by the cell (**endocytosis**) or exported from it (**exocytosis**). The transport of large molecules inside the cell also takes place in this kind of vesicle (e. g. protein transport from the RER to the Golgi apparatus). Two different forms of endocytosis can be distinguished. One form, **pinocytosis**, is the continuous, nonspecific uptake of extracellular fluid via relatively small vesicles, by which means the cell is provided with "sips" of the external fluid. At the same time, large or small dissolved molecules are indiscriminately conveyed into the cell.

The second form of endocytosis requires the presence of specific **receptors** on the outside of the cell membrane. One and the same cell may possess many different types of receptors (about 50 on a fibroblast). As many as 1000 such receptors are often concentrated at sites on the membrane (so-called circulating proteins) where the inside of the membrane is lined (*coated pits*; → **E 2**) with special proteins (mainly **clathrin**). Since this is where endocytosis starts, the endocytotic vesicles are temporarily coated with clathrin (*coated vesicles*). This so-called **receptor-mediated endocytosis** (= adsorptive endocytosis) is **specific** since the receptors only recognize certain substances, and only these can be endocytized and taken up selectively by the cell.

Subsequent to endocytosis, the "coat" is removed intracellularly; this is often followed by fusion with **primary lysosomes** (→ **B**), whose hydrolytic enzymes "digest" the endocytized substance in the **secondary lysosomes** that resulted from the fusion. Small products of digestion, such as amino acids, sugars, and nucleotides, are transported across the lysosomal membrane into the cytoplasm, where they are available for cellular metabolism.

Such transport mechanisms of the lysosomal membrane are specific. If, for example, the one responsible for the amino acid L-cystine is defective (genetically), cystine (the solubility of which is low) accumulates and precipitates in the lysosomes, finally damaging the cells and the entire organ (*cystinosis*).

The phospholipids of the cell membrane taken up in the process of endocytosis to form the vesicular wall are reincorporated into the cell membrane, together with the receptor proteins and the clathrin, via largely unexplained **recirculation** processes. Certain other membrane proteins, such as the ion pumps (so-called noncirculating proteins) are for the most part not involved in this rapid, continuous process of recirculation.

An example of receptor-mediated endocytosis is the uptake of **cholesterol** and its esters. They are transported via the plasma in the low-density type of lipoprotein (**LDL**; → p. 222) to extrahepatic cells. When these cells require cholesterol (e. g. for membrane synthesis or for steroid hormone production), they incorporate into their cell membrane more LDL receptors that recognize and bind the LDL (apo-)proteins, which in turn sets off endocytosis of the LDL. A single 22 nm-sized LDL particle can take up about 1500 cholesterol ester molecules in this manner. Patients with a genetic defect in this receptor mechanism develop an elevated plasma cholesterol level, which results in premature arteriosclerosis.

Other examples of receptor-mediated endocytosis are the cellular uptake of transferrin-bound **iron**, hemopexin-bound **heme**, haptoglobin-bound **hemoglobin** (**Hb**) (→ p. 62), and **cobalamines** bound to their different transport proteins (→ p. 226).

Phagocytosis of pathogenic organisms (recognition of antigen-antibody complexes) and endogenous cell debris, for which the neutrophil granulocytes and the macrophages are specialized (→ p. 66 ff.), is also usually receptor-mediated (see also opsonization; p. 66 ff.). The quantitative aspects of phagocytosis are illustrated by the fact that about 10 billion senile erythrocytes undergo phagocytosis in only one hour (→ p. 60), and that macrophages, for example, phagocytize about 125 % of their own volume and 200 % of their cell membrane hourly.

When the hormone **insulin** binds to the receptors on the surface of its target cell, the hormone-receptor complex migrates into the coated pits and is endocytized (is "internalized"; → p. 248). In this way, the density of the receptors available for binding the hormone is thus lowered ("down regulation" of the receptors by high levels of hormone).

Exocytosis is also a regulated process. Hormones (e.g. in the posterior lobe of the pituitary; → p. 240), neurotransmitters (e.g. acetyl choline; → p. 54), and enzymes (e.g. in the pancreatic acini; → p. 212ff.) are available as "prepacked" vesicles, ready for release in response to a rise in the intracellular Ca^{2+} concentration. The Ca^{2+} is probably bound to **calmodulin** as an intermediary step (→ p. 17).

Transcellular transport of macromolecules (proteins, hormones) also can take place by cytosis. Endothelia, for example, take up the molecule on one side of the cell and release it, unchanged, on the other side (**transcytosis**).

Cytosis as a Means of Locomotion

In principle, most cells within the organism are capable of active locomotion, although normally only a few cell types make use of it. The only cells equipped with a special mechanism for locomotion are sperms; each sperm can swim at a speed of about 35 μm/s by lashing its whiplike tail. Other cells also move, although much more slowly, such as fibroblasts, which move at about 0.01 μm/s (→ **E1**); for example, in the event of injury they migrate into the wound and contribute to scar formation. Other examples are neutrophilic granulocytes and the macrophages, which, attracted chemotactically, penetrate the vascular wall and migrate in

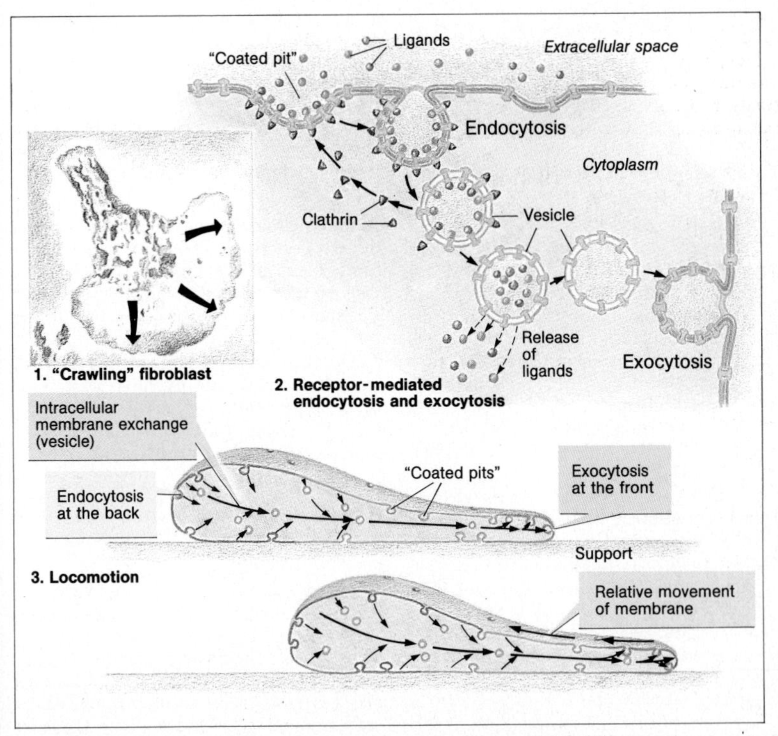

E. Endocytosis and exocytosis as a means of cell locomotion (Explanation see text)
(After *M.S. Bretcher*, Spektrum der Wissenschaft 2/88, p. 56–62)

the direction of the invading bacteria (migration, → p. 66ff.). Lastly, the "degenerate" tumor cells exert their disastrous effects by invading different tissues of the body.

The migration of such cells is achieved by crawling, like ameba, over a firm base (→ **E**). This is done by **endocytosis** of part of the cell membrane on the side furthest from its goal, the intracellular transport of the endocytotic vesicles across the cell, and their reincorporation into the cell wall near the goal by **exocytosis** (→ **E3**). Fibroblasts turn over about 2% of the cell membrane per minute in this way.

The cycle is completed by obligatory displacement from "front" to "back" by caterpillar-like movements of the parts of the cell membrane not involved in cytosis. Since the cell membrane of fibroblasts mainly adheres to the fibronectin of the extracellular matrix, the cell moves forward relative to the matrix (→ **E3**). Adhesion of the cell also requires the presence of specific receptors (i.e. for fibronectin in the case of fibroblasts).

The Development of Electric Potentials in Transport Processes

The transport of ions involves a movement of electrical charges that results in the development of an *electric potential*. If, for example, K^+ ions diffuse out of the cell, a **diffusion potential** arises, the outside of the cell being positive with respect to the cell interior. The diffusion potential can now drive the outwardly diffusing K^+ ions (diffusion along a chemical gradient) back into the cell (potential-driven transport, → p. 10). Diffusion of the K^+ ions continues until the two gradients are equal and opposite, that is, until their sum, or the **electrochemical gradient**, is zero. At this point, the ratio prevailing between the concentration (instead of activity; see comment below) of the ions outside the membrane and their concentration within the cell is the **equilibrium concentration**, and the potential obtained is the **equilibrium potential** of the ion species involved, e.g. K^+. The potential caused by diffusion of one ion species drives also other cations and anions through the membrane if it is permeable to these ions, too.

For the ion "x" the equilibrium potential E_x between the inner (i) and outer (o) surface of the cell membrane can be calculated from **Nernst's equation** as follows (see also p. 24):

$$E_x = R \cdot T \cdot (F \cdot z_x)^{-1} \cdot \ln([x]_a/[x]_i), \qquad (5)$$

where R is the gas constant ($= 8.314\,J \cdot K^{-1} \cdot mol^{-1}$); T, the absolute temperature (310 K in the body); F, the Faraday constant, that is, the charge per mol ($= 9.65 \times 10^4\,A \cdot s \cdot mol^{-1}$); z_x, the number of charges on the ion (+1 for K^+, +2 for Ca^{2+}, −1 for Cl^- etc.); ln, the natural logarithm; and x, the molal concentration of the ion (if the activity coefficient is equal in both compartments, concentrations can be used instead of activities). At body temperature (310 K), $R \cdot T/F = 0.0267\ V^{-1}$. If $\ln([x]_o/[x]_i)$ is rewritten as $-\ln([x]_i/[x]_o)$ and ln is converted to log(ln = 2.3 · log), and substituted in equation (5), the Nernst equation becomes

$$E_x = -61 \cdot z_x^{-1} \cdot \log([x]_i/[x]_o)\,[mV]. \qquad (6)$$

If 'x' is K^+ and $[K^+]_i = 150$ mmol/kg H_2O and $[K]_o = 5$ mmol/kg H_2O, the K^+ equilibrium potential $E_k = -90$ mV (see also p. 24).

At the equilibrium potential, just as many of the ions concerned are moved by the chemical gradient in one direction as are pushed back by the potential in the other direction. The sum of these two, the **net ion flux**, is thus zero. It will deviate from zero when the actual membrane potential (E_m) deviates from the equilibrium potential (E_x). Thus the **driving potential** for the net ion flux (I_x) is $E_m - E_x$. **Ohm's Law** for the ion flux (→ p. 10, equation (4)) thus becomes:

$$I_x = g_x(E_m - E_x). \qquad (7)$$

(Units: conductance g_x/membrane area $[S \cdot m^{-2}]$; ion current I_x/membrane area $[A \cdot m^{-2}]$; E[V].)

In the resting cell (→ p. 24). E_m is approximately −70 mV, E_K is approximately −90 mV, and E_{Na} approximately +70 mV. Thus for Na^+ the driving potential is roughly 140 mV; for K^+ (in the reverse direction), about 20 mV. At rest, a larger I_{Na} is avoided only because g_{Na} is very small (approximately 1/10 − 1/100 of g_K). Should, however, the Na^+ pores open briefly (activation of the Na^+ channels; see below) during an action potential (→ p. 26), there is a very rapid influx of Na^+ into the cell on account of the high electrochemical driving potential.

For Na^+, Ca^{2+}, K^+, Cl^-, and other ions, the "conductance" of the cell membrane is more than just a simple physical event. At many cell membranes the passive inflow of Na^+ is mediated by a **carrier**, which can cotransport glucose in the same direction or carry H^+ ions in the opposite direction (→ e.g. p. 128). In addition, the cell membranes (e.g. nerve, muscle) possess special **channels** for the different types of ions, whereby the conductance of the channels can be regulated (see below).

If, in the *active transport* of ions (e.g. Na^+) employing the same *carrier*, an ion (e.g. Cl^-) with the opposite charge is transported in the same direction, or an ion (such as H^+) with the same charge is transported in the opposite direction and the ratio of the charges is 1 : 1, the term **electroneutral transport** is applied. If, on the other hand, 3 Na^+ ions, for example, are transported in the one direction and 2 K^+ ions are transported simultaneously in the other direction (e.g. by Na^+-K^+-ATPase), the excess Na^+ ion gives rise to a potential: **electrogenic** or **rheogenic transport**.

Control of the Permeability of Membranes to Ions

In certain cells the conductance of the membranes for ions can be varied by the opening or closing of **channels** or **pores** (usually *specific* for a particular ion or groups of ions) by a kind of **gate** via which the ions enter or leave the cell. Such a gate can be only open or closed. Opening of the gates is not synchronized, but is a statistical event whereby the *mean* number of open gates can be regulated by the *magnitude of the cell potential* (e.g. in excitable cells like nerve und muscle) or by a *chemical substance* (e.g. the postsynaptic effect of acetyl choline, → **F** and p. 54).

The excitation of the **cardiac muscle**, for example, causes the (relatively slow) opening of Ca^{2+} channels, at the same time as the conductance to K^+ drops. In this case also, Na^+ channels open very quickly but briefly. Excitation (depolarization, → p. 26) causes the gate of the Na^+ channels to open (*activation*) and, immediately thereafter, the open gate to shut (*inactivation*). In the brief period at the beginning of the action potential during which the gate is open (high Na^+ conductance, → p. 26), Na^+ rushes into the cell.

In presynaptic **nerve endings** an incoming action potential opens Ca^{2+} pores, and the inflowing Ca^{2+} ions activate the release of neurotransmitters (→ e.g. p. 54 ff.) or hormones (e.g., in the posterior lobe of the pituitary gland, → p. 240).

In **exocrine cells**, too (e.g. in the pancreas), *exocytosis* is regulated by Ca^{2+} influx (see below). A further example: the opening of the Ca^{2+} pores in the longitudinal tubules of **skeletal muscle** in response to an action potential (AP) initiates the contraction of muscle (see below and p. 36 ff.).

It is thus a general principle that the organism builds up electrochemical gradients (e.g. cell interior is low in Na^+) relatively slowly by *actively pumping ions* (in this case by Na^+-K^+-ATPase: approximately 1 $\mu mol \cdot m^{-2} \cdot s^{-1}$) and then exploits this electrochemical gradient for rapid ion fluxes by regulating the passive membrane permeability (pores) (in this case Na^+ inflow on arrival of an AP: approximately 1000 $\mu mol \cdot m^{-2} \cdot s^{-1}$).

The Role of Ca^{2+} Ions in the Regulation of Cellular Processes

About 50% of the plasma Ca^{2+} is bound to proteins and to complexing agents. The remainder is present in its free form (→ p. 151). In the interstitial fluid most of the Ca^{2+} is in the free form (about 2.9 mmol/kg H_2O) on account of the low protein concentration. Inside the cell, in the *cytoplasm*, on the other hand, the concentration of free Ca^{2+} is *several orders of magnitude lower* (about 0.1–0.01 μmol/kg H_2O), as a result of the continual removal of Ca^{2+} by active transport. A primary active Ca^{2+} transport has been demonstrated in addition to a

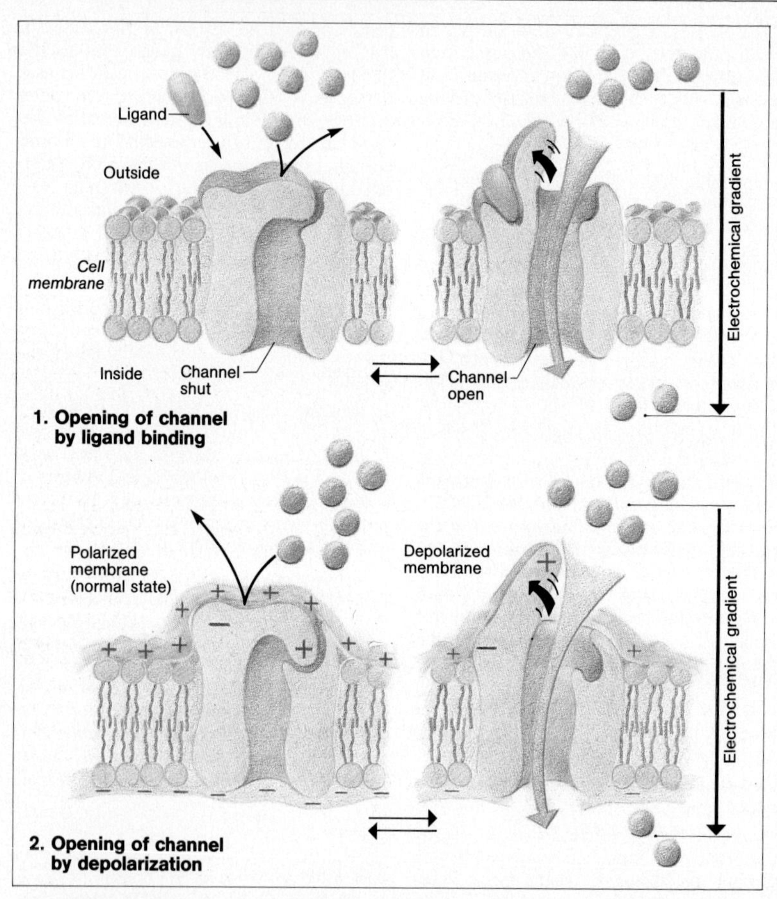

F. Opening or "gating" of ion channels (1) by means of binding a ligand, for example, a neurotransmitter such as acetylcholine at a synapse, or a hormone, or (**2**) by depolarization of the cell membrane (e.g., of nerve-, muscle-, and epithelial cells). The membrane channels are not opened or closed synchronously, but the probability of the open or closed state of the channel is raised or lowered. The driving force for the passage of ions is the sum of the electric gradient (membrane potential) and chemical gradient (concentration difference) of the ion across the membrane, collectively known as the electrochemical gradient. (After Albertus B, et al. Molecular biology of the cell. 1st ed. New York and London: Garland, 1983: 299.)

secondary active countertransport driven by the Na^+ gradient.

The intracellular Ca^{2+} concentration is **regulated** by higher or lower passive **influx of Ca^{2+}** from the extracellular space (ECS; see previous section) or from intracellular Ca^{2+} reservoirs. The influx can be elicited by *APs* (→ p. 164) or by the second messengers (→ p. 242ff.) of transmitters or hormones for which receptors are present on the outer surface of the cell membrane (e.g. for acetylcholine, → **F**).

In **skeletal muscle** the arrival of an action potential causes Ca^{2+} to flow out of the longitudinal tubuli (special bladderlike Ca^{2+} reservoirs) into the muscle cell, where it binds to *troponin C* and sets off the contraction of the muscle (electromechanical coupling, → p. 36ff.).

The Ca^{2+} that streams into **cardiac muscle** cells not only comes from the longitudinal tubuli but also from the ECS. The resulting rise in intracellular Ca^{2+} during excitation is the reason why the action potential (AP) of the myocardium has a particularly long-lasting *plateau* (200–500 ms), which is the basis for the "all or none" type of contraction of the heart (→ p. 164). The intracellular Ca^{2+} concentration also influences the strength of myocardial contraction.

In **smooth muscle** the inflow of Ca^{2+} can be elicited by an AP, by a transmitter, or by a hormone. In this case (as in many other cells), the primary intracellular **receptor for Ca^{2+}** is probably **calmodulin.**

Calmodulin is a protein with a molecular weight of 16700 and is very similar in structure to troponin C (→ p. 36ff.); it binds 4 mol Ca^{2+} per mol. In collaboration with another protein, the Ca^{2+} calmodulin complex forms an enzyme (**m**yosin **l**ight **c**hain **k**inase of MLCK) that phosphorylates the light chain of myosin (light meromyosin, → p. 34). The conformation of myosin is changed in the process, allowing actin to activate the ATPase of the myosin, which leads to contraction. A drop in Ca^{2+} and dephosphorylation of the myosin by another enzyme (**m**yosin **l**ight **c**hain **p**hosphatase or MLCP) terminates contraction, and the smooth muscle relaxes.

The role of calmodulin in the contraction of smooth muscle thus seems to be similar to that of troponin C in skeletal muscle, although its mode of action is somewhat different (→ p. 36ff.). Ca^{2+} and calmodulin also initiate the movements of *sperm.*

Exocytosis from secretory cells (e.g. pancreas or salivary glands) can probably be regarded as a primitive type of motor activity. The inflow of Ca^{2+} (mainly from the RER; → p. 4) and the calmodulin-Ca^{2+} complex influence the *microtubule-microfilament system* via which, in some way not yet understood, exocytosis is regulated. In this case, again, what triggers the Ca^{2+} inflow is as a rule the binding of *hormones* to extracellular hormone receptors. Thus Ca^{2+} plays the role of a **second (or third) messenger** in the target cell of the hormone (**first messenger**). Cyclic adenosine monophosphate, cAMP, (→ p. 242) plays a very similar role in the action of peptide hormones and catecholamines. In many cells, one hormone initiates the inflow of Ca^{2+} while another hormone triggers the formation of cAMP. The two second messengers then exert either an antagonistic or a synergistic effect on cellular metabolism. The antagonistic effect might partly be due to the Ca^{2+}-calmodulin complex activating the enzyme phosphodiesterase, which is responsible for the breakdown of cAMP (→ p. 242). In addition, the Ca^{2+}-calmodulin complex is involved in *cell growth* and also regulates the activity of a number of other *enzymes* in liver, kidney, heart, pancreas, brain, sperms, thrombocytes, and so on.

In a wide variety of cells, extracellular hormone-receptor interaction converts phosphoinositides of membranal origin to **inositol trisphosphate** and **diacylglycerol**, which in turn act as second messengers within the cell (→ p. 242f.). This bifurcating signal not only leads to protein phosphorylation and other intracellular events, but also mediates (via inositol trisphosphate) Ca^{2+} release (e.g. in the pancreas; see above). Here, Ca^{2+} is the **third messenger**, which in turn influences the cell functions, for example, by binding to calmodulin.

The Production and Transformation of Energy

Without a supply of energy, life cannot go on. Plants harness the energy from the sun to convert CO_2 from the air into O_2 and organic compounds. These compounds can be directly utilized (mainly as hydrocarbons) by the human and animal organism to supply their energy needs, or they can be stored in the form of fossil fuels (coal and oil). Thus, energy can be converted from one form into another. If we consider such a transformation as taking place in a **closed system** (exchange of energy, but not of material, with the surroundings), energy can neither appear spontaneously nor disappear. This is described in the **First Law of Thermodynamics**, which states that the change in internal energy (= change in energy content ΔU) of a system (e.g. a chemical reaction) is equal to the sum of work absorbed (+ W) or done (− W) and the heat evolved (− Q) or absorbed (+ Q), respectively.

$\Delta U = Q - W$ [J] (work done, heat absorbed)

$\Delta U = W - Q$ [J] (work absorbed, heat evolved).

(By definition, the signs indicate the direction of flow with respect to the system under consideration.)

For biological systems, the most important consequence of this law is that the heat produced in the transformation of one substance into another is always the same, *whatever the pathway taken*. In other words, whether glucose undergoes direct combustion with O_2 in a calorimeter (→ p. 198), with the production of CO_2 and H_2O, or whether these same end products are formed from glucose via devious metabolic pathways, the energy production is the same (i.e. the physical and physiological caloric values are equal, → p. 198). They are not equal if, as in the case of amino acids, the end products formed in the calorimeter (here, CO_2, etc.) are different from those formed in human metabolism (here, urea). However, even in this situation the chemical energy taken up by the body (in this case, amino acids) is always equal to the heat liberated + mechanical work + energy content of the metabolic products excreted (in this case, urea).

If, as in the organism, the pressure (p) remains constant, part of the energy is used for the change in volume (ΔV); the volume work ($p \cdot \Delta V$) is not available and has to be allowed for: ΔU (see above) $+ p \cdot \Delta V = \Delta H$, where ΔH is termed the change of **enthalpy**. (ΔV is usually low in the organism so that $\Delta H \approx \Delta U$.)

In order to determine which part of this enthalpy ΔH is freely available, the **Second Law of Thermodynamics** has to be taken into consideration. According to this law, the disorder or randomness, called **entropy**, of a closed system increases ($\Delta S > 0$) if, for instance, one form of energy is converted into another. The rise of entropy times the absolute temperature ($\Delta S \cdot T$) equals the heat dissipated during such a process. Thus, the **free enthalpy** ΔG (= *Gibbs' energy*) is calculated as follows:

$$\Delta G = \Delta H - T \cdot \Delta S.$$

This relationship also defines the conditions under which, for instance, spontaneous **chemical reactions** may occur. If $\Delta G < 0$, the reaction is called **exergonic** and may occur spontaneously; if $\Delta G > 0$, the reaction is **endergonic** and cannot, without the expenditure of additional free energy, take place spontaneously.

ΔG is obtained by taking the *actual* concentrations of the compounds involved into consideration. If the reaction is $A \rightleftharpoons B + C$, the *standard free enthalpy* $\Delta G'_0$ (where the concentrations A, B, and C are 1 molal, and the pH = 7) is converted into ΔG as follows:

$$\Delta G = \Delta G'_0 + R \cdot T \cdot \ln \frac{[B] \cdot [C]}{[A]}; \qquad (8)$$

or (at 37 °C):

$$\Delta G = \Delta G'_0 + 8.31 \cdot 310 \cdot 2.3 \cdot \log \frac{[B] \cdot [C]}{[A]} \; [J \cdot mol^{-1}].$$

Assuming, for instance, that $\Delta G'_0$ of a reaction is $+20\ kJ \cdot mol^{-1}$ (endergonic), ΔG becomes < 0 (exergonic) if $[B] \cdot [C]$ is, for example, 10^4 times smaller than [A]:

$$\Delta G = 20000 + 5925 \cdot \log 10^{-4} = -3.7\ kJ \cdot mol^{-1}.$$

In this case B and C are formed (reaction runs to the right).

If in the same example the ratio [B] · [C] to $[A] = 4.2 \cdot 10^{-4}$, ΔG becomes 0 and the reaction is in equilibrium (no net reaction). This ratio therefore is called the **equilibrium constant K** of the reaction. Using equation (8), K can be converted into $\Delta G'_0$, and vice versa, as follows:

$$0 = \Delta G'_0 + R \cdot T \cdot \ln K; \text{ or}$$
$$\Delta G'_0 = -R \cdot T \cdot \ln K; \text{ or}$$
$$K = e^{-\Delta G'_0/(R \cdot T)}.$$

If, finally, the ratio [B] · [C] to $[A] > 4.2\ 10^{-4}$, then $\Delta G > 0$ and the net reaction runs backward, that is, A is formed.

It is evident from these considerations that whereas $\Delta G'_0$ indicates the *equilibrium* point of a reaction, ΔG is a measure for *how far away* the reaction is from equilibrium.

ΔG, however, does not give any indication of the **rate of a reaction**. Even if $\Delta G < 0$, the reaction may be extremely slow. Its rate depends on the amount of energy that is *transiently* needed for producing intermediate products of the reaction, for which ΔG is larger than that of the initial substance or the end products. The additional energy needed here is called threshold or **activating energy** ΔG^{+}. Catalysts ("activators") or, in biology, enzymes increase the rate of reaction by lowering ΔG^{+}. This is illustrated in Fig. **G**.

According to *Arrhenius*, the rate constant k (s^{-1}) of a unimolecular reaction is proportional to $e^{-\Delta G^{+}/(RT)}$. If, therefore, the activation energy ΔG^{+} of a unimolecular reaction is halved by an enzyme (e. g. from 126 to 63 kJ · mol^{-1}), the rate constant increases at 310 K (37 °C) by the factor of ($e^{-63\,000/(8.31 \cdot 310)}/e^{-126\,000/(8.31 \cdot 310)}$ =) about $4 \cdot 10^{10}$. In other words, the time after which 50 % of the substance is metabolized (t/2) is lowered from 10 years to 7 ms in this case! (Rate constant [s^{-1}] times concentration of the starting substance [$mol \cdot l^{-1}$] = unidirectional rate of reaction [$mol \cdot l^{-1} \cdot s^{-1}$].)

A reaction can also be accelerated by increasing the **temperature**. A rise of 10 °C usually results in a 2- to 4-fold increase of the reaction rate (i. e. the $\mathbf{Q_{10}}$ **value** is 2 to 4).

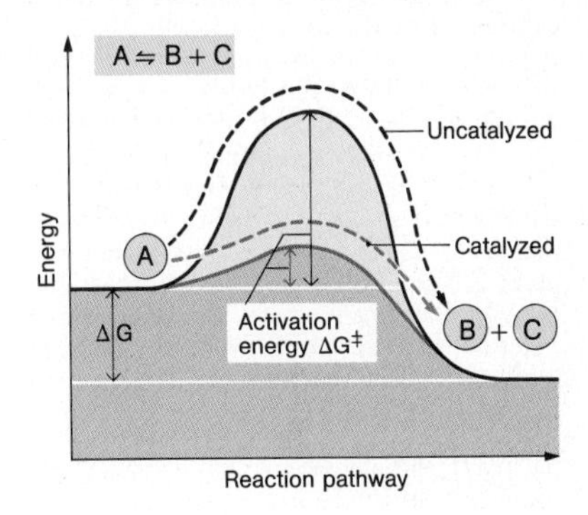

G. Activation energy and catalyzed reactions. Substance A can achieve a lower, more favorable state of energy by reacting to give B + C. This reaction can only take place if enough activation energy can be acquired by A. Catalysts and, in biology, enzymes lower the activation energy and thereby facilitate the reaction (red lines)

The second law of thermodynamics (see above) also states that in a closed system – and the universe is a closed system in this sense – a continuous dissipation of free energy takes place, the sum of all systems moving towards increasing randomness and disorder. The organism, however, is an **open system**, and as such it can take up energy and deliver end products. The entropy of the closed system consisting of organism and surroundings will increase, whereas the open system "organism" considered alone, not only keeps its *entropy* constant, but even *decreases* it by the expenditure of free enthalpy. Examples are seen in the establishment of osmotic gradients or of pressure differences within the body. Thus, whereas a closed system is characterized by maximum entropy, true reaction equilibrium (with reverse reactions), and by performing work only once, the body, as an open system, is capable of continuous work with the production of a *minimum of entropy*. Very few processes within the body attain true equilibrium (e.g. CO_2 transport); most of them (e.g. enzyme reactions, cell potential) are in flow equilibrium or **steady state** and usually *irreversible* (e.g. due to excretion of end products). The reverse of the "reaction" germ cell to adult is obviously impossible, too. In the steady state, the *rate* of the reaction and not its equilibrium is what matters. **Regulation** can be achieved by influencing the rate of reaction.

The universal "currency" of free enthalpy (or free Gibb's energy) in the organism is adenosine triphosphate, **ATP**. It is a product of the cellular breakdown of nutrients.

ATP is formed by the **oxidation** of biological molecules such as glucose. Oxidation in this sense refers to the **removal of electrons** from the relatively electron-rich (= reduced) carbohydrates. The endproducts of this reaction are CO_2 and H_2O. This oxidation (or shift of electrons) occurs in several steps and makes it possible for part of the energy set free in oxidation to be coupled to the formation of ATP (**coupled reactions**; → **H**).

The standard free enthalpy $\Delta G_0'$ of **ATP**

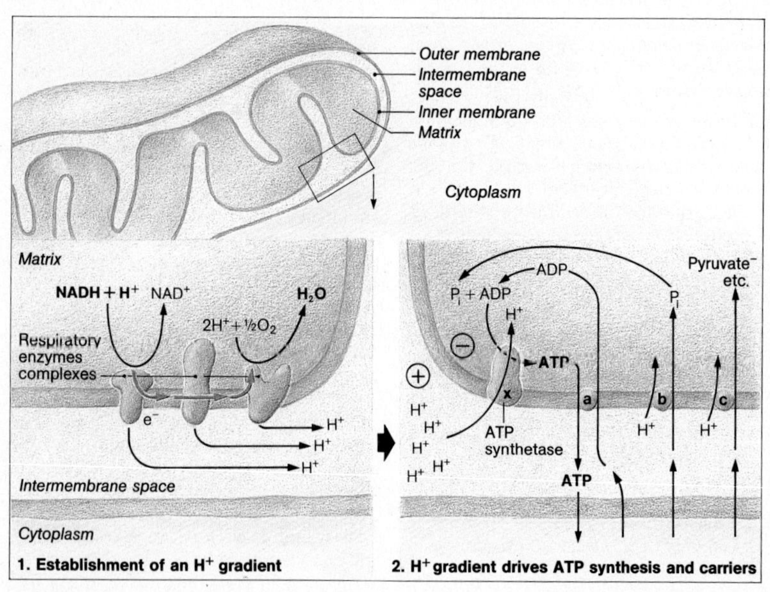

H. Energy transfer in oxidative phosphorylation and transport at the inner membrane of mitochondria. A high-energy electron (**e**$^-$) passes along the electron transport chain to lower energy levels, thus pumping H^+ out of the matrix space. The resulting high electrochemical gradient for H^+ (electrical potential + chemical gradient) across the inner membrane drives the H^+ back into the matrix via ATP synthetase (**x**). Here, the energy of the H^+ gradient is utilized for ATP production. (If ATP concentration in the matrix is high, the reaction is reversible.) Carriers exist which (**a**) exchange ATP for ADP, and co-transport (**b**) P_i and (**c**) pyruvate (and other substrates) together with H^+ into the matrix space. Ca^{2+} also is driven by the electrical potential in this direction

hydrolysis amounts to $-30.5\ kJ \cdot mol^{-1}$. As can be seen from equation (8), the real free enthalpy ΔG increases if the ratio of $([ADP] \cdot [P_i])/[ATP]$ is smaller than the equilibrium constant K of ATP hydrolysis (ADP = adenosine diphosphate). The high ATP level in the cells leads to a ΔG of about -46 to $-54\ kJ \cdot mol^{-1}$.

Substances with a significantly higher $\Delta G'_0$, like creatine phosphate ($-43\ kJ \cdot mol^{-1}$), can *produce* ATP from ADP and P_i. On the other hand, the universal energy "currency" of ATP can be used for the formation of other high-energy compounds with a somewhat lower energy content than ATP, for example, UTP, GTP, glucose-6-phosphate, and so on.

The energy set free during ATP hydrolysis drives hundreds of reactions (e.g. active transport through membranes, synthesis of proteins, muscle contraction). Finally, all these reactions *generate order* in the cells and in the whole body. As outlined above, the price of maintaining life by this decrease of entropy in the body is paid by increasing the entropy of the universe.

Control and Transmission of Information

A system as complicated as the human organism can only function if its individual activities are subject to **regulation**. This requires a two-way exchange of *information* between the control center and the controlled variable.

The information concerning the controlled variable is *fed back* to the control center and if the initial "commands" have not been adequately carried out they can be adjusted. *A regulatory system with feedback* is termed a **control system**.

Control systems play a very important role in the organism. Muscular movements, equilibrium, blood volume, blood pressure, oxygen content of the blood, pH value, body temperature, blood sugar level, and many other values are regulated within the body. The control system can be situated in the organ concerned (*autoregulation*) or can include a higher level (CNS, endocrine glands). The regulating signals are sent from the control center (which determines the *set value*) to the effector organ or part of an organ that carries out the commands received. The situation is registered by *receptors* and the results are reported back to the center for comparison with the original **set value**. If the result is unsatisfactory, the signals are adapted.

Information is conveyed between the components of the control system via nerves, or the hormones in the blood (→ p. 232). Over short distances within an organ information can also be conveyed by diffusion.

Nerve Cell and Synapse

An excitable cell responds to a stimulus by changing the electrical properties of its cell membrane. Human beings possess two types of excitable cells:

a) **nerve cells**, to transmit an impulse; and

b) **muscle cells**, which respond to the stimulus by contracting.

The human nervous system contains more than 10^{10} nerve cells or **neurons**. The neuron is both the *structural* and *functional unit* of the nervous system. A typical motor neuron (→ **A**) consists of the **soma** or cell body, an **axon** and **dendrites**. The neuron contains the usual intracellular organelles including mitochondria (→ "Mi" in **C**), as well as neurotubuli and neurofibrils. The dendrites furnish a large area for contact with other neurons (up to 0.25 mm^2). They are *afferent* or centripetal fibers because they receive signals from other neurons and transmit them *to* the soma. The axon and its branches, the **collaterals**, are *efferent* or centrifugal fibers because they carry signals *from* the soma to other neurons muscles, and gland cells. They split up and terminate in swellings, the **synaptic knobs**, or **terminal buttons** (→ **A**), which contain vesicles that store **neurotransmitters** (→ **B**). At the synapse (see below), the terminal buttons make contact with the soma, dendrites, or axon of the next neuron. There are thousands of these contact sites on a single neuron, covering up to 40% of the total neuron surface.

From the soma towards the ends of dendrites and axon (in some cases, in the opposite direction) **axoplasmic transport** of proteins, amino acids, transmitter substances, and so on (at a speed of about 200–400 mm/day) takes place along the *neurotubules*. Neither the mechanism nor the role of this transport system are definitely understood (possible role in cell nutrition, growth, and long-term changes in excitatory properties). Neurotransmitters are packed in vesicles for their transport along the outside of the neurotubules to the periphery; actin of the tubules and adenosine triphosphatase (ATPase) from the vesicles are known to play an important role in this process.

The cell membrane of the soma extends along the axon as the **axolemma** (→ **A**; and "AL" in **C**). The axon is surrounded in the central nervous system (CNS) by **oligodendrocytes** and in the peripheral nervous system by **Schwann cells** (→ **A**; and "SC" in **C**). Axon + sheath = nerve fiber. In some neurons the Schwann cells envelope the axon in multiple concentric layers to form the **myelin sheath** (→ **C**), which serves as an *insulation* for ion currents (hydrophobic lipoprotein). Along the axon the myelin is interrupted at 1.5 mm intervals at the **nodes of Ranvier** (→ **A**). The speed of conduction in the **myelinated nerve fibers** is relatively *high* as compared with the relatively *slow* conduction in the *unmyelinated fibers* (→ **C**). In addition the smaller the *diameter* of the nerve fiber, the slower the conduction (→ p. 29, C).

The **synapse** (→ **A, B**) is the area where signals are transmitted from the axon, and its collaterals, of one neuron to the axon, dendrite or soma of another, or to muscle cells (→ p. 32) or gland cells. In mammals there is no actual contact at the synapse (with rare exceptions). The **synaptic cleft** (10–40 nm) separates the two neurons and acts as an insulator. For transmission of a signal the electrical impulse that reaches the **presynaptic membrane** must release a chemical transducer, the **neurotransmitter**, into the synaptic cleft. The transmitter(s), set free from presynaptic vesicles by exocytosis, diffuses to the **postsynaptic membrane** and initiates a new electric signal.

There are several transmitters: acetylcholine, norepinephrine, γ-aminobutyric acid, dopamine, glycine, glutamate, substance P, and so on. Although it had been generally assumed that each neuron has only one neurotransmitter, the release of a second transmitter ("cotransmitter") can be frequently observed.

Since there is no neurotransmitter release from the postsynaptic membrane, the synapse functions as a *one-way valve*, transmitting only from the presynaptic to the postsynaptic neuron. It is also the site where neuronal signals can be modified by other nerves. The transmitter may produce either an excitatory or an inhibitory response depending on its nature (→ p. 30).

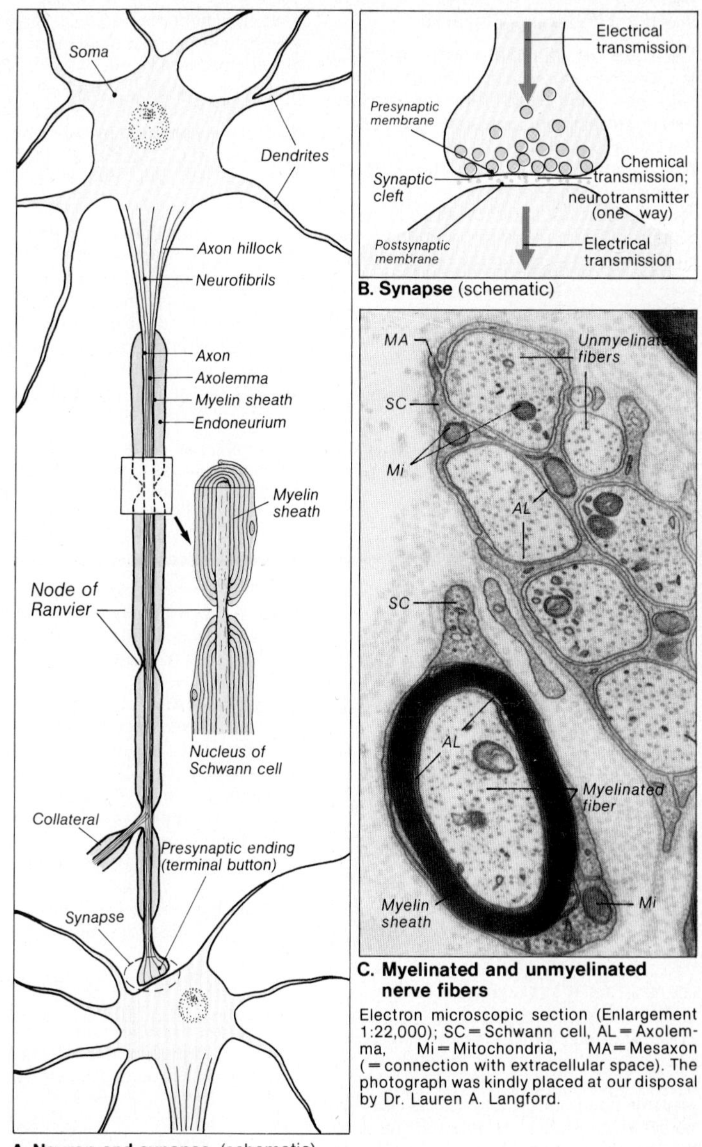

A. Neuron and synapse (schematic)

B. Synapse (schematic)

C. Myelinated and unmyelinated nerve fibers

Electron microscopic section (Enlargement 1:22,000); SC = Schwann cell, AL = Axolemma, Mi = Mitochondria, MA = Mesaxon (= connection with extracellular space). The photograph was kindly placed at our disposal by Dr. Lauren A. Langford.

Resting Membrane Potential

An electric potential difference (→ p. 14) can be recorded across the membrane of living cells. According to cell type this so-called **resting (membrane) potential** amounts to **50–100 mV** (*cell interior is negative*) in muscle and nerve cells. The resting potential is due to the *uneven distribution of ions* (→ **B**) between the inside and the outside of the cell membrane. The following phenomena are involved in establishing the cell potential:

1. By means of **active transport** (→ p. 11), Na^+ is continuously "pumped" out of the cell and K^+ "pumped" into it (→ **A 2**), so that the K^+ concentration in the intracellular fluid (ICF; → **A, B**) is about 35 times higher and the Na^+ concentration 20 times lower than in the extracellular fluid (ECF). The "pump" is identical with the membranal Na^+-K^+-ATPase (→ p. 11).

Since the anionic proteins and phosphates of the ICF (→ p. 65, B) cannot leave the cell, the diffusible ions would also be *unequally distributed* to some extent by purely passive means (*Gibbs-Donnan distribution*; → **A 1**):

$[K^+ + Na^+]_{ICF} > [K^+ + Na^+]_{ECF}$
$[Cl^-]_{ICF} < [Cl^-]_{ECF}$.

2. Under resting conditions the permeability of the cell membrane to $\mathbf{Na^+}$ ions is low (i.e. its Na^+ conductance is low), so that the Na^+ gradient (→ **A 3–5**) cannot be levelled under these conditions by passive back diffusion.

3. The cell membrane is of extremely *low permeability* for the negatively charged **proteins** and organic **phosphates** (→ **A 4, 5**).

4. The conductance of the resting cell membrane for $\mathbf{K^+}$ is relatively high ($g_K \gg g_{Na}$). On account of the steep concentration gradient (point 1) K^+ ions diffuse from the ICF into the ECF (→ **A 3**). Due to the positive charge of the K^+ the diffusion of only a few such ions is sufficient to produce a potential across the membrane, since the greater part of the intracellular anions do not follow and no significant inward diffusion of Na^+ is possible (see points 2 and 3). This **diffusion potential** continues to increase until the further efflux of K^+ ions (driven by the concentration gradient) is prevented by the rising potential. Since the conductance of the cell membrane for $\mathbf{Cl^-}$ is relatively high (see below), the rising potential at the same time drives the Cl^- out of the cell against its chemical gradient (→ **A 4**). Thus, the diffusion of K^+ down a chemical gradient is increasingly opposed by the rising potential, while that of Cl^- (driven by the potential) is opposed by the rising chemical Cl^- gradient.

Finally, an **equilibrium potential** is attained for K^+ (E_K) as well as for Cl^- (E_{Cl}). At E_K the driving force for outward K^+ diffusion (the chemical gradient) is exactly equal to the inwardly directed force due to the potential (electric gradient), i.e. the electrochemical gradient for K^+ is 0. The same holds true for Cl^-. E_K can be calculated using the **Nernst equation** (→ p. 14).

Although the permeability of the resting membrane to $\mathbf{Na^+}$ is very low as compared with K^+, Na^+ ions diffuse continuously into the cell (high electric and chemical gradients, → **A 5** and **B**). The resting membrane potential is consequently usually somewhat **less negative than E_K**.

Due to the relatively high conductance of many cell membranes for $\mathbf{Cl^-}$, the Cl^- is usually so distributed between ICF and ECF that the Cl^- equilibrium potential equals the resting membrane potential. On account of its negative charge, the Cl^- gradient is in the opposite direction to the K^+ gradient (→ **B**). If, however, the E_{Cl} calculated from the Cl^- distribution (Nernst equation with $z = -1$) deviates from the resting membrane potential (as in certain cells), it can be concluded that Cl^- is being transported against an electrochemical gradient, that is, (usually secondary) active Cl^- transport is taking place (→ e.g. p. 132).

All living cells exhibit a (resting) membrane potential, but the excitable cells (nerve, muscle) are also able to alter the ion permeability of their membranes, which leads to considerable changes in potential (→ p. 24).

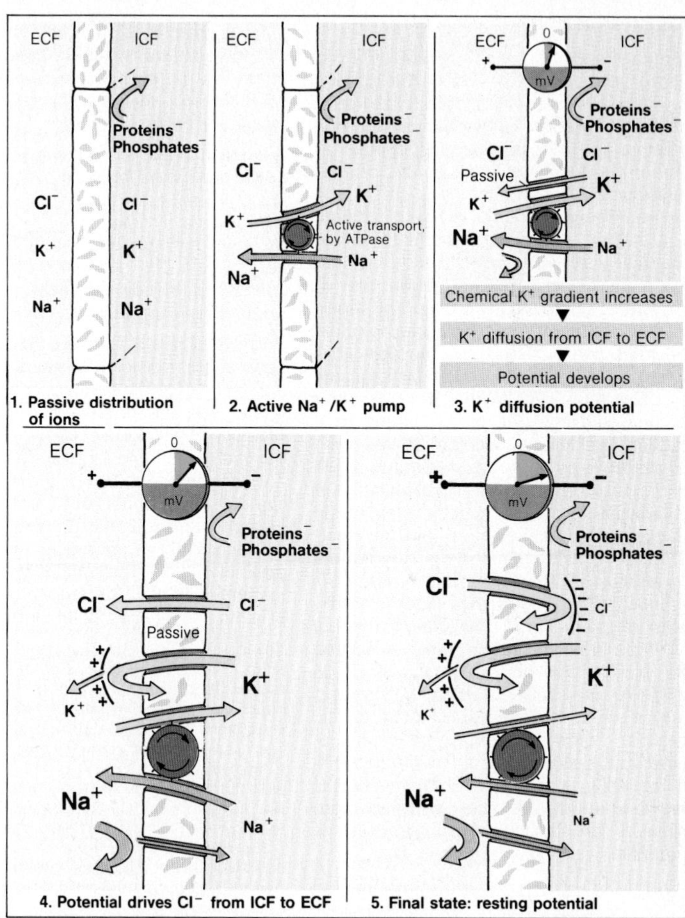

A. Causes and consequences of the resting potential

	"Effective" concentration (mmol/kg H_2O)		Equilibrium potential
	Interstitium (ECF)	Cell (ICF)	
K^+	4.5	160	− 95 mV
Na^+	144	7	+ 80 mV
H^+	$4 \cdot 10^{-5}$ (pH 7.4)	10^{-4} (pH 7.0)	− 24 mV
Cl^-	114	7	− 80 mV
HCO_3^-	28	10	− 27 mV

B. Typical "effective" concentrations and equilibrium potentials of important ions in skeletal muscle (37° C) (after Conway)

Action Potential

Maintenance of a membrane potential is a property of all living cells. **Excitability**, however, is shown only by specialized cells, nerve, and muscle. These cells respond to a stimulus by producing transient changes in the ion conductances and potential of their membranes. If the stimulus is sufficiently strong an action potential (**AP**) is generated, which in the case of the nerve is the signal that is transmitted along the nerve cell and in the muscle leads to contraction. An AP consists of the following events: The stimulus reduces the resting membrane potential (→ p. 24) to a less negative value (**depolarization**). When it reaches a critical voltage, the so-called **threshold potential**, Na^+ channels become **activated** (→ p. 15) leading to a sudden tremendous **increase in the Na^+ conductance** (g_{Na}; → **A2** and **B**) and a fast Na^+ influx into the cell. During this **depolarization phase**, the negative state inside the cell is not only reversed, but the membrane potential even reaches positive values (**overshoot**; → **B**). *Before* the overshoot is attained the g_{Na} decreases again (**inactivation** starts after < 0.1 ms), coupled with a slow **rise in K^+ conductance** (g_K; → **A3** and **B**), which allows positively charged K^+ to diffuse to the outside of the cell membrane, and leads to the reestablishment of the negative resting membrane potential (**repolarization phase**). For a few milliseconds before the g_K returns to its resting value, the membrane potential may even be more negative than the original resting membrane potential (*hyperpolarization*; → **B**).

Since the Na^+-K^+-ATPase is usually electrogenic (3 Na^+ are exchanged for 2 K^+), the ion pump may also bring about hyperpolarization following a high frequency of APs.

Below the threshold potential, a weak stimulus (e.g. an electric current) leads to (passive) changes of the resting potential. Symmetrical hyperpolarization or depolarization occurs, depending on the direction of the stimulating current (**electrotonic potentials**; see also p. 28). Near (but still below) the threshold potential, these local potentials become higher during depolarization than during hyperpolarization. This **local response** (or **local excitation**) following a depolarization already involves a slight activation of the Na^+ channels which, however, is not sufficient to elicit an AP.

Once the threshold potential is attained the cell responds with an **all-or-none** depolarization, that is, the cell responds in its own *characteristic manner, irrespective of the magnitude of the stimulation.*

A large number of APs can be elicited in quick succession because the quantity of ions traversing the membrane during one AP is extremely small (roughly 1/100000 of the total ions present within the cell). In addition, the Na^+-K^+ pump (→ p. 24) is in continuous action to help restore the original ion concentrations.

For a brief period following the depolarization phase, the nerve or muscle cannot be excited by even the strongest stimulus. This is the **absolute refractory period** and is succeeded by a **relative refractory period** (at the end of the repolarization phase, → **B**), during which an AP of a lower amplitude can be elicited, but only by a stimulus greater than the initial threshold stimulus. As the membrane potential returns to its original state, the threshold stimulus and the amplitude of the AP also return to their original value (→ p. 45, for instance).

The increase of g_{Na} (activation) and therefore the influx of Na^+ (I_{Na}) is dependent upon the magnitude of the potential *preceding* excitation and *not* the duration of depolarization. I_{Na} is at a maximum at an initial potential of approximately − 100 mV, which means that for a resting membrane potential of approximately − 60 mV, the I_{Na} is only some 60% of the maximum and that in mammalian cells Na^+ channels cannot be activated beyond a potential of about − 50 mV. This phenomenon explains the *absolute* and *relative refractory periods* and the nonexcitability of the cells if substances or events causing long-lasting depolarization, such as *succinyl choline* (→ p. 32) or hypoxia (→ p. 102), lead to a resting membrane potential more positive than roughly − 50 mV. **Ca^{2+}** influences the connection between potential and Na^+ influx. An increase of the Ca^{2+} concentration, for instance, makes the Na^+ channels better prepared for activation, and at the same time the threshold potential becomes less negative. On the other hand, a lack of Ca^{2+} leads to a lower threshold and thus to increased excitability (muscular cramps, *tetany*; → p. 114).

The AP of nerve and striated muscle differ only slightly, whereas the **AP of cardiac muscle** has certain unique characteristic features (→ p. 31 A, and pp. 42 and 164).

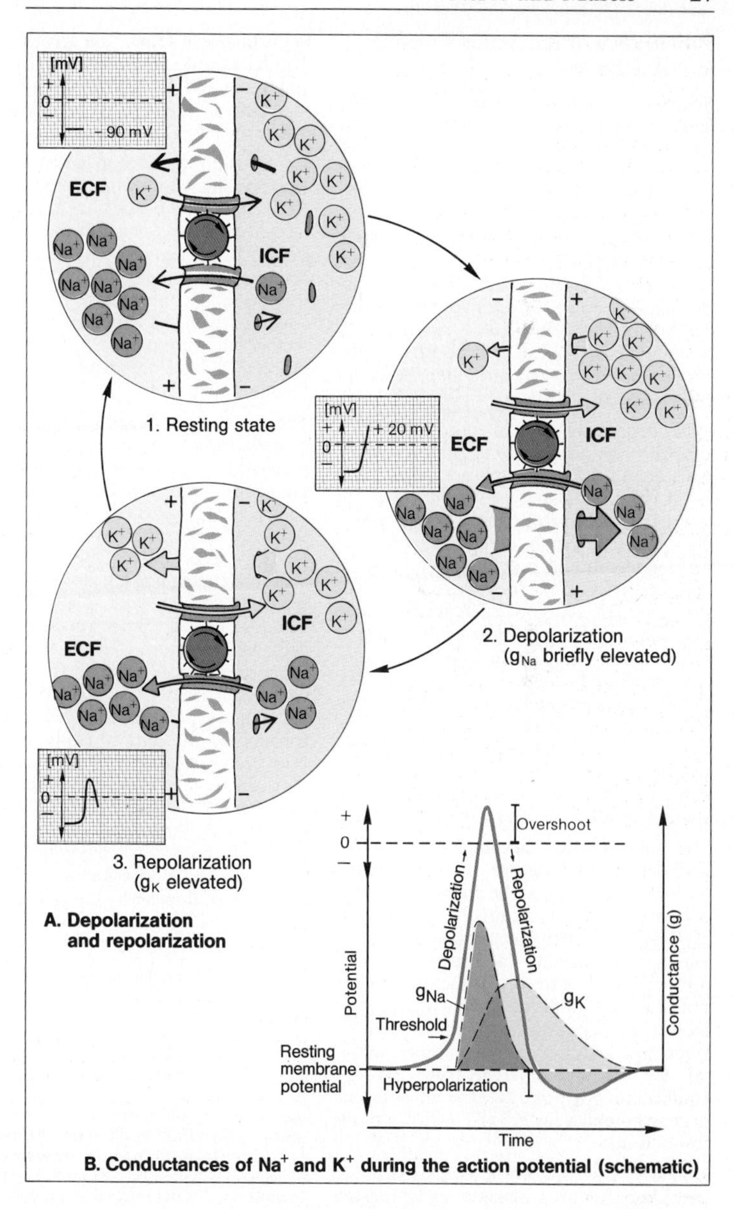

A. Depolarization and repolarization

B. Conductances of Na^+ and K^+ during the action potential (schematic)

Propagation of the Action Potential in Nerve Fibers

When a voltage is applied to an electric **cable**, current flows; since the metal core of the cable has only a small resistance, an electric impulse can be transmitted over very large distances in this way. In contrast, the (longitudinal) resistance of a **nerve fiber** is very high and its insulation (especially if it is unmyelinated) is far from being perfect, so that such a cablelike or **electrotonic transmission** soon comes to a stop. But before this can happen, the impulse — in order to ensure its propagation — is boosted by the further production of an action potential (AP; → p. 26).

What happens is the following (→ **A**): Once the AP is set going, a very brief inrush of Na^+ across the cell membrane takes place (→ **A1a**), that is an *ionic current* (→ p. 14) flows across the membrane. The previously charged nerve membrane (outside positive; resting membrane potential; → p. 24) is *depolarized* (the outside even becomes negative; → p. 27, B). The difference in charge between neighboring segments of the nerve membrane results in an equilibration of charges along the fiber (inside and outside), that is, in an **electrotonic flow of current**. The withdrawal of charges from the neighborhood causes its **depolarization**, and if the *threshold potential* (→ p. 27, B) is reached a new AP builds up (→ **A1b**) whilst the spatially preceding AP is already subsiding.

The influx of Na^+ accompanying an AP leads to a local discharge of the membrane (*capacitor property* of the membrane). This discharge thus initially involves a *capacitative* (in this case, depolarizing) *current*, which, with distance (a) becomes smaller and (b) rises less steeply. If the threshold is not attained (and no AP triggered) an increasing outward flow of K^+ follows (repolarizing) since the membrane potential, E_m, is now different from the K^+ equilibrium potential, E_k, thus allowing the driving potential for K^+ ($E_m - E_k$; → p. 14) to build up.

A new AP can thus only arise distally to the preceding AP at a distance at which the capacitative current can **rapidly depolarize to the threshold value**. At a greater distance, depolarization (a) is inadequate in its amplitude and (b) above all, it is so slow that the Na^+ channels are inactivated again before an AP is produced.

An AP is normally propagated *in only one direction* since, following the passage of an AP, each section of the fiber goes through a refractory phase (→ p. 26) during which it cannot be excited, or only with great difficulty. If a stimulus should, however, be transmitted in the reverse direction (*antidromic*), which can happen if the nerve fiber is externally stimulated (→ p. 30), it cannot proceed beyond the next synapse (valve function; → p. 22).

The consecutive triggering of APs in closely neighboring sites of the nerve fiber keeps the signal traveling, but is nevertheless a very time-consuming process. For example, in *unmyelinated nerve fibers* transmitting in this way, the speed of transmission is about 1 m/s (→ **B2**, and **C**, type C). *Myelinated fibers* can transmit much more rapidly (→ **C**, types A and B), with a speed of up to 120 m/s. Since the latter are insulated like a cable by their myelin sheath, the depolarizing electrotonic discharge along the fiber can span a greater distance (approximately 1.5 mm; → **A2**). In this case the AP is transmitted in "jumps" (**saltatory conduction**) from *node to node* (→ p. 23). The length that can be jumped is limited because the *equilibrating current* (1–2 nA) becomes weaker with increasing distance (→ **B1**). Before it becomes subthreshold the AP signal has to be refreshed at the noninsulated **node of Ranvier** by the production of a new AP, which involves a time loss of 0.1 ms (→ **B1**).

The **velocity of conduction** (→ **C**) also depends upon the **diameter** of the nerve fiber; the greater the diameter, and hence the greater the cross section of the fiber, the smaller the *longitudinal resistance* of the axon. The electrotonic equilibrating current and thus local depolarization (→ **A**) can therefore be transmitted over larger distances. This means that per length of fiber less APs have to be built up, which, again, enhances the speed of transmission. In thicker nerves, however, the **membrane capacity** is higher. Although this reduces the velocity of conduction, it is outweighed by the positive effect of the lower longitudinal resistance.

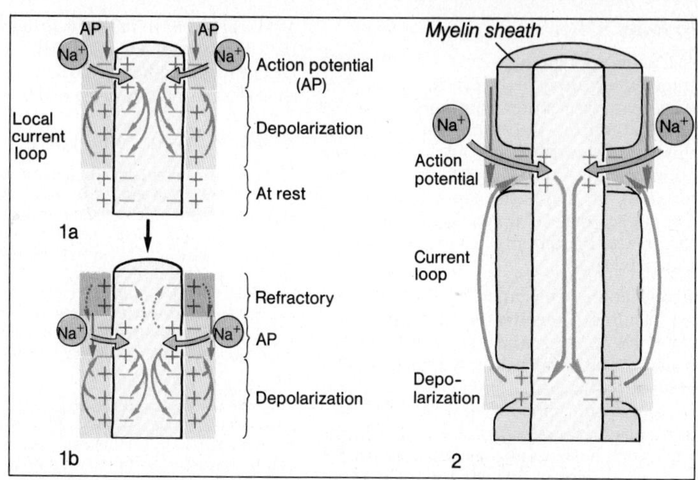

A. Serial (1a, 1b) and saltatory (2) conduction of action potentials

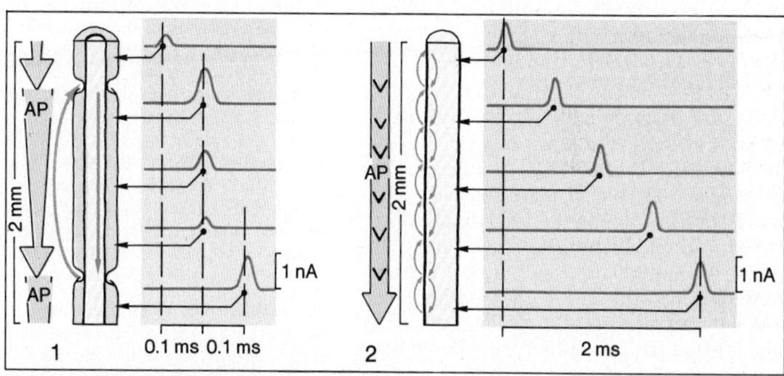

B. Impulse propagation (action currents) in myelinated (1) and unmyelinated (2) fibers

Fiber	Function (examples)	Diameter (µm)	Conduction velocity (m/s)
Aα	Afferents - muscle spindle, tendon organ. Efferents - skeletal muscle	15	70–120
Aβ	Afferents - touch	8	30– 70
Aγ	Efferents - muscle spindle	5	15– 30
Aδ	Afferents - temperature, fast pain	3	12– 30
B	Sympathetic, preganglionic	3	3– 15
C	Sympathetic, postganglionic afferents - slow pain	1 (unmyelinated)	0.5– 2

C. Classification of neurons

(after Erlanger and Gasser)

Synaptic Potentials

The action potential (AP; → **A1** and p. 26) transmitted along a (presynaptic) neurite releases a **transmitter substance** from the terminal button. Depending upon its type, this substance can bring about depolarization (excitation) or hyperpolarization (inhibition) of the postsynaptic membrane. The higher the AP frequency in the axon, the greater the amount of transmitter substance released.

Acetylcholine, substance P, and glutamate are examples of **excitatory transmitters** that raise the g_{Na}, g_{Cl}, and g_K (→ p. 9) of the postsynaptic membrane in the synapse (*subsynaptic*). Due to the high electrochemical Na^+ gradient, influx of Na^+ predominates and **depolarization** takes place; this is the **excitatory postsynaptic potential** or **EPSP** (maximum approx. 20 mV; → **C**). The EPSP does not begin until 0.5 ms after the arrival of the AP at the presynaptic terminal button (→ **C**). This *synaptic delay* or *latency* results from the relatively slow liberation and diffusion of the transmitter.

Although a single EPSP is usually insufficient to generate a postsynaptic AP, the excitability of the neuron is enhanced by the *local depolarization* so that several simultaneous EPSPs are able to depolarize cells to the threshold potential (spatial and temporal **summation**), thus setting off a transmitted AP. Unlike the AP, the EPSP is *not* an "all or none response"; its magnitude depends upon the strength of the stimulus (see above and **D**).

If a prolonged burst of APs arrives presynaptically, the EPSP response increases with each AP: **synaptic potentiation.** The reason for this is that at higher AP frequencies (approximately 30 Hz) the **presynaptic Ca^{2+} concentration** can no longer drop to its resting value between two APs (increased release of transmitter!).

Inhibitory transmitter substances such as glycine and γ-aminobutyric acid (GABA) increase only g_{Cl} and g_K, but not the g_{Na} of the subsynaptic membrane. The membrane is thus **hyperpolarized** and the excitability of the cell is lowered: This is an **inhibitory postsynaptic potential** (**IPSP**; maximum approx. 4 mV; → **D** and p. 280). EPSP and IPSP can occur *at the same time* in the *same cell*; the sum of all the EPSPs and IPSPs determines whether or not an AP is transmitted postsynaptically (→ **D**).

Artificial Nerve Stimulation

If an electric stimulus is applied externally to a nerve fiber, a current flows from the positive stimulating electrode (*anode*) into the interior of the neuron and leaves again at the negative electrode (*cathode*). The nerve is *depolarized* below the cathode, and if the threshold value is attained an AP is set off (→ p. 26). Usually, an undesirable *hyperpolarization* occurs below the anode. This effect can be greatly reduced by employing a very large electrode, that is, the *indifferent electrode.*

The stronger the stimulus, the shorter the duration required before an AP is elicited (*stimulus / response curve*; → **B**). The **excitability of a nerve** can be characterized by (a) the weakest stimulus (current strength) that can fire APs if applied for infinite time (**rheobase**) and (b) by the **chronaxie**, or the duration of time required to generate APs if the stimulus is twice the rheobase (→ **B**). Chronaxie provides a **measure of the excitability** of the nerve *without a knowledge of the absolute current strength* at the cell membrane of its composing fibers. Therefore, chronaxie can be determined by the application of skin electrodes and is of clinical importance. The same is true for the **measurement of conduction velocity** of a nerve. After excitation by skin electrodes the APs are recorded at a certain distance, and the travelling time is measured. Note that a nerve contains fibers varying in function and diameter and, consequently, in their conduction velocity (*compound AP* of a mixed nerve).

In **accidents with electric current**, the important factor is the amount of current passing through the body. For a given voltage, the lower the *resistance* the greater the current strength. This is why wet skin, which is a good conductor, or bare feet are especially dangerous if electric appliances are being handled (e.g. in bathrooms). *Direct current* (DC) can, in most cases, only act as a stimulus when being switched on or off. *Alternating current* (AC) of *low frequencies* (usually 50 or 60 Hz) can even elicit lethal *ventricular fibrillation* (→ p. 174). On the other hand, high voltage (usually 2000–4000 V DC) is therapeutically applied for **defibrillation** of a fibrillating heart. By this means all the myocardial fibers are depolarized at once, and regular excitation and contraction of the heart can continue. *High-frequency AC* (> 15 kHz) is unable to depolarize nerves and muscles; it does, however, warm up the tissues and is employed therapeutically in **diathermy**.

(Text to **A1–A3** → p. 26 and p. 40)

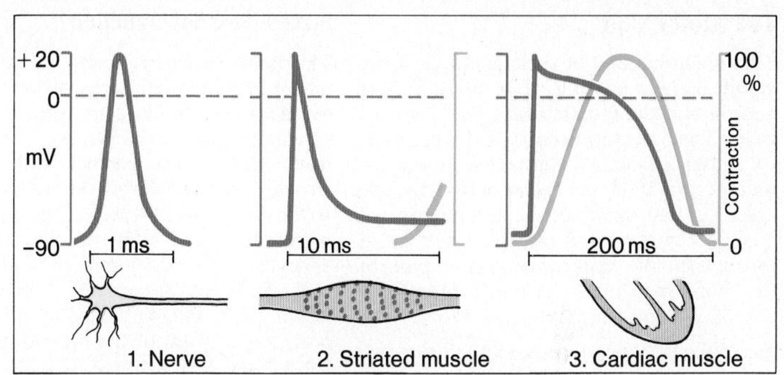

A. Action potential - nerve and muscle

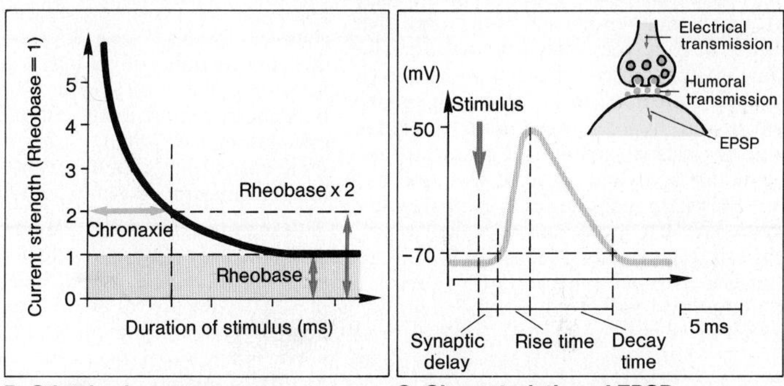

B. Stimulus/response curve

C. Characteristics of EPSP

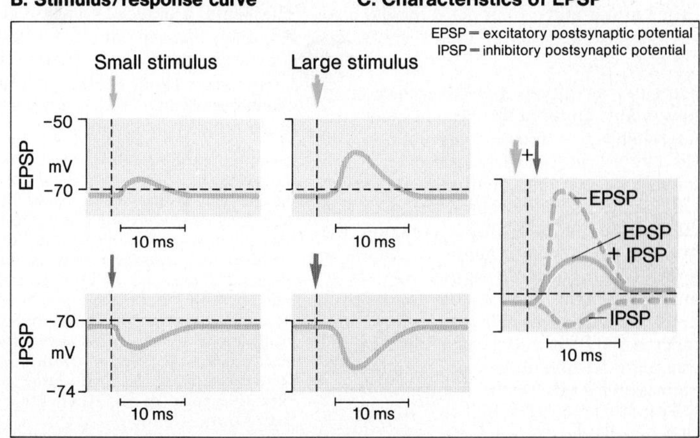

D. Graded response of EPSP and IPSP to stimuli

The Motor Unit

The functional unit of skeletal muscle is the **motor unit**, consisting of a single motor neuron and the muscle fibers that it innervates. The innervation ratio varies in different muscles, from 5 (external eye muscle) to more than 1000 (M. temporalis) muscle fibers per neuron. Since the muscle fibers of a single motor unit may be distributed throughout the entire muscle, the axon of the motor neuron is split into many *collaterals* for their supply.

Two different types of motor unit, **fast twitch** and **slow twitch units**, can be distinguished. To which type a unit belongs is determined around birth by properties of the respective motor neuron (frequency of impulse?). The slow twitch muscles have a more active oxidative metabolism (→ p. 46), are more sensitive to hypoxia, have more capillaries and myoglobin (acting as a small but immediately accessible store of O_2), and are more resistant to fatigue than the fast twitch units. The latter predominate in the *"white" muscles* (e.g. M. gastrocnemius), which are engaged in *rapid movements* (in this case, running and walking). *"Red" muscles,* which are specialized for *maintaining posture* (e.g. the soleus muscle, involved in standing), contain predominantly slow twitch motor units. In humans there is also an intermediate form of motor unit (*fast-"red"*).

Muscular activity is **graded** according to how many motor units become activated (**recruitment**). A muscle may contain 100 (M. lumbricalis) or nearly 2000 (external eye muscles) motor units. Recruitment of more units increases the tension developed. The more motor units a muscle contains, the finer the graduation of muscle activity. The **order of recruitment** of slow and fast or large and small units depends on the type of motor activity (phasic or tonic contraction, reflex activity or voluntary effort). In addition, the tension of each motor unit can be increased by raising the frequency of the neuronal impulses (state of **tetanus** in skeletal muscles; → p. 41, B).

Neuromuscular Junction

The neuromuscular junction between the motor neuron and the muscle, i.e. the **motor end plate** (→ **A**), has a transducer function similar to that of the synapse. The neurotransmitter is **acetylcholine, ACh**, which is stored in the vesicles of the nerve ending. In the *active zones* (→ **A3**) of the presynaptic membrane, the ACh vesicles empty by exocytosis (→ p. 12) into the subsynaptic cleft. Each vesicle contains a certain *quantum* of ACh. Facing the active zone are the *postsynaptic folds* of the muscle membrane (→ **A2** and **A3**), with the **ACh receptors** (→ **A3**). When a molecule of ACh reacts with a receptor, the corresponding membrane channels for Na^+ (and K^+) (→ p. 16, F) open and an influx of Na^+ takes place (2 pA for roughly 0.2–1 ms; → **B1**). One quantum of ACh opens more than 2000 such channels over an area of about 1 μm^2, that is, the ionic current amounts to several nA for a few ms (**miniature endplate current**; → **B2**).

Although single ACh quanta discharge spontaneously, this is insufficient to elicit muscle contraction, which can only take place after the arrival of APs via the motor neuron. The resulting influx of Ca^{2+} in the nerve ending (→ p. 54) causes hundreds of quanta of ACh to empty simultaneously. The neurally induced **endplate current** (→ **B3**) then fires an AP in the muscle and contraction takes place. ACh is very rapidly *broken down* again in the synaptic cleft by **cholinesterases** (→ p. 54), thus making possible rapid repolarization.

A number of **poisons** or **drugs** block neuromuscular transmission (→ also p. 54). *Botulin toxin* inhibits the emptying of the vesicles. *Curare* displaces the ACh from its binding site (*competitive inhibition*) but is *not depolarizing* itself. Curare-like substances are used for muscular **relaxation** during operations. *Cholinesterase inhibitors* reverse this inhibition by locally raising the concentration of ACh. At an intact synapse, the cholinesterase inhibitors bring about a *long-lasting depolarization*, with inactivation of the Na^+ channels (→ p. 26) and muscular paralysis. *Succinyl choline* has the same effect. It depolarizes like ACh but is broken down more slowly.

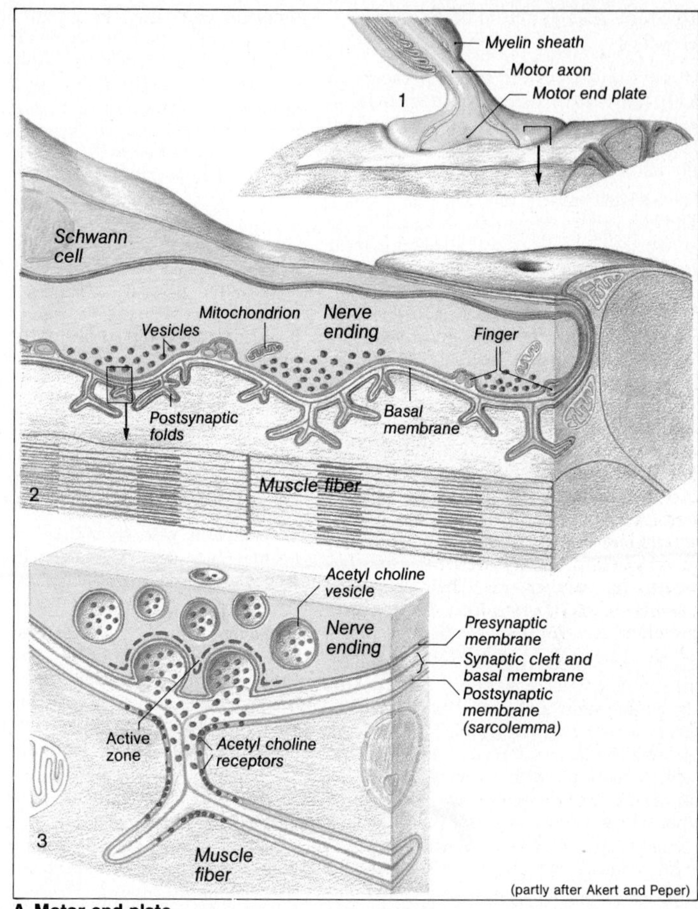

(partly after Akert and Peper)

A. Motor end-plate

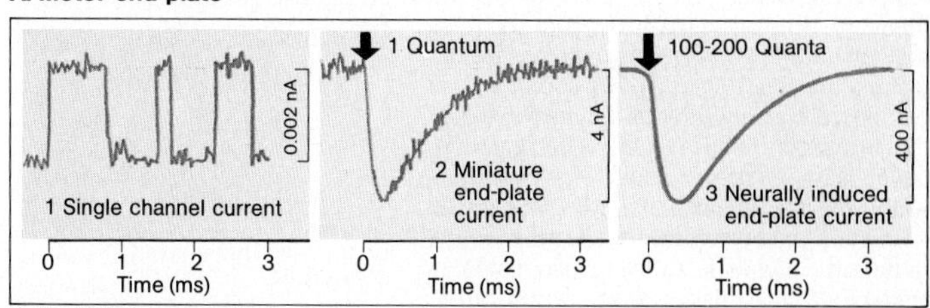

B. End-plate currents

(after Neher and Sakmann(1), after Peper and co-workers (2))

Structure and Function of Skeletal Muscle I

Muscle transforms the **chemical energy** of **ATP** (→ p. 19f.) directly into **mechanical energy** (and heat energy), a process in which enzymatic and structural elements are equally involved.

The **striated muscle cell** is a **fiber** with a diameter of 10–100 μm and a length of up to 20 cm (→ **A 2**). What we call fibers in meat are in fact *bundles of fibers* (diameter approximately 0.1–1 mm; → **A 3**). The cell membrane of the muscle fiber is called *sarcolemma*, and encloses the *myofibrils*, *sarcoplasm* (cytoplasm), several cell nuclei, *sarcosomes* (mitochondria), lysosomes, fat droplets, glycogen granules, and various other organelles and inclusions. Among the many substances dissolved in the sarcoplasm are glycogen, myoglobin, glycolytic enzymes, creatine phosphate, and amino acids. Each muscle fiber contains several hundred **myofibrils** (→ **A 1**), each of which is subdivided by the Z-plates into compartments, the **sarcomeres**. Their length is roughly 1.5 to 3.0 μm, depending on prestretching of the muscle (→ **B**).

Under the microscope (two-dimensional), the sarcomeres of a myofibril (→ **A**) show alternating light and dark *bands* and *lines* (hence *striated* muscle), which result from the arrangement of the (thick) **myosin filaments** and (thin) **actin filaments** (→ **B**). A sarcomere lies between two *Z-lines*, or (three-dimensional) *Z-plates* (→ **B**), which are in fact platelike protein structures. The *actin filaments* are inserted through and attached to the Z plates in such a way that half of the chain of each of the 2000 actin filaments is in one sarcomere, and its other half is in the adjacent sarcomere. In the vicinity of the Z plates, the sarcomere is composed entirely of actin filaments; this is the so-called *I band* (→ **B**). The region in which actin and myosin filaments overlap is visible as the *A band*; the *H zone* is the portion of the sarcomere that consists solely of *myosin filaments* (about 1000 per sarcomere), which thicken in the center of the sarcomere to give the *M line*.

The **myosin molecule** (→ **B, C**) consists of a long tail (consisting of a pair of strands twisted around each other) joined by a neck region to a double head, with each half containing an ATPase (→ p. 36ff.). The attachment of the tail (light meromyosin) and the head to the neck (head and neck = so-called heavy meromyosin; one pair per myosin molecule; → **C**) is *joint-like*, permitting a certain degree of movement. A myosin filament is a bundle of about 150–360 such molecules. The ability of the head-neck parts of the molecule to swivel enables the myosin to form a reversible complex with actin (*actomyosin complex*; → p. 38) and makes it possible for the actin and myosin filaments to slide along one another (*sliding filament mechanism*; → p. 36 and p. 38).

Actin is a globular protein molecule (G actin), 400 of which form a chain of beads, namely, actin F. Two such chains, intertwined, constitute the **actin filament** (→ **B**).

Tropomyosin, also threadlike, winds itself around the actin filament whereby one **troponin** molecule is attached every 40 nm (→ **B**).

Troponin is made up of three subunits: (a) troponin-C combines with Ca^{2+}; (b) troponin-T combines troponin with tropomyosin; (c) troponin-I inhibits the formation of bridges between myosin and actin in the resting state. Its inhibitory effect is cancelled when troponin-C is saturated with Ca^{2+}.

During contraction the tropomyosin "thread" settles into the groove between two actin F-chains, thus leaving the binding sites free for the myosin. This switch-over is brought about by the Ca^{2+}-sensitive troponin (→ pp. 36–39).

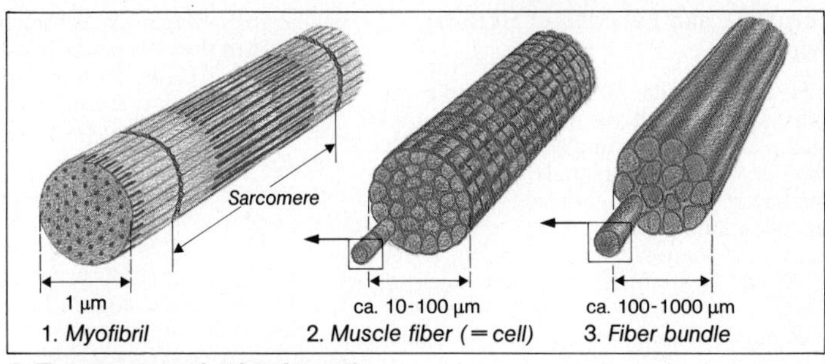

A. Fine structure of striated muscle

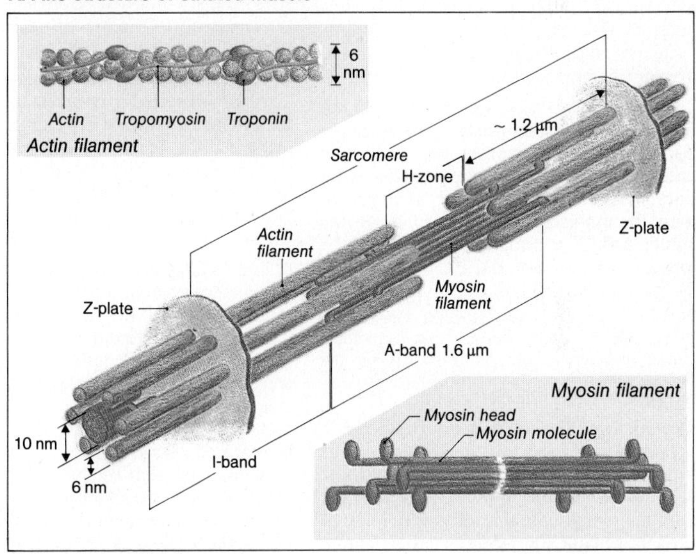

B. Structure of the sarcomere

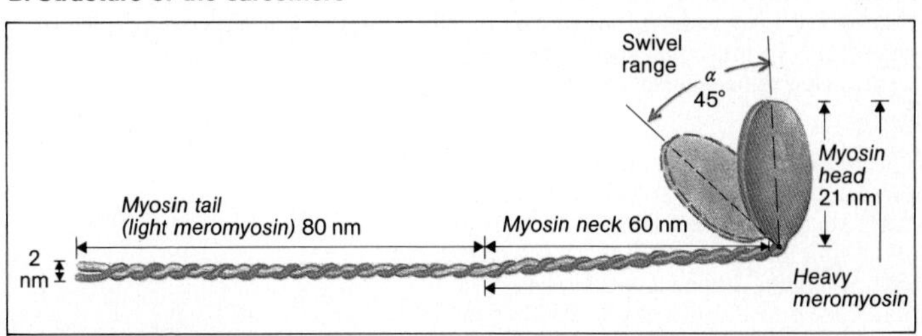

C. Myosin molecule

Structure and Function of Skeletal Muscle II

The motor units of skeletal muscle are normally activated via their motor neuron (indirect muscle stimulation; → p. 32). Neuronal action potentials release acetyl choline at the motor endplate, inducing an endplate current that spreads electrotonically to a limited extent (→ p. 28). If the threshold is attained, an action potential (AP) is set off and spreads over the sarcolemma of the entire muscle fiber (= muscle cell). In many places, the sarcolemma is deeply invaginated perpendicular to the muscle fibrils, constituting a system of **transverse tubules** known as the **T system** (→ **A**).

The endoplasmic reticulum (→ p. 4) of the muscle cell, the **sarcoplasmic reticulum** (→ **A**), is specially organized into a blindly ending **longitudinal tubular system** lying parallel to the myofibrils and unconnected with the extracellular space (ECS) or the sarcoplasm. They form a **reservoir for Ca^{2+} ions**.

The T-system runs in very close proximity to and between the ends of two adjacent longitudinal tubules. The so-called *triads* seen in a microscopic section are composed of one transverse tubule and the two bulbous ends of the longitudinal tubules of the sarcoplasmic reticulum (→ A).

The AP rapidly spreads throughout the T system, which is part of the ECS, and thus quickly reaches even the innermost parts of the muscle fiber; Ca^{2+} is released from the adjoining longitudinal tubules. Elevation of the **intracellular Ca^{2+} concentration** from a resting value of about 0.01 µmol/l to 1–10 µmol/l triggers a series of events that culminates in muscular contraction; this is termed **electromechanical coupling** (→ **B** and p. 38).

The myosin and actin filaments of a sarcomere (→ p. 34) are so arranged that they can slide over one another so as to overlap and bring about shortening of the muscle, that is, **sliding filament mechanism**. The Z plates are pulled towards one another as the degree of overlap of thick and thin filaments increases; I bands and H zones become shorter. (The length of each filament is unchanged!) Maximum contraction is attained when the ends of the thick filaments butt against Z plates. At this stage the ends of the thin filaments (actin) overlap at the center of the sarcomere (→ p. 41, C).

ATP (→ p. 19f.) is essential for the sliding of the filaments (and thus for muscle contraction), and the *myosin heads* (→ p. 35, C) possess the ATPase activity that breaks down ATP. After Ca^{2+}-troponin interaction (→ p. 34), the myosin heads become attached to the thin filaments at a particular angle (also called **cross-bridges**); following a structural alteration in the myosin molecule the heads tilt to a different angle, drawing the thin filament along with them (→ **C** and p. 38).

This pull is exerted in opposite directions at the two ends of the myosin filament with the result that the region of overlap of actin and myosin increases on both sides of the Z line. Shortening of the sarcomere thus takes place at both ends of the myosin bundle (→ p. 35).

A single **sliding cycle** or "**rowing stroke**" (→ **C**) shortens a sarcomere by 2 × 8 nm, which is barely 1 % of the sarcomere length of 2 µm. This means that the entire muscle fiber (with a maximum length of 20 cm), which consists of serially arranged sarcomeres, also shortens by 1 %. In order to achieve 50 % shortening of a muscle, the cycle has to be repeated a large number of times, that is, frequent repetition of the cycle of attachment of the myosin heads – tilting and sliding – detachment of the heads – assumption of the original angle – attachment to the next portion of the actin filaments, and so on (→ **C1–C3**).

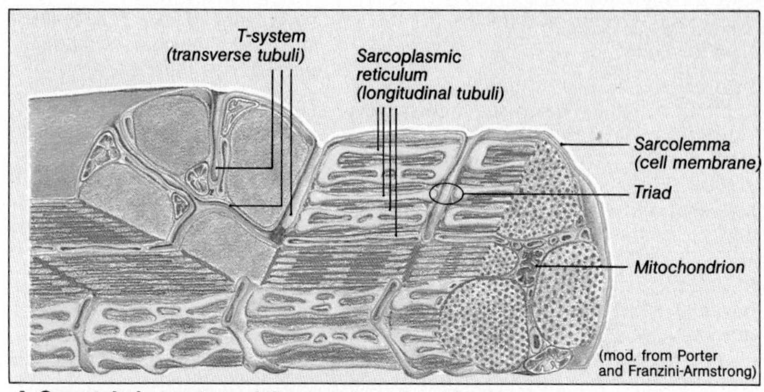

A. Sarcotubular system of the muscle cell (= fiber)

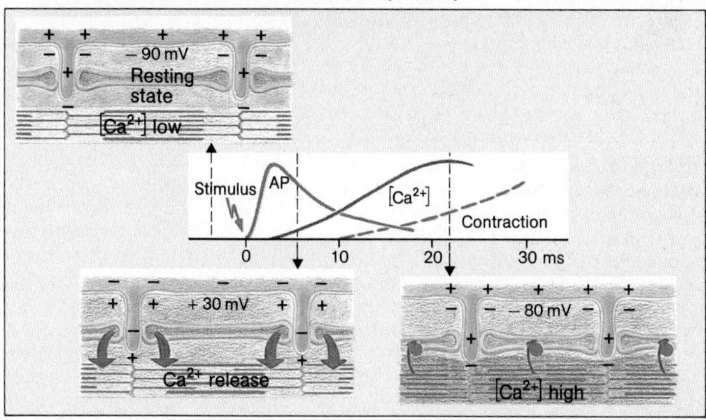

B. Ca^{2+} as mediator between electrical stimulation and contraction

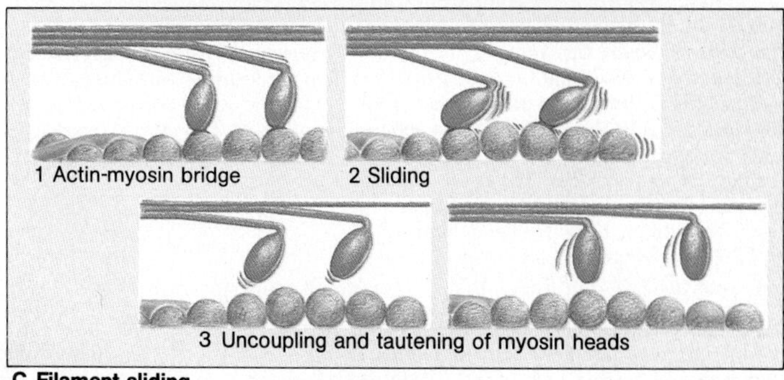

C. Filament sliding

Molecular Events in Muscle Contraction

In order to contract, a muscle requires, in addition to actin and myosin, among other substances, Ca^{2+}, Mg^{2+}, ATP, and ATPase. **Ca^{2+}** is stored in high concentrations in the longitudinal tubuli of the sarcoplasmic reticulum (→ p. 36). An incoming action potential (AP) spreads throughout the muscle fiber via the T system and renders the longitudinal tubuli temporarily permeable to Ca^{2+}. As a consequence, the Ca^{2+} concentration in the muscle cell rises by about 1000-fold. This Ca^{2+} binds with troponin, and tropomyosin loses its inhibitory action on the formation of the actin-myosin (A-M) complex (→ **A** and p. 34ff.). The Ca^{2+} set free is immediately pumped back into the longitudinal tubuli (active transport; → **A** and p. 11), with one molecule of **ATP** being consumed for the transport of two Ca^{2+}-ions.

Each of the two myosin heads (**M**) of a myosin molecule also binds one **ATP**. In this form (M-ATP complex), their heads hang at an angle of 90° to their necks (→ p. 35, C). At high intracellular concentrations of Ca^{2+} the myosin heads bind to the actin (**A**), the actin activates the **ATPase** of the myosin head, and the ATP of the complex is split up into ADP and inorganic phosphate (**P_i**; → **A 1**). Three mmol/l **Mg^{2+}** ions are required for this process, which results in the formation of an A-M-ADP-P_i complex (→ **A 1**). When the P_i dissociates from the complex the myosin heads tip from 90° to 50° (→ **A 2 a**) with the result that the actin fibers and myosin fibers slide along one another. With the loss of ADP the myosin heads achieve their end position of 45°, which puts a stop to the sliding (→ **A 2 b**). The A-M complex that is left is stable (*rigor complex*) and can only be broken by a renewed binding of ATP to the myosin heads: *Plasticizing or relaxing effect of ATP* (→ **A 3**).

The ready extensibility of muscle in the resting state (since the A and M filaments are free to slide past each other) is important, for instance, for the passive diastolic filling of the heart or for movement, since shortening of one muscle mostly requires the simultaneous lengthening of its antagonistic muscles.

In the dead organism no more ATP is formed, Ca^{2+} cannot be pumped back into the longitudinal tubuli, and the A-M complex cannot be split. This is the basis for the development or *rigor mortis*, which persists until the actin and myosin molecules are broken down by decay processes.

In the presence of ATP, the myosin heads resume their original angle of 90° upon dissociation from the actin (45° → 90°; → **A 4**). If the intracellular Ca^{2+} concentration is still sufficiently high, which is mainly dependent upon the frequency of incoming APs, the cycle A 1 – A 4 begins again (up to about 50 times for full shortening of the muscle). The cycle does not proceed simultaneously in all of the myosin heads; if it did, muscular contraction would be jerky. Although at any one moment only part of the myosin heads are tilted the total number is the same, thus ensuring smooth, continuous contraction. When the intracellular Ca^{2+} drops below roughly 1 µmol/l, the sliding cycle is ended (*resting state*; → **A**).

The sliding cycle as described above applies to an *isotonic* muscle contraction, that is, with shortening of the muscle. In a strictly *isometric* contraction (length of muscle remains constant, tension is generated), the tipping of the myosin heads is largely prevented and the A-M-ATP complex (→ **A 3**) is probably immediately converted to the A-M-ADP-P_i complex. Most of the isometric muscular tension is generated by the *attempt* of the heads to tip. This explanation implies that the *serial elastic component* (→ p. 40; and p. 41, A) of the muscle is primarily situated in the neck-head region of myosin.

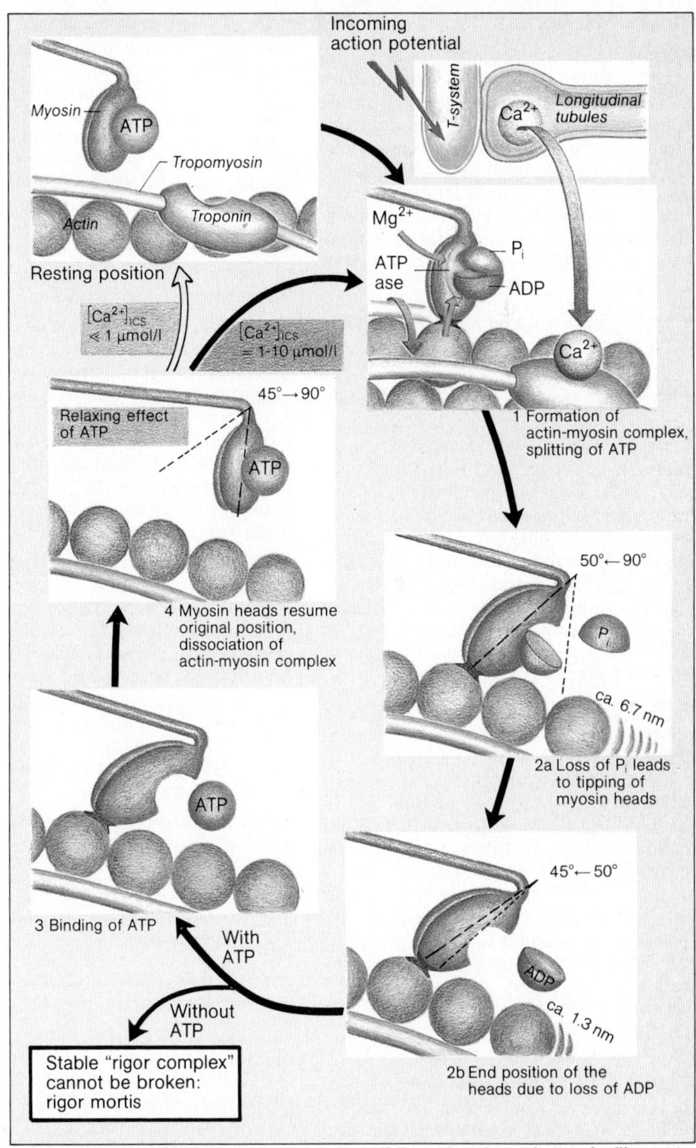

A. Molecular mechanisms involved in the sliding of actin and myosin filaments (isotonic contraction)

Muscle Mechanics I

When an endplate potential (→ p. 32) exceeds the threshold value, an action potential (AP) is initiated (max. depolarization after about 2 ms; → p. 31, A2) and spreads rapidly (2 m/s) over the muscle fiber into the T system. The maximal intracellular Ca^{2+} concentration is attained after, at most, 10 ms; the maximal muscle contraction, after 10 ms (external eye muscles) to more than 100 ms (M. soleus) (see, for example, p. 37, B).

Muscular force is varied by (a) the number of motor units activated (→ p. 32): *recruitment* of more motor units increases the muscle force developed, and (b) the *frequency* of the APs.

A single stimulus invariably causes a maximal release of $\mathbf{Ca^{2+}}$ and thus always produces a maximal *single* twitch of the striated muscle fiber (**all-or-none rule**). Nevertheless, a single stimulus does not elicit the maximal possible shortening of the muscle fiber since it is too brief to maintain the relatively slow filament sliding until its end position is reached. Further shortening can only be achieved if a *second stimulus* arrives *during* the first twitch.

In this way repeated stimuli lead to a superposition of single twitches called **mechanical summation** (→ **B**). If the frequency of the stimuli is increased so that the intervals between the APs becomes less than 1/3 of the time required for a single twitch (20 Hz in slow muscles, up to 60–100 Hz in fast muscles; → p. 32), the twitches fuse and a maximal possible contraction of the motor unit results. This is known as **tetanus** (→ **B**). In tetanic contraction the force generated can be as much as four-fold that arising from a single twitch. Whereas the $\mathbf{Ca^{2+}}$ **concentration** sinks again between the individual stimuli in incomplete summation, it remains elevated during tetanus.

Another kind of prolonged shortening of the muscle, apart from tetanus and rigor (→ p. 38), is known as **contracture**. This is a state of contraction *not* caused by a conducted AP but evoked either by a *sustained local depolarization*, as for example in connection with elevated extracellular K^+ concentrations (*K^+ contracture*), or by pharmacologically induced *release of Ca^{2+}* inside the cell, for example, by caffeine. The contraction of the so-called **tonus fibers** (certain fibers of the external eye muscles and the muscle spindles; → p. 278) is also a type of contracture. Such fibers respond to a stimulus in proportion to the extent of an *AP-independent depolarization* and therefore *not* with an all-or-none contraction. The degree of contraction depends upon the changes in the intracellular concentration of Ca^{2+}.

In contrast, the general or **reflex tone** of the skeletal musculature is maintained by normal APs in the individual motor units. Single twitches are not detectable, due to the asynchronous stimulation of the various motor units. The postural muscles, especially when apparently at rest, are in a state of involuntary tension that is enhanced, for example, by increased awareness and is under reflex control (→ p. 278ff.).

Muscle contraction can be measured under two extreme conditions: (a) **isometric:** the *length* of the muscle remains *constant* and the tension changes; (b) **isotonic:** the length changes and the *tension* remains *constant* (→ **A**). If both tension and length change simultaneously, the contraction is termed **auxotonic**. If an isometric contraction is followed by an isotonic or auxotonic contraction, it is called an **afterloaded contraction** (e.g. → p. 182).

Muscle contains elastic components that are arranged either in parallel or in series with the sarcomeres (→ **A**): (1) The *parallel elastic component* (**PEC**), provided by the membrane of the muscle fibers (sarcolemma) and the connective tissue (fasciae), prevents the filaments from falling apart when stretched in the resting state. The effect of the PEC is shown quantitatively by the resting tension curve (→ p. 43, A and B). (2) The *serial elastic component* (**SEC**) is found in isometric twitches in which the total length of the muscle remains unchanged as the short sliding of the filaments stretches the connective tissue slightly (tendons, etc.). The second part of the SEC is due to the attempt of the filaments to slide and is a property of the neck region of myosin (→ p. 38).

Text to **C** → p. 42).

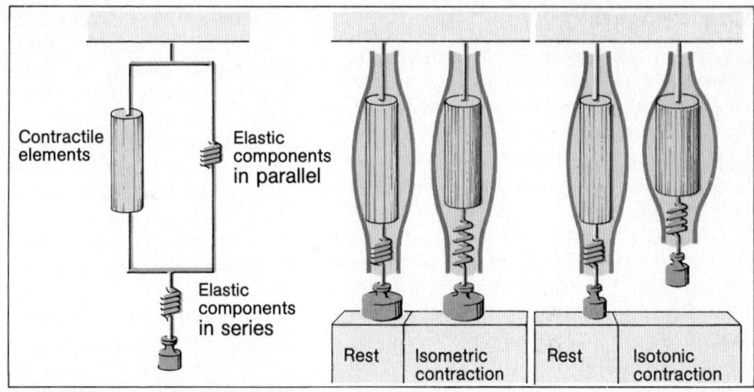

A. Model of muscle contraction

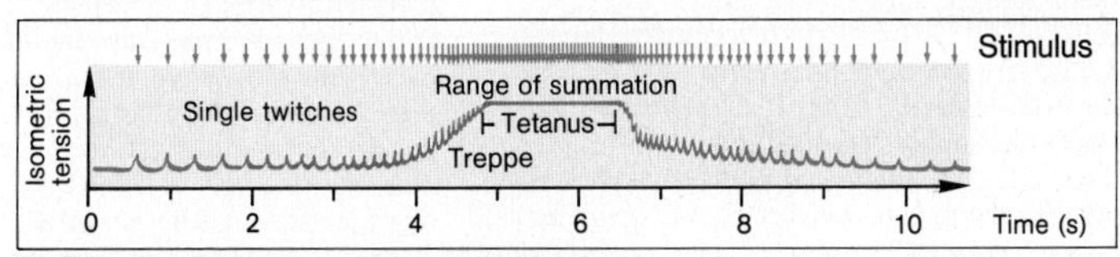

B. Stimulus frequency and muscle tension

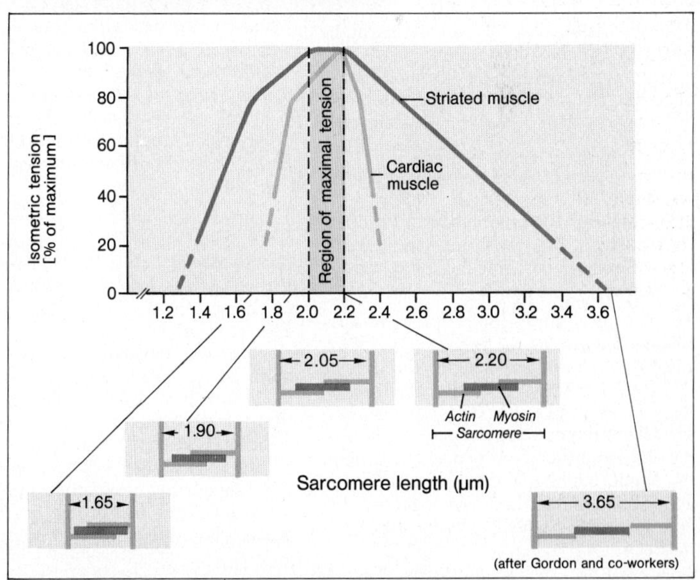

C. Isometric tension relative to sarcomere length

Muscle Mechanics II

Length (L) and **tension** (T) of a muscle are closely related (→ **B**, and p. 41, C). The *total tension* generated by a muscle is the sum of the active and resting tensions.

Active tension is determined by the number of actin-myosin cross-bridges and varies with the initial fiber or sarcomere length (*prestretching*; → **A**). The *maximal* active isometric tension (T_o) can develop at the **resting length** (L_{max}; sarcomere length about 2–2.2 µm; → p. 41, C), at which the maximal number of cross bridges can form. As the sarcomere shortens ($L < L_{max}$), the thin filaments begin to overlap and the tension developing is smaller than T_o (→ p. 41, C). At $L = 70\%$ of L_{max} (sarcomere length 1.65 µm), the thick filaments butt against the Z plates and T falls off even more. When the sarcomere is stretched to lengths greater than L_{max}, the degree of overlap decreases, fewer cross-bridges can form and, again, less tension than the maximum can develop (→ p. 41, C).

The *length-tension* (L-T) *relationship* can be modified by the intracellular Ca^{2+} concentration and permits the fine modulation of muscle response; it plays its greatest role in the cardiac muscle (**homeometric regulation**).

Resting tension is developed when the fibril is stretched in the resting state ($L > L_{max}$). If a fibril is stretched to more than 130% of the resting length, the resting tension accounts for the larger part of the **total tension** developed (→ **A** and **B**).

The L-T curve corresponds to the **pressure-volume diagram of the heart** (→ p. 182f.). Instead of the muscle length, the cardiac volume is measured; instead of the tension, the ventricular pressure. The diastolic pressure corresponds to the resting tension of the cardiac muscle and is dependent upon the degree of filling, so that the end-diastolic volume determines the stroke volume of the heart: **Frank-Starling mechanism** (→ p. 182f.).

Main differences between skeletal and cardiac muscle (see also p. 45):

1. Striated muscle can be stretched to greater fiber lengths than cardiac muscle, that is, when stretched to the same extent the passive resting tension of cardiac muscle is greater than that of skeletal muscle.
2. Skeletal muscle normally operates on the *plateau* of the L-T curve whereas cardiac muscle has no plateau and operates on the ascending limb below L_{max} (→ **B**), thus furnishing the heart with a *reserve* that allows the tension to increase when the heart is stretched by a greater diastolic filling (Frank-Starling mechanism).
3. Cardiac muscle has a longer-lasting AP (→ p. 31, A 3) because after the usual fast inactivation of Na^+ channels, g_K drops transiently and g_{Ca} becomes elevated for 200 to 500 ms. Thus, the resulting "slow" Ca^{2+} influx into myocardial cells causes a **plateau in their AP**, that is, the refractory period continues until the contraction has almost ceased (→ p. 45). Cardiac muscle can therefore **not be tetanized**.
4. There are no motor units in the heart (→ p. 32). Unlike the situation in skeletal muscle, the excitation spreads over the entire atrial or ventricular myocardium: *all-or-none contraction*.
5. The force of contraction of cardiac muscle can be varied by the length of the AP, which is in turn regulated by changes in the Ca^{2+} influx into the cells (→ p. 166).

The greater the **load** (force), the smaller the **speed** of an isotonic contraction (**force-velocity curve**; → **C**). The *maximal force* or *tension* (+ little heat) is developed if there is no shortening, as when pulling or pushing. The *maximal velocity* (e.g. M. biceps brachii: 7 m/s) of shortening (V_{max}) occurs (and more heat is produced) when little or no load is involved, as when playing an instrument or throwing an object. The force-velocity curve shows that light weights can be lifted faster than heavy weights (→ **C**). The total energy consumption (mechanical work and heat) in isotonic contraction is greater than in isometric contraction.

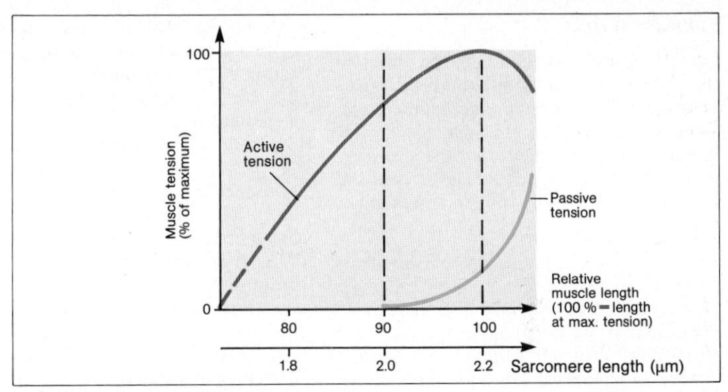

A. Active and passive tension of the muscle

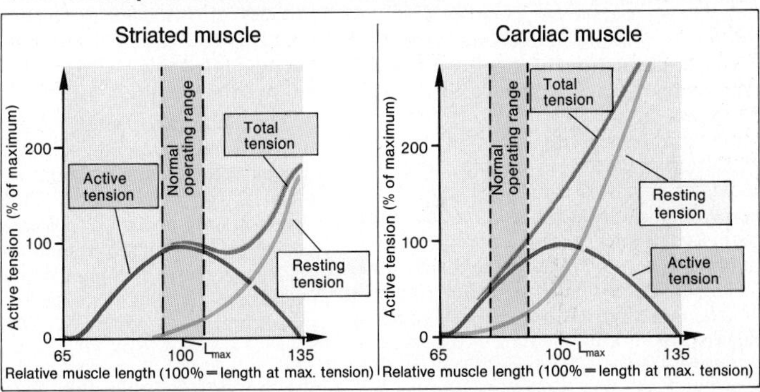

B. Length-tension curves for striated and cardiac muscle

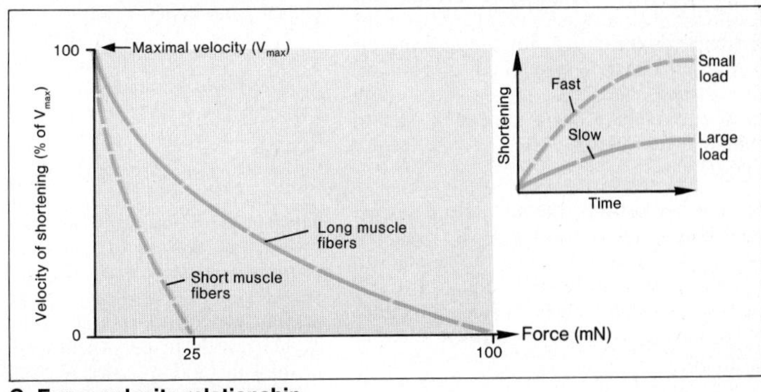

C. Force-velocity relationship

Smooth Muscle

Smooth muscle includes all the different types of muscle that do not have cross-striations. They are of great clinical significance since they are involved in the function of many organs (stomach, intestines, bladder, uterus, bronchi, eyes, etc.) and control blood flow by influencing vascular resistance.

Smooth muscle contains actin F-filaments (→ p. 35) and a kind of myosin, but thick filaments are not common and there is no organization into sarcomeres (hence the absence of cross-striations). The tubular system (→ p. 36) and the triads encountered in striated muscle are not found in smooth muscle. The **membrane potential** of smooth muscle is often unstable and subject to **rhythmic fluctuations** of low frequency and amplitude (e.g. gastrointestinal tract 3–15 min^{-1} and 10–20 mV). If depolarization connected with the slow wave of excitation exceeds a certain threshold, bursts of **action potentials (spikes)** are triggered off. Their number and frequency rises with the degree of slow spontaneous depolarization. A spike is followed after 150 ms by a slowly increasing and declining contraction that reaches its peak approximately 500 ms after the start of the spike (→ **A**, left diagram). The duration of the contraction depends upon the number of spikes. Compared with skeletal muscle, a very low frequency of spikes suffices for fusing of the twitches (**tetanus**; → p. 40). Smooth muscle is thus in a constant state of partial contraction ("**tone**"; see below). In some smooth muscles the spike has a plateau similar to that of the cardiac AP (→ **A**, center diagram).

As in other muscles, the membrane potential is largely determined by the K^+ gradient (→ p. 24). The contraction of the smooth muscle is set off by the influx of $\mathbf{Ca^{2+}}$ (mainly from the ECS). **Calmodulin** probably plays a role comparable to that of troponin in skeletal muscle (→ p. 34).

On the basis of the type of excitation, **two kinds of smooth muscle** can be distinguished:

1. The smooth muscle of the *visceral organs* such as the stomach and intestines (→ p. 210), urinary bladder, ureter, and uterus belong to the **single-unit type**. Their muscle cells are interconnected by cellular bridges or **gap junctions** (→ p. 7, D), which have a high conductance for ions. As in the heart (→ p. 206), some of the cells depolarize spontaneously (**pacemaker potential**), and the excitation spreads via the gap junctions throughout the entire smooth musculature of the organ involved (hence *single-unit* type of muscle). Contraction of this type of muscle is thus largely *independent of external nervous impulses* and can persist over longer periods of time: **myogenic tone**. **Stretching** of the muscle causes a depolarization and enhances the existing tone. The muscles of the *smaller blood vessels* are mainly of this type, and their contraction as a result of stretching is involved in autoregulation of the blood flow (→ p. 176).

2. The **multi-unit type** of smooth musculature is also present in most blood vessels, as well as in the epididymus und vas deferens, the iris, and ciliary body. The excitation arises mainly from the **autonomic nerves** rather than from the muscle itself: **neurogenic tone**. Gap junctions are rare. Thus, the excitation remains localized within the motor unit (→ p. 32) involved (hence *multi*-unit type of muscle).

Besides **Acetylcholine** and **norepinephrine** from the autonomic nerve endings (→ p. 54ff.), **hormones**, reaching the cell via the circulation, exert an influence on smooth muscle. For example, the *uterine musculature* reacts to estrogens, progesterone, and ocytocin (→ p. 262ff.); the musculature of the *blood vessels* reacts to histamine, angiotensin II, adiuretin (vasopressin), serotonin, bradykinin, etc.

The *length-tension* curve for smooth muscle shows that tension at a particular length progressively decreases if the muscle is kept stretched. This property of **plasticity** is responsible, for example, for the large capacity of the urinary bladder. The tension in its wall (and its internal pressure) only rises to any considerable degree when the bladder is almost full, so that not until this point is the urge to urinate experienced.

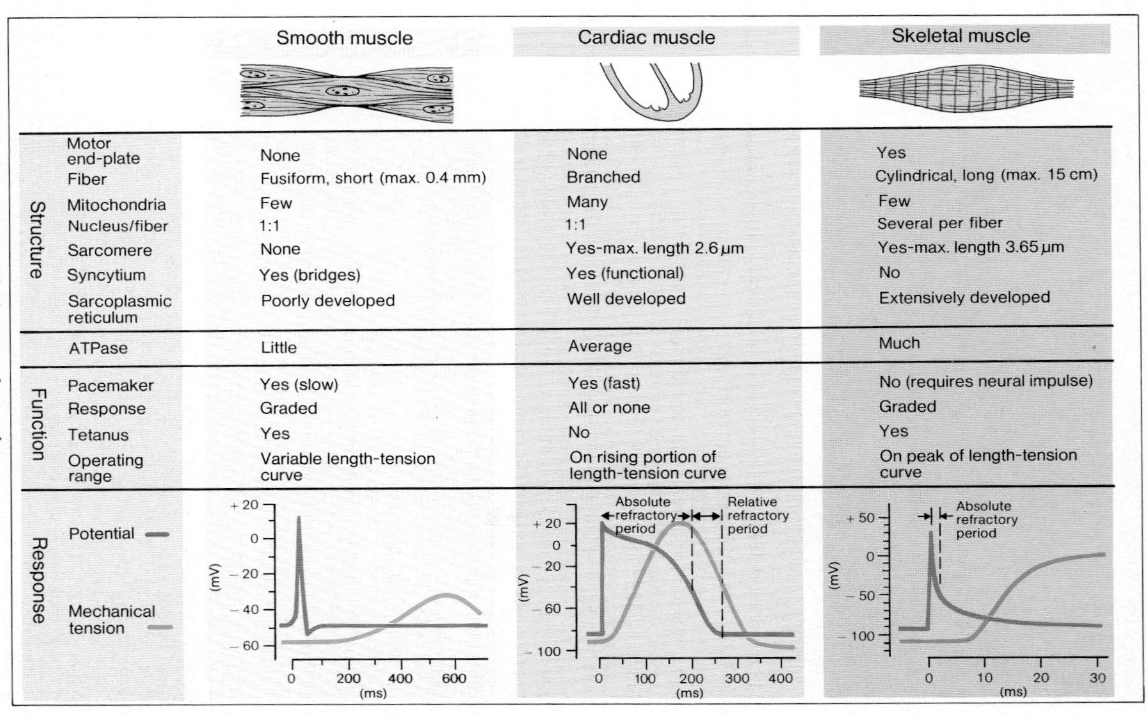

		Smooth muscle	Cardiac muscle	Skeletal muscle
Structure	Motor end-plate	None	None	Yes
	Fiber	Fusiform, short (max. 0.4 mm)	Branched	Cylindrical, long (max. 15 cm)
	Mitochondria	Few	Many	Few
	Nucleus/fiber	1:1	1:1	Several per fiber
	Sarcomere	None	Yes-max. length 2.6 µm	Yes-max. length 3.65 µm
	Syncytium	Yes (bridges)	Yes (functional)	No
	Sarcoplasmic reticulum	Poorly developed	Well developed	Extensively developed
	ATPase	Little	Average	Much
Function	Pacemaker	Yes (slow)	Yes (fast)	No (requires neural impulse)
	Response	Graded	All or none	Graded
	Tetanus	Yes	No	Yes
	Operating range	Variable length-tension curve	On rising portion of length-tension curve	On peak of length-tension curve
Response	Potential / Mechanical tension			

A. Structure and function of 3 classes of muscle

Energy for Muscle Contraction

The mechanical energy of muscle contraction comes directly from **chemical energy**, that is, from high-energy **ATP** (→ p. 20). It is split into ADP and inorganic phosphate (→ **A**) with the release of energy (ΔG; → p. 18) that is used for filament sliding (→ p. 38). *No O_2* is needed for breaking down ATP to ADP, that is, the process of muscular contraction can take place *anaerobically*. ATP is immediately regenerated by the three following processes: (1) Breakdown of creatine phosphate (**CrP**; → **A**); (2) anaerobic glycolysis, that is, breakdown of glycogen or glucose to lactic acid; and (3) aerobic oxidation of glucose to CO_2 (→ also p. 198).

CrP is present in the muscle and represents a rapidly available energy reserve. Its high-energy phosphate bond can be transferred to ADP for the (anaerobic) regeneration of ATP (→ **B1**). The approx. 5 µmol ATP/g muscle are sufficient for only about 10 contractions, whereas with the help of the roughly 25 µmol CrP/g muscle about 50 more contractions are possible before this reserve is also exhausted. The energy of CrP is sufficient to support a *maximal effort* (100 m sprint) over a *brief period* (10–20 s).

Anaerobic glycolysis begins slightly later than CrP breakdown (maximum after about 0.5 min). The **glycogen** reserves of the muscle are broken down to **lactic acid** via glucose-6-phosphate, with the production of 3 mol ATP per mol glucose residue (→ **B2**). In *light work*, this type of ATP formation is succeeded after about 1 min by an aerobic oxidation of glucose that is more productive of energy. If, however, in *heavy work*, the aerobically produced energy is insufficient to meet the energy requirements, anaerobic glycolysis proceeds at the same time. *Blood glucose* is now broken down to lactic acid (energy gain in this case is only 2 mol ATP/mol glucose, since 1 ATP is required for the phosphorylation of glucose). A limit to this type of energy production is set by the *accumulation of lactic acid*, which is buffered to lactate (→ p. 110f.).

Prolonged muscular activity is only possible if the energy is produced **aerobically** from glucose (2 + 34 mol ATP/mol glucose!) and fats (→ **B3**). The muscular blood flow, cardiac performance, respiration, and so on, have to be increased until they meet the requirements of muscle metabolism (pulse frequency becomes constant; → p. 49, B).

Several minutes elapse before this "steady state" is achieved, during which time the gap is bridged, on the one hand, by anaerobic energy production (see above) and, on the other hand, by tapping the short-lived oxygen reserve of the muscle (*myoglobin*) and by increased oxygen extraction from the blood. The transition from the one phase to the other is often experienced as a phase of exhaustion and fatigue.

The **limit for prolonged high performance** in top athletes amounts to about 370 W and depends primarily upon the speed of the O_2 supply and of the aerobic breakdown of glucose and fat. If this limit is exceeded, equilibrium between metabolism and cardiovascular function is not attained (the pulse, for example, continues to rise; → p. 49, B). Although the energy gap can be bridged temporarily by *continuation of anaerobic glycolysis* (see above) the lactic acid formed **lowers pH**, both locally in the muscle and systemically. Consequently, the chemical reactions necessary for muscle contraction are inhibited; insufficient ATP is available, **fatigue** sets in, and the work has to be terminated.

In CrP breakdown and in anaerobic glycolysis the organism accumulates an **O_2 debt**, which enables it to sustain for about 40 s a performance approx. threefold that achieved with the relatively slower aerobic oxidation of glucose. In the ensuing phase of recovery this O_2 debt (max. more than 20 l) has to be paid off, which explains why the O_2 consumption remains high for some time even though the body is at rest. The increased cardiac and respiratory work during the phase of recovery is one of the reasons why considerably more O_2 has to be repaid than has been borrowed. The elevated energy turnover during recovery serves mainly to replenish the CrP, O_2, and glycogen reserves, and employs some of the accumulated lactate.

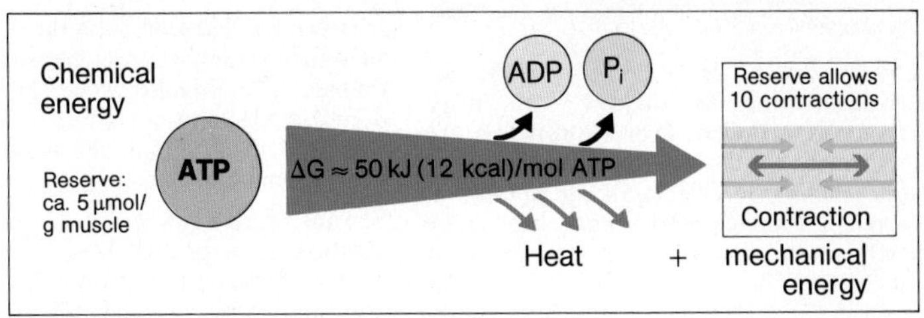

A. ATP as direct energy source

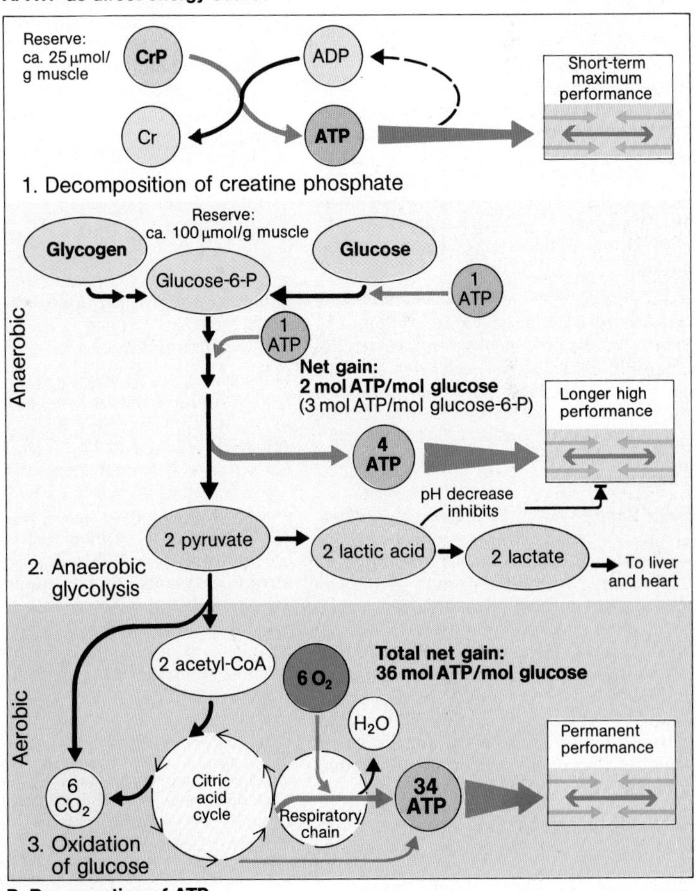

B. Regeneration of ATP

Exercise

The muscular work performed by the organism can be divided into three categories: 1. **Positive dynamic work**; work is performed by the alternately contracting and relaxing muscle (walking uphill). 2. **Negative dynamic work**; work has to be performed to counteract the stretching of the muscle fibers, in alternation with contraction under zero load (walking downhill). 3. **Static (postural) work** (e. g. standing still).

Many activities involve a combination of two or three of the above types of work.

External mechanical work is performed in dynamic, rhythmic work, but not in purely postural work since in this case force times distance (→ p. 327) = 0. In both cases there is an expenditure of chemical energy, which, in postural work, is completely transformed into the *heat of maintenance* at a rate proportional to the muscle tension times the duration of postural work.

In heavy muscular work the muscles have to be supplied with up to 500 times more $\mathbf{O_2}$ than when the body is in a resting state. At the same time, larger quantities of the metabolic breakdown products, CO_2 and lactate, have to be removed (→ p. 46). To meet these needs, considerable adjustments in the cardiovascular and respiratory systems take place.

Blood flow (→ **A**): The increased blood flow in the muscles themselves is achieved by **local chemical means**, that is, the rising P_{CO_2}, the sinking P_{O_2}, and the drop in pH due to the accumulation of lactic acid etc., cause **local vasodilatation** (→ p. 46). In purely postural work, increased blood flow of this nature is partially hindered by the pressure exerted on its own vessels by the continuously contracted muscle. This is the reason why **fatigue** sets in sooner when standing than when performing rhythmic dynamic work.

Heart: During maximal muscular work the blood flow to the muscles alone amounts to about 25 l/min, or 4–5 times the total resting cardiac output (→ p. 154). The drastic increase in **cardiac output** needed for maximal muscular work (up to 30 l/min) is achieved by **increased pulse rate** (→ **B**) and an approximately 1.2-fold **increase in stroke volume**. The **systolic blood pressure** (→ p. 160) **climbs** to values of more than 25 kPa (185 mmHg), although the diastolic pressure remains unchanged.

The increased cardiac output provides, in addition to a greater blood flow to the skeletal musculature, more blood for the **skin** (heat loss; → p. 192ff.) and for the cardiac muscle. At the same time, the blood flow to kidneys and gastrointestinal tract drops to values below the resting level (→ **A**). The blood flow to the skin (at rest about 0.5 l/min) increases to nearly 2 l/min in heavy work but drops to resting levels during maximal exercise. The consequent rise in **body temperature** is one of the factors limiting maximal muscular work to a short period of time.

In *light* and *average work* the heart rate and lactate level soon attain new, constant values (no fatigue), whereas *maximal work* has to be interrupted after a brief period since the heart cannot keep up the high performance required of it (→ **B**).

During exercise, **pulmonary ventilation rises** from a resting value of 6–8 l/min to values of up to 100 l/min (→ **C1, 3**). This is achieved by a rise in both respiratory frequency (→ **C2**) and tidal volume. The combination of increased ventilation and a greater cardiac output makes it possible for the O_2 uptake to be increased from about 0.3 l/min (rest) to 4–5 l/min (→ **C4**). The **oxygen extraction** in the capillaries rises since the metabolic acidosis (→ p. 114) caused by the increased lactic acid formation (→ p. 46) and the rise in temperature (→ p. 101) displace the oxygen dissociation curve to the right.

Training: Not only has a trained athlete more muscle mass and more know-how, he also has more mitochondria in his muscles, enabling him to break down more glucose by oxidative pathways. This is the main reason why his blood lactate rises later and less than that of an untrained individual. Training increases stroke volume and tidal volume, which leads to very low cardiac and respiratory frequencies during rest, but permits higher cardiac output and ventilatory rate during the performance of work than in the untrained.

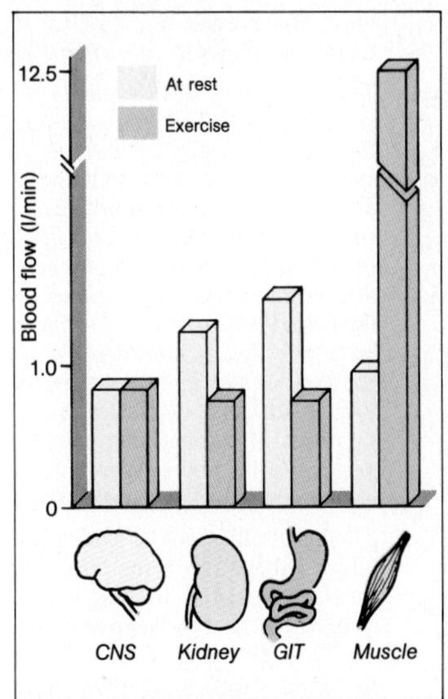

A. Blood flow response to exercise

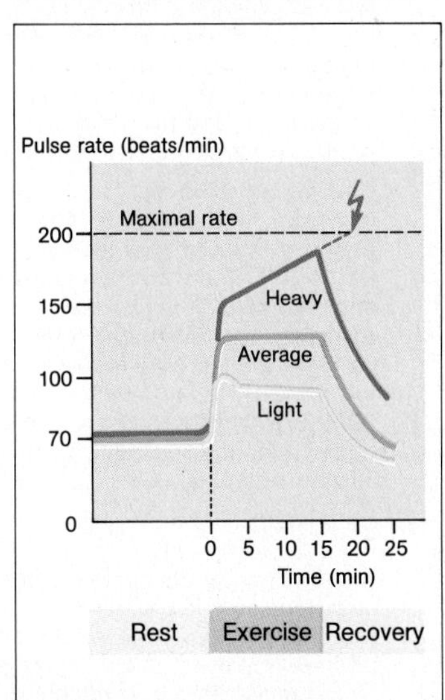

B. Work load and pulse rate

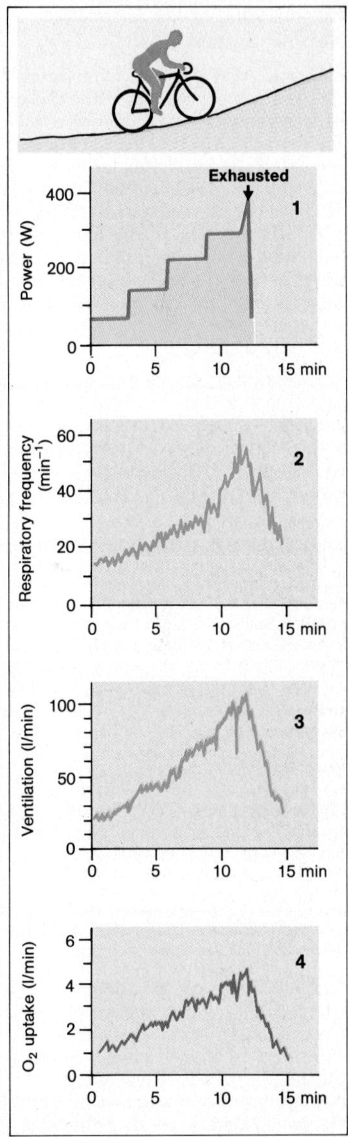

C. Respiration during exercise

(after J.Stegemann)

Organization of the Autonomic Nervous System

The *somatic* nervous system mediates afferent (from sense organs, etc.) and efferent (to skeletal muscles) communication with the environment. Much of the somatic nervous activity is under voluntary control and accessible to consciousness. The **vegetative** or **autonomic nervous system (ANS)**, on the other hand, serves mainly to regulate the functions of the internal organs, adapts them to the needs of the moment, and thus controls the **internal milieu** of the body. Such activities are for the most part not subject to voluntary control.

In the periphery of the organism the autonomic and somatic nervous systems are functionally and anatomically almost entirely separate, whereas in the central nervous system (CNS) intimate connections exist between the two (→ p. 232 and p. 290).

The peripheral ANS is *efferent*. Nevertheless, autonomic nerves often contain afferent fibers. They come from receptors in the inner organs (gastrointestinal tract, lung, heart, urinary bladder, etc.) and are therefore usually called **visceral afferents**. Sometimes they obtain their name from the nerve they run with (e.g. vagal afferents).

The ANS is primarily involved in **reflex arcs** (→ p. 278ff.), with visceral or somatic afferent and autonomic and somatic efferent limbs. **Afferent fibers** convey the stimuli from pain receptors, and from mechanoreceptors and chemoreceptors of the heart, lungs, gastrointestinal tract, urinary bladder, blood vessels, etc. The ANS provides the **efferent fibers** conveying the reflex response to such afferent information, causing contraction of the *smooth muscle* (→ p. 44) of the different organs (blood vessels, eyes, lungs, gastrointestinal tract, bladder, genital organs, etc.), and influencing the function of the heart and glands. The efferent limbs of these reflexes also involve the somatic nervous system (e.g. in coughing or vomiting).

Simple reflexes can be completed entirely within the organ concerned (→ e.g. p. 210), whereas more complex mechanisms are controlled by higher **autonomic centers** in the spinal cord and in the brain (→ **A**). The principle locus of integration of the ANS is the **hypothalamus** (→ p. 290), where the functions of the ANS are coordinated into the **programs** involving other divisions of the nervous system and the endocrine system. The **cortex** is a further level of integration of ANS with other systems.

The **peripheral ANS** consists of two divisions, for the most part anatomically and functionally distinct (→ **A** and p. 52f.): the **sympathetic** and **parasympathetic** divisions. The centers for the sympathetic divisions are situated in the *thoracic* and *lumbar* regions of the spinal cord; and those for the parasympathetic division in the *sacral* region of the spinal cord (for the bladder, parts of the large intestine, genital organs; → **A**), and in the *brain stem* (for the eyes, glands, and organs supplied by the vagus nerve). Both divisions of the peripheral ANS consist of *preganglionic* fibers, switching to *postganglionic* fibers in the **ganglia**.

The cell bodies of the **sympathetic** preganglionic fibers are in the lateral horn of the thoracic and lumbar spinal cord and terminate either in the *paravertebral ganglionic chain*, in the *cervical* or *abdominal ganglia*, or in the so-called terminal ganglia. Within the ganglion, transmission of the stimulus to the postganglionic fiber is *cholinergic*, the transmitter substance being **Acetylcholine (ACh**; → p. 54). With the *exception of the sweat glands*, stimulation of the end organs is *adrenergic*, and the transmitter substance is **norepinephrine** (→ **A** and p. 56).

The **parasympathetic** preganglionic fibers run from the brain stem with the cranial nerves to muscles and glands in the head (III, VII, IX) and to the organs in thorax and abdomen (X). The sacral fibers reach the pelvic organs with the pelvic nerve. The *parasympathetic ganglia* are situated close to, or even within, the effector organ. The transmitter substance of the parasympathetic division, in the ganglion as well as in the effector organ, is **ACh** (→ p. 54).

Most organs are innervated by fibers from both divisions of the ANS. The response of the two types of fiber may be either *antagonistic* (e.g. as in the heart) or almost *parallel* (e.g. salivary glands; → p. 204). The **adrenal medulla** is a combination of ganglion and gland; preganglionic sympathetic fibers release **epinephrine** and **norepinephrine** into the *systemic blood circulation* (→ p. 58).

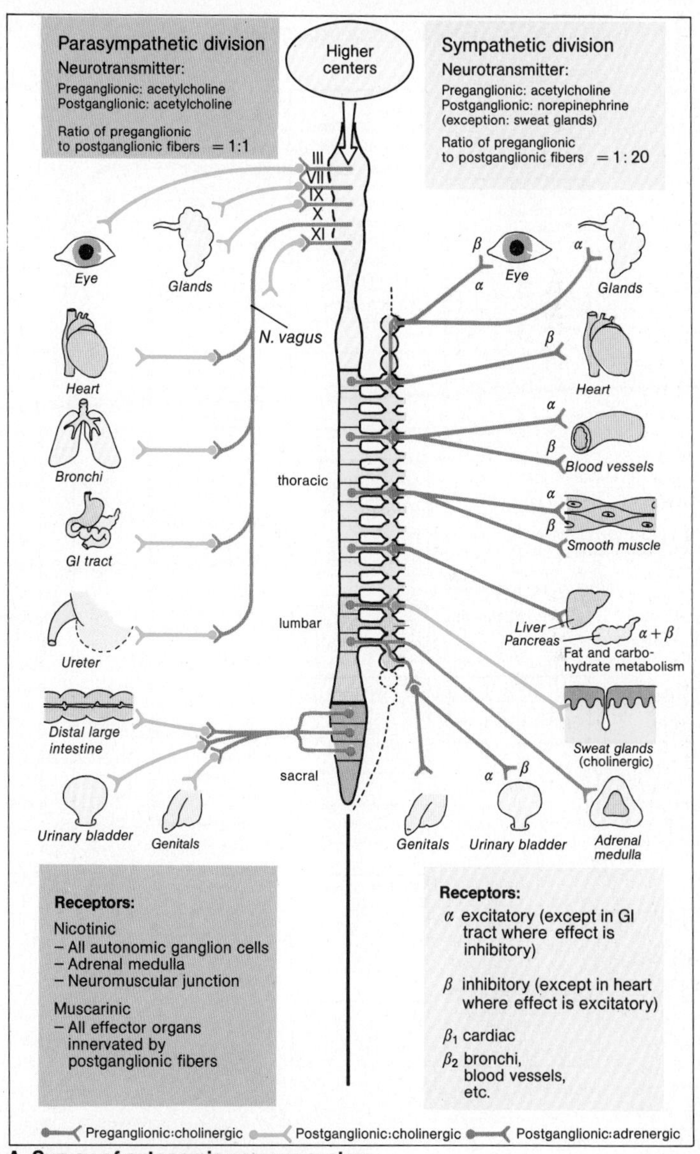

A. Survey of autonomic nervous system

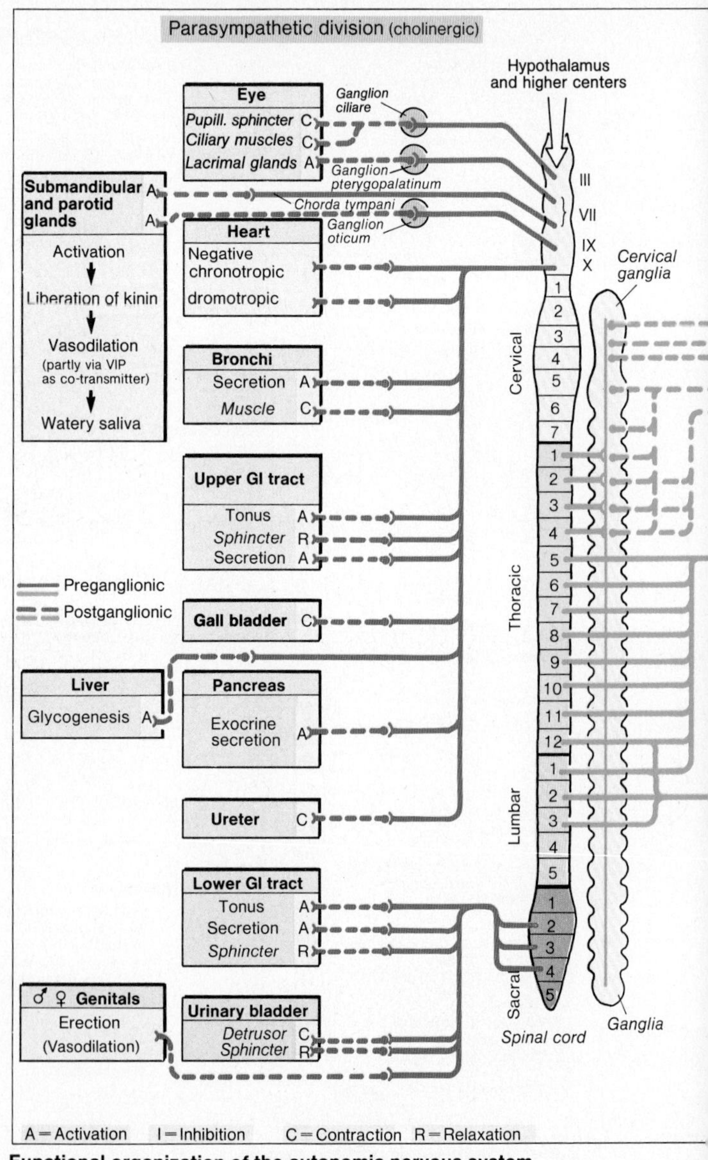

Functional organization of the autonomic nervous system

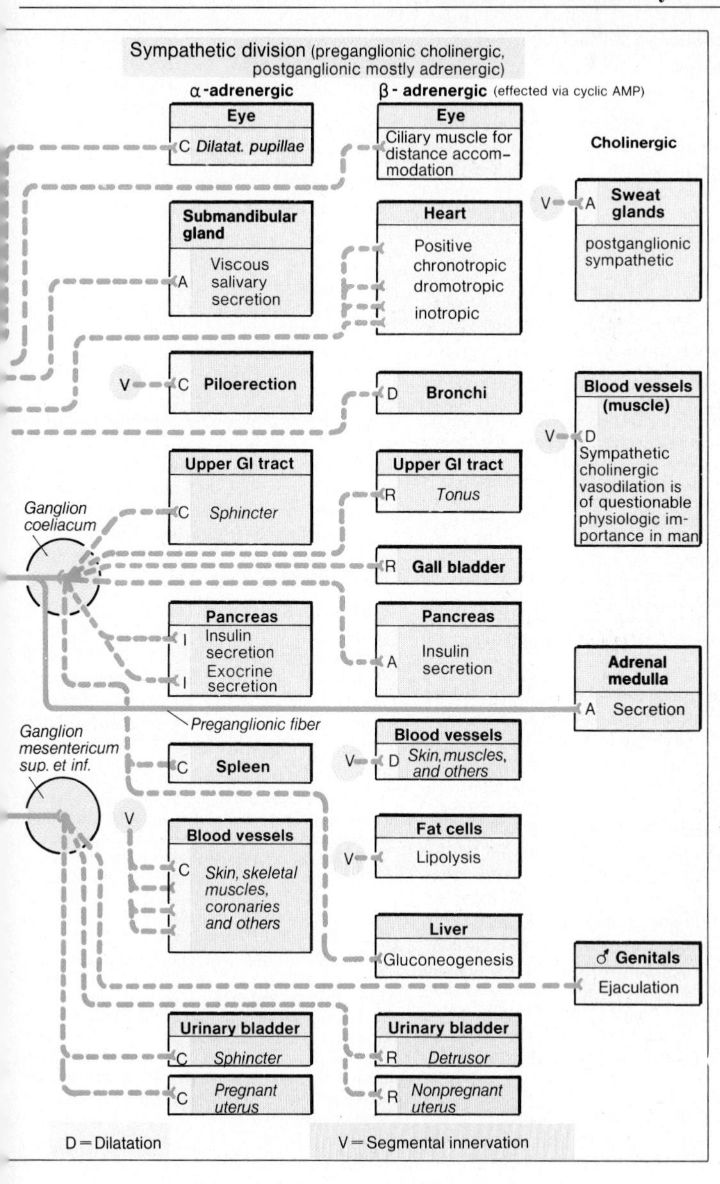
Sympathetic division (preganglionic cholinergic, postganglionic mostly adrenergic)
α-adrenergic
β-adrenergic (effected via cyclic AMP)
Cholinergic
Eye
C Dilatat. pupillae
Eye
Ciliary muscle for distance accommodation
V
A
Sweat glands
postganglionic sympathetic
Submandibular gland
A
Viscous salivary secretion
Heart
Positive chronotropic dromotropic inotropic
V
C
Piloerection
D
Bronchi
Blood vessels (muscle)
V
D
Sympathetic cholinergic vasodilation is of questionable physiologic importance in man
Upper GI tract
C
Sphincter
Upper GI tract
R
Tonus
Ganglion coeliacum
R
Gall bladder
Pancreas
I
Insulin secretion
I
Exocrine secretion
Pancreas
A
Insulin secretion
Adrenal medulla
A
Secretion
Preganglionic fiber
Ganglion mesentericum sup. et inf.
C
Spleen
Blood vessels
V
D
Skin, muscles, and others
V
Blood vessels
C
Skin, skeletal muscles, coronaries and others
Fat cells
V
Lipolysis
Liver
Gluconeogenesis
♂ Genitals
Ejaculation
Urinary bladder
C
Sphincter
Urinary bladder
R
Detrusor
C
Pregnant uterus
R
Nonpregnant uterus
D = Dilatation
V = Segmental innervation

Acetylcholine – Cholinergic Transmission

Acetylcholine (ACh) is the neurotransmitter at (1) all autonomic preganglionic nerve endings, (2) all parasympathetic and (3) some sympathetic (→ p. 53) postganglionic nerve endings, (4) at the neuromuscular junction (→ p. 32) and (5) at several synapses in the central nervous system (CNS). Some cholinergic innervations *initiate* organ activity (e.g. skeletal muscle), while others *modify* the intrinsic activity of the organ (e.g. smooth muscle and the conducting system of the heart).

The **synthesis of ACh** occurs in the cytoplasm of the nerve endings. *Acetylcoenzyme A* (*ACoA*) is synthesized in the mitochondria, and its acetyl group is transferred to *choline* in the presence of the enzyme *choline acetyltransferase* (ChAT). ChAT is synthesized in the soma of the nerve cells and is transported by *axoplasmic flow* (→ p. 22) to the nerve endings. Choline itself cannot be synthesized in nerves and must be accumulated from the extracellular fluid by a sodium-dependent *secondary active transport* process (→ p. 11). This is the rate-limiting step in ACh synthesis. When ACh is broken down in the synaptic cleft 50 % of the choline produced is taken up again by the nerve cell (→ **A**).

Storage and release of ACh: ACh is stored in **vesicles** in the nerve endings. The total amount stored is kept constant by continual matching of the amount synthesized to the amount released. The ACh *quantum* (→ p. 32) stored in and released from a single vesicle contains roughly 4000 ACh molecules. The arrival of a presynaptic action potential (AP) causes an *influx of* Ca^{2+} that leads to the release of several hundred quanta, sufficient to produce an **EPSP** (→ p. 30 f.). This postsynaptic change in potential results from alterations in membrane properties: *ACh raises the membrane permeability* to, or conductance for, $\mathbf{Na^+}$, $\mathbf{K^+}$, and $\mathbf{Ca^{2+}}$ (→ **A** and p. 11, F); in the heart, ACh only raises the K^+ conductance (→ p. 166).

Termination of ACh action is achieved by the hydrolysis of ACh with the help of the enzyme **acetyl cholinesterase (AChE)**.

Since the frequency of APs in somatic motoneurons can amount to many hundred Hz, the breakdown of the ACh at the motor endplate (→ p. 32) has to be completed within a millisecond in order to allow for repolarization between two APs (→ p. 26 ff.). The hydrolysis of ACh at the autonomic cholinergic nerve endings (with a lower discharge frequency) is much slower.

There are **two classes of receptors** for ACh:

1. **Nicotinic receptors** (*autonomic ganglia, motor endplates of skeletal muscle, adrenal medulla*, and in certain places in the *CNS*). In addition to responding to ACh, these receptors are stimulated by nicotine, which nevertheless has an inhibitory effect at high concentrations. Nicotinic receptors do not seem to be a homogeneous group; some cholinergic stimulating and blocking agents influence cholinergic transmission via such receptors, for instance in ganglion cells and in the motor endplate, quite selectively, although the inhibitory effect of the *curare* derivative d-tubocurarine on both is identical.
2. **Muscarinic receptors** (in parts of the *CNS* and at the *parasympathetic effector organs*) are stimulated by *muscarine* (in addition to ACh). This substance has no effect on nicotinic receptors. *Atropine* inhibits the muscarinic receptors in the heart, smooth muscle, CNS, and so on.

Carbachol and *pilocarpine* are employed therapeutically to stimulate the parasympathetic system (**direct parasympatheticomimetics**) because their breakdown by ACh esterase is slower than that of ACh. **Indirect parasympatheticomimetics** (*neostigmine*, among others), in contrast, act via *inhibition of AChE*. Potentially, such anti-AChE agents can (1) stimulate muscarinic receptor sites in autonomic effector organs (e.g. long-lasting contraction of bronchi), and (2) stimulate and subsequently paralyse (a) the nicotinic receptors in autonomic ganglia and in skeletal muscle, and (b) the muscarinic receptors in the CNS. Whereas therapeutic doses of such agents only cause some of these effects, nearly all effects are noted at toxic or lethal doses. Some insecticides act in this way, for example, *paraxon*, which is the active metabolite of *parathion* (E 605).

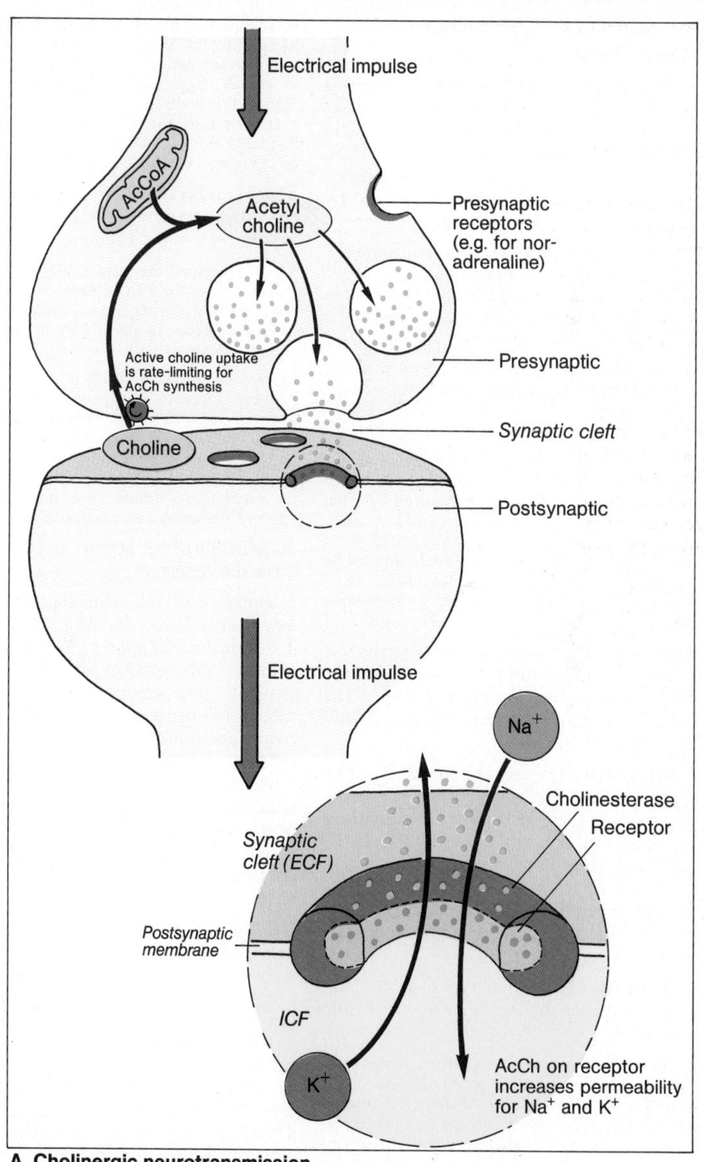

A. Cholinergic neurotransmission

Norepinephrine – Adrenergic Transmission

Norepinephrine (**NE**) or **noradrenaline** is the transmitter substance at most *sympathetic postganglionic* nerve endings and at some synapses in the central nervous system (CNS), especially in the hypothalamus. **Epinephrine** (**E**) or **adrenaline** is released by the adrenal medulla (→ p. 58).

The terminal branches of the nonmyelinated, sympathetic postganglionic nerve fibers exhibit *varicosities* or swellings, giving them the appearance of strings of beads (→ **A**, top left). These swellings form the synaptic contact with the effector organ and are also the site of synthesis and storage of NE.

NE synthesis: The nerve cell membrane of these swellings actively takes up the amino acid L-*tyrosine*, which is the starting substance in the chain of reactions leading to the synthesis of NE (→ **A**). The rate-limiting step in this synthesis, that is, the conversion of L-tyrosine to L-dihydroxyphenylalanine or *L-dopa* by tyrosine hydroxylase, is accelerated by Na^+ and Ca^{2+} and inhibited by the end product NE (negative feedback). **NE is stored** in large granular **vesicles** in a *micellar complex* (→ p. 218).

Release of NE occurs by exocytosis (→ p. 12) when an action potential reaches the synapse and causes an influx of Ca^{2+}. The exact mechanism of NE release is not completely understood.

Adrenoceptors (→ see also p. 59, B): Two main classes, α- and β-receptors, are distinguished by their sensitivity to the three substances E, NE, and isoproterenol (IPR) (the latter does not occur physiologically). α-Receptors are most sensitive to NE; β-Receptors react most strongly to IPR; and the response to E occupies an intermediate position.

α-Receptors are of two types, α_1 and α_2, distinguishable by specific agonists and inhibitors.

α_1-Receptors dominate in the salivary glands (promotion of K^+- and H_2O secretion) and in the smooth muscle: *contraction* of arterioles, uterus, bronchioles, sphincters in the urinary bladder and gastrointestinal (GI) tract, vas efferens, M. dilator pupillae, and so on. The second messenger is **inositol trisphosphate**; a subordinate messenger role is played by Ca^{2+} and cyclic GMP (→ p. 244 ff.).

α_2-Receptors occur e.g. in the CNS, kidneys, uterus, parotid gland, pancreas, mast cells (degranulation), and on thrombocytes (aggregation), as well as on certain presynaptic membranes (see below), for example on cholinergic neurons of the gastrointestinal tract. The binding of E and NE to α_2-receptors leads, via the G_i-protein, to *inhibition* of the adenyl cyclase (→ p. 242 ff).

There are also two types of **β-receptors**, both of which employ **cAMP** at the second messenger (→ p. 242 ff.).

Via **β_1-receptors**, the intracellular Ca^{2+} concentration can be *raised*. In the **heart**, for example, this has positive chronotropic dromotropic, and inotropic effects; in the **kidney**, it causes an increased release of renin (→ p. 59, B).

Employing specific agonists (e.g. Fenoterol) it is possible to localize the **β_2-receptors**, on which β_1-agonists (e.g. NE) have only a relatively weak effect. The β_2-receptors function by *decreasing* the Ca^{2+} concentration, and in this way can bring about dilatation of blood vessels and bronchioles, the stimulation of insulin release, and an increase in lipolysis (fat cells) and glycogenolysis (liver).

Termination of NE action can be effected via three mechanisms:

1. *Diffusion* of NE from the synaptic cleft *into the capillaries* (→ **A 1**);
2. *Extraneuronal uptake of NE* (in the heart, glands, and smooth muscle), with subsequent intracellular breakdown by catechol-O-methyltransferase (**COMT**) and by mitochondrial monoamine oxidase (**MAO**; → **A 2**);
3. *Reuptake* of NE (70%) into the presynaptic nerve endings (→ **A 3**) by active transport) following which the free NE in the cell is inactivated by mitochondrial MAO and its breakdown completed by dehydrogenases.

NE in the synaptic cleft also stimulates **presynaptic α_2-receptors**, thus inhibiting the release of further NE vesicles (→ p. 59, A). Presynaptic α_2-receptors of this nature are also found at cholinergic nerve endings (e.g. in the heart atria). Conversely, *presynaptic* (muscarinic) *acetyl choline receptors* are found at adrenergic nerve endings. This form of reciprocal influence makes possible a kind of "peripheral regulation" of sympathetic and parasympathetic tonus.

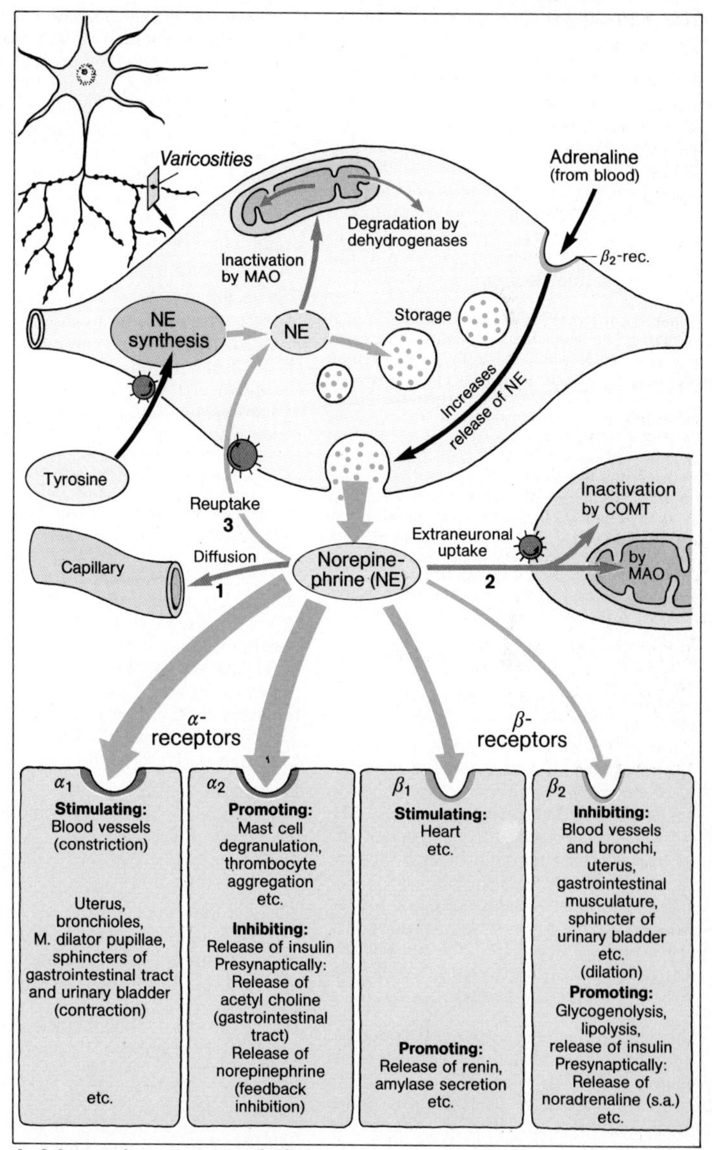

A. Adrenergic neurotransmission

The Adrenal Medulla

The **adrenal medulla** is a neuroendocrine transducer. Electrical *nervous impulses* (sympathetic preganglionic fibers; → p. 51 f.) are transformed into *hormonal signals:* **epinephrine (E)**, or **adrenaline, norepinephrine (NE)**, or **noradrenaline**, collectively termed **catecholamines**. The *neurotransmitter*, as at all preganglionic nerve endings, is *acetylcholine (ACh)*, which elicits the release of NE and E (exocytosis, → p. 12) at the "presynaptic" membrane.

Synthesis and **storage** of catecholamines is similar to that at the sympathetic postganglionic nerve endings (→ p. 56), but, due to the presence of an additional enzyme, *phenyl ethanolamine-N-methyltransferase*, the major part of the NE is converted into E. The precursor of NE, *dopamine* (from L-tyrosine via L-dopa), is taken up into granules. In about 15% of the granules, the sequence ends here and NE is stored. In 85% of the granules, NE leaks into the cytoplasm where it is partly broken down by MAO (→ p. 56) and partly converted to E, which is taken up again into granules. The uptake of dopamine, NE, and E is an active process. The granules contain, besides lipids, proteins (*chromogranin*) and other substances, 1 mol *ATP* per mol catecholamine.

Regulation of catecholamine synthesis: Synthesis is accelerated in acute sympathetic discharge by a decrease of the NE level, which in turn relieves the feedback inhibition of tyrosine hydroxylase (→ p. 56). Chronic stimulation also increases the activity of the enzyme converting dopamine to NE. **Adrenocorticotropic hormone (ACTH**; → p. 240) is involved in these processes. In contrast, **cortisol** (coming in high concentrations directly from the adrenal cortex, → p. 260) specifically induces the N-methyl transferase, thus increasing the ratio E: NE. The ratio in which NE and E are released also differs according to species and sympathetic activity (see below).

The adrenal medulla normally functions at a low basic level of activity, releasing only small quantities of E and NE. In situations involving **physical** or **psychic stress**, much larger quantities are released. This means that even cells with no sympathetic innervation become involved in the **alarm reaction**.

The **liberation of catecholamines** in the adrenal medulla is elicited via enhancement of sympathetic activity, for example by physical work, cold, heat, hypoglycemia (low blood sugar), pain, O_2 deficiency, a drop in blood pressure, fear, and anger ("stress"). In severe hypoglycemia, for instance, plasma E rises from the basal level of some 0.16 to 8.2 nmol/l and NE from 1.2 to 4 nmol/l. The *hypothalamus* is the principal organ concerned with the regulation of these reactions (→ p. 290).

The essential **role** of the catecholamines liberated in alarm situations (→ p. 290) is to mobilize (*lipolysis, glycogenolysis*) reserve chemical energy (fat, glycogen), to promote the *uptake of glucose* by the cells (→ p. 247), and in this way to provide the necessary fuel (fatty acids, glucose) for the extra work performed by the muscles. Catecholamines also influence the K^+ balance (→ p. 148).

In *skeletal muscle* the catecholamines activate, via cyclic AMP (→ p. 242), enzymes that promote the breakdown of glycogen and the formation of lactate (→ p. 46). The positive inotropic and chronotropic effects (β_1 receptors; → **B** and p. 56) of the catecholamines on the *heart* cause a rise in the heart rate and stroke volume, and thus in the cardiac output and blood pressure. At the same time, blood is shunted from the gastrointestinal tract to the skeletal musculature (→ **B** and p. 46).

Epinephrine also reinforces *adrenergic transmission* (→ p. 56). By binding to β_2-receptors of the varicosities of *postganglionic sympathetic fibers*, it increases NE release by the latter.

Whilst the alarm reaction is still in progress, catecholamines are already stimulating the hypothalamus to release hormones that set off the refilling of the depleted energy reservoirs. The levels of these hormones in the blood are at their highest about four hours after the onset of the alarm reaction.

(Text to **A**: → p. 56)

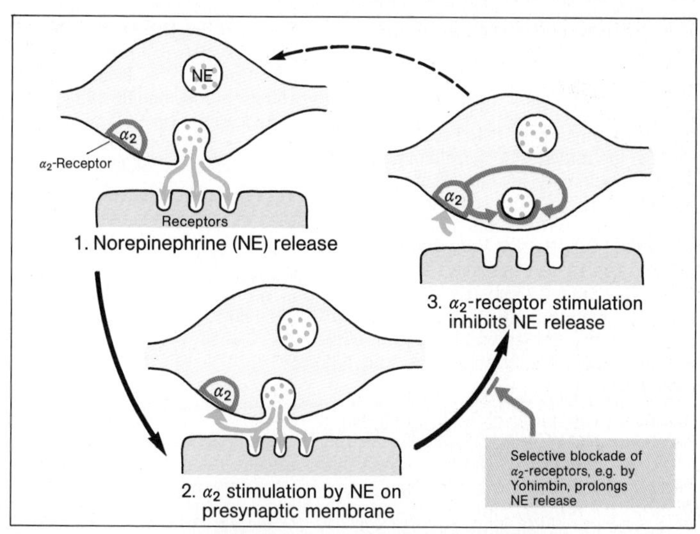

A. Feedback control of NE release via presynaptic α_2-receptors

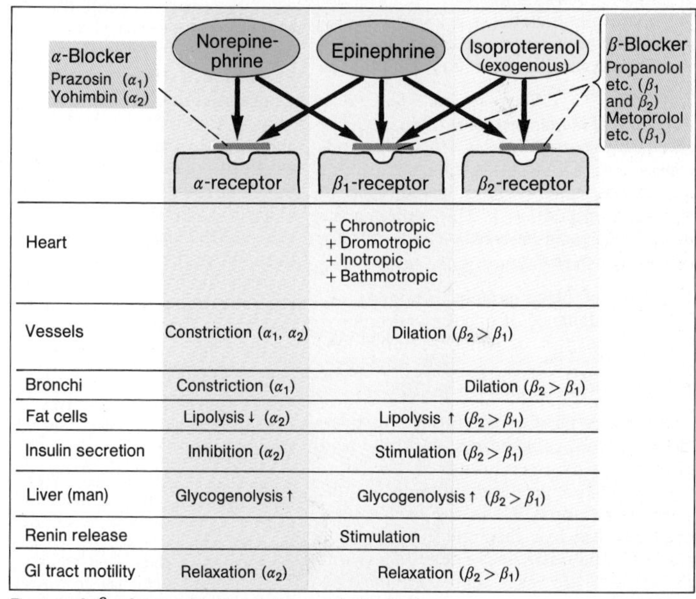

	α-receptor	β_1-receptor	β_2-receptor
Heart		+ Chronotropic + Dromotropic + Inotropic + Bathmotropic	
Vessels	Constriction (α_1, α_2)	Dilation ($\beta_2 > \beta_1$)	
Bronchi	Constriction (α_1)		Dilation ($\beta_2 > \beta_1$)
Fat cells	Lipolysis ↓ (α_2)	Lipolysis ↑ ($\beta_2 > \beta_1$)	
Insulin secretion	Inhibition (α_2)	Stimulation ($\beta_2 > \beta_1$)	
Liver (man)	Glycogenolysis ↑	Glycogenolysis ↑ ($\beta_2 > \beta_1$)	
Renin release		Stimulation	
GI tract motility	Relaxation (α_2)	Relaxation ($\beta_2 > \beta_1$)	

B. α and β adrenergic stimulation and blockade

Composition and Function of the Blood

The **blood volume** of the adult human comprises 6 to 8% of the body weight; 1 l of blood contains about 0.46 l red blood cells (RBCs) in males; 0.41 l in females. This value, which may also be expressed as a percent (46% in males), is the **hematocrit** (→ p. 65, A). In 1 µl (= 1 mm^3) of blood there are about $5 \cdot 10^6$ RBCs (**erythrocytes**) in males, $4.5 \cdot 10^6$ in females, 4000–10000 **leukocytes**, and 0.15 to $0.3 \cdot 10^6$ **thrombocytes** or platelets. A normal *differential count* of leukocytes includes about 67% granulocytes, 27% lymphocytes, and 6% monocytes.

The fluid phase of the blood is the **plasma**; it has an osmolality (→ p. 335) of 290 mosm/kg H_2O and contains 65–80 g protein/l. The various **proteins** in the plasma include albumin (57%), α_1-globulin (4%), α_2-globulin (8%), β-globulin (12%), γ-globulin (16%), and fibrinogen (3%). When the blood clots (→ p. 74ff.), fibrinogen is used up; the remaining fluid is **serum**. Serum and plasma are similar except for the fibrinogen content.

Functions of blood. Transport of dissolved or chemically bound matter (O_2, CO_2, nutrients, metabolites, etc.); transport of heat for heating and cooling; **transmission of signals** (hormones, → p. 232); **buffering** the body fluids (acid-base balance, → p. 110); **defense** (→ p. 66ff.).

Functions of blood proteins. Immune defense; maintenance of oncotic pressure (→ p. 335); transport of water-insoluble substances (e.g. lipids; → lipoproteins on p. 222); protein binding (→ p. 10).

Erythrocytes (RBCs) are formed in bone marrow. Iron, cobalamines, and folic acid are some of the prerequisites. In the fetus, erythrocytes are produced in the spleen and liver. Immature RBCs in the marrow are nucleated but lose their nuclei when they reach the bloodstream. Their dimensions (7.5 × 2 µm) are such that they have to squeeze their way through the small capillaries, which facilitates the exchange processes with the surrounding tissues.

The chief **function of the RBC** is to carry O_2 and CO_2 between lungs and tissues; this is accomplished by **hemoglobin**, **Hb**, (male 160 g/l blood, female 145 g/l) and *carbonic anhydrase* (→ pp. 96f. and 144f.). One RBC contains 28–36 pg Hb (**m**ean **c**orpuscular **h**emoglobin = **MCH** = [Hb]/RBC count).

The high *intra*cellular Hb concentration (roughly 300 g/l erythrocytes = **MCHC** = [Hb]/hematocrit) is a major factor in cellular osmolality and requires that the intracellular electrolyte concentration be maintained at a lower level than that in the plasma. To accomplish this, the RBC membrane contains an active transport system for Na^+ and K^+. The ATP needed for this purpose comes from anaerobic glycolysis (→ p. 47, B).

As in other cells (e.g. → p. 24f.), the membrane potential (caused by K^+ accumulation in the RBC) expels Cl^- from the cell interior (ICS), which results in the desired drop of intracellular osmotic pressure (→ p. 335) to isotonicity. In addition, (secondary) active Cl^- uptake increases $[Cl^-]_{ICS}$ above its equilibrium concentration (→ p. 24). Consequently, the ratio of active uptake to passive efflux of Cl^- determines $[Cl^-]_{ICS}$. Thus, by varying this ratio, the *volume* of RBC (and other cells) can be *regulated* (the **m**ean **c**orpuscular **v**olume of one RBC = **MCV** = 93 fl).

Production of RBCs is mainly controlled by a hormonal mechanism. Cellular O_2 deficiency is the initiating event in the production and release of the hormone **erythropoietin**, which stimulates red cell production in the bone marrow (→ **A**, top). More than 90% of erythropoietin is produced in the glomeruli of the *kidney*; the rest, mainly in the liver. When the red cell mass increases and corrects the hypoxia, synthesis of erythropoietin subsides after some hours (negative feed-back). Erythropoiesis is also influenced by the CNS, which can stimulate the marrow to discharge stored RBC.

The **life span of the RBC** is about 120 days. Aged RBC are removed from the blood in sinuses of the spleen and are degraded. The fragments are broken down by macrophages in the *mononuclear phagocytotic system* or **MPS** (also called reticuloendothelial system or **RES**) of the spleen, liver, bone marrow, and so on (→ **A**). When the membrane of the RBC ruptures (*hemolysis*), Hb is set free and is metabolized to globin and bilirubin (→ p. 216). The iron of Hb is recycled. In spherocytic anemia, the *osmotic fragility* of the RBC is greater than normal, and its life span is reduced. Removal of the spleen can bring some degree of relief.

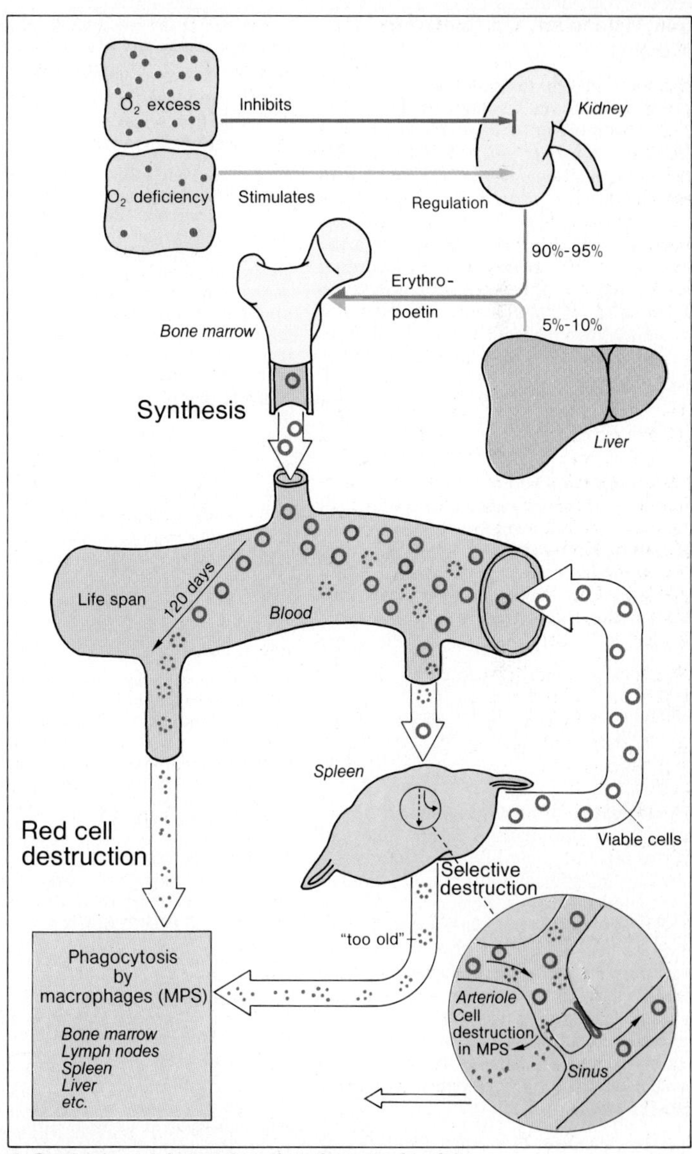

A. Synthesis and destruction of erythrocytes in adults

Iron Metabolism, Erythropoiesis, and Anemias

The total quantity of iron (**Fe**) in the body amounts to about 45 mmol in the female and 60 mmol in the male (1 mmol = 55.8 mg). Of this, 60–70% is bound in **Hb**, 10–12%, the *functional iron*, is present in myoglobin and in iron-containing enzymes (e.g. catalase), and 16–29% is *stored iron (ferritin, hemosiderin*; → **A**). Although the quantity of Fe in the food varies according to the diet, in the mean it amounts to 0.2 mmol/d in the female and about 0.33 mmol/d in the male. Of this, only about 6% (male) to 12% (female) is absorbed in the duodenum (→ **A**, **B**). Fe absorption *varies according to the body's needs* and in cases of iron deficiency may increase to as much as 25% of the quantity ingested.

Iron absorption. Dietary Fe bound to heme and other lipophilic substances is absorbed more by diffusion, whereas free Fe (mainly Fe [II]) is *actively* taken up by the intestinal mucosa. Normal iron absorption depends (1) upon the *gastric hydrochloric acid* (which releases Fe from its complexes and stimulates Fe [III] absorption in the upper duodenum ("pH 3" in **B**); (2) upon the *availability of Fe* [II] (more soluble at neutral pH than Fe [III]; → **B**); and possibly (3) upon glycoproteins of the gastric mucosa (*"gastroferrin"*), which can bind large quantities of Fe [III] ions.

The **regulation of Fe absorption** is not fully understood, although it seems that the availability of *apotransferrin*, the iron-transporting protein in the plasma, is involved (apotransferrin + Fe = *transferrin*; → **A, B**). Some of the excess Fe absorbed into the intestinal mucosa is bound to ferritin and taken up by the lysosomes of intestinal mucosa where it is stored for several days in a readily available form until it reenters the intestinal lumen with the contents of disintegrating mucosal cells (→ **B**). Thus for a certain length of time it is again available for possible reabsorption.

If Fe is administered by injection, thus bypassing the gastrointestinal tract, the apotransferrin capacity may be exceeded (maximum about 0.2 mmol) and free Fe can lead to **iron poisoning** (impaired coagulation of the blood results in hemorrhage, circulatory shock, and so on).

Ferritin (intestine, spleen, liver, bone marrow, heart, muscle, etc) represents a rapidly available **iron reserve**, whereas the iron in **hemosiderin** is less readily mobilized.

Most of the body's Fe is incorporated into *erythrocytes* in the bone marrow (about 0.54 mmol/d). The iron of defective RBCs (about 1/3) is set free by the macrophages of the bone marrow and is immediately reavailable (→ **A**). Aged erythrocytes are also phagocytized by macrophages (→ p. 60, and p. 66ff.): their heme and Hb-bound Fe is conveyed in the plasma to the liver cells, where it is taken up by **endocytosis**. In each case the iron is reutilized (roughly 97% *Fe "recycling"*; → **A**). Liver cells, erythroblasts, and other cells have special *receptors for transferrin*, which is also taken up by endocytosis.

Iron requirements (= losses) are normally 18 µmol/d, but are higher during menstruation (→ **A**), and especially in the second half of pregnancy (fetal growth) and following delivery (blood losses). Since the fetus takes up about 60 µmol Fe/d in the ninth month of pregnancy, the maternal diet at this time should contain an additional 0.5 mmol Fe/d (approximately 12% absorption).

In **anemias**, either the erythrocyte content or the Hb concentration in the blood is lowered. In addition to anemias resulting from *hemorrhage* or *Fe-deficiency* (defective absorption, pregnancy, chronic loss of blood, infections), other types are caused by deficiency of **cobalamines** (vitamin B_{12}) or of **folic acid** (→ **C**). The latter types of anemia are characterized by enlarged erythrocytes and by the fact that the decrease in their number in the blood is greater than the decrease in Hb concentration (*hyperchromic anemia*; MCH increased). The most common causes of such anemias are insufficient secretion of *intrinsic factor* (essential for the absorption of cobalamines); or an autoimmune reaction, which renders it ineffective; or malabsorption of folic acid (see also p. 226). On account of the large stores of cobalamines present in the body, decreased absorption only leads to signs of deficiency after some years, whereas insufficient uptake of folic acid results in anemia after as little as 3–4 months (→ **C**).

The *folic acid antagonists* (e.g. Methotraxat) are often employed in tumor therapy as cytostatic agents (inhibiting cell formation), at the same time reducing the numbers of erythrocytes (*aplastic anemia*) and other rapidly dividing cells.

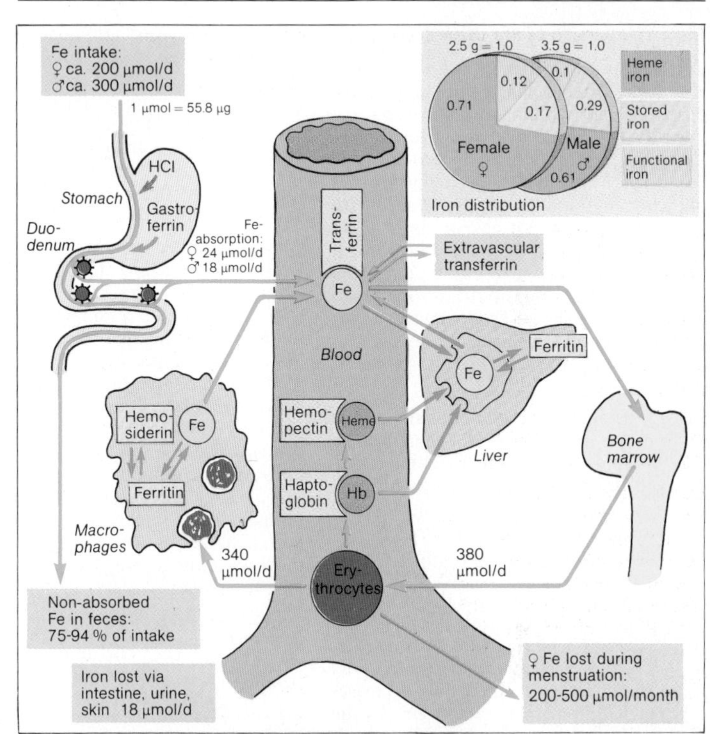

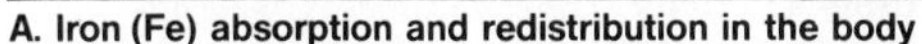
A. Iron (Fe) absorption and redistribution in the body (after Munro and Linder)

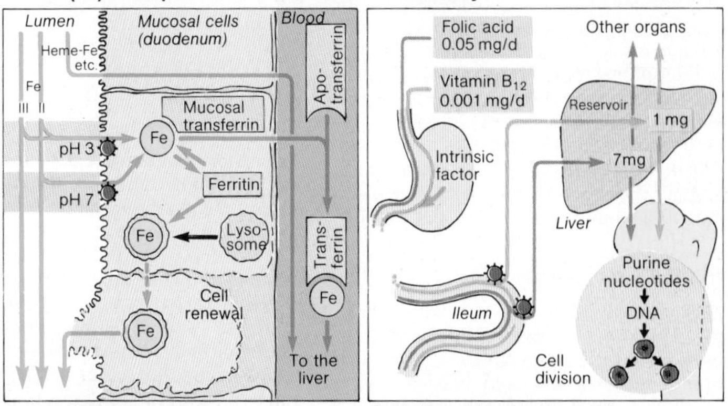

B. Iron (Fe) absorption

C. Folic acid and vitamin B_{12} (cobalamine)

Flow Properties of the Blood

Erythrocytes are extremely flexible, deformable *anuclear* cells that behave much like drops of fluid within the blood. The low **viscosity** (= 1/*fluidity* = shear force τ/velocity gradient between adjacent layers γ [Pa · s]) of their contents, the fluid properties of their membrane and their large surface to volume ratio are responsible for the fact that the blood, particularly when fast-flowing, behaves more like an *emulsion* than a cell suspension. The viscosity of *flowing* blood is, with a value of approximately four relative units (r.U), therefore only about double that of the plasma (2 r.U.; water: 1 r.U. = 0.7 mPa · s) at 37 °C.

Passage through fine blood capillaries or the pores of the spleen (→ p. 61), which have a considerably smaller diameter than the free-floating red blood cells, presents no problems to a normal, healthy, highly deformable erythrocyte. The slowness of flow in the small vessels, however, causes an increase in the viscosity, which is partially compensated for by the erythrocytes swimming in the center of the blood stream ($\eta\downarrow$: **Fahraeus-Lindqvist effect**). However, the viscosity of the blood may rise critically if (a) the velocity of blood flow decreases too much (circulatory shock) or (b) the fluidity of the erythrocytes drops as a result of environmental hyperosmolality (the erythrocytes shrink and appear crenated), cell inclusions (sickle cell anemia), changes in the cell membrane (e.g. in "old" RBCs) etc. Under these conditions, due to *aggregation* (formation of "stacks of coins"), the blood acquires the properties of a *suspension with high viscosity* (up to 1000 r.U.), which, in small vessels, can rapidly bring the blood flow to a standstill (→ p. 156 and p. 186).

Plasma Components

Plasma is what is left after the cellular components (→ p. 60) of blood that has been rendered incoagulable (→ p. 74) have been centrifuged off (→ **A**). Plasma consists of *water* in which high-molecular-mass *proteins* (→ **A**), small uncharged molecules (glucose, urea, etc), and **ions** are dissolved. The sum of these particles gives the osmolal concentration (osmolality) of the plasma (→ p. 136). The major *cation* (positively charged ion) contributing to this osmolality is Na^+; the major anions (negatively charged) are HCO_3^- and Cl^-. Although the proteins carry a high net anionic charge (→ **B** and p. 24), their osmotic effect is relatively lower because this depends upon number and not size or charge of the particles.

Only very limited quantities of plasma proteins can leave the blood vessels. The amount varies from organ to organ. The liver capillaries, for example, are relatively much more permeable than those of the brain. The composition of the *interstitial fluid* (→ p. 138 f.) therefore differs to a greater or lesser extent from that of the plasma (→ **B**), particularly in its protein content. The *intracellular fluid* has a totally different composition: the predominant cation is K^+, and phosphates (organic and inorganic) and proteins constitute the major part of the anions (→ **B**). The proportions of these constituents may vary from one cell type to another.

The **plasma proteins** (→ **A**, right) consist of nearly 60% of albumin (35–45 g/l), which serves as a vehicle for substances such as bilirubin (→ p. 216) and hormones (→ p. 234 ff.). Albumin is also largely responsible for the *colloidal osmotic (oncotic) pressure* (→ pp. 158 and 335) and can replace dietary protein in cases of deficiency. Proteins play a role in the transport of lipids (*apoproteins*; → p. 220 ff.), hemoglobin (*haptoglobin*; → p. 63), iron (*transferrin*; → p. 62 f.), cortisol (*transcortin*; → p. 260), cobalamines (*transcobalamines*; → p. 226) and many other substances. Most of the plasma factors involved in coagulation and fibrinolysis (→ p. 74 f.) are proteins, too.

The majority of **immunoglobulins** (**Ig**, → **C**) are γ-globulins. They are the defense proteins of the plasma (antibodies, → p. 66 ff.). Of these, IgG is present in the relatively highest concentration (7–15 g/l) and is the immunoglobulin capable of crossing the placental barrier most easily (from mother to fetus, → **C**). Immunoglobulins each consist of two group-specific heavy protein chains (IgG: γ, IgA: α, IgM: μ, IgD: δ, IgE: ε) and two light protein chains (λ or κ) linked together in a characteristic Y-form (→ p. 67) by (-S-S-) bridges.

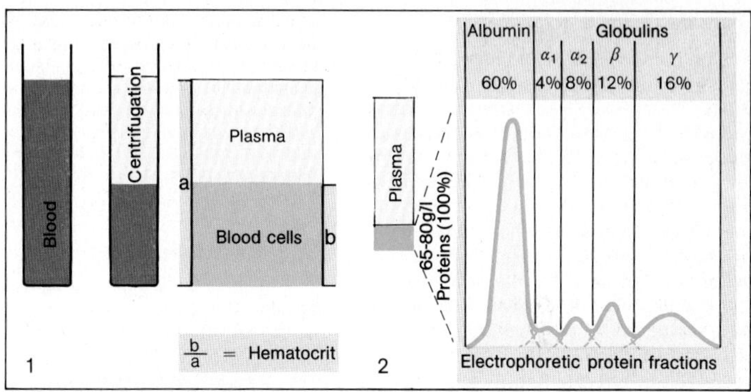

A. Composition of the blood

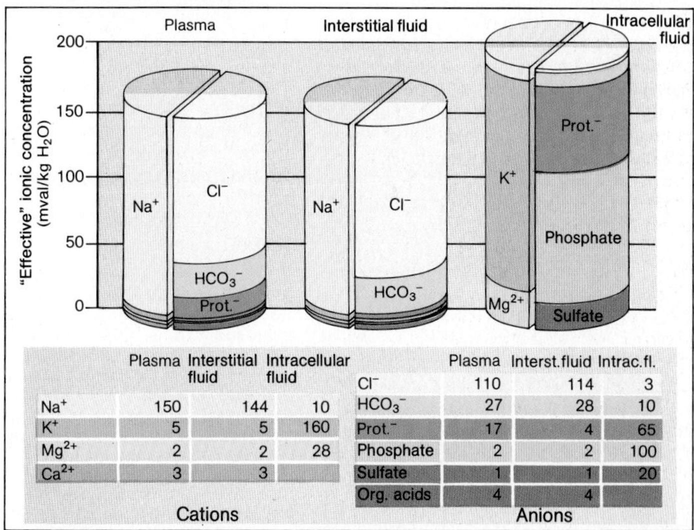

Cations

	Plasma	Interstitial fluid	Intracellular fluid
Na^+	150	144	10
K^+	5	5	160
Mg^{2+}	2	2	28
Ca^{2+}	3	3	

Anions

	Plasma	Interst. fluid	Intrac. fl.
Cl^-	110	114	3
HCO_3^-	27	28	10
$Prot.^-$	17	4	65
Phosphate	2	2	100
Sulfate	1	1	20
Org. acids	4	4	

B. Ionic composition of the body fluids

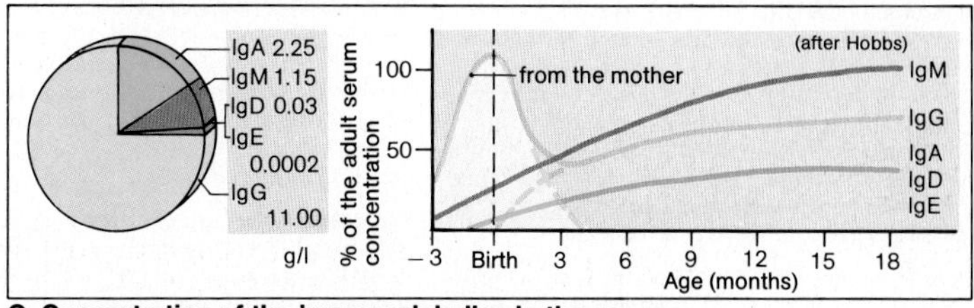

C. Concentration of the immunoglobulins in the serum

Immune Defense

The body is constantly endangered by infectious microbes from the environment (bacteria, viruses, fungi, parasites). In order to combat these intruders, the organism is equipped with a defense system that provides it with a considerable degree of immunity. There are two kinds of immunity: (pathogen-) **unspecific**, **inborn immunity** and (pathogen-) **specific**, **acquired** (adaptive) **immunity** (= immunity in its stricter sense). The two systems are closely interwoven, and both involve the participation of mobile cells and soluble factors.

If pathogens succeed in invading the body, the unspecific defense system comes into action. At the same time, the invaders represent **antigens** to which the specific defense system reacts by producing **antibodies**. With the help of these antibodies, the organism can usually combat the pathogen much more efficiently, in addition to "memorizing" it (*immunological memory*).

Whether or not an antigenic substance (e.g. cell protein) is to be treated as "foreign" is "learned" by the immune system around the time of birth. Substances with which the organism is confronted at this time are usually recognized for the rest of its life (**immunological tolerance**), and all others are treated as intruders. If this distinction between "foreign" and "self" breaks down, the result is **autoimmune disease**, a condition in which the organism produces antibodies to its own proteins.

A child coming into contact for the first time with the measles virus, for example, sickens with measles. Although the unspecific system is fully active, it normally is unable to prevent the multiplication and spread of the virus in the body. During the illness, however, in addition to producing the T-killer cells (→ p. 72) the body also produces antibodies to the virus (**primary response** or **sensibilization**); the virus is thereby rendered harmless and the patient recovers. This antibody production can be rapidly reactivated at need ("memory"), which is what happens if the measles virus again enters the organism at a later date. This time, however, antibody production gets going at once (**secondary response**) and immediately renders the virus harmless, thus preventing another attack of measles. The organism is now **immune**.

In order to prevent the initial outbreak of a disease altogether, **vaccination** with a harmless form of the microorganism concerned can be carried out (e.g. smallpox), or with the dead microorganism itself (e.g. tetanus), or with synthetic components thereof (peptides). If repeated several times, antibodies are produced in sufficient quantities (**active immunization**) to provide complete protection. A patient who is already ill can be successfully treated with the *serum*, or its *γ-globulin fraction*, of animals that have produced antibodies against the pathological organism (**passive immunization**, e.g. with diphtheria serum).

Unspecific Defense

The unspecific defense against foreign substances (bacteria, viruses, inorganic particles, etc.), and in certain cases also against the breakdown of the body's own substance (e.g. erythrocyte debris), is provided by certain **dissolved substances** such as proteins (e.g. *lysozyme, complement factors*), signal substances (e.g. *lymphokins and monokins*, together termed *interleukins*), and aggressive groups (e.g. *O_2 radicals*), as well as **phagocytes** (i.e. monocytes/macrophages; → p. 68) and the **neutrophilic granulocytes**. The latter are leukocytes and are formed in the bone marrow (life span approximately 1 day). Granulocytes are active not only in the blood and tissues but also on the surface of mucosal epithelia (e.g. in the mouth).

If, as an example, bacteria succeed in entering the body tissues (→ **A 1**), the neutrophilic granulocytes are attracted by foreign chemical substances or the complement factor C5a (*chemotaxis*; → **A2**). Granulocytes from the blood attach themselves to the inner walls of the blood vessels (*margination*), from which they then escape and move towards the site of attack (*migration*). Here they surround and ingest the invaders by endocytosis: **phagocytosis** (→ **A3**). These events, combined with increased blood flow (reddening!) and enhanced capillary permeability for proteins (swelling!), constitute **inflammation**.

Organic substances are "digested" in the granulocytes. This is preceded by a merging of the phagocytized pathogen (*phagosome*) with the enzyme-containing *lysosomes* in the granulocytes to form the *phagolysosome* (heterophagosome), in which the invader is finally broken down (→ **A4**). "Indigestible" particles (e.g. coal dust in the lungs) are permanently retained in the body.

Although microorganisms can be bound directly to phagocytes, the "appetite" of the latter can be greatly enhanced if the surface of the foreign substance is "labelled" (**opsonization**) with the complement factor C3b (unspecific), or with antigen-specific immunoglobulins (IgM, IgG), or,

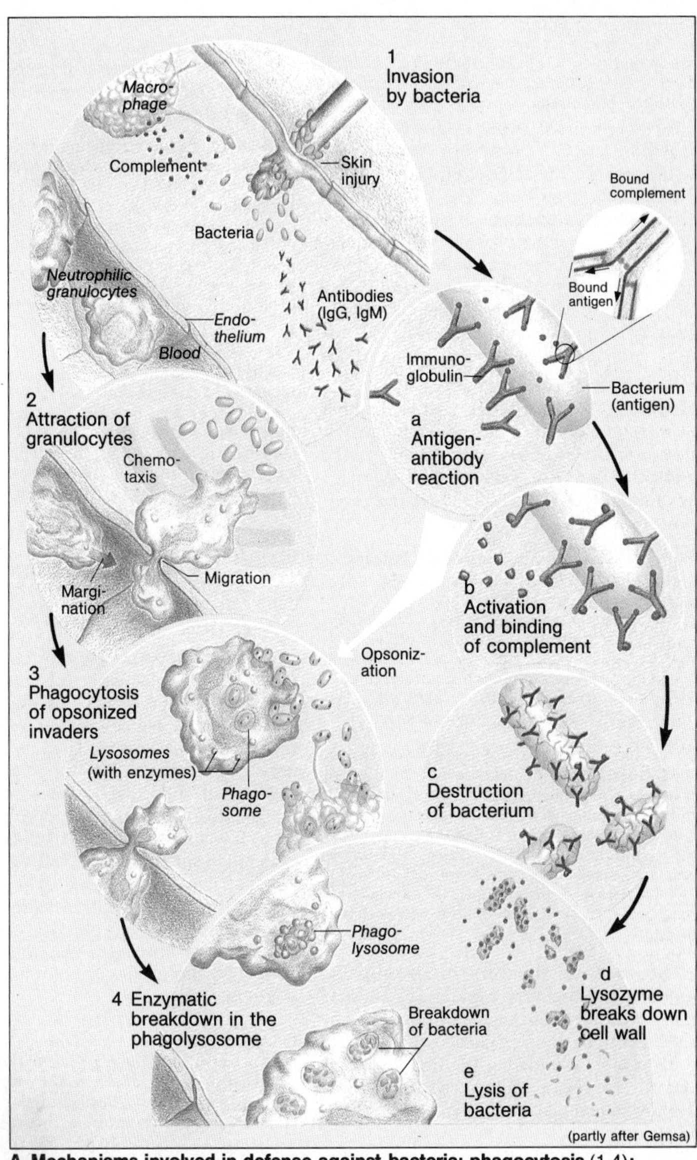

A. Mechanisms involved in defense against bacteria: phagocytosis (1-4); **extracellular lysis** (1, a-e)

still better, with both, since the phagocytes have special receptors for these opsonins. A condition for opsonization with Ig is that the organism must have previously been in contact with the pathogen (acquired immunity!), whereas C3b and other opsonins bind a wide range of pathogens relatively unspecifically.

Pathogens are also attacked (unspecifically) outside the phagocytes. The reaction cascade of the **complement system** (see textbooks of biochemistry) ends with the perforation of the outer wall of (Gram-negative) bacteria. At the same time, **lysozyme** (in plasma, lymph, and secretions) breaks down the bacterial wall enzymatically, which eventually leads to its dissolution (**lysis**; → **A, a–e**).

For killing phagocytized pathogens, the neutrophilic granulocytes not only have at their disposal the *enzymes* stored in the lysosome, but also **oxidants** like *hydrogen peroxide* (H_2O_2) and *oxygen radicals* (O_2^-; $^1O_2$).

Normally, the concentration of such oxidants is kept at a low level by the reducing enzymes, such as *catalase* and *superoxide-dismutase*, in order to prevent premature self-destruction of the granulocytes. This "restraint" is abolished during invasion by foreign substances in order to permit maximum development of the bactericidal effect of the oxidants, whereby granulocytes and even other of the body's own cells may suffer damage.

Disturbances in the process of phagocytosis result in an increased susceptibility to infection. Two examples are: (1) the *lazy leukocyte syndrome*, in which migration is defective; and (2) *chronic granulomatous disease*, in which the formation of H_2O_2 in the granulocytes is defective. More often, migration and phagocytosis are disturbed for unknown reasons, for example, in diabetes mellitus, by ethanol abuse, and by corticosteroid therapy.

Although large numbers of neutrophilic granulocytes are very rapidly available at the site of infection, their powers of chemical defense are soon exhausted and their life span is brief. Following the first "wave of attack" by the granulocytes, the task of defense is taken over by the **macrophages**. They originate in the **monocytes** that are circulating in the blood and are also capable of phagocytosis. Although the speed of migration of the macrophages is less than that of the granulocytes, they have, for example, a significantly *longer life span*, are capable of a longer-lasting *synthesis of enzymes*, and are able to *secrete complement*, among other properties.

In addition to the circulating monocyte/macrophages there are **local wandering macrophages**, as for example in the liver (*stellate Kupffer cells*), in the pulmonary alveoli, on the intestinal serosa, in the spleen sinuses, in lymph nodes, in the skin, in the joints (*synovial A-cells*), and in the brain (*microglia*), as well as endothelial-bound macrophages (e.g. in the renal glomeruli). These cells are collectively known as the **MPS** or the **RES** (→ p. 60).

The natural killer cells (**NKCs**) (= 5% of the leukocytes in the blood) are specialized for unspecific defense against *viruses*. They detect changes on the surface of virus-infected cells, then collect on their surface and kill them. This not only deprives the viruses of the possibility to reproduce (the cell's enzyme apparatus!), but also makes them vulnerable to the rest of the defense system. The NKCs are activated by **interferons**, which are usually produced and released by the virus-infected cell itself. Interferons also induce an increased resistance to the virus in the uninfected cells.

Specific Defense

Although phagocytes are effective against a whole range of bacteria, other pathogenic organisms have "learned" in the course of evolution to defend themselves against them. Some pathogens, such as mycobacteria, for example, can suppress the formation of phagolysosomes or prevent phagocytosis altogether, or, once phagocytized (e.g. streptococci and staphylococci), even kill the granulocytes. Pathogens of this kind and most viruses can only be combated successfully by the **specific defense system**, in which macrophages, humoral antibodies (immunoglobulins; → p. 64), and different types of lymphocytes act in close *cooperation* (see below).

Lymphocytes originate in the bone marrow (→ **B**). During fetal life and infancy some of them migrate into the **thymus**, where they acquire their specificity (*immune competence*): **T-lymphocytes**. Other lymphocytes are "imprinted" as **B-lymphocytes**, in the **B**ursa Fabricii in birds, and in the *equivalent of the bursa* in humans, that is, in the **b**one marrow. Later in life, both kinds of lymphocytes are chiefly found in the *spleen* and in different regions of the

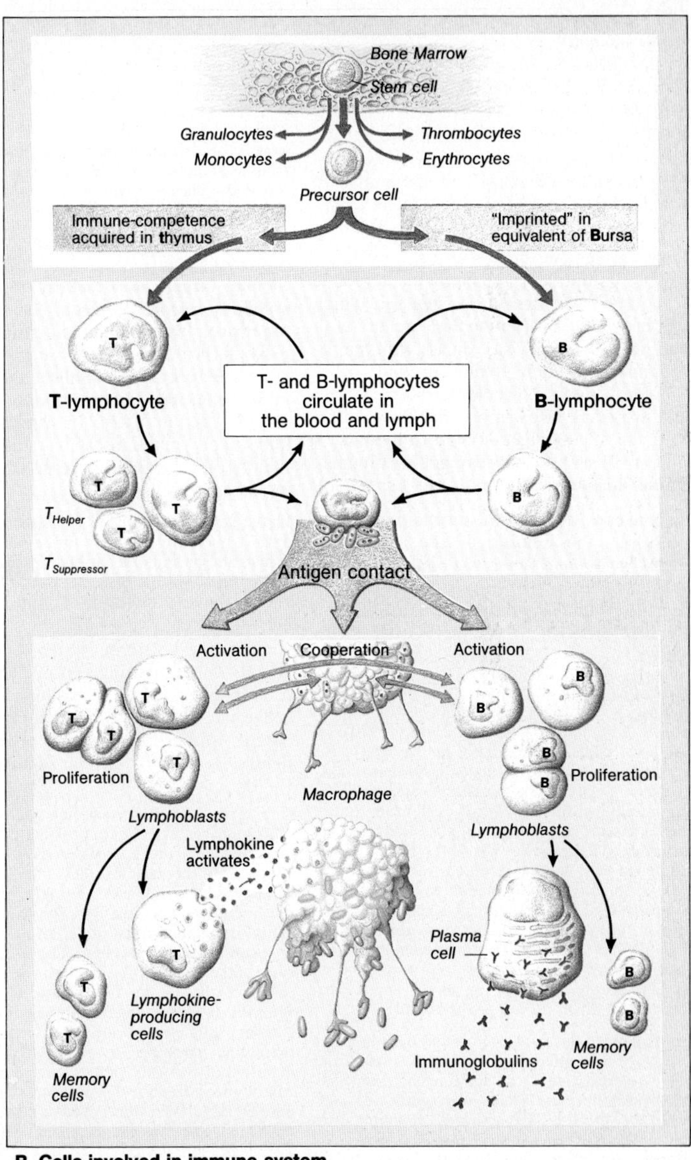

B. Cells involved in immune system

lymph nodes, whence they proceed to the *lymph vessels* and *blood vessels* in which they circulate, ready to play their role in immune defense. The individual life span can amount to years.

The **first contact** with antigen (antigen-specific receptors on the lymphocyte surface) results in **activation** for both types of lymphocytes (see below), in the course of which they are transformed into *lymphoblasts* (→ **B**).

Humoral immune defense (which also has a cellular component; see below) is a specific reaction of the immune system, in which **immunoglobulins (antibodies)** react with **antigens** (→ p. 67 A 1 a), that is with microorganisms or their toxins, or with other "foreign" macromolecules (> 4000 Dalton, e. g. proteins). If low-molecular compounds (e. g. therapeutic agents) bind to the body's own protein they, too, can act as antigens. These substances are known as *haptenes*.

The release of humoral antibodies is preceded by the binding of the antigen to *membrane-bound immunoglobulins* on the surface of the B-lymphocytes (acting here as receptors) and by **presentation** of the antigen by macrophages or by the B-cells themselves.

After phagocytosis of the pathogen by **macrophages** (→ **C 1**), phagolysosomes are formed in which the pathogen is broken down (→ **C 2**). Its antigenic peptides are bound, in the cell, to the proteins of the class II **major histocompatibility complex (MHC)**. The resulting antigen-MHC complex is incorporated into the cell membrane of the macrophages (→ **C 3**) and *presented* to the **T-helper-(T_H) lymphocytes** (→ **C 4**); these cells are equipped with specific receptors for the antigen-MHC complex. In a similar manner, the **B-lymphocyte**, which has already internalized and broken down the (in this case soluble) antigen together with the membrane-immunoglobulin (antigen receptor), is also able to present the class II MHC-protein (typical for B-cells and macrophages) to which the antigen is bound.

In response to this double information mediated by B-cells or macrophages, that is, in the presence of an (1) antigen and (2) an immune-competent cell, the T_H cell liberates lymphokins, which now activate the B-cells. The activated B-lymphocytes multiply (*clonal selection* or *expansion*), giving rise to *memory cells* and **plasma cells**. These cells are specialized to produce (and release by exocytosis) the specific **immunoglobulin** required for each antigen (→ **B; C 5, 6**). This production of antibodies starts up much more quickly on *repeated contact* with the pathogen since, at the first response, the **memory cells** stored the information eliciting the response.

The vast variety of antibodies (10^6-10^9?) is not genetically coded in the germ cells; instead, certain sections of the gene (V, D, J, C) are recombined and mutated during the development of the lymphocytes.

Immunglobulins are incapable of directly destroying pathogens, but merely *mark* them as targets for attack by other defense systems (opsonization, complement system; see above). In serum and interstitial fluid (→ p. 64) in the primary response, first IgM and later IgG appear in roughly equal amounts; in the faster secondary response, the release of IgG predominates. In the case of the lungs, lachrymal fluid, saliva, and intestine, mainly IgA is formed in response to invading foreign bodies.

In the first months of life an **infant** is protected against microorganisms by its unspecific defense system and by humoral antibodies acquired partly before birth, via the placenta, from the maternal plasma (IgG; → p. 65, C), or from its mother's milk.

The humoral defense system is not fully effective in combating certain types of microorganisms (viruses, mycobacteria, Brucellae, etc.) because some of them manage to escape intracellular killing. This gap in defense is filled by **cellular immune defense** (→ **D**).

One of its reactions, proceeding relatively slowly (maximum reached after approximately 2 days), is carried out by the T_H cells: *delayed immune response*. The other reaction, that of the T-killer cells, is responsible for killing *virus-infected cells* and *tumor cells*, and for the *rejection of transplanted organs* originating in an organism with "foreign" histocompatibility proteins. The following are some of the cells involved in cellular defense: the **T_H** cells (see above); the **T-suppressor (T_s)** cells, responsible for regulating the immune response; the **T-killer cells** (see below); and the antigen-presenting **macrophages**, of which a subgroup, mainly in the thymus, spleen, and skin, is particularly specialized for this purpose. Finally, intercellular signal substances mediating this **cooperation** are **lymphokins** and **monokins** released by the T-cells and monocytes.

The initial reaction in cellular immune defense is again phagocytosis by **macrophages** (→ **D 1**).

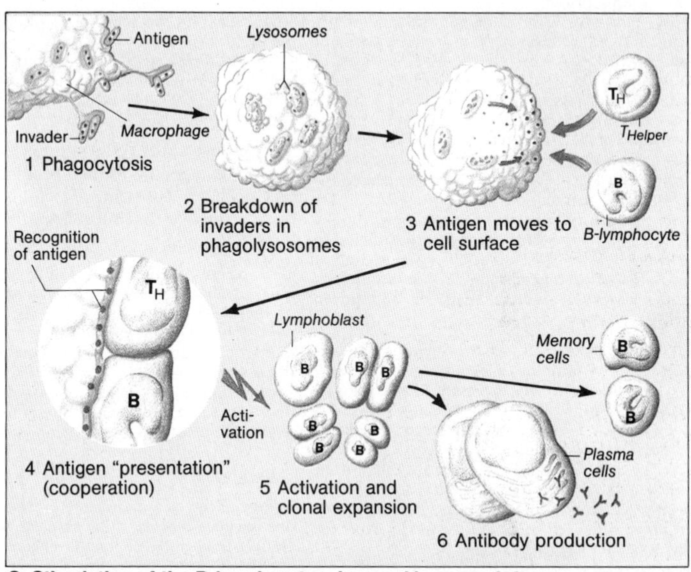

C. Stimulation of the B-lymphocytes: humoral immune defense (partly after Gemsa)

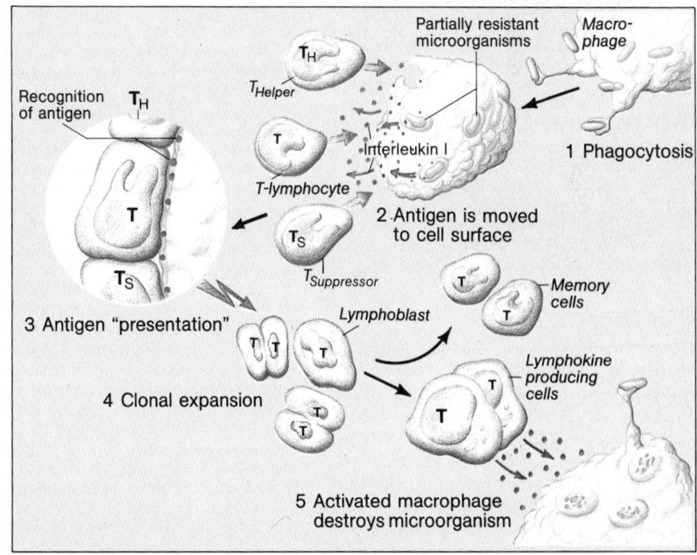

D. Stimulation of the T-lymphocytes and activation of macrophages (partly after Gemsa)

Although most of the phagocytized microorganisms survive intracellularly, a small amount of antigen that, together with the MHC proteins of class II (for activation of the T_H-cells) and class I (activation of T-killer cells), is *presented* (→ **D2**) on the cell surface of the T-cells is sufficient for **activation of the T-cells**. The monokin *interleukin 1* from the macrophages, as well as the *interleukin 2* brom the T_H-cells, participate in this activation (→ **D3**). Once activated, the T-cells "responsible" for this specific defense reaction multiply rapidly (*clonal selection*, → **D4**), giving rise, in addition to *memory cells*, to T-cells whose *lymphokins* lead to **activation of the macrophages**. The latter are then able to cope with almost all pathogens and foreign cells.

Cells attacked by viruses thus present on their surface the virus antigens together with the **class I MHC-proteins**, which are present in all nucleated cells of the body. For the MHC-antigen complex, the cytotoxic **T-killer (T_C) cells** possess receptors. These receptors recognize and bind only *virus-infected* cells. This ensures that healthy cells are not killed and that the receptors are not rendered ineffective by binding free viruses. Binding results in the diseased cell being killed (consequences for the virus: see above; NKC).

In spite of these antiviral defense mechanisms, some viruses manage to survive for years in the body (e.g. hepatitis and Herpes viruses). This type of **slow virus infection** is nevertheless relatively "conventional" with regard to virus structure and immune response. The "unconventional viruses" responsible for scrapie, kuru, and Creutzfeldt-Jakob disease elicit no immune response at all and lead to slowly progressive degeneration of the CNS. The **h**uman **i**mmune deficiency **v**irus (**HIV**) causing **a**cquired **i**mmune **d**eficiency **s**yndrome (**AIDS**) decimates the T_H- (T_4-)cells, which play a central role in the immune system (see above). As a result, the life of AIDS patients is endangered by innumerable, otherwise harmless infections.

Blood Groups

The erythrocytes, too, possess antigenic properties known as blood groups. **ABO system: A** (antigen A on the erythrocytes; antibody anti-B in serum); **B** (B, anti-A); **O** (neither A nor B, but anti-A + anti-B); **AB** (A + B, neither anti-A nor anti-B) (→ **G**). These ABO antibodies belong to the IgM-class. With their high-molecular mass of 900000 Daltons, they are normally unable to cross the placental barrier to any great extent.

If, in a **blood transfusion**, for example, A accidentally comes into contact with anti-A or B with anti-B, the erythrocytes clump together (*agglutination*) and burst (*hemolysis*; → **G**). Thus, the blood group of donor and of recipient have to be known and the compatibility of the blood confirmed by a **cross test** before a transfusion is given.

In contrast to the ABO system, the C, D, E, c, and e types of **Rhesus properties** of the erythrocytes (present, Rh^+; absent, rh^-) only give rise to antibodies after previous sensitization.

The Rhesus antibodies are able to cross the placental barrier relatively easily since they belong to the IgG class (→ pp. 64 and 65, C). Persons with an rh^- blood group may form antibodies to Rh^+ erythrocytes (anti-Rh^+) as, for example, following the wrong transfusion, or an Rh^+ child with an rh^- mother. Subsequent contact of the same kind leads to a more severe antigen-antibody reaction with resultant agglutination and hemolysis (→ **H**).

Allergies

A true allergy is a **disturbance in the regulation of the immune system**. For example, a harmless antigen (such as pollen) may erroneously be judged by the body to be "dangerous," and elicits an overreaction. Allergic reactions may be of the anaphylactic type (s to min) or the delayed type (days), and are mediated by the humoral and the cellular immune defense respectively. In the **anaphylactic type**, the antigen (= **allergen**) sensitizes the B-lymphocytes, and on second contact plasma cells rapidly release large quantities of **IgE** (normally, IgE makes up only 0.001 % of the Ig). Binding of the allergen to 2 IgE, which are attached to IgE receptors of the **mast cells** (→ **E**), leads to exocytosis of histamine, serotonin, lymphokins, to name only a few. They act chiefly on blood vessels (*dilatation*, *edema*), mucus glands (*hay fever*) and sensory nerve endings (*itching*). In addition, synthesis and release of prostaglandins and leukotrienes (→ p. 235) are stimulated. The latter are components of SRS-A (slow reacting substance of anaphylaxis); SRS-A has a powerful constricting effect on the bronchi (*asthma*). The **delayed type** of allergy can be elicited by mycobacteria (e.g. Tbc), fungi, contact allergens (chrome compounds, poison ivy), and many other substances. In the wider sense, *serum disease*, which results from very high levels of antigen (e.g. in passive immunization), is also an allergy. After about 6 days the concentration of the antigen-antibody complexes in the blood rises dramatically (→ **F**); as a result they are deposited in the region of blood capillaries and thus elicit the symptoms of the disease.

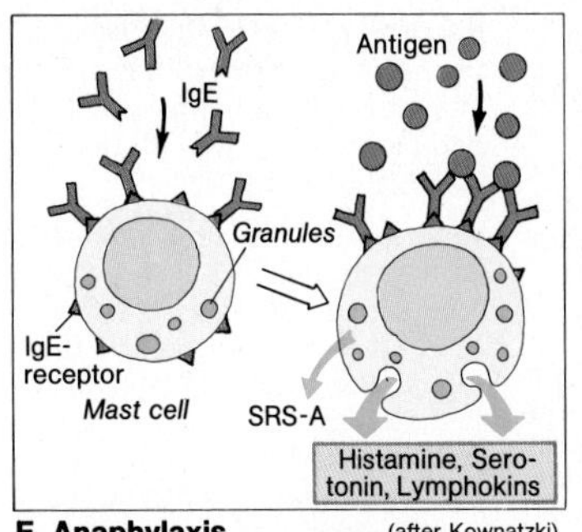

E. Anaphylaxis (after Kownatzki)

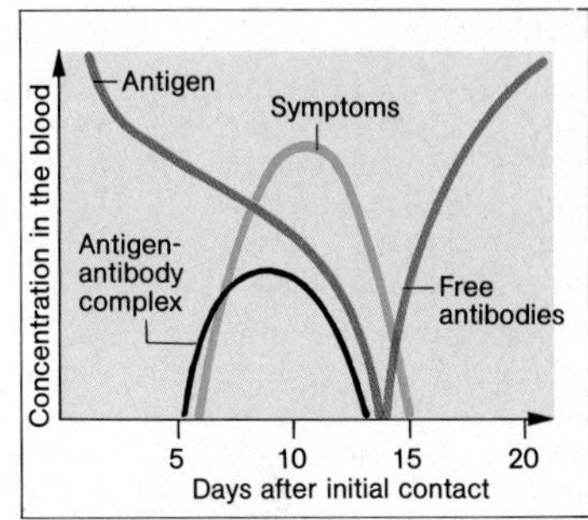

F. Serum desease (after Kownatzki)

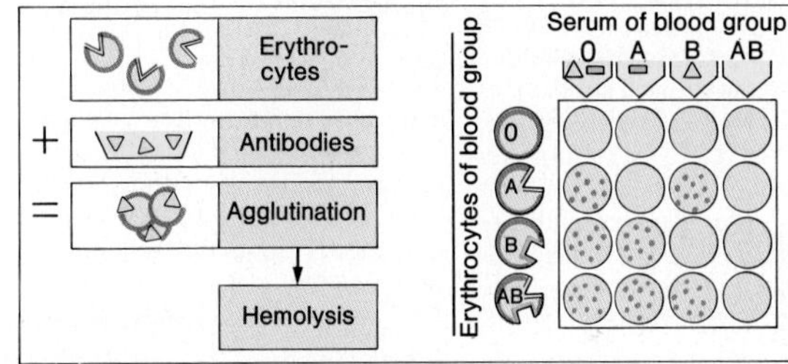

G. AB0 blood group compatibility

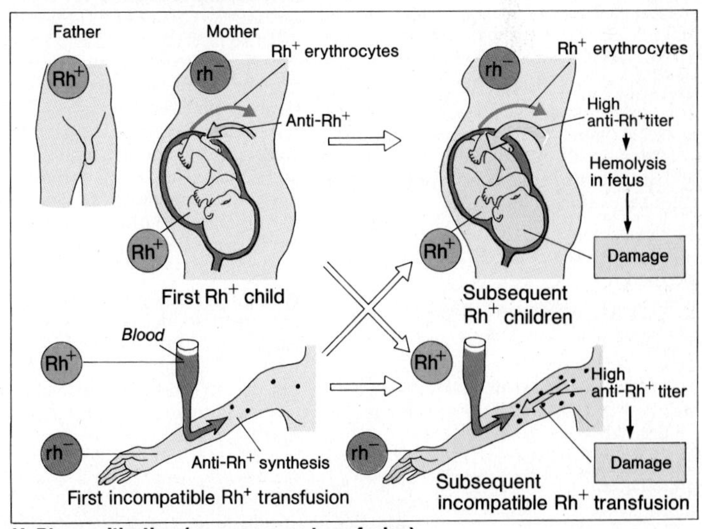

H. Rh sensitization (pregnancy or transfusion)

Hemostasis

Hemostasis or arrest of bleeding involves the interaction of *plasma-* and *tissue-factors* with the blood **platelets (TC = thrombocytes)** and the *vessels*. The mechanism ensures that leaky vessels are sealed within the space of a few minutes.

If the innermost lining of a blood vessel (*endothelium*) is defective (e.g. due to injury), the blood comes into contact with the underlying *collagen fibers*. The TC adhere at the site of injury, with the help of the *von Willebrand factor (vWF)*. This process, known as **adhesion** (→ **A 1**), **activates the TC**, which then change their shape (*metamorphosis* of platelets to spheres with pseudopodia) and expel, by exocytosis, substances contained in vesicles (*granules*): **secretion**. Of these substances, *ADP*, for example, stimulates aggregation; *vWF* and *fibronectin* promote adhesion (→ p. 14); *serotonin* (→ **A 2, B**), mitogen, and *PDGF* (= **p**latelet-**d**erived **g**rowth **f**actor), among others, have a vasoconstrictor effect. In addition to this, activated TC release *Thromboxan A_2* (vasoconstrictor) and *PAF* (**p**latelet-**a**ctivating **f**actor). As well as activating more TC, PAF also activates *phagocytes* (→ p. 66 ff.). The final result is a massive accumulation of TC: **aggregation**.

This **thrombocyte plug** (*white thrombus*), especially in the case of minor defects, leads to a provisional stopping of the leak, aided by local **vasoconstriction** and rolling up of the inner epithelial layer.

At the same time, the actual process of **clotting** (→ **A 3**) is set off by two further mechanisms:

a) an **"extrinsic" system**, triggered by tissue factors set free when a tissue is damaged (→ p. 76 ff.).

b) an **"intrinsic" system**, activated by contact between clotting factor XII and the collagen fibers (→ p. 76 ff.).

Both systems can activate, either singly or in combination, plasma factor **X**, which, together with other factors (→ p. 76 ff.), converts **prothrombin** (factor II) to **thrombin**. This in turn converts **fibrinogen** (factor I) to **fibrin** (→ **B**).

Phospholipids on the surface of lipoproteins are essential for several steps in coagulation. They are released from the TC (thrombocyte factor 3, TF 3, → **B**) and from injured tissue (tissue factors).

Fibrin consists of fibers that form a network enmeshing the TC and erythrocytes, thus producing the **definitive "mixed"** or **red clot**. The further sealing goes through three subsequent stages: (1) *retraction* of the clot, mediated by contractile proteins of the thrombocytes; (2) *organization*, during which fibroblasts proliferate and connective tissue forms; and (3) *scar formation* and regeneration of the endothelium in the lumen of blood vessels (→ **B**, bottom).

Free **Ca^{2+}** ions ("Factor IV") are also essential for several steps in coagulation (see below; → p. 77). The addition of *citrate* or *oxalate* to freshly withdrawn blood binds the Ca^{2+} ions and prevents clotting. This is a necessary procedure in a number of blood and clotting tests.

Vitamin K is required for the formation of the coagulation factors **prothrombin (II), VII, IX, and X** in the liver (→ p. 76 ff.). It is an essential cofactor for the γ-carboxylation of these factors that takes place at some of the N-terminal glutamyl residues of the protein chains, to give *γ-carboxyglutamyl residues* (posttranslational modification; → p. 3). With the help of the latter residues the coagulation factors can bind to Ca^{2+}, which in turn attaches to phospholipids (formation of "complexes"). Thus, important reactions in the process of blood coagulation can take place on the surface of lipoproteins (→ p. 220 ff.).

Normally, vitamin K is provided by intestinal bacteria, but if the gut flora has been damaged, as for example by oral administration of antibiotics, vitamin K deficiency can occur. The same is true in disorders of fat digestion and absorption, since vitamin K is fat soluble.

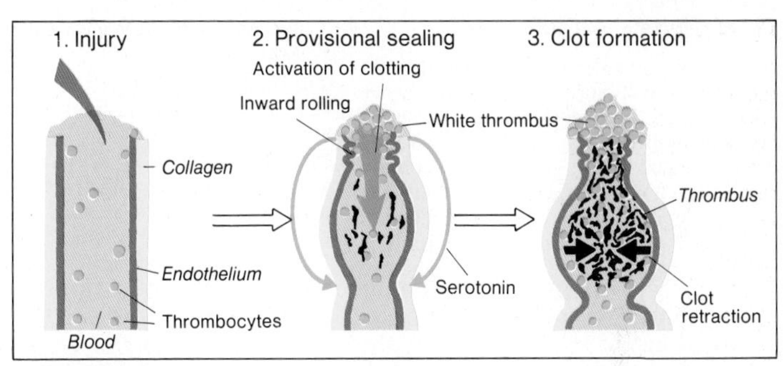

A. Hemostasis

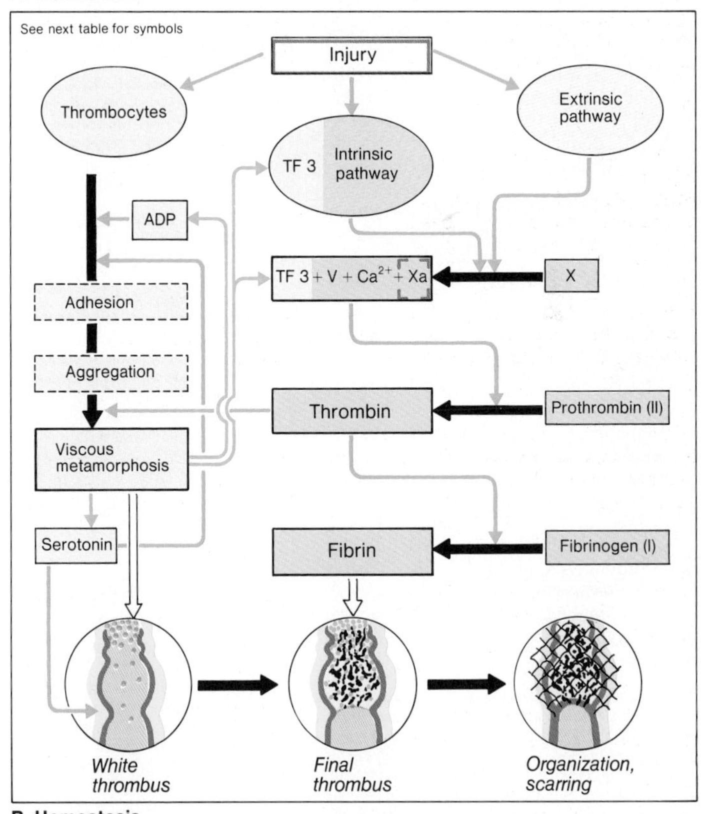

B. Hemostasis

Blood Clotting and Fibrinolysis

Minor defects in the vascular endothelium are mainly dealt with by the thrombocytes (→ p. 74) and the **intrinsic system**. The latter becomes activated if plasma factor XII is itself activated (XII a, → **A**, top) by coming into contact with surfaces other than the vascular epithelium (e.g. with collagen). Kininogen and kallikrein are cofactors. The intrinsic pathway involves a thrombocyte factor (TF 3), plasma factors, and Ca^{2+} (→ **A**, top). In the case of widespread damage involving tissues as well as blood vessels, the **extrinsic system**, including tissue factors (*tissue thrombokinase*), plasma factor VII, and Ca^{2+}, comes into play. Both pathways, either individually or together, activate **factor X** (→ p. 75) to factor Xa. Factor Xa, together with phospholipids (from thrombocytes [TF 3] or tissues), plasma **factor V**, and Ca^{2+}, converts prothrombin to thrombin (→ **A**, middle). **Thrombin** has three major effects: (1) it catalyzes the conversion of fibrinogen to **fibrin** for clot formation; (2) it activates the fibrin-stabilizing **factor XIII** to XIII a (→ **A**, middle); and (3) it influences the thrombocytes during hemostasis (→ **B** and p. 75).

The single (monomeric) fibrin fibers form a loose network of $fibrin_s$, which is polymerized by factor XIII a (→ **A**, middle) to $fibrin_i$.

Once the process of clotting has begun, some mechanism must come into operation to prevent it from spreading throughout the vascular system (**thrombosis**). In addition to antithrombin 3 (see below), **plasmin** plays an important role by causing the dissolution of fibrin (**fibrinolysis**; → **A**) and other clotting factors. **Fibrin fragments** resulting from fibrinolysis inhibit the formation of fibrin and thus offer one safeguard against unlimited clotting.

Plasmin is derived from inactive *plasminogen*, in a process that requires a number of blood and tissue factors, probably including factor XII a. *Streptokinase* and *urokinase* are employed therapeutically as plasminogen-activating agents in dissolving freshly formed clots within the blood vessels. Excessive fibrinolysis, on the other hand, is hindered physiologically by *antiplasmins*. The same purpose can be served by therapeutic administration of ε-aminocaproic acid, aprotinin, and similar substances (→ **A**, below).

The most important plasma protein involved in protection against thrombosis is **antithrombin 3**. It forms complexes with thrombin, factor X a, IX a, XI a, XII a, and kallikrein, and thus prevents their further activity. The formation of such complexes can be promoted by naturally occuring (e.g. from the mast cells or from the endothelium) or administered **heparin**. *Deficiency of antithrombin 3* leads to thrombosis.

Therapeutic reduction of the clotting powers of the blood (**anticoagulant therapy**) is desirable if danger of thrombosis exists, that is, if a blood clot threatens to obstruct important blood vessels. *Heparin* (see above) inhibits thrombin and other factors indirectly, whereas *dicoumarol* and similar substances inhibit the vitamin-K-mediated γ-carboxylation (→ p. 74) of prothrombin, and factors VII, IX, and X in the liver. *Acetyl salicylic acid* (aspirin, etc.) and other substances inhibit the aggregation of thrombocytes (→ p. 74) by blocking their prostaglandin metabolism (→ p. 235).

A tendency to **hemorrhage** due to an abnormal reduction in the ability of the blood to clot can be caused by
1) congenital lack of a factor (e.g. factor VIII; leads to *hemophilia A*),
2) acquired deficiency of a factor (*liver damage, vitamin K deficiency*),
3) elevated consumption (utilization) of a particular factor (*consumption coagulopathy*),
4) too few or diseased thrombocytes (*thrombocytopenia, thrombocytopathia*),
5) certain vascular diseases,
6) excessive fibrinolysis, and so on.

Factor		half-life in vivo (h)
I	fibrinogen	96
II	prothrombin	72
III	thromboplastin, thrombokinase, tissue factor	
IV	ionized Ca^{2+}	
V	accelerator globulin	20
VII	proconvertin	5
VIII	antihemophilic globulin (A)	12
IX	plasma thromboplastic component (PTC), Christmas factor	24
X	Stuart-Power factor	30
XI	plasma thromboplastin antecedent (PTA)	48
XII	Hageman factor	50
XIII	fibrin-stabilizing factor	250

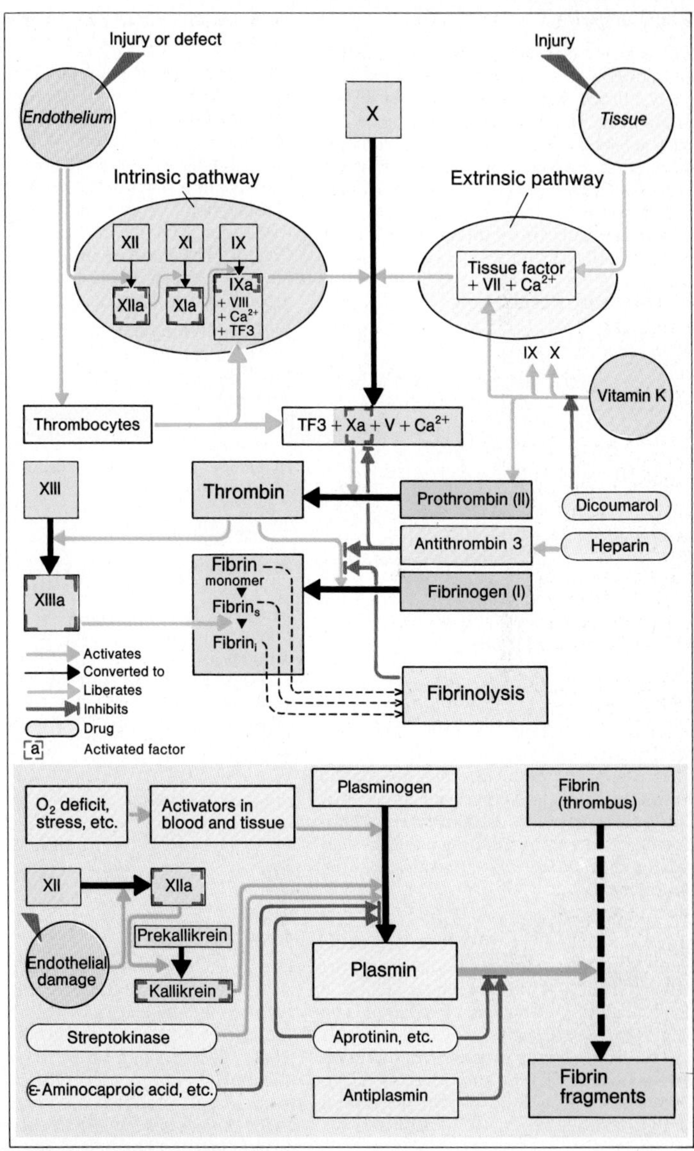

A. Blood clotting and fibrinolysis

The Lungs

The primary task of the lungs is *respiration*. In addition, they play a role in *metabolism*. They convert, for example, angiotensin I to angiotensin II (→ p. 152) and remove certain substances (e.g. serotonin) from the circulation. Further, the pulmonary circulation *buffers* the *blood volume* (→ p. 160 and p. 184) and traps small blood clots (emboli) before they can cause damage in the arterial pathways (heart, brain).

Functions of Respiration

Respiration in its narrowest sense, that is, "external" respiration, is **gaseous exchange** between organism and environment ("internal respiration" = oxidation of nutrients → p. 198). In contrast to unicellular organisms, in which distances for the *diffusion* (→ p. 8 f.) of O_2 and CO_2 between cell and environment are sufficiently short, the mutlicellular human organism requires a special convective transport system for gaseous exchange, that is, the respiratory tract and the circulatory system (→ p. 154 ff.).

O_2 in the air that is inspired in the course of respiratory movements reaches the *alveoli* of the lungs (*ventilation*), where it diffuses into the blood. The O_2 is transported in the blood to the tissues, where it diffuses to the mitochondria inside the cells. The CO_2 formed here moves in the reverse direction. The respiratory gases are thus alternately transported by **convection** over longer distances (ventilation, circulation) and by **diffusion** across thin limiting membranes (gas/fluid in the alveoli, blood/tissue at the periphery).

About 300 million thin-walled vesicles, the **alveoli** (diameter of about 0.3 mm), form the endings of the terminal branches of the bronchial tree. They are enmeshed in a dense network of **pulmonary capillaries**. The total surface area of the alveoli is about 100 m^2. Here, owing to the enormous area, gaseous exchange can take place by diffusion, that is, CO_2 diffuses into the alveoli and O_2 diffuses out of the alveoli into the blood of the pulmonary capillaries (→ p. 92 ff.). In this way the oxygen-deficient ("venous") blood in the pulmonary artery is "arterialized" and, via the left ventricle, once again reaches the periphery.

When the body is at rest the heart pumps about 5 l/min blood (**cardiac output**) through the lungs and subsequently through the systemic circulation. With this blood flow roughly 0.3 l/min O_2 are conveyed, at rest, from the lungs to the periphery (**oxygen consumption** $\dot{V}_{O_2}$) and approximately 0.25 l/min CO_2 in the reverse direction ($\dot{V}_{CO_2}$). (These are *net values*, i.e. the difference between l/min transported in the arteries and in the veins.)

In order to bring this volume of O_2 from the environment into the alveoli, and to expire the CO_2, a **total ventilation** ($\dot{V}_T$) of about **7.5 l/min** is necessary. This is achieved by breathing in and out a **tidal volume** (V_T) of roughly **0.5 l** about **15 min^{-1}** (**respiratory frequency f**). **Alveolar ventilation** ($\dot{V}_A$) is smaller than $\dot{V}_T$ because dead space ventilation ($\dot{V}_D$) makes up a significant fraction of $\dot{V}_T$ (see also p. 86).

In a mixture of gases, the **partial pressure** of each gas (the pressure that each gas exerts) equals the total pressure of the gas mixture times the relative fraction (fractional "concentration" F; → p. 328) of the individual gas. The partial pressure of the individual gases adds up to the total gas pressure (*Dalton's Law*). At sea level, air has a mean barometric pressure of 101.3 kPa (= 760 mmHg).

Composition of dry air

Gas	F (l/l)	*P* at sea level (kPa)	(mmHg)
O_2	0.209	21.17	158.8
CO_2	0.0003	0.03	0.23
N_2 + inert gases	0.791	80.1	601
dry air	1.0	101.3	760

During its passage through the respiratory channels (mouth, nose, throat, bronchial system) the inhaled air becomes totally saturated with **water**, so that the P_{H_2O} rises to the constant value (at 37 °C) of 6.37 kPa (47 mmHg). This causes a drop in the P_{O_2} from about 21.2 kPa (= 159 mmHg) to 19.9 kPa (149 mmHg), and a corresponding drop in the P_{N_2}. The partial pressures in the alveoli, arterial, and venous blood, and in the expired air, are shown in **A**.

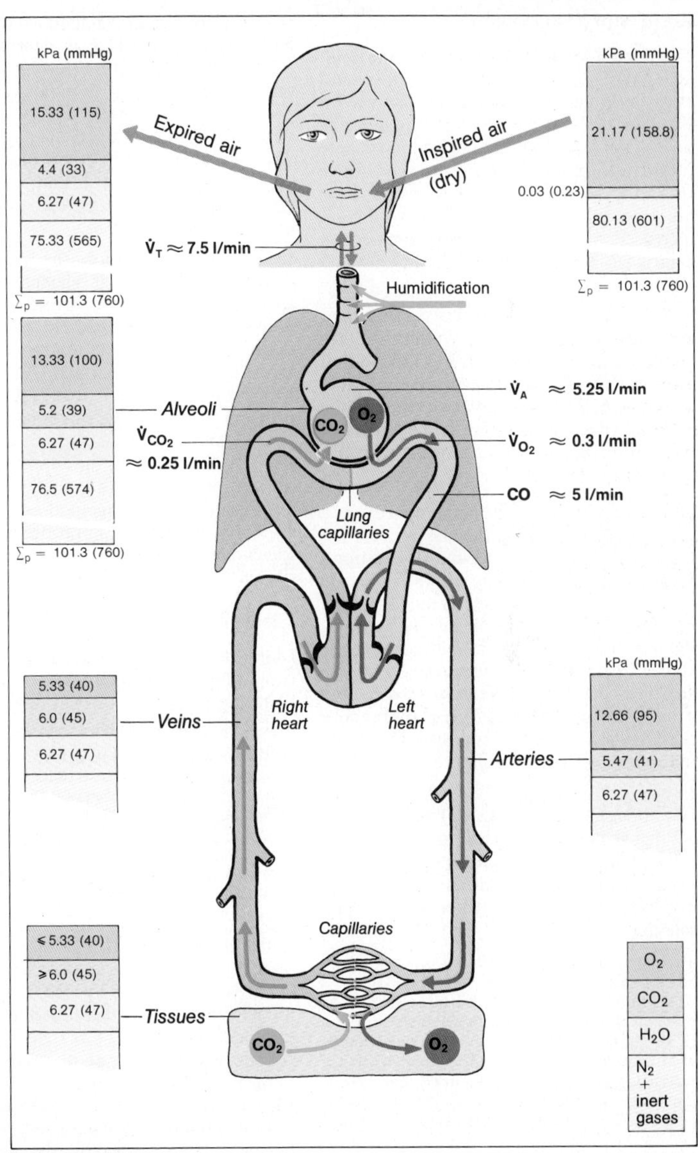

A. Respiration

Respiratory Mechanics

The **driving force for ventilation** is the pressure difference between the atmosphere and the **intrapulmonic pressure** in the alveoli (P_{pul}; → **B**). For inspiration, P_{pul} must be below the external atmospheric pressure (negative), for expiration it must be above (positive; → **B**). These pressure gradients are established when the lung volume is increased on inspiration and decreased on expiration by action of the diaphragm and thorax (→ **A**).

Inspiration is an *active* process. Muscular contraction increases the volume of the chest, the lungs inflate, and P_{pul} falls so that air flows into the lungs. At the end of quiet inspiration, the lungs and chest recoil to the positions they occupied at the beginning of inspiration. Thus, quiet expiration is largely a *passive* process.

Respiratory muscles. The **diaphragm** exerts a direct influence on lung volume by contracting (inspiration) and relaxing (expiration). It accounts for 75 % of the volume change in quiet inspiration. It moves as much as 7 cm on deep inspiration. More indirect influences on inspiration are exerted when the thorax enlarges by contraction of the **scalene** and **external intercostal muscles** and other accessory muscles. The diaphragm *or* the external intercostal muscles alone can maintain adequate ventilation at rest.

Expiration occurs chiefly by passive recoil (see above). It can be assisted by contraction of the **abdominal muscles**, which increase intra-abdominal pressure, pushing the relaxed diaphragm towards the chest cavity, and by contraction of the **internal intercostal muscles**.

The external intercostal muscles run downward and forward from rib to rib (→ **A**). The attachment of the muscle to the upper rib (B) is closer to the pivot point (A) than the attachment to the lower rib (C') is to its joint (A'); the lower rib has a longer lever arm (A'–C' > A–B). Therefore, when the external intercostal muscles contract, they *elevate* the lower rib, which pivots on its joint at the vertebra. This action pushes the sternum outward and increases the anteroposterior diameter of the rib cage. The internal intercostal muscles are oppositely oriented (→ **A**), and their contraction causes the ribs to pivot in the opposite direction.

In order for the movements of the thorax and diaphragm to help in ventilation, the lungs must follow these movements without, however, being completely fixed to thorax and diaphragm. This is made possible by the presence of a thin layer of fluid between the two layers of the **pleura**, one of which covers the lungs (*pleura pulmonalis*), while the other covers the surrounding organs (*pleura parietalis*).

In their natural position, the lungs tend to "shrink" due to their elasticity and the surface tension of the alveoli (→ p. 90). Since the fluid in the pleural space cannot expand, the lungs therefore remain in contact with the inner surface of the thorax, which results in suction, that is, to a negative pressure relative to the surroundings (**intrapleural pressure** or **intrathoracic pressure** [P_{pl}]; → **B**). When the thorax expands during inspiration the suction becomes stronger, decreasing again at expiration (→ **B**). Only in forced respiration, assisted by the expiratory muscles (see above), can P_{pl} become positive.

Cleansing of the Inspired Air

Many dirt particles in the inspired air are caught by the **mucus** covering the nasal and pharyngeal cavities, as well as the trachea and the bronchial tree.

In the *bronchial tree* (more than 20 successive branchings) the total cross-sectional area of a set of branches is greater than that of the respective stem from which it arises. Therefore, the air flow produced by the changes in P_{pul} comes to a stop in the terminal branches of the bronchioles and with it any remaining particles of dust from the outside air. (O_2 and CO_2 cover the remaining few mm to or from the alveoli by diffusion.)

In the bronchial tree, the particles are either phagocytosed on the spot or are returned towards the glottis by the *cilia* of the tracheobronchial epithelium (**mucociliar escalator**). Cilia beat 12–20 times/s and propel the mucous film at a speed of about 1 cm/min. Mucus is produced at a rate of 10–100 ml/day, depending on local irritants (e.g. smoke) and on vagal stimulation. The mucus is usually swallowed and its fluid resorbed in the gastrointestinal tract.

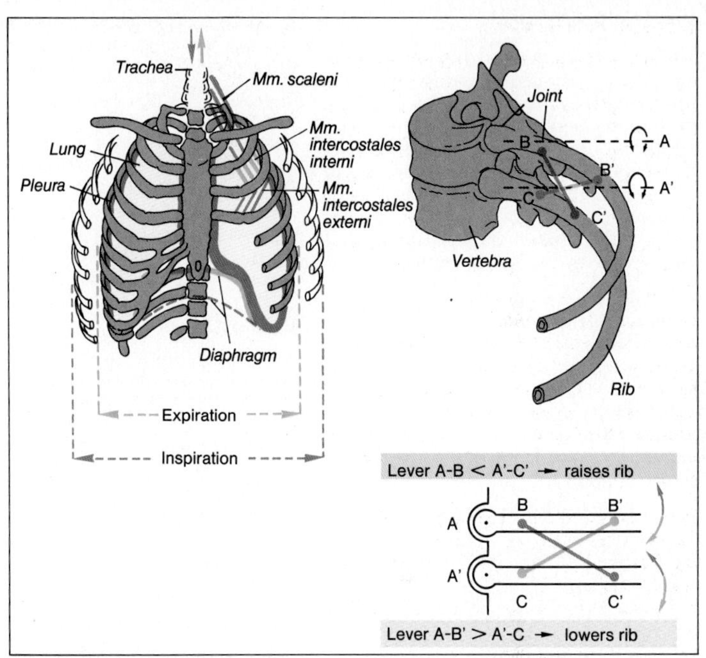

A. Respiratory musculature

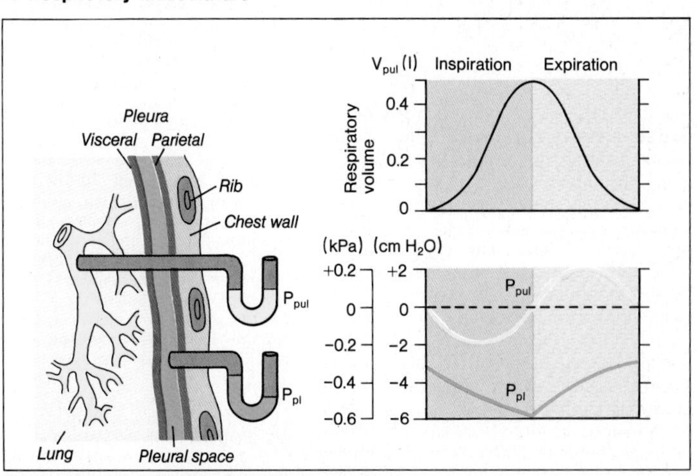

B. Intrapleural and intrapulmonary pressures

Artificial Respiration

Artificial respiration is necessary when spontaneous ventilation is insufficient or fails completely. In these cases, artificial respiration should always be attempted because the heart continues to function even after respiratory stimuli have ceased. Lack of oxygenation for seconds causes loss of consciousness and for only a few minutes causes irreversible damage to the central nervous system and death (→ p. 102).

Mouth-to-mouth respiration is an emergency measure until spontaneous respiration can be restored (→ **A**). In all attempts at artificial respiration, the air passage should be cleared. The subject is turned onto the back with the nose held closed. The first-aid assistant blows into the subject's mouth. This process raises the P_{pul} (→ p. 81, B) above atmospheric pressure and inflates both the lungs and the chest. Breaking the mouth-to-mouth contact allows expiration as the chest collapses passively by elastic recoil. Pressure applied to the chest can increase the rate and completeness of expiration. The helper inspires fresh air and repeats the cycle about 15 times/min. The O_2 content of the helper's expired air (F_{EO_2}) provides the patient with adequate O_2. The *success of artificial respiration* can be recognized by a change in the patient's skin color from blue (cyanotic) to pink.

A similar principle is involved in **mechanical positive-pressure respiration**, which is employed (during narcosis) in operations in which the patient's respiratory musculature is deliberately paralyzed (curarelike substances). Air is blown in (inspiration) by a pump (→ **A, left**). The expiration and inspiration pathways in the machine must be well separated (control valve → **A, top**), since the dead space (→ p. 86) would otherwise be too large. This kind of artificial respiration can be performed with a constant volume (*volume controlled*) or at a constant pressure (*pressure controlled*). Each method has its advantages and disadvantages, but in either case respiration must be continuously monitored (i.e. concentration of expired gases, blood-gas composition, and so on).

The conventional respirator used for the treatment of chronic respiratory insufficiency, as may occur after bulbar poliomyelitis, is the "iron lung" (→ **A**), which provides **mechanical negative-pressure respiration**. The patient is enclosed to the neck in an airtight chamber. Air is pumped out of the chamber to lower the pressure and to permit inspiration. When air is allowed to flow into the chamber again, expiration takes place.

One deficiency of these systems is that they hinder *venous return* to the heart (→ p. 184), but this can be partly avoided if expiration as well as inspiration is assisted.

Pneumothorax

Pneumothorax occurs when air enters the pleural space (→ p. 81). In **open pneumothorax**, when the chest wall is penetrated, the lung on the open side collapses by elastic recoil and does not contribute to ventilation (→ **B**). Air moves in and out of the pleural space when the subject breathes. Gas exchange in the healthy lung is also compromised because (1) air is exchanged in part with that in the collapsed lung instead of with the external atmosphere, (2) the weight of the collapsed lung prevents full ventilation, and (3) atmospheric pressure in the pleural space presses the mediastinum into the healthy side. Respiration is stimulated but distress can be severe. In **tension pneumothorax** (→ **B**), the air entering the pleural space with each respiratory movement is unable to escape (e.g. a flap of tissue on the wound may act as a valve). Pressure in the pleural space on the damaged side becomes positive; the resulting hypoxia elicits a progressive increase in the total ventilatory rate, with the result that pressure may rise to as much as 30 mmHg in the pleural space on the damaged side. The mediastinum shifts to the unaffected side. Cardiac filling (→ p. 162) is reduced, and peripheral veins distend. Cyanosis develops, and the condition may become fatal if pressure is not relieved by withdrawing air. However, if the hole becomes sealed, P_{pl} stabilizes, the healthy lung retains its function, and anoxia does not develop. After 1 to 2 weeks, the air pocket is completely absorbed. **Closed pneumothorax** (the most common type of pneumothorax) may develop spontaneously, especially in emphysema when the lung ruptures through the visceral pleura, so that there is a direct connection between bronchial system and pleural space. Forced (mechanical) positive pressure respiration or too rapid surfacing from a dive (→ p. 106) can also result in closed pneumothorax.

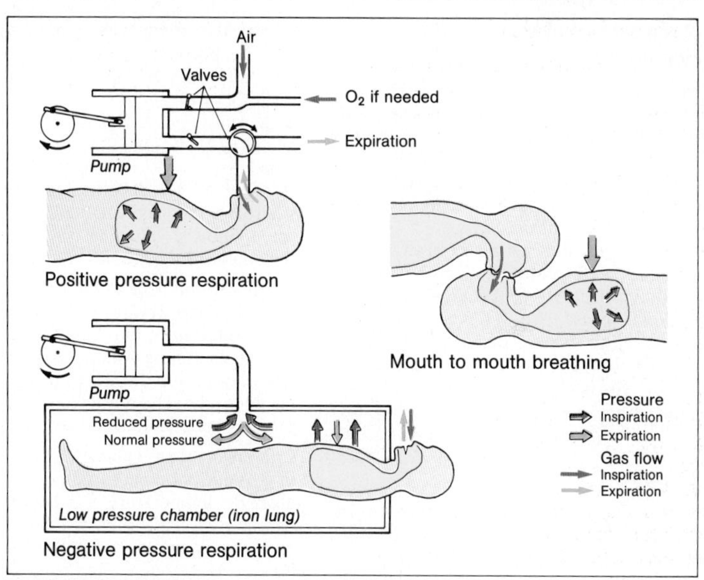

A. Artificial respiration

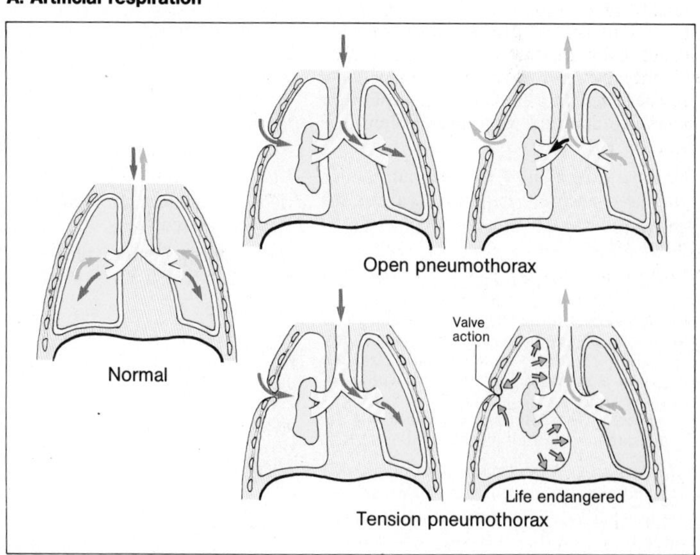

B. Pneumothorax

Respiratory Volumes and the Spirometer

In quiet (passive) expiration, the thorax comes to rest at the **resting expiratory level**. Quiet breathing at rest allows a ventilatory exchange of about 0.5 l, the *tidal volume* (V_T). An active maximal inspiratory effort allows intake of about 2.5 l of air in excess of the resting V_T, the *inspiratory reserve volume*. Beyond the resting expiratory level, however, still more air can be expired: maximum roughly 1.5 l (*expiratory reserve volume*). These two reserve volumes are called upon when, during exertion (→ p. 48 f.), the resting tidal volume is inadequate for gas exchange. Even after maximal expiration, some air remains in the lungs; this is the *residual volume*. Sums of these volumes in various combinations are called capacities.

The **vital capacity** is the air expired up to maximal expiration following maximal inspiration, and is the sum of the tidal volume + inspiratory reserve volume + expiratory reserve volume (about 4.5 to 5.7 l in a young man 1.8 m in height). The *total lung capacity* includes, in addition to the above three volumes, the residual volume; the *functional residual capacity* is the sum of the expiratory reserve volume and the residual volume (→ **A** and p. 86).

With the exception of the residual volume and the capacities including it, all of the above values can be measured with the **spirometer** (→ **A**).

It consists of an inverted chamber in a water seal. A subject breathes into and out of the chamber, thus moving an indicator that depicts the magnitude of the corresponding volume change. These movements are recorded as the **spirogram** on a drum moving at a fixed rate, allowing calculations of ventilatory rates as well as volumes, and of values incorporating these data (e. g. compliance, → p. 88; O_2 utilization, dynamic function tests; → p. 90).

These volumes and capacities vary according to height, age, sex, and physical training. The normal value for vital capacity ranges between 2.5 and 7 l in the absence of pulmonary disease.

In order to take into account at least some of the factors that affect the above values, empirical formulae have been introduced for the purpose of standardization. The normal values for VC of Europeans, for instance, are

male:
$VC\ (l) = 5.2\,h - 0.022\,a - 3.6\ (\pm 0.58)$
female:
$VC\ (l) = 5.2\,h - 0.018\,a - 4.6 \pm 0.42)$,
where *h* is the height in meters, *a* the age in years, and the values in brackets represent the standard deviations.

Even with the help of these formulae, only relatively large deviations from the norm are detectable. More informative are the measurements of lung volumes which, if measured repeatedly in the same subject, reflect for instance the changes accompanying the course of a pulmonary disease.

Conversion of gas volumes. The volume V [l] of a quantity of gas n [mol] depends on the absolute temperature T [K] and the total pressure P [kPa], that is, the barometric pressure P_B minus water vapor pressure P_{H_2O}:

$V = n \cdot R \cdot T/P$,

where R = general gas constant = 8.31 $J \cdot K^{-1} \cdot mol^{-1}$.

A distinction is made between the following conditions:

STPD: **S**tandard **T**emperature **P**ressure **D**ry (273 K; 101 kPa; $P_{H_2O} = 0$)

ATPS: **A**mbient **T**emperature **P**ressure H_2O-**S**aturated (T_{amb}; P_B; P_{H_2O})

BTPS: **B**ody **T**emperature **P**ressure **S**aturated (310 K; P_B; P_{H_2O} = 6.25 kPa)

and hence:
$V_{STPD} = n \cdot R \cdot 273/101$
$V_{ATPS} = n \cdot R \cdot T_{amb}/(P_B - P_{H_2O})$
$V_{BTPS} = n \cdot R \cdot 310/(P_B - 6.25)$.

V_{STPD}/V_{BTPS}, for example, therefore amounts to $\frac{273}{310} \cdot \frac{P_B - 6.25}{101}$.

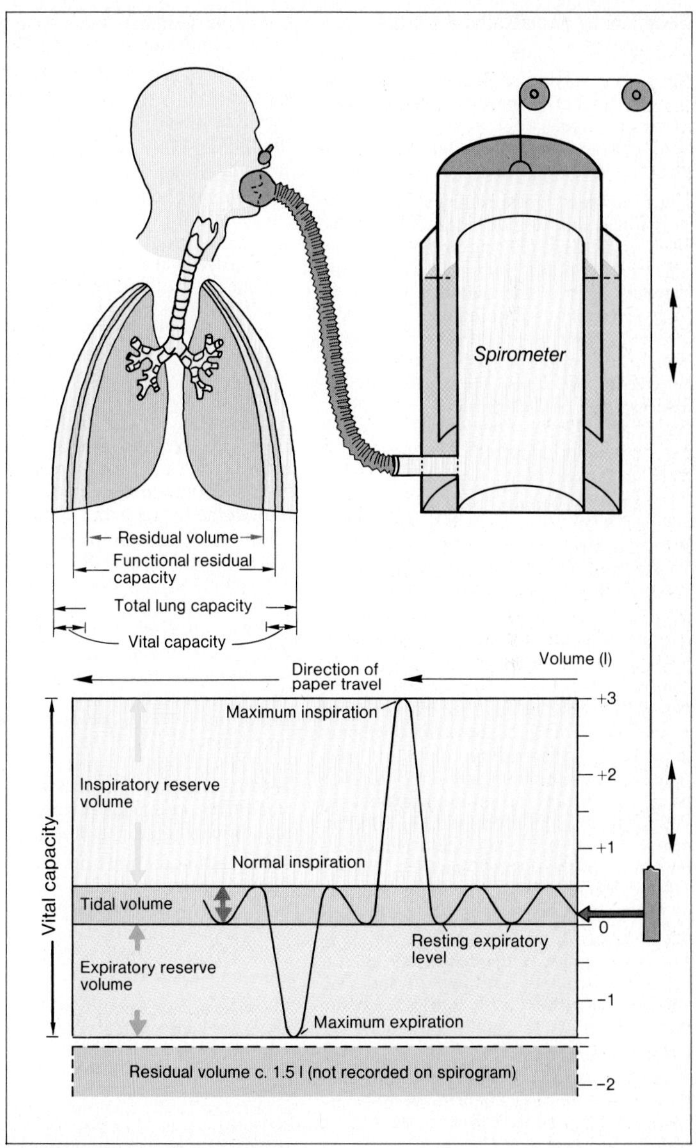

A. Measurement of lung volumes

Dead Space and Residual Volume

Gas exchange in the respiratory system occurs only in the alveoli. Therefore, the expiratory tidal volume (V_T) is composed of two parts, one (V_A) coming from the alveolar space, the other (V_D) from the **dead space** (→ **A**).

The **dead space** is the total volume of the airchannels that conduct inspired air to the alveoli; it does not take part in gas exchange. The mouth, nose, pharynx, trachea, and bronchi comprise the **anatomic dead space (ADS)**, which has a volume equivalent to the body weight in pounds expressed in ml (roughly 150 ml). The ADS functions as an air conduit in which the air is simultaneously freed of dust particles, humidified, and warmed before it reaches the alveoli. It also contributes to speech as a sound box that determines characteristics of the voice (→ p. 324). The ADS is normally roughly equivalent to the **functional dead space**; however, when gas exchange in part of the alveoli is restricted, the functional dead space exceeds the ADS (→ p. 92).

The dead space (V_D) can be calculated from the CO_2 content of alveolar gas, and the V_T by means of the **Bohr equation** (→ **A**). V_T is comprised of the gas that equilibrated in the alveoli, V_A, and the gas that did not (dead space), V_D. In each of these three volumes there is a corresponding equivalent fractional CO_2 concentration (→ **A**): the fractional CO_2 concentration in the expired air (F_{ECO_2}) in V_T; the alveolar CO_2 concentration (F_{ACO_2}) in V_A; and the original concentration in the inspired air (F_{ICO_2}) in V_D. The product of each volume component and its corresponding fractional CO_2 concentration gives the volume of CO_2 for each. The CO_2 volume in the expiratory volume ($F_{ECO_2} \times V_T$) equals the sum of the CO_2 volume in V_A and V_D (→ **A**, II). However, the term F_{ICO_2} is small enough to be disregarded.

To determine V_D, three magnitudes must be measured: (1) V_T is determined with the spirometer (→ p. 84); (2) F_{ACO_2} is determined in the terminal portion of the expired air, that is, in the alveolar air; and (3) F_{ECO_2}. When $V_T = 0.5$ l, $F_{ACO_2} = 0.06$ l/l (6 vol%), and $F_{ECO_2} = 0.045$ l/l (4.5 vol%), V_D amounts to 0.15 l. The ratio V_D/V_T is an index of "wasted" ventilation (at rest about 0.2 to 0.3; → also p. 92).

Residual volume and **functional residual capacity** (→ p. 84) must be measured indirectly because they are not represented on the spirograph.

A *test gas*, such as the N_2 that is normally present, must be used (→ **B**). The fractional concentration of N_2 in the lungs, F_{LN_2}, is constant at about 0.80 l/l. The subject rebreathes several times from a bag of known volume (V_B) containing an N_2-free gas. The N_2 that is in the lungs is expired and inspired into and from the bag several times until it reaches equilibration between the lungs and the bag. In this process the volume (amount) of N_2 remains nearly unchanged, but its concentration in the expired air will decrease by virtue of the larger volume of distribution. Thus, the amount of N_2 at the beginning of the determination (N_2 that was only in the lungs) is the same amount that ultimately is distributed between lung and bag volume at the end of the determination. The pulmonary residual volume, V_L, can then be calculated (→ **B**). V_B and F_{LN_2} are known values, and only the fractional volume of N_2 in the bag at the end of the determination, F_{XN_2}, need be determined. If the procedure is begun at the extreme expiratory level, V_L measures residual volume (about 1.5 l); if begun at the resting expiratory level, V_L measures functional residual capacity (about 3 l; → p. 84). Instead of N_2, an exogenous mixture of helium and O_2 can be used as test gas.

These techniques only measure the part of the lungs reached by air, whereas the use of *whole body plethysmography* for determining residual capacity ensures that encapsulated air spaces (such as cysts) in the lungs are also measured.

The value that is clinically significant is the ratio residual volume/total lung capacity (→ p. 84) (normally 0.25). In obstructive pulmonary disease, this ratio may be more than 0.55 and indicates the severity of the disease process. The residual volume is increased (and forced vital capacity [FVC] is decreased) in these diseases mainly because the airways close prematurely at an abnormally high lung volume.

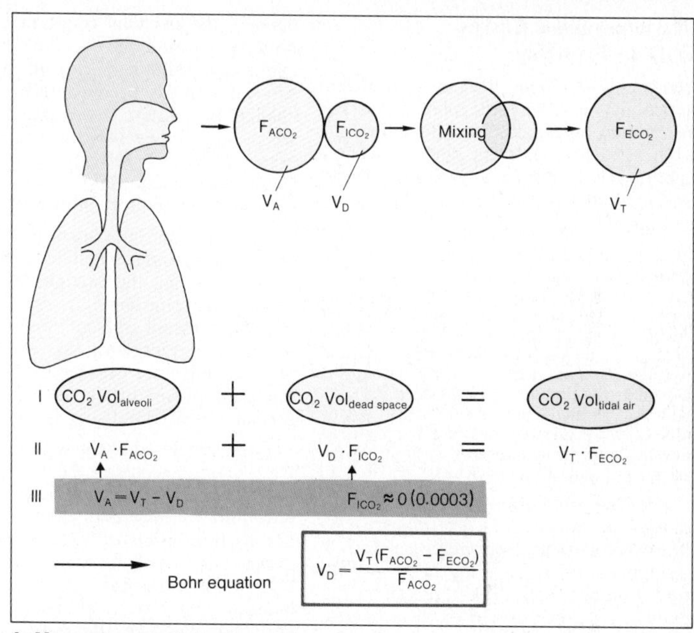

A. Measurement of dead space

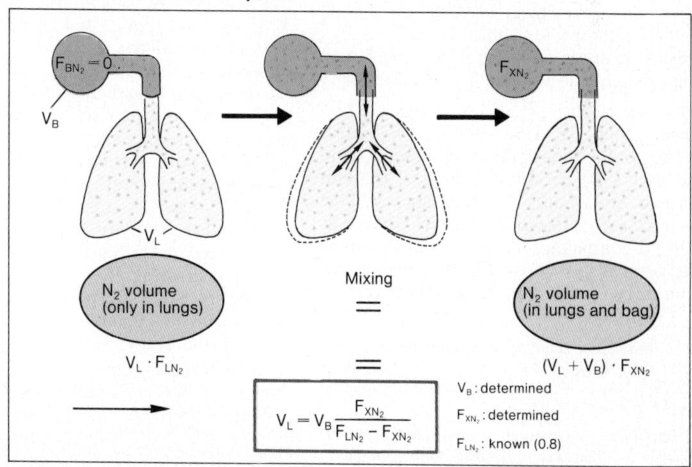

B. Measurement of residual volume or functional residual capacity

Pressure-Volume Curves, Work of Breathing

The pressure volume (PV) curve is determined by measuring the pressure in the airchannels (intrapulmonary pressure, P_{pul}; → p. 81, B) at different stages of chest inflation (V_{pul}) during a respiratory cycle. The relationship is shown graphically by plotting V_{pul} against P_{pul} (→ **A**).

To determine the PV characteristics, V_{pul} at the resting expiratory level is often set at zero. Also, $P_{pul} = 0$ relative to the atmospheric pressure (→ **A, a**). From this starting point, small measured volumes of air are inspired ($+V_{pul}$) or expired ($-V_{pul}$). At the end of each stepwise stage, the valve to the spirometer is closed and P_{pul} is measured. (Respiratory muscles must be relaxed.) Note that during these measurements, V_{pul} is respectively compressed or enlarged if compared to the volume originally measured by means of the spirometer (→ **A**, diagonal arrows).

Under these *static conditions*, the **PV curve for lung and thorax** can be determined (→ **A**, blue curve **c–a–b**). Inspiration generates a rise in pressure (V_{pul} and $P_{pul} > 0$; → **A, b**); expiration generates a fall in pressure (V_{pul} and $P_{pul} < 0$; → **A, c**). These pressures arise from the passive elastic recoil (→ **A**, blue arrows) of the lungs and the chest wall, which attempt to return to the original resting position (→ **A, a**). The recoil forces become greater as more V_{pul} deviates from 0.

The slope of the static PV curve ($\Delta V_{pul}/\Delta P_{pul}$) describes the **static compliance** (→ **B**), or stretchability, of the unit "lungs and thorax". Compliance will be small towards the ends of the PV curve. The largest compliance is observed in the range between $V_{pul} = 0$ to 1 l, the normal respiratory range (normal = 1 ml/Pa or 0.1 l/cm H_2O).

The compliance can also be determined *separately* for the **thorax** ($\Delta V_{pul}/\Delta P_{pl}$), or the **lungs** ($\Delta V_{pul}/\Delta(P_{pul} - P_{pl})$), where P_{pl} = the intrapleural pressure (→ p. 81, B).

Compliance is increased in emphysema and reduced in pulmonary fibrosis and pulmonary congestion.

The PV diagram can also be determined during maximal active participation of the respiratory muscles (→ **A**, red and green curves), the **maximal inspiratory and expiratory pressure** curves. At an extreme expiratory position ($V_{pul} \ll 0$), the expiratory muscles can produce only very small changes in pressure (→ **A, g**); in contrast, when V_{pul} is positive, the pressure may exceed 15 kPa (= 150 cm H_2O) (→ **A, e**). The reverse relationships are found in regard to the inspiratory muscles (→ **A, d, f**).

If the PV relationships are measured during breathing (*dynamically*), different values are obtained during the two phases of inspiration and expiration (*hysteresis*, → **C**). The "driving pressure gradient" on the abscissa of this plot represents, for example, the pressure difference between the mouth and the atmosphere during artificial positive pressure respiration (→ p. 83, A).

The areas enclosed by the two curves (→ **C**, $A_{Rinsp.}$, $A_{Rexp.}$) represent the **work of breathing** done *against frictional resistance* to air flow and to lung and chest movements during inspiration or expiration, respectively. They have the dimension pressure times volume ($Pa \cdot m^3 = J$, → p. 327). The hatched area (→ **C**) is the *elastic work* ($A_{elast.}$) required to stretch the lungs and chest wall. Total **work of inspiration** is $A_{Rinsp.} + A_{elast.}$ (pink + hatched area). Total **work of expiration** is $A_{Rexp.} - A_{elast.}$ (greenish *minus* hatched area). Because at rest the elastic energy stored during inspiration is larger than the expiratory work against frictional resistance ($A_{elast.} > A_{Rexp.}$), expiration does not need extra energy. In forced respiration the greenish area becomes larger than the hatched area ($A_{Rexp.} > A_{elast.}$). Thus, muscle activity becomes necessary and is needed (1) for faster air flow and (2) for decreasing V_{pul} below the resting respiratory level. Respiratory work may increase up to 20 times above its resting value during heavy exercise.

Work of breathing done against frictional resistance of air flow and lung (but not of chest wall) and against elastic forces of the lung alone can be measured during spontaneous breathing. Instead of the "driving pressure gradient" (→ **C**), P_{pl} is measured by a pressure probe in the esophagus. In this case, in diagram **C**, "0 kPa" has to be replaced by P_{pl} at the resting respiratory level (– 0.3 kPa; → p. 81, B), and the highest pressure after inspiration by the most negative P_{pl} (→ p. 81, B).

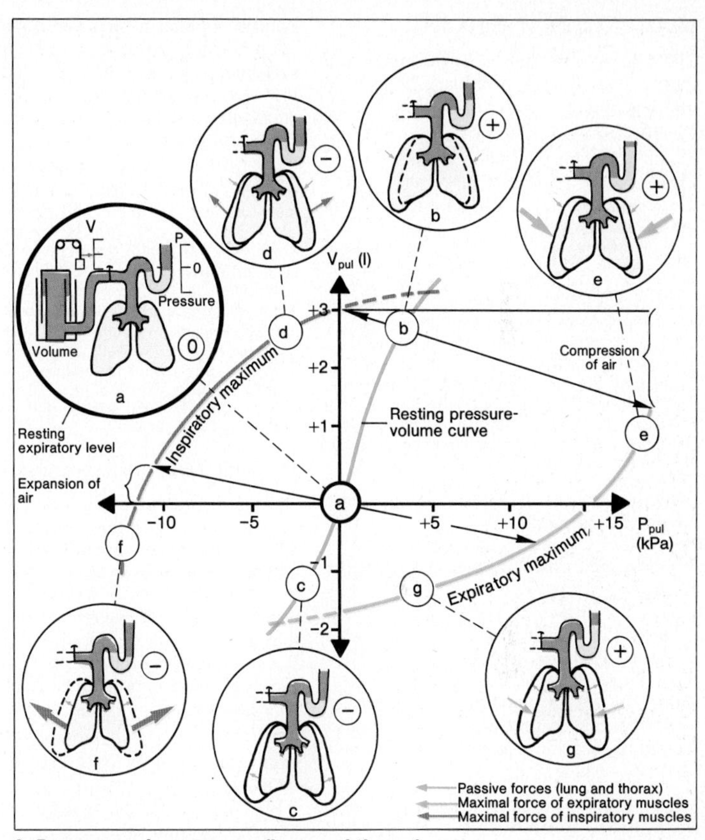

A. Pressure-volume curves (lung and thorax)

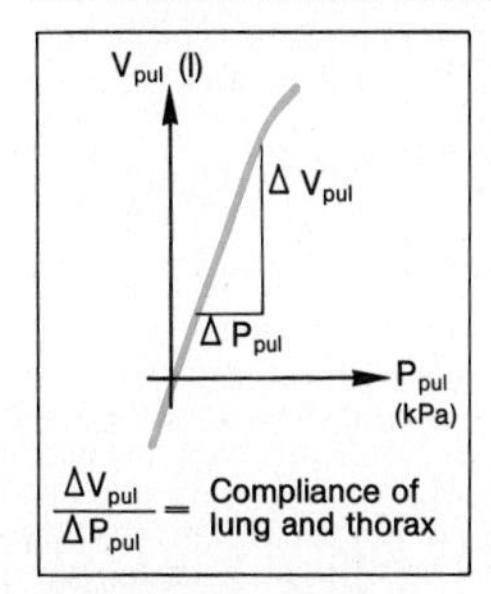

B. Static compliance

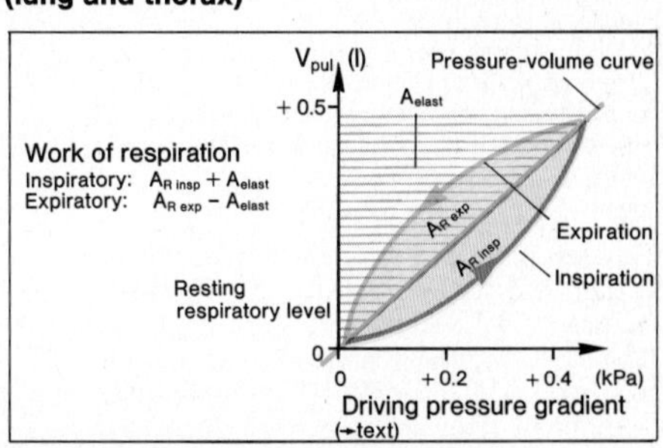

C. Dynamic pressure-volume curves

Surface Tension of the Alveoli

The passive compliance of the lungs and thorax (as a unit; → p. 88) largely depends upon *surface tension*. It arises at the boundary between a gas and a liquid, in this case on the nearly 100 m^2 of the *gaseous exchange surface of the alveoli.*

The influence of surface tension can be demonstrated by measuring the intrapulmonary pressure while attempting to fill a fully collapsed lung. When air is used to fill the lung there is, especially at the outset, a high resistance to opening, just as in attempting to inflate a new balloon. In a newborn baby, during the first breath of life, this resistance is overcome only by an intrapleural pressure (→ p. 81, B) of as much as -6 kPa ($= -60$ cm H_2O), whereas in later life only about -0.6 kPa are necessary for expansion of the lungs. In contrast, when the lung is filled with a saline solution, the resistance is only one fourth as large because saline reduces the surface tension to nearly zero.

The surface tension is defined by the *Laplace relationship* (see also p. 156): the tension in the wall of a spheric chamber is proportional to the transmural pressure times the radius. When a gas bubble is surrounded by a fluid, the surface tension (γ) (dimensions: N/m) of the fluid increases the transmural pressure (Δp) across the wall of the bubble. $\Delta p = 2\gamma/\text{radius}$. The γ is normally a constant for a given fluid (for plasma $= 10^{-3}$ N/m), so that Δp is influenced by the radius, becoming larger as the air space becomes smaller.

These relationships may be exemplified by a soap film across the opening of a cylinder (→ **A 1**), where r is large and pressure is small. (In a soap bubble, two gas/fluid interfaces are involved: $\Delta p = 4\gamma/r$.) To blow a bubble, a large "opening pressure" (Δp) is required because the radius decreases (→ **A 2**). As the bubble enlarges further (r increases again; → **A 3**), less pressure is required to enlarge it until, finally, the walls of the bubble are no longer able to contain the pressure and the bubble breaks.

The alveolus responds similarly, but rupture at high pressures is resisted by the tissue elasticity. Two other features are apparent: (1) below a critical pressure, the bubble collapses (→ **A 2**) and (2) when two bubbles of different radii are connected, there is a pressure difference between the smaller and the larger bubble (→ **A 4**); gas will flow to the larger bubble at the expense of the smaller bubble. In the alveoli, these effects are resisted by a *surface-active substance* or **surfactant**, a phospholipid film in the alveoli that lowers surface tension; it is more effective in smaller alveoli, whereby (1) and (2) are largely prevented.

Surfactant is a complex of protein and phospholipids, of which the chief component is dipalmitoyl lecithin. One of the two types of cells lining the alveoli, the type II cells, secrete surfactant by exocytosis. Surfactant deficiency lowers compliance, leads to a collapse of alveoli (*atelektasis*), and contributes to the development of pulmonary edema in selected conditions and to the development of pulmonary abnormalities after bronchial occlusion or after breathing air with a high P_{O_2} (→ p. 96). In some newborn infants, surfactant is insufficiently effective (*hyaline membrane disease*); many alveoli are collapsed and the opening pressure cannot be achieved.

Dynamic Tests of Respiratory Function

The total ventilation $\dot{V}_T$ is the product of tidal volume V_T [l] and respiratory frequency f (min^{-1}). By raising both V_T and f, $\dot{V}_T$ can be increased from about 7.5 to as much as 120–170 l · min^{-1}: **maximal breathing capacity, MBC**. Measurement of this value may provide, for example, a clinical tool (→ **B**) for monitoring the course of diseases of the respiratory musculature (e.g. myasthenia gravis).

The Tiffeneau test determines the **forced expiratory volume in the first second** (FEV_1), which is usually expressed as part of the FVC, that is, as the *relative first second capacity* (normal value > 0.7; → **C**). (FVC is the volume, after maximum inspiration, that can be expired as rapidly and forcefully as possible. FVC is often smaller than VC; → p. 84). Like VC, the relative first second capacity is also corrected for age and sex by means of empirical formulas. The **maximum expiratory flow rate** achievable is roughly 10 l/s.

The clinical use of such tests makes it possible, for example, to distinguish **restrictive respiratory disease** (reduction of the functional pulmonary volume, e.g. due to pulmonary edema or inflammation, or to impeded inflation of the lungs, e.g. as a result of scoliosis) from **obstructive respiratory disease** (narrowing of the respiratory channels as, for example, in asthma, bronchitis, emphysema, paralysis of the vocal cords) (→ **C**).

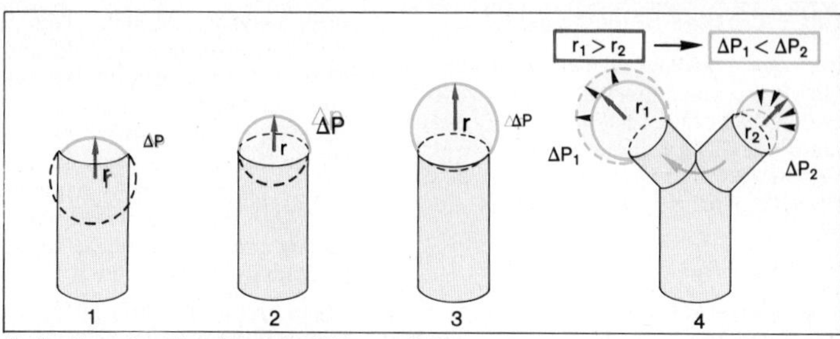

A. Surface tension (soap bubble model)

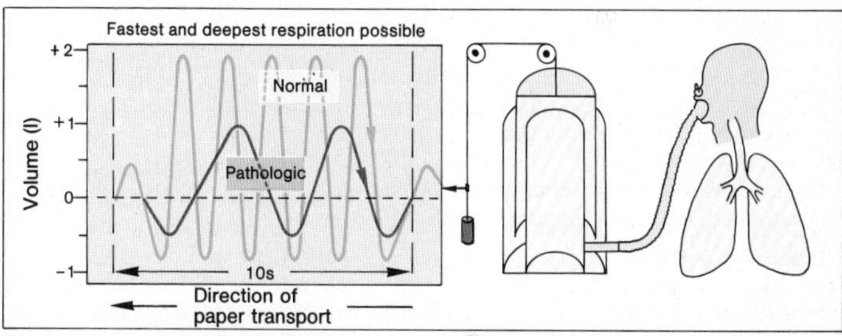

B. Maximal voluntary ventilation

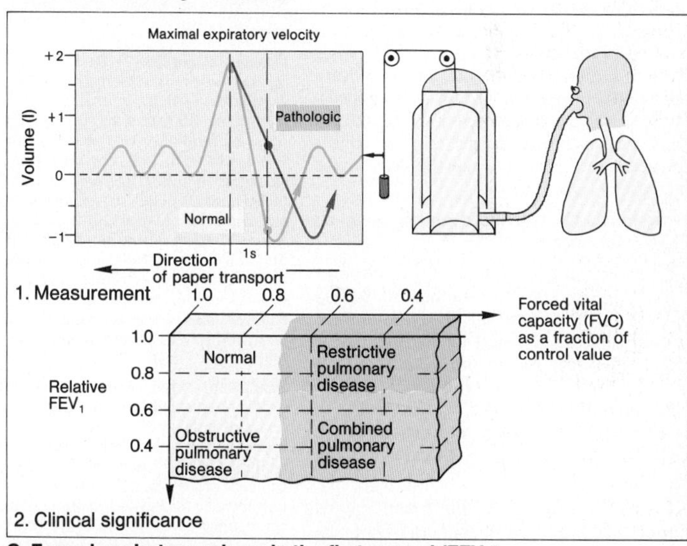

C. Forced expiratory volume in the first second (FEV_1)

Gas Exchange in the Lung

For gas exchange between alveoli and blood, the lungs must be ventilated. However, not all of the tidal volume (V_T) reaches the alveoli. The volume reaching the alveoli (V_A) is less than V_T because of the dead space (V_D) in the upper airchannel (→ p. 86). Thus,

$V_T = V_A + V_D$.

The **total ventilation**, $\dot{V}_T$ (l/min), $= V_T \cdot f$ (f = frequency). Similarly, the **alveolar ventilation**, $\dot{V}_A$, and the **dead space ventilation**, $\dot{V}_D$, can be derived from V_A and V_D. At rest, $\dot{V}_A$ normally amounts to about 70% of $\dot{V}_T$ or 5.25 l/min. In rapid shallow breathing, f is increased. If $\dot{V}_T$ remains unchanged, V_T is decreased and $\dot{V}_D$ will increase ($= V_D \cdot f$) because the magnitude of V_D is fixed by the anatomy of the upper airchannel. As a consequence, the functionally significant value, $\dot{V}_A$, will decrease.

Example: Normal $V_T = 0.5$ l, $f = 15$ breaths/min, $V_D = 0.15$ l, $V_A = 0.35$ l. Then $\dot{V}_T = 7.5$ l/min, $\dot{V}_D = 2.25$ l/min, $\dot{V}_A = 5.25$ l/min. In rapid shallow ventilation: $V_T = 0.375$ l, $f = 20$ breaths/min, $V_D = 0.15$ l (constant), $V_A = 0.225$ l. Then $\dot{V}_T = 7.5$ l/min (unchanged), $\dot{V}_D = 3$ l/min (elevated), $\dot{V}_A = 4.5$ l/min (reduced). Because of the lower alveolar ventilation ($\dot{V}_A$), gas exchange decreases. The consequences are the same when the dead space increases (→ p. 106).

How much O_2 is **utilized** and how much CO_2 is **formed**?

The inspired air contains an O_2 fraction of 0.21 (F_{IO_2}) and a CO_2 fraction of 0.0003 (F_{ICO_2}). In the expired air, the fractions are 0.17 (F_{EO_2}) for O_2, and 0.035 (F_{ECO_2}) for CO_2. The rate of inspiration of $O_2 = \dot{V}_T \times F_{IO_2}$ and of expiration $= \dot{V}_T \times F_{EO_2}$. The difference of the rates, $\dot{V}_T (F_{IO_2} - F_{EO_2})$, is the rate of oxygen utilization ($\dot{V}_{O_2}$). CO_2 formation ($\dot{V}_{CO_2}$) is calculated from $\dot{V}_T (F_{ECO_2} - F_{ICO_2})$.

At rest, $\dot{V}_{O_2}$ is about 0.3 l/min and $\dot{V}_{CO_2}$ is about 0.26 l/min. Both $\dot{V}_{O_2}$ and $\dot{V}_{CO_2}$ increase about tenfold on heavy exertion. Normally, when exercise is mild, $\dot{V}_{O_2}$ is 1 l/min; when moderate, 1.5 l/min; and when severe, 2 l/min. The relationship $\dot{V}_T/\dot{V}_{O_2}$ is constant in mild and moderate exercise (approximately 25); in severe exercise it increases to roughly 35. In the lungs, $\dot{V}_{CO_2}/\dot{V}_{O_2}$ is the respiratory exchange rate (R); in the cells, it is the respiratory quotient (RQ) (range 0.75 to 1.0 depending on diet; → p. 198). R = RQ only at steady state; e.g. during breath holding, R (lungs) and RQ (cells) have quite different values.

Mean partial pressures in the **alveoli** and in **"venous" blood** (A. pulmonalis) are (see also **A**):

	P_{O_2} kPa (mmHg)	P_{CO_2} kPa (mmHg)
Alveoli	13.33 (100)	5.33 (40)
"Venous" blood	5.33 (40)	6.13 (46)
Δp:	8 (60)	0.8 (6)

These *pressure differences* ΔP constitute the *driving force for diffusion of gas across the alveolar membrane* (→ p. 9). The **diffusion path** from the alveolus is approximately 1 μm to the plasma and an additional 1 μm to the red cell. The path is short enough to allow equilibration of gas pressure during the time the blood is in contact with the alveoli (about 0.75 seconds at rest: → **A**).

Gaseous exchange (→ **B, 1**) can be *hindered* in several ways: by reducing flow in the alveolar capillaries (→ **B, 2**); by thickening of the alveolar membrane (diffusion barrier; → **B, 3**); and by reducing ventilation of alveoli (→ **B, 4**). In the first two cases (→ **B, 2, 3**), the *functional dead space* is enlarged (→ p. 86); in the second and third cases blood is insufficiently oxygenated or "arterialized" (see also p. 94). At a moderate diffusion barrier, the curves in **A** are flatter but diffusional equilibrium may still be reached at rest. If, however, the contact time is shortened, as during exercise, equilibration (i.e. full oxygenation), is no longer possible. CO_2 equilibration is less sensitive to such an event because the alveolar permeability for CO_2 is 20fold higher than for O_2.

Insufficient oxygenation can also result from arteriovenous **shunts** in the lung. A physiological (extra-alveolar) AV shunt also occurs: the bronchial veins of the lung and the thebesian veins of the heart carry unsaturated blood into the left ventricle of the heart. For this reason the mean P_{O_2} falls from 13.3 kPa (= 100 mmHg) in the pulmonary capillaries to 12.7 kPa (= 92 mmHg) in the aorta (changes in P_{CO_2} are analogous but opposite).

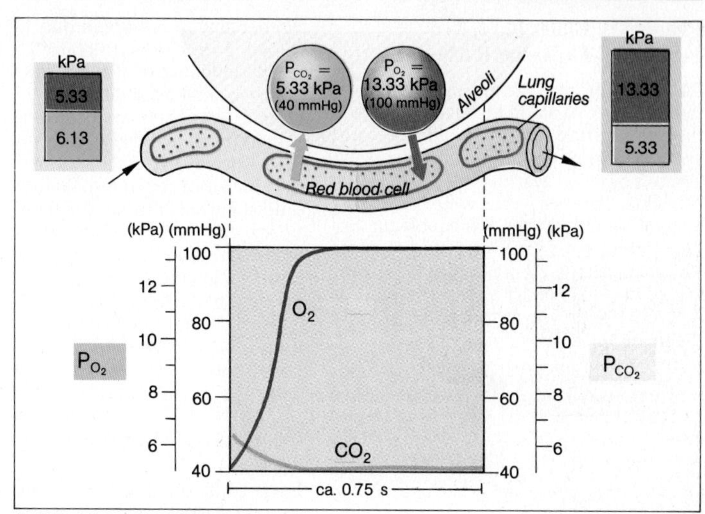

A. Alveolar gas exchange

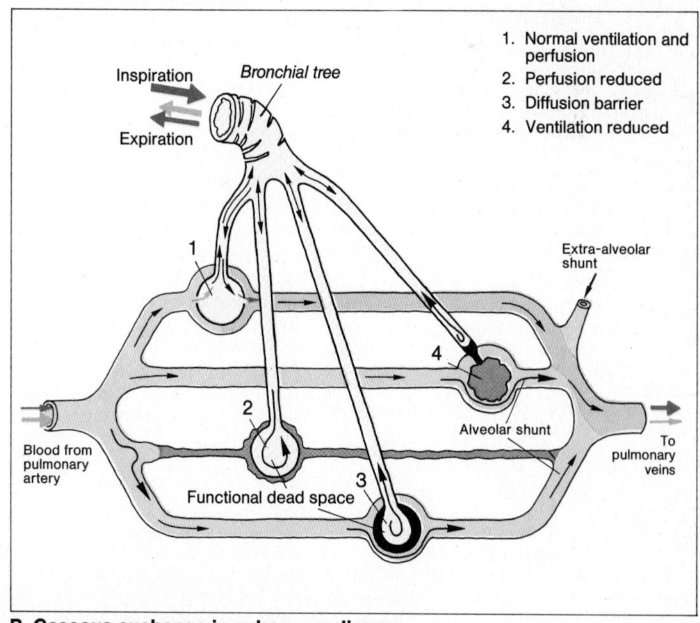

B. Gaseous exchange in pulmonary disease

Pulmonary Blood Flow, Ventilation-Perfusion Ratio

The right ventricle pumps, on average, the same amount of blood through the lungs as the left ventricle sends through the entire systemic circulation. Ignoring the small amount of blood reaching the lungs via the bronchial arteries, the *mean pulmonary blood flow* or *perfusion* ($\dot{Q}$) equals the cardiac output (= 5 l/min at rest; → p. 154). The **pulmonary blood pressure** at the beginning of the pulmonary artery amounts to 3.33 kPa (25 mmHg) at systole and 1.07 kPa (8 mmHg) at diastole.

In the pulmonary artery, the **mean pressure** ($\bar{P}$) is 2 kPa (15 mmHg), which drops to an estimated 1.6 kPa (12 mmHg) at the beginning of the pulmonary capillaries ($\bar{P}_a$) and to about 1.07 kPa (8 mmHg) at the end of the capillary bed ($\bar{P}_v$). These values apply to all regions of the lungs that are at the same level as the pulmonary valve.

In vessels below this niveau (towards the base of the lung) $\bar{P}$ and $\bar{P}_a$ are higher (with an *upright thorax*) because of the additional hydrostatic pressure of the blood column (up to nearly 2 kPa), whereas in regions of the lungs higher than the pulmonary valve (towards the tip) $\bar{P}$ is lower (→ **A**). The pressure at the arterial end of the capillaries can even drop below 0 so that the atmospheric pressure prevailing in the alveoli ($\bar{P}_A$) compresses the capillaries ($\bar{P}_A > \bar{P}_a > \bar{P}_v$: → **A**). Thus, at most, blood flows briefly through this *zone 1* during systole (→ **A**). In the central regions of the lungs (*zone 2*, → **A**) there may be an occasional narrowing of the capillaries, ($\bar{P}_a > \bar{P}_A > \bar{P}_v$), but at the base of the lungs (*zone 3*, → **A**) blood flow is continuous ($\bar{P}_a > \bar{P}_v > \bar{P}_A$). Thus the pulmonary blood flow ($\dot{Q}$) per unit volume of the lung increases from the tip to the base (→ **B**, red line). For other reasons the ventilation of the alveolar space ($\dot{V}_A$) also increases in this direction, although less steeply (→ **B**, orange line), so that the **ventilation-perfusion ratio $\dot{V}_A/\dot{Q}$** drops from the tip to the base of the lung (→ **B**, green line).

The *average* P_{O_2} in the alveoli amounts to 13.33 kPa (100 mmHg), and the P_{CO_2} averages 5.33 kPa (40 mmHg). In the O_2-deficient blood of the pulmonary artery the P_{O_2} is 5.33 kPa (40 mmHg) and the P_{CO_2} is 6.13 kPa (46 mmHg). The latter two values are equilibrated by alveolar gaseous exchange with the pressures in the alveoli (→ p. 92). These mean values for the whole lung apply for an average ventilation ($\dot{V}_A$) of approximately 5.25 l/min and a perfusion (blood flow, $\dot{Q}$) of roughly 5 l/min. The mean ventilation/perfusion ratio ($\dot{V}_A/\dot{Q}$) is in this case therefore 5.25/5 ≈ 1 (→ **C 2**). If, in an extreme case, ventilation ceases altogether (functional **shunt**), then $\dot{V}_A/\dot{Q} = 0$ (→ **C 1**), whereas in another extreme case (i.e. the absence of perfusion), atmospheric conditions prevail in the alveoli involved (functional dead space, → p. 86), then $\dot{V}_A/\dot{Q}$ approaches infinity (∞, → **C 3**). This means that in extreme cases $\dot{V}_A/\dot{Q}$ can vary in different parts of the lung from 0 to ∞, that is, P_{O_2} can vary from that in the blood of the right ventricle to that in the (humified) outside air (→ **D**). In the upright, resting lung, $\dot{V}_A/\dot{Q}$ drops sharply from tip to base (3.3–0.63, → **D**, → **B**: green line). During physical exertion these differences are less marked.

Large differences in $\dot{V}_A/\dot{Q}$ values from region to region reduce the efficiency of the lung in gaseous exchange. On account of the shape of the dissociation curve (→ p. 101) the relatively high alveolar P_{O_2} in the tip of the lung (→ **D**, right) cannot, in practice, compensate for the relatively low alveolar P_{O_2} at the base of the lung. In the event of a total shunt, even O_2 inhalation cannot help the affected part of the lung since the administered O_2 cannot reach the capillary bed at the afflicted site (situation **C 1**).

However, a mechanisms does exist to prevent the occurence of extreme $\dot{V}_A/\dot{Q}$ values: this is so-called **hypoxic vasoconstriction**, which regulates alveolar blood flow. Very low alveolar P_{O_2} values stimulate receptors in the alveoli that, by unknown pathways (tissue hormones?), lead to constriction of the afferent blood vessels. In this way the parts of the lung receiving little or no oxygen are bypassed (shunts), and the better-functioning regions receive relatively more blood for gaseous exchange.

Many pulmonary diseases are accompanied by very pronounced deviations of the $\dot{V}_A/\dot{Q}$ values from the normal. In the shocked lung, for example the shunt can amount to as much as 50% of $\dot{Q}$. If this is accompanied by edema (→ p. 158) of the lungs, or some other hindrance to diffusion, or by a defect in the surfactant (→ p. 90), a dangerous respiratory insufficiency can rapidly develop.

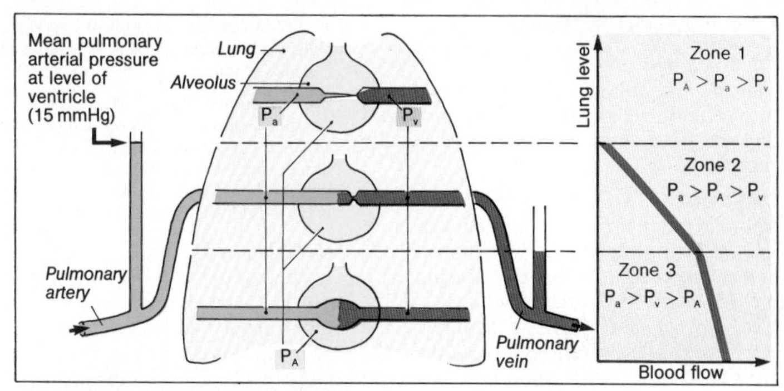

A. Regional pulmonary blood flow

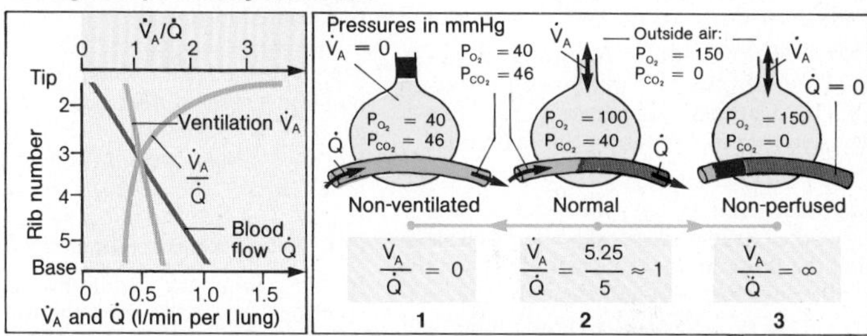

B. Perfusion and ventilation of the lung regions

C. Influence of the ventilation/perfusion ratio ($\dot{V}_A/\dot{Q}$) on the partial pressures in the lung

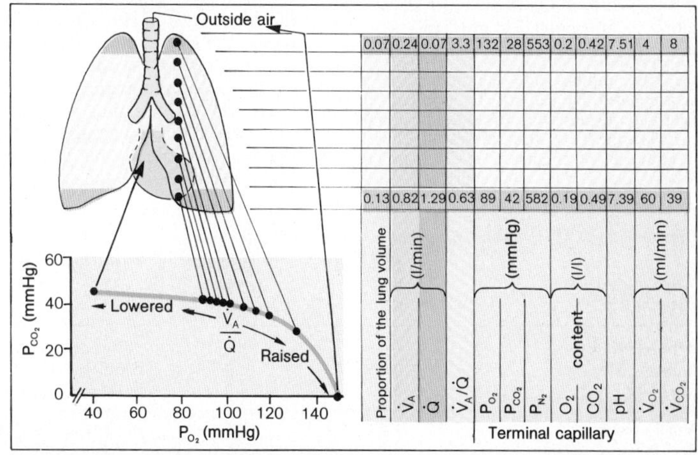

D. Regional parameters of lung function

(A, B, C, D, after West and collaborators)

CO_2 Transport in Blood

The main end products of energy metabolism are CO_2 and H_2O. The CO_2 that is formed in the cells dissolves in cell water and diffuses through the interstitial space to the venous blood.

Dissolved CO_2. CO_2 is 20 times more soluble in blood than is O_2; there is always more CO_2 in solution than O_2. The dissolved CO_2 in the plasma equilibrates, by diffusion, with the cell water in the red blood cells, RBC. The P_{CO_2} of arterial blood supplying the cells is 5.33 kPa (= 40 mmHg). At the cells, CO_2 diffuses into the blood and raises the P_{CO_2} to 6.27 kPa (= 47 mmHg) in venous blood.

HCO_3^-. The CO_2 diffusing into the RBC (→ **A**, blue arrow) is rapidly converted to HCO_3^-. **Carbonic anhydrase (CA)** plays a decisive role in catalyzing the reaction:

$CO_2 + H_2O \rightleftarrows HCO_3^- + H^+$ (→ p. 145, A). In the absence of CA, the reaction rate is slow; the enzyme accelerates the reaction so that the short contact time with RBCs in the capillaries (< 1 s) is sufficient for the conversion to H^+ and HCO_3^- of about 80% of the CO_2 taken up. The H^+ is buffered within the RBC, chiefly by Hb; the HCO_3^- formed diffuses out of the RBC into the plasma along its concentration gradient.

Chloride shift. About 70% of the HCO_3^- formed in the RBC diffuses into the plasma. It is not accompanied by cations. To maintain electric neutrality, an anion (Cl^-) must diffuse into the cell to replace the HCO_3^- that diffuses out. Thus, RBCs in venous blood contain more Cl^- than RBCs in arterial blood. The intracellular Cl^- increases osmolality and motivates water movement into the cells, which increases their volume to some extent (→ also p. 60).

Carbamino bondings (→ **A, B**). Some of the CO_2 combines with free NH_2 groups of Hb in the RBC to form carbamino bondings (= carbamate). A much smaller amount (about 1/10) forms carbamino bonds with plasma proteins. The reaction is $Hb-NH_2 + CO_2 \rightleftarrows Hb-NH-COO^- + H^+$.

Buffering of H^+ ions. Both the carbamino reactions and the HCO_3^- formation generate H^+, a portion of which lowers the pH of blood from 7.40 (arterial) to 7.36 (venous). A much larger portion is buffered, principally by Hb. The binding capacity of desoxygenated Hb for H^+ is greater than that of oxygenated Hb (→ **B**). When O_2 is given up at the tissues, oxygenated Hb is converted to desoxygenated Hb; thus, Hb is a better buffer just at the site where H^+ production is greatest. By improving its buffering of H^+, Hb lowers the concentration of H^+ in the RBC, drives the reaction $CO_2 + H_2O \rightleftarrows HCO_3^- + H^+$ to the right, and allows more CO_2 to be taken up (*Haldane effect*).

In the pulmonary capillaries, all of these reactions are reversed by the lowered P_{CO_2} (→ **A** and **B**, red arrows). HCO_3^- diffuses back into the RBC, accepts H^+, and forms CO_2 and H_2O. This reaction is supported by the liberation of H^+ ions, which accompanies the oxygenation of Hb in the lung (*Haldane effect*; → **B**); additionally, the carbamino bondings are reversed due to the lower pH. CO_2 is thus generated by two processes. It diffuses along its concentration gradient into the alveoli where the P_{CO_2} is lower and exits with the expired air.

Distribution of CO_2 in blood (approximate values in mmol/l *blood* [§]):

	Dissolved	**HCO_3^-**	**Carbamino**	**Total**
Arterial plasma*)	0.7	13.2	0.1	14.0
Arterial RBC**)	0.5	6.5	1.1	8.1
Arterial blood	*1.2*	*19.7*	*1.2*	*22.1*
Venous plasma*)	0.8	14.3	c. 0.1	15.2
Venous RBC**)	0.6	7.2	1.4	9.2
Venous blood	*1.4*	*21.5*	*1.5*	*24.4*
A-V difference blood	*0.2*	*1.8*	*0.3*	*2.3*
(% of total A-V diff.)	(9%)	(78%)	(13%)	

*) 0.55 l/l blood; **) 0.45 l/l blood; [§]) 1 mmol = 22.26 ml CO_2

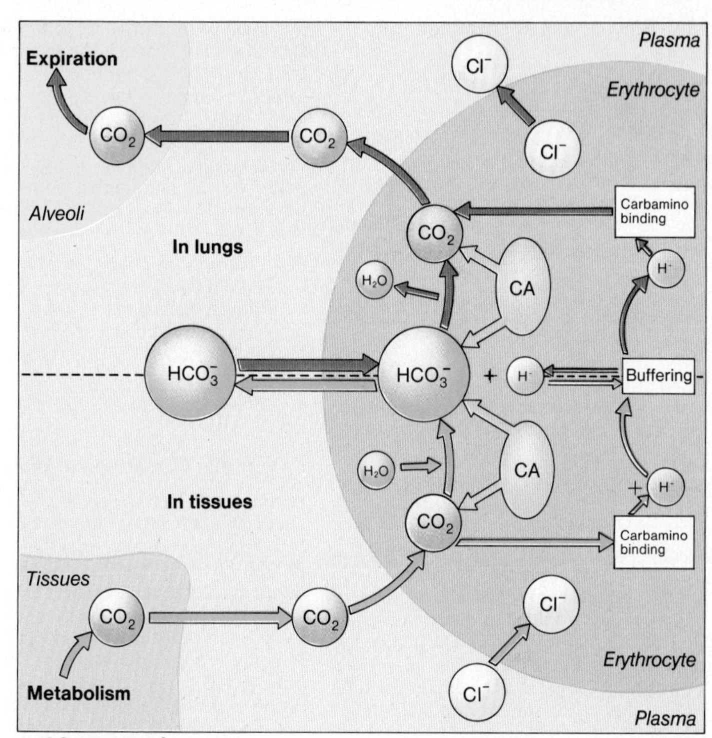

A. CO_2 transport

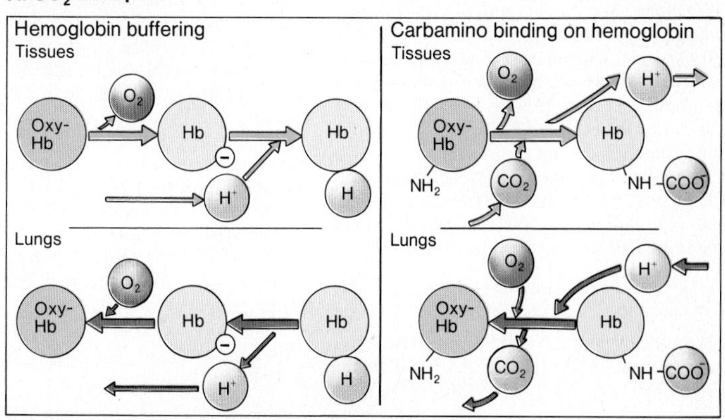

B. Buffering and carbamino binding in erythrocytes

CO_2 Binding and Distribution in Blood

The **blood content of CO_2** in mixed venous blood both as dissolved CO_2 and as chemically bound "CO_2" is 24 to 25 mmol/l, in arterial blood 22–23 mmol/l. The actual content at any time is more or less dependent on the partial pressure of CO_2 (P_{CO_2}). The distribution of CO_2 in arterial blood and in venous blood is shown in the table on p. 96.

The concentration of **dissolved CO_2**, $[CO_2]$, in blood is linearly dependent on the P_{CO_2} and is given by

$$[CO_2] = \alpha \cdot P_{CO_2} \text{ (mmol/l)}$$

where α represents the solubility coefficient; for CO_2 in plasma at 37 °C:

$$\alpha_{CO_2} = 0.225 \text{ mmol} \cdot l^{-1} \cdot kPa^{-1}$$
$$= 0.03 \text{ mmol} \cdot l^{-1} \cdot mmHg^{-1}.$$

The curve for dissolved CO_2 is therefore a straight line (→ **A**, green curve). In contrast, the curve for **chemically combined CO_2** relative to P_{CO_2} is curvilinear. The shape of this curve is determined by the limited buffer sites of Hb and by the limited number of carbamino bondings that can be formed with the available Hb. The curves for **total CO_2** represent the sum of dissolved and of bound CO_2 (→ **A**, red and blue curves).

Two curves are shown, one for CO_2 binding when Hb is fully saturated with O_2 (100% HbO_2) and one when Hb is fully desaturated (0% HbO_2). The capacity of the blood to "bind" CO_2, that is, to bind H^+ during formation of HCO_3^- and carbamino bondings, depends on the state of Hb (→ p. 96). For a given P_{CO_2}, blood fully saturated with O_2 binds *less* CO_2 than O_2-free blood (→ **A**, red and violet curves). The physiological meaning of this difference is twofold:

(1) venous blood (low in O_2, high in CO_2) can carry more CO_2 away from the tissues;

(2) in the pulmonary capillaries, when Hb is saturated with O_2, release of CO_2 is facilitated by formation of HbO_2 (*Haldane effect*).

Mixed venous blood is about 70% saturated with O_2 (70% HbO_2). The dissociation curve corresponding to this extent of saturation lies between the two curves for 0% and 100% saturation. The P_{CO_2} for mixed venous blood is 6.27 kPa (= 47 mmHg) (→ **A**, point v). In arterial blood, Hb saturation is close to 97% and the P_{CO_2} is 5.33 kPa (= 40 mmHg) (→ **A**, point a). The line joining points a and v represents the "physiological dissociation curve" for CO_2.

The relationship between the concentrations of HCO_3^- and dissolved CO_2 varies according to the pH of the blood. At pH 7.4, in plasma, the molal concentration ratio HCO_3^-/CO_2 is 20 : 1. These values compare with the normal pH in the red blood cell of 7.2, which corresponds to a ratio of 12 : 1 (→ pp. 110–119).

CO_2 in the Cerebrospinal Fluid

In contrast to HCO_3^- and H^+, CO_2 diffuses relatively easily across the barrier between blood and the cerebrospinal fluid (CSF; → p. 272), enabling the P_{CO_2} of the CSF to adjust quickly to **acute changes in the P_{CO_2}** of the blood. Since changes in the pH, if caused by CO_2 (so-called respiratory pH changes), can only be buffered by the non-bicarbonate buffers, NBBs (→ p. 116), whose concentration in the CSF is low, acute fluctuations in the P_{CO_2} of the CSF lead to relatively large changes in its pH. These are registered by **central chemoreceptors** and result in the appropriate changes in respiratory activity (→ p. 104). The blood, on the other hand, is well-supplied with NBBs so that a drop in its pH due to CO_2 can be effectively buffered. Consequently, the actual HCO_3^- concentration in the blood (→ p. 118) rises to higher values than in the CSF, and HCO_3^- diffuses (relatively slowly) into the CSF. This brings about a rise in the pH of the latter and (via chemoreceptors) a reduction in the stimulus to respiration; the process is enhanced by renal compensation (rise in pH, → p. 116). In this way, a kind of "adaptation" to **chronic increase of P_{CO_2}** builds up (→ also p. 104).

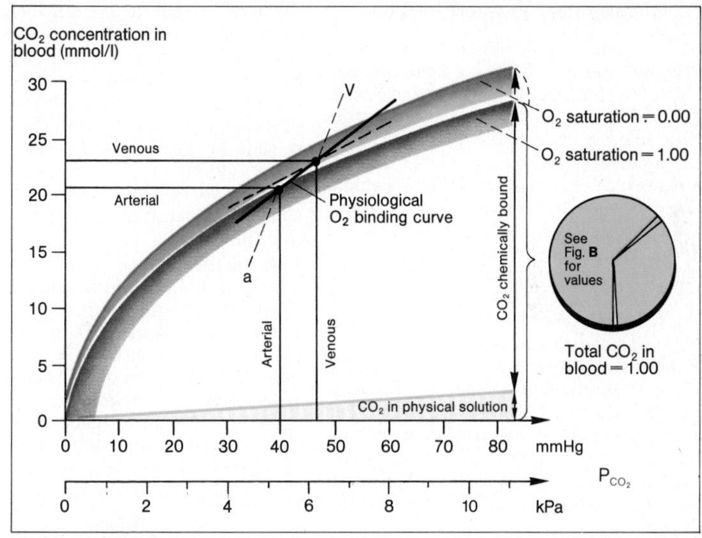

A. CO_2 binding curve

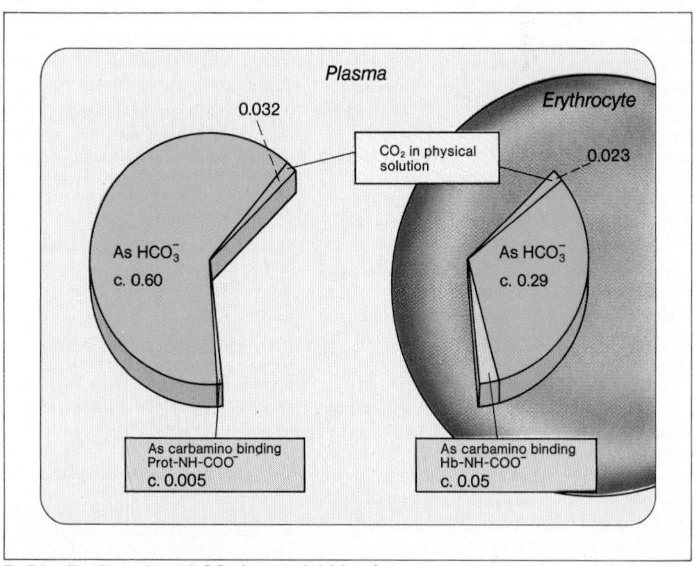

B. Distribution of total CO_2 in arterial blood

O_2 Binding and Transport in Blood

Hb is a protein of 64800 Dalton with four subunits, each containing a **heme** moiety. Heme is a complex of porphyrin and **Fe(II)**. Each of the four Fe(II) combines reversibly with one O_2 molecules. This process is **oxygenation**, not oxidation. The **O_2 dissociation curve** (→ **A**, red curve) has a sigmoid shape because of varying affinities of the heme groups for O_2. When O_2 binds to the first group, affinity of the second is enhanced; binding to the second enhances the third, and so on. When fully saturated with O_2, 1 mol Hb binds 4 mol O_2 (4 × 22.4 l) or 1 g Hb can carry 1.38 ml O_2. Virtually full saturation occurs at a P_{O_2} of about 20 kPa; O_2-carrying capacity cannot be increased significantly by elevating P_{O_2} further. The average normal concentration of Hb in blood is 150 g/l; at this Hb content, the **O_2 capacity** of blood, or the maximal potential O_2 concentration, = 0.207 l/l or 9.24 mmol O_2/l blood.

O_2 content is the concentration of O_2 *actually* in the blood. The ratio *O_2 content/O_2 capacity* represents **O_2 saturation** of Hb. Because of alveolar and extra-alveolar atrioventricular shunts (→ p. 92), **P_{O_2} of arterial blood** is smaller than alveolar P_{O_2} and amounts to 12.7 kPa (95 mmHg) in young adults, resulting in an arterial O_2 saturation of about 0.97, whereas saturation in mixed venous blood is about 0.73 (P_{O_2} = 5.33 kPa, = 40 mmHg). The arterial value drops with age; the venous value differs from organ to organ because the **O_2 extraction** depends on the type of organ and the demands made upon it.

The **O_2 dissociation curve** is sigmoid (see above) and rises steeply at low values of P_{O_2} but becomes almost flat at high values of P_{O_2}, when it approximates to the O_2 capacity. The curve can be displaced by various influences. Since the amount of available Hb affects O_2 content, the curve may be shifted upward in Hb excess and downward in Hb deficiency. This shift can be seen if *concentration* of O_2 or of Hb-O_2 (→ **A, B**, yellow and purple curve), but not if O_2 *saturation* (→ **C**) is plotted against P_{O_2}. In contrast, shifts of the curve to the left or right occur without a change in the O_2 capacity. A **shift to the left** increases the slope of the rising portion of the curve because the affinity of Hb for O_2 has been increased (→ **A**, blue curve). In effect, (a) for a given P_{O_2}, more Hb-O_2 is formed or (b) for a given O_2 content, the P_{O_2} is less. A **shift to the right** represents decreased affinity of Hb for O_2. Three principal factors influence the affinity: pH (Bohr effect), P_{CO_2} as such, temperature, 2,3-bisphosphoglycerate (BPG).

1. At the tissues, P_{CO_2} is increased and pH is decreased, both of which reduce affinity of Hb for O_2, allowing more O_2 to dissociate. In the lung capillaries CO_2 is given off, the pH rises, and Hb can carry more O_2 (affinity increases). The effect of pH is known as the **Bohr effect**. A "physiological dissociation curve" (→ **B**) can be made for O_2 (as for CO_2, → p. 99) in order to take into account changes of pH and P_{CO_2} in mixed venous blood compared to arterial blood.

2. A reduction in temperature shifts the O_2 dissociation curve to the left.

3. 2,3-BPG is generated by glycolysis; it binds to Hb but not to Hb-O_2. High concentrations of BPG in the red blood cell reduce affinity of Hb for O_2 and shift the dissociation curve to the right.

A simple measure of the displacement to left or right is the **O_2-half saturation pressure (P_{50})**, the pressure at which the Hb is 0.5 loaded (50%) with O_2. The P_{50} value (at pH 7.4 and 37 °C) usually amounts to 3.46 kPa = 26 mmHg (can be read off the blue curve in **C**).

Other O_2-carrying proteins, **myoglobin** (monomeric → nonsigmoid curve) in muscles and **fetal Hb**, differ in detail from Hb (→ **C**). They have a steeper rising portion of the curve (shift to the left). Other gases can bind with Hb. **Carbon monoxide (CO)** has a much steeper dissociation curve and therefore, even in low concentrations, displaces large amounts of O_2 from Hb, causing the high *toxicity of CO*. **Methemoglobin** is formed when the ferrous iron of the normal Hb is oxidized to ferric. Met-Hb has no affinity for O_2 (→ **C**).

The concentration of **dissolved O_2** is very small and is linearly dependent on the P_{O_2} (→ **A**, orange line). The concentration is given by $\alpha \times P_{O_2}$, where α represents the solubility coefficient. For O_2 in plasma at 37 °C, $\alpha = 0.01\ \text{mmol} \cdot \text{l}^{-1} \cdot \text{kPa}^{-1}$. In arterial blood ($P_{O_2}$ = 12.7 kPa), dissolved O_2 amounts to about 0.13 mmol/l. This value is only 1/70 of the O_2 combined with Hb.

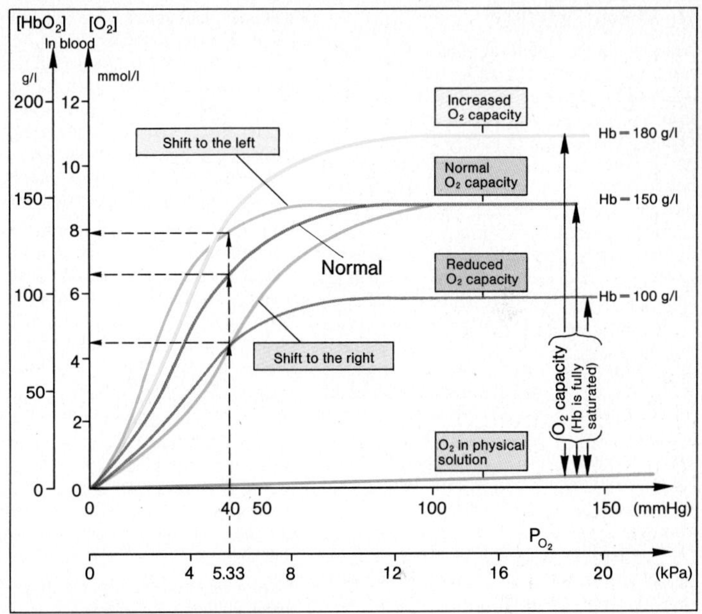

A. O_2 binding curve of the blood

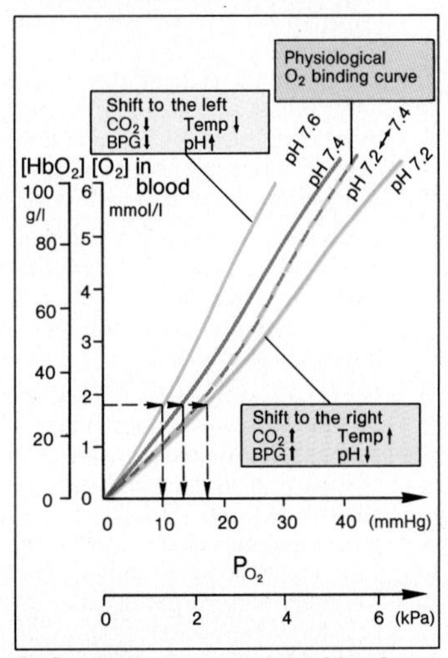

B. O_2 binding curve of the blood (detail)

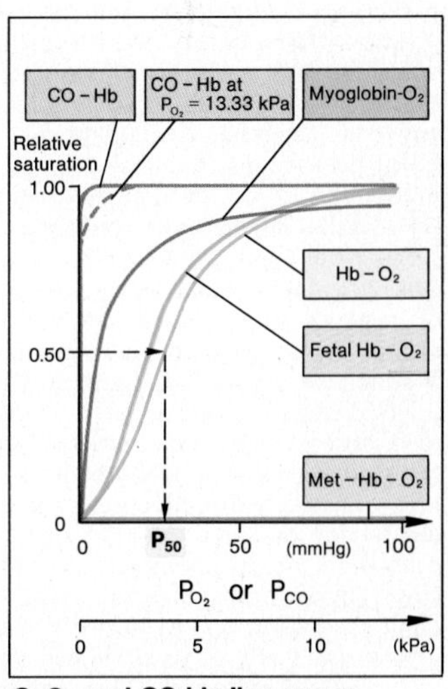

C. O_2 and CO binding curves

O_2 Deficit – Hypoxia

Anoxia, the complete absence of O_2, does not actually occur clinically, but the term is often used. **Hypoxia** describes the situation in which the cell is provided with too little O_2, and its partial pressure at the site of the end-consumers, the **mitochondria**, drops below a **critical P_{O_2}** of 0.1–1 mmHg. Hypoxia can develop because only a limited cylindric space surrounding a capillary can be adequately supplied with O_2 by diffusion processes. The radius of such a cylinder depends upon the P_{O_2}, the O_2 diffusion coefficient, the O_2 utilization of the tissue, and the rate of capillary flow. For a maximally functional muscle, 20 µm is a realistic radius for the cylinder (diameter of capillary lumen is 6 µm); tissue cells at a distance greater than 20 µm from the capillary will be undersupplied with O_2 (hypoxia; → **A, 4**). Four types of hypoxia are recognized clinically:

Hypoxic hypoxia (→ **A, 1**) occurs when the O_2 availability to the RBC from the atmosphere is reduced (low P_{O_2}). **Causes:** (a) low atmospheric P_{O_2}, as at high altitudes (→ p. 108); (b) hypoventilation, as in paralysis of respiratory muscles, in depression of respiratory control in the medulla, in high external pressure on the thorax (→ p. 106), in airway obstruction, and in atelectasis (collapsed lung); (c) alveolar-capillary diffusion block, as in pulmonary edema, pneumonia, or fibrosis; (d) ventilation-perfusion imbalance (→ p. 94), as in emphysema.

Anemic hypoxia (→ **A, 2**) occurs because the O_2 capacity of the blood is lowered. The arterial P_{O_2} is normal, but the amount of Hb available to carry O_2 is reduced. At rest, hypoxia due to anemia is rarely severe, but in exercise it can become restrictive. **Causes:** (a) reduced erythrocyte count from blood loss, or decreased production or increased destruction of RBC; (b) reduced Hb concentration in spite of normal erythrocyte count (hypochromic anemia), as can result from Fe deficiency (→ p. 62); (c) synthesis of abnormal Hb, as in sickle cell anemia; (d) reduced binding of O_2, as in carbon monoxide poisoning (→ p. 100) or chemical alteration of Hb (methemoglobinemia).

Ischemic, circulatory, or stagnant hypoxia (→ **A, 3**) occurs during shock, heart failure, or intravascular obstruction. The P_{O_2} at the lungs and the Hb concentration are normal, but delivery of O_2 to the tissues is inadequate, O_2 extraction is increased, and the local P_{O_2} falls to low levels. The arteriovenous difference for O_2 is increased as a consequence of the increased extraction.

Unlike hypoxemic and anemic hypoxias, in ischemic hypoxia the *removal of metabolic products is also impeded.* In this case even anaerobic glycolysis (→ p. 46) is of little help, since the lactic acid remains in the ischemic area and cell metabolism rapidly comes to a stop due to local acidosis.

Histotoxic hypoxia (→ **A, 5**) occurs when the tissues cannot utilize O_2 for oxidative processes. Delivery of O_2 is adequate, but the cells cannot make use of the O_2 supplied to them. This circumstance occurs in cyanide poisoning, for example, in which cytochrome oxidase (among others) is inactivated.

Effects of hypoxia depend upon the tissue affected because tissue sensitivities to hypoxia differ. In general, the **brain** is most sensitive (→ **B**). Anoxia can cause unconsciousness in 15 seconds, irreversible cell damage in about 2 minutes, and cell death in 4 to 5 minutes. Lesser degrees of hypoxia, as in heart failure or chronic pulmonary disease, may be manifested clinically by confusion, disorientation, and bizarre behavior.

Cyanosis occurs when desoxygenated Hb in the capillaries exceeds 50 g/l. Since it has a dark color, nail beds, lips, ear lobes, and areas where the skin is thin take on a dusky purplish coloration.

Because development of cyanosis depends on the *absolute* concentration of desoxygenated Hb, hypoxia may be severe without cyanosis (as in anemia), and cyanosis may be seen without significant hypoxia (as in polycythemia). In the latter case, polycythemia might be an adaptive response to, for example, a previous hypoxic hypoxia (→ p. 108).

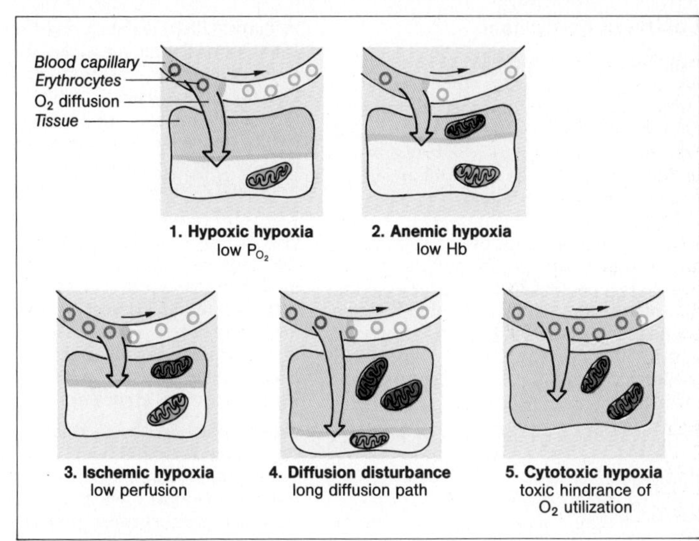

A. Types of hypoxia

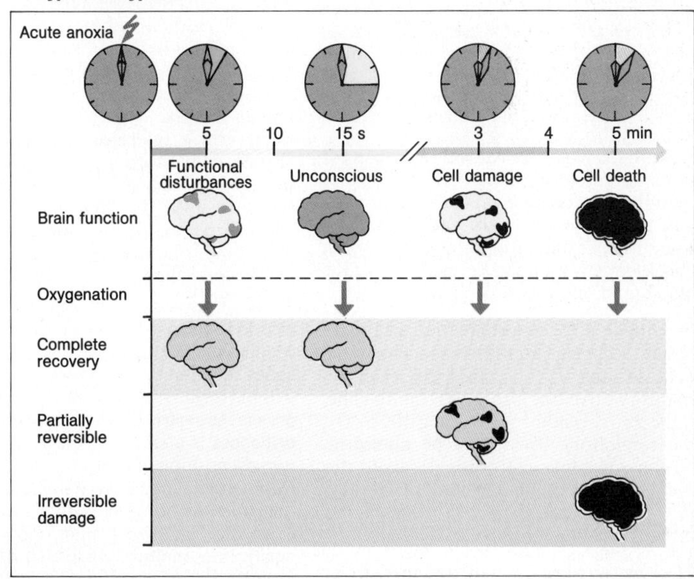

B. Effects of anoxia on the CNS

Control of Respiration

Respiration is controlled in the central nervous system (CNS); voluntary respiration is governed by the cortex, and automatic respiration by structures in the medullopontine region (→ **A**). The respiratory muscles (→ p. 80) are supplied by nerves from the cervical medulla (C IV–VIII) and from the thoracic medulla (Th I–VII). Regulation of respiration adjusts ventilation to maintain appropriate blood levels of P_{O_2}, P_{CO_2}, and pH whereby the P_{CO_2} and the pH of the blood are closely interconnected (→ p. 110ff.). There are several sensors for afferent input to the CNS, chemoreceptors, mechanoreceptors, and others.

Peripheral chemoreceptors are found in the carotid and aortic bodies. In humans, the primary O_2 sensory organ is the **carotid body**. Impulses from these sensors increase as P_{O_2} falls below about 13.3 kPa (= 100 mmHg). Output of impulses cannot be sustained below 4 kPa (= 30 mmHg). The increased ventilatory response to a fall in P_{O_2} is potentiated by a rise in P_{CO_2} or in concentration of H^+. Responses to P_{CO_2} are linear above 5.3 kPa (= 40 mmHg) and to H^+ from pH 7.7 to 7.2.

A rise in CO_2 and consequently a drop in pH in the cerebrospinal fluid (**CSF**) stimulates **central chemoreceptors** on the anterior medulla oblongata (→ p. 98 and 272). This stimulus reinforces respiratory activity with the aim of lowering the elevated P_{CO_2} of the blood (and hence also of the CSF).

In **chronic CO_2 retention**, the medullary center becomes insensitive to changes in P_{CO_2} so that P_{O_2} becomes the chief respiratory drive (see also p. 98). In this state, if the P_{O_2} is raised by breathing 100% O_2, the respiratory drive may be abolished, resulting in coma and death. To avoid this event, patients with chronically elevated P_{CO_2} should receive only O_2-enriched air rather than 100% O_2 (→ p. 108).

Mechanoreceptors occur in the upper airway and in the lungs. They are of several types and have various functions. In the lungs, the principal receptors are the **pulmonary stretch receptors** (**PSR**) of the Hering-Breuer reflex. Inflation of the lung stretches the PSR and initiates impulses that are carried to the CNS by large myelinated fibers in the vagus (X). They increase respiratory time and reduce frequency. They are also involved in reflexes that bring about bronchoconstriction, tachycardia, and vasoconstriction.

Control of automatic breathing by the CNS is governed by so-called respiratory centers in the pons and the medulla. These centers modulate the depth of inspiration and the cutoff point that terminates inspiration. The medullary center is important for establishing respiratory rhythm and for the Hering-Breuer reflex, which inhibits inspiration when the lung is stretched.

Other inputs to the medullary center include: **proprioceptors** (→ p. 278), which coordinate muscular activity with respiration; **body temperature**, which for instance increases respiratory rate during fever; **pressoreceptors** or **baroreceptors** (→ p. 179), which send afferents to the medullary center as well as to the cardioinhibitory area in the medulla; in the reverse direction, respiratory activity affects blood pressure and pulse rate; the effects are small; **higher CNS centers** (cortex, hypothalamus, limbic system), which influence respiration during anxiety, pain, sneezing, yawning, etc. **Voluntary breath holding** inhibits automatic respiration until the **breakpoint** is reached when the rise in P_{CO_2} overrides the voluntary inhibition. The breakpoint can be delayed by previous hyperventilation (→ p. 106).

Terms of respiratory activity are: **hyperpnea** and **hypopnea**, which describe mainly the deepness, while **tachypnea**, **bradypnea**, and **apnea** describe the frequency of respiration without regard to efficiency or needs; **dyspnea** is the awareness of shortness of breath; **orthopnea** is a severe dyspnea requiring an upright position of the thorax for breathing; **hypoventilation** or **hyperventilation** describe the situations in which alveolar ventilation is smaller or higher than the metabolic needs, thus leading to elevated or lowered alveolar P_{CO_2}, respectively.

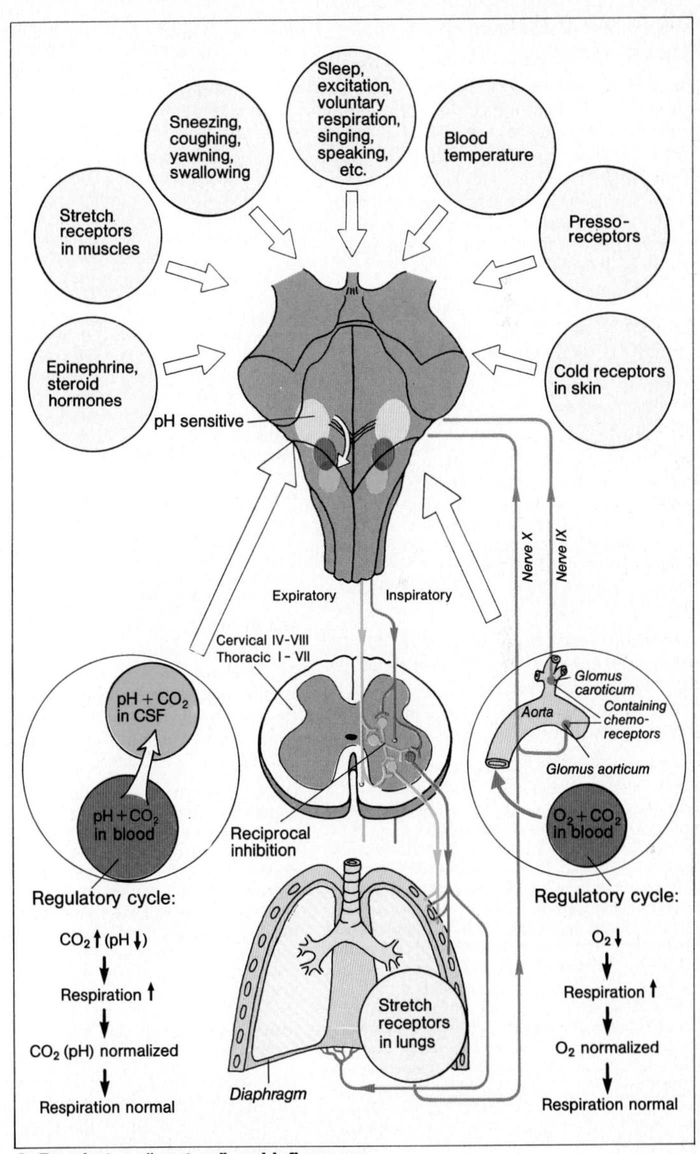

A. Respiratory "centers" and influences

Respiration at Elevated Barometric Pressure: Diving

Even if air is supplied, diving causes respiratory problems since the barometric pressure increases by 98 kPa (= 1 at = 735 mmHg) for each 10 m depth in water: Swimming below the water surface is possible when the upper airchannel is lengthened by a **snorkel**. However, there are several limits to the depth at which snorkel diving is practical:

1. The snorkel increases functional dead space (→ **B**). When the dead space (→ pp. 86, 92) approximates the ventilatory volume, fresh air no longer reaches the alveoli. Reducing the diameter of the snorkel to reduce its dead space volume is not practical because the airchannel resistance increases.
2. Water pressure on the chest increases inspiratory effort. The maximum inspiratory pressure is approximately 11 kPa (112 cm H_2O) (→ p. 89), so at a depth of 112 cm inspiration is impossible and the chest is fixed in expiration (→ **B**).
3. Additionally, on account of the high external pressure, blood from the periphery is pressed into the lungs (venous return, → p. 184), causing pulmonary edema.

To permit ventilation at greater depths, respiration must be assisted by an **apparatus** that adjusts the pressure at which gas is delivered from a gas bottle to the pressure that exists at the diving depth.

The elevated barometric pressure that exists at depth increases the amount of dissolved N_2 and other gases in the plasma by elevating their partial pressure (→ **A**). At 60 m, seven times more N_2 is dissolved in plasma than at sea level. If return to sea level is gradual, the P_{N_2} is reduced by diffusion and expiration of N_2; no ill effects occur. If return is not slow enough, the relatively rapid decompression allows N_2 bubbles to form in plasma and in tissues (similar to opening a bottle of carbonated drink). In the tissues, bubbles cause severe pain; in the plasma, they may occlude small vessels and may cause damage to the heart or CNS. Treatment for decompression sickness (the bends) is immediate recompression and slower decompression. Although N_2 is physiologically nonreactive at atmospheric pressure, higher levels of P_{N_2} are toxic and depress the nervous system in the same manner as common anesthetic gases. At a pressure of 400–500 kPa (30 to 40 m depth), N_2 may produce euphoria; at greater depths and at longer bottom times it produces narcosis ("rapture of the deep"), which resembles alcoholic intoxication. These effects can be prevented by substituting helium for N_2, since helium has a much weaker anesthetic effect. The normal fractional O_2 concentration in air (0.21 l/l or 21 Vol%) becomes fatal at a depth of 100 m (325 ft) because inspiratory P_{O_2} rises to the absolutely toxic value of 220 kPa (→ p. 108).

Holding the breath during a dive without apparatus leads to elevation of P_{CO_2} because the CO_2 produced in the tissues is not eliminated by expiration. When the P_{CO_2} rises sufficiently, chemoreceptors are stimulated and the breakpoint is reached (→ p. 104). It is possible to delay the rise in P_{CO_2} by hyperventilation prior to diving. This process lowers the level of P_{CO_2} at the start of a dive (→ **C**) and allows a longer period of time before P_{CO_2} rises to the breakpoint. Figure **C** shows partial pressures for CO_2, O_2, and N_2 during hyperventilation and during a dive at 10 m depth for 40 seconds. The high pressure during the dive increases alveolar P_{CO_2}, P_{O_2}, and P_{N_2}. The gases diffuse from alveoli to blood. When P_{CO_2} rises to the breakpoint, the diver returns to the surface. During the ascent, P_{O_2} in the blood falls sharply because pressure is reduced and because tissues continue to utilize O_2. Eventually, alveolar exchange ceases. It is possible by hyperventilation to prolong the rise in P_{CO_2} to such an extent that P_{O_2} falls to intolerable levels during decompression and the diver loses consciousness before surfacing (→ **C**, dotted line).

Barotrauma. In diving, the gas-filled spaces of the body (lungs, middle ear) are compressed (to 1/2 at 10 m depth; to 1/4 at 30 m) by the rising pressure during descent unless the reduction in volume is made good. By using a diving apparatus, this is performed automatically for the lungs.

On the other hand, the connection between the middle ear and the throat via the Eustachian tubes is only occasionally open (swallowing) or not at all (colds). If volume equilibration does not take place during descent the increasing water pressure in the outer ear passage pushes the ear drum inwards, which is painful, and can even cause it to burst. In this case the influx of cold water causes a unilateral stimulation of the organ of balance (→ p. 298), producing nausea, dizziness, and difficulties in orientation. This can be prevented by equalizing the air volumes at regular intervals by actively pressing air from the lungs into the middle ear by holding the nose and pressing.

During resurfacing the gas spaces expand, but if surfacing takes place too quickly (> 18 m/min), without the release of air at regular intervals, the lungs may tear (pneumothorax; → p. 82), which often leads to lethal hemorrhage and air embolies.

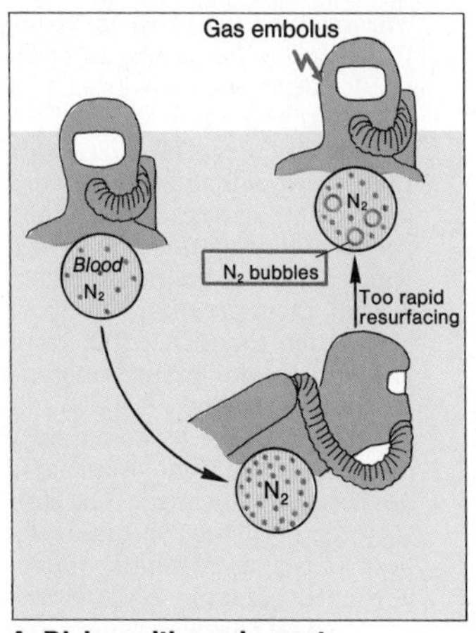

A. Diving with equipment

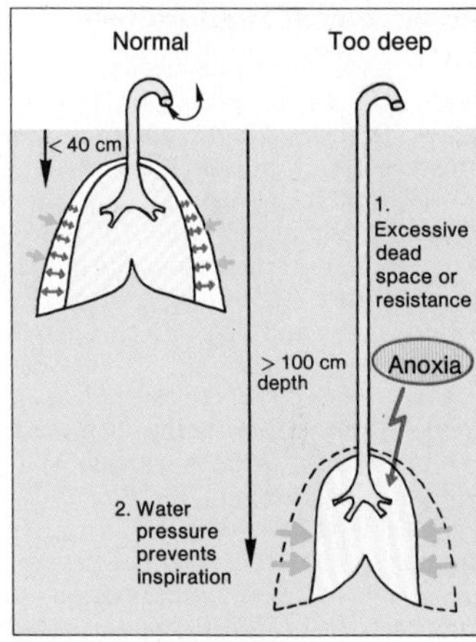

B. Diving with snorkel

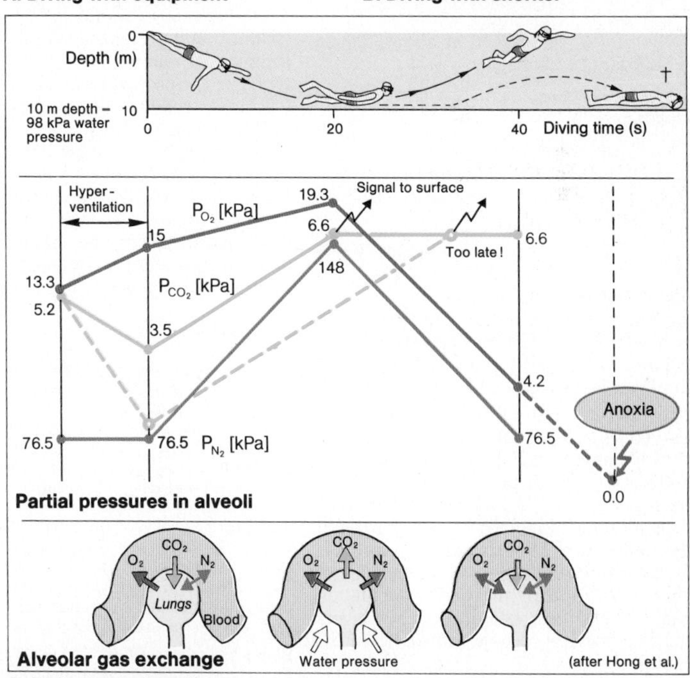

C. Diving unassisted

Respiration at High Altitude

At sea level, the mean barometric pressure (P_B) is 101.3 kPa (= 760 mmHg). O_2 comprises 21% of the air and has a partial pressure (P_{IO_2}) at sea level of 21.33 kPa (= 160 mmHg) (→ p. 78). P_{O_2} in alveoli (P_{AO_2}) is roughly 13 kPa (= 100 mmHg) (→ **A**, column 2). At increasing altitude, P_B falls and therefore P_{IO_2} and P_{AO_2} also fall (→ **A**, columns 1 and 2). Even at 3000 m (10000 ft), P_{O_2} is low enough (8 kPa = 60 mmHg) to stimulate an increase in ventilation via the chemoreceptors (→ p. 104). At 4000 m (13000 ft) (→ **A**, column 2, dashed line), the P_{AO_2} falls below a **critical level** of 4.7 kPa (= 35 mmHg) and **hypoxia** (→ p. 103) becomes severe. Progressively increasing stimulation of the chemoreceptors allows a gain in ventilation (→ **A**, column 4) and makes it possible to reach still higher altitudes before the critical value of P_{AO_2} is reached at about 7000 m (23000 ft) ("altitude gain"; → **A**).

As can be seen from the **alveolar gas equation**,

$$P_{AO_2} = P_{IO_2} - P_{ACO_2}/RQ,$$

at the given barometric pressure any drop in P_{ACO_2} (caused by hyperventilation) leads to a rise in alveolar P_{O_2}.

Above 7000 m, consciousness is usually lost. However, breathing **O_2 from a gas bottle** allows higher elevations to be achieved (→ **A**, column 1, green curve). The critical P_{AO_2} is displaced to 12000 m (39500 ft) (→ **A**, column 3), but if $\dot{V}_T$ is also increased, the limit can be pushed to 14000 m (46000 ft). Already at an altitude of 10400 m (34000 ft), P_B falls below 25 kPa (= 187 mmHg), and even when breathing 100% O_2 it is not possible to achieve the normal P_{AO_2} of 133.3 kPa (= 100 mmHg). This altitude cannot be exceeded without development of hypoxia. Modern airliners fly below this limiting P_B so that, in case of depressurization of the cabin, ventilation can be sustained with an O_2 mask and hypoxia does not become critical. Below the critical P_{AO_2} (4.7 kPa = 35 mmHg) or the critical P_B of 16.3 kPa (= 122 mmHg), a pressurized suit or pressurized cabins are essential for survival.

The maximal increase in ventilation rate (roughly 3× the resting rate) during *acute* O_2 deficiency is relatively small when compared to the increase during heavy work at normal altitudes. The explanation for this is that hyperventilation at elevated altitudes appreciably reduces P_{CO_2} in the blood and produces a *respiratory alkalosis* (→ p. 116), which decreases the respiratory drive at the central chemoreceptors (→ p. 104). However, after some time the respiratory alkalosis becomes partially compensated by increased excretion of HCO_3^-, thus returning the pH of the blood almost to normal. The respiratory stimulus due to O_2 lack can now exert its full effect. The stimulation of O_2 receptors at high altitudes also causes the heart to beat more rapidly, thus ensuring an adequate O_2 supply to the tissues as a result of increased cardiac output (→ p. 154).

Erythropoiesis is stimulated at high altitudes. After a prolonged period of **acclimatization**, the number of erythrocytes in the blood rises (polycythemia) as a result of increased secretion of erythropoietin (→ p. 60). A limit to this feature of acclimatization is set by the accompanying rise in viscosity of the blood (→ p. 64 and p. 156). When arterial P_{O_2} rises, erythropoietin secretion subsides. In the High Andes (5500 m = 18000 ft) the inhabitants have polycythemia and low P_{AO_2} but at least the younger ones appear to live normally.

O_2 Toxicity

If P_{IO_2} rises above normal (> 22 kPa or 165 mmHg), *hyperoxia* develops. The P_{O_2} may be high due to an increased O_2 concentration (O_2 therapy) or due to an elevated total pressure (P_B) at a normal O_2 content (diving; → p. 106). O_2 toxicity is a function of P_{IO_2} (critical level about 40 kPa or 300 mmHg) and time of exposure. *Lung damage* with a decrease of surfactant (→ p. 90) develops after days if $P_{O_2} \approx$ 70 kPa (0.7 at) and within 3 to 6 hours if $P_{O_2} \approx$ 200 kPa (2 at). *Coughing* and *pain during breathing* are the first symptoms. At $P_{O_2} >$ 220 kPa (2.2 at), corresponding to 100 m depth during diving with compressed air, *convulsions* and *unconsciousness* develop quickly.

Babies become *blind* if exposed to a $P_{IO_2} \gg$ 40 kPa for a longer period of time in an incubator. Under such conditions, the vitreous body becomes irreversibly opaque.

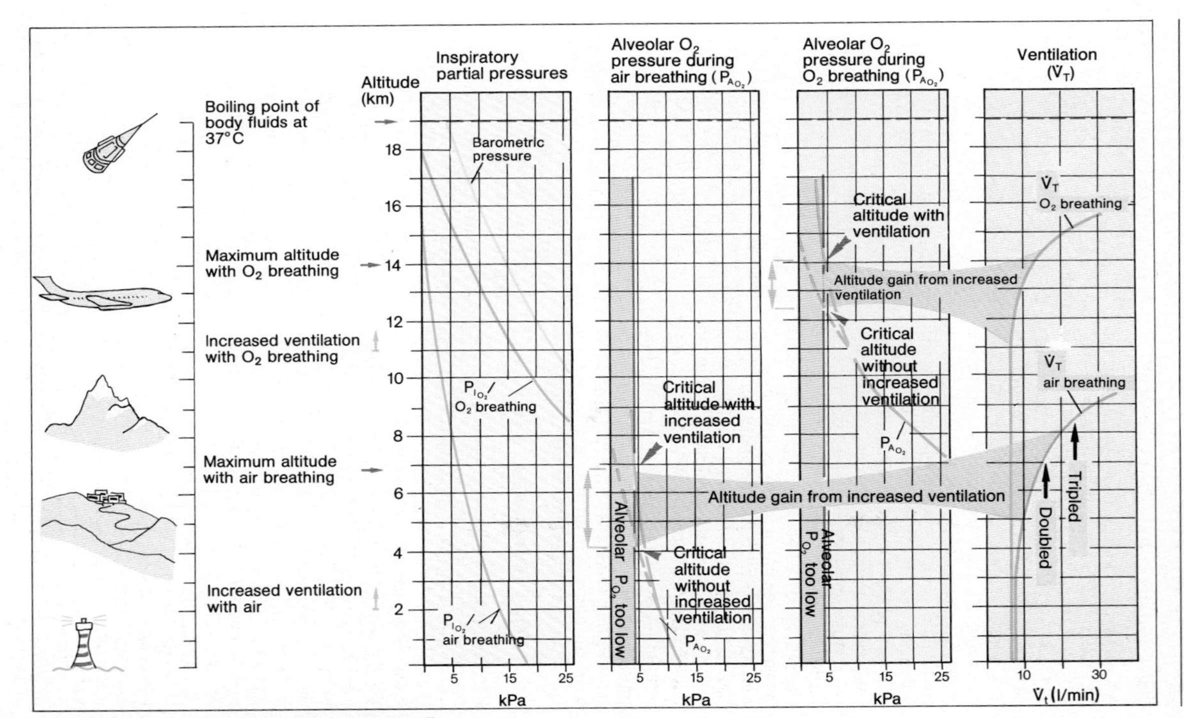

A. Respiration at altitude (not acclimatized)

pH – Buffers – Acid-Base Balance

The **pH value** is a logarithmic measure of the "effective" H^+ ion concentration or H^+ activity ($= f_H \cdot [H^+]$; → p. 334), where **pH** $= -$ **log** ($f_H \cdot$ **[H^+]**). The pH value of the **blood** averages about 7.4 (normal range; → p. 114), which corresponds to a H^+ activity of roughly 40 nmol/l. Except for local deviations (gastric juice, urine), the pH is similar in all fluid compartments whereby the pH value of the ICF is about 7.0–7.2. A **constant pH** is of vital importance for the organism; the molecular form of the proteins, for example, and thus the normal structure of the cell constituents is pH dependent. Consequently, larger deviations from the normal lead to disturbances in metabolism, in the permeability of membranes, in electrolyte distribution, and so on. Blood pH values outside the range of 7.0–7.8 are incompatible with life.

Constancy of the pH value is maintained by **buffers** (→ p. 334), which act by partially neutralizing the acids and bases that arise from the diet and from metabolism. A main buffer of the body fluids is the system

$$\mathbf{CO_2 + H_2O \rightleftarrows HCO_3^- + H^+}.$$

For any given pH value in a solution, the concentration ratio for each buffer "base" (e.g. $[HCO_3^-]$) to the corresponding buffer "acid" (in this example $[CO_2]$) is fixed (Henderson-Hasselbalch equation; → **A** and p. 113).

The significance of the CO_2-HCO_3^- buffer system in the blood is that it not only can buffer H^+ ions, like the other buffer systems, but its components can be *varied* independently of one another by certain organs; CO_2 is produced continually as an end product of metabolism, and its concentration or partial pressure (P_{CO_2}) in the blood is principally controlled by the **lungs**. P_{CO_2} can be kept constant by varying the ventilation, even when it is produced in excess or when CO_2 is consumed in buffering base. The **kidney** (in collaboration with the **liver**) controls HCO_3^- (→ **A**).

Another very important buffer is the **hemoglobin (Hb)** in the erythrocytes:

$HbH \rightleftarrows Hb^- + H^+$;
$HbO_2H \rightleftarrows HbO_2^- + H^+$

The relatively acid HbO_2 takes up less H^+ ions and gives up more of them than the less acid desoxygenated form of Hb (see also pp. 96 and 98). Thus if, for example, Hb is oxygenated in the lungs to HbO_2, H^+ ions are set free. They partially compensate for the rise in pH caused here by the removal of respiratory CO_2.

In the blood, a buffer action is also exerted by *plasma proteins, inorganic phosphates* ($H_2PO_4^- \rightleftarrows H^+ + HPO_4^{2-}$), and (in the erythrocytes) *organic phosphates*. Additionally, the cell interior in the different tissues can be drawn upon for buffering purposes.

The **total buffering capacity** (→ p. 112) of the blood at pH 7.4 and constant P_{CO_2} is about 75 mmol $\cdot\, l^{-1} \cdot (\Delta pH)^{-1}$ and the **total buffer base** concentration is normally **roughly 48 meq/l** (→ pp. 114 and 118). The latter is the sum of the concentrations of all buffer forms that can take up H^+ ions (HCO_3^-, Hb^-, HbO_2^-, BPG, plasma proteinate$^-$, HPO_4^{2-}, etc.).

For evaluation of the buffering ability of blood in the organism, the buffer base concentration is usually preferred to the buffering capacity because the latter value is dependent on P_{CO_2}.

The pH of plasma and red cell, and thus of the blood, can be changed by a variety of factors (→ **A** and p. 114f.):

a) **H^+ ions** can be directly *added*, e.g., from metabolism (hydrochloric acid, lactic acid, keto acids, sulfuric acid, etc.) or they can be *removed from the blood*, e.g., by H^+ excretion by the kidneys (→ p. 144ff.) or H^+ loss by vomiting (→ p. 114).

b) **OH^- ions** can be added, e.g., with the (basic) salts of weak acids (vegetable diet).

c) **The concentration of CO_2** ($= [CO_2]$) can be changed, e.g., by a change in CO_2 production in **metabolism** or in CO_2 expiration in the **lungs**. If the CO_2 concentration drops, the pH value rises and *vice versa* (→ **A**).

d) The **bicarbonate concentration** ($= [HCO_3^-]$) can change primarily, e.g., due to excretion of HCO_3^- by the **kidneys** or HCO_3^- losses due to diarrhea (→ pp. 114 and 146). A rise (or drop) in $[HCO_3^-]$ results in a rise (or drop) in pH.

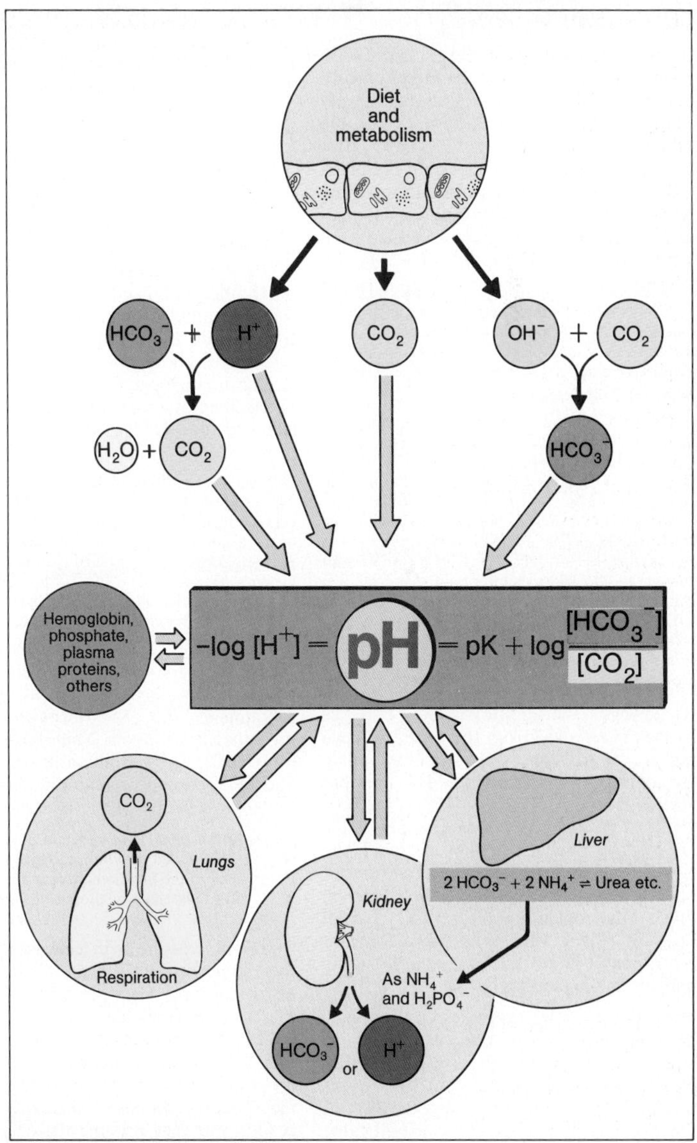

A. Factors influencing pH in blood

Bicarbonate/CO_2 Buffer

In any buffered solution, the pH is linked to the **concentration ratio** of the buffer pair. In a bicarbonate solution, therefore, the **pH value** is determined by the ratio of the bicarbonate concentration (= **[HCO_3^-]**) to the concentration of the physically dissolved CO_2 (= **[CO_2]**; → **A, top**, Henderson-Hasselbalch equation). According to this equation, the pH of a bicarbonate buffer is 7.4 when [HCO_3^-] : [CO_2] = 20 : 1 (→ **A**).

Changing the absolute concentrations of HCO_3^- and CO_2 will not change the pH as long as the ratio remains 20 : 1. As an example, the pH = 7.4 no matter whether [HCO_3^-] : [CO_2] = 24 : 1.2 mmol/l (normal) or 10 : 0.5 (low) or 34 : 1.7 (high), since the ratio in each case is 20 : 1.

If H^+ ions enter a buffered solution, they are bound by the buffer base (in this case HCO_3^-), with the formation of the buffer acid: $HCO_3^- + H^+ \rightarrow CO_2 + H_2O$. In a **closed system**, exactly the same amount of buffer acid is formed as buffer base consumed (the reverse obtains for the addition of OH^- ions). The values (see above) of 24 : 1.2 mmol/l for [HCO_3^-] : [CO_2] (→ **A, top**) change to 22 : 3.2 on addition of, for example, 2 mmol/l H^+-ions, and the pH drops to 6.93 (→ **A, left**). This means that the buffer capacity (see below) of the HCO_3^-/CO_2-buffer is small in a closed system, since the pK value of 6.1 is far from the desired pH value of 7.4 (→ p. 334f.).

If, however, the additional CO_2 that has formed is *removed* from the solution (**open system**; → **A, right**), only the bicarbonate concentration changes when H^+ ions are added (2 mmol/l). The ratio [HCO_3^-] : [CO_2] (= 22 : 1.2) and thus also the pH value (7.36), drop much less in this case than in the closed system. In the human organism an open system for bicarbonate buffering arises due to the regulation of the **CO_2 partial pressure (P_{CO_2})**, and thus the **CO_2 concentration** ([CO_2] = $\alpha \cdot P_{CO_2}$; → p. 98) in the plasma, by **respiration** (→ **B**).

Normal metabolic processes continuously generate CO_2 at a rate of 15000 to 20000 mmol/day, but the P_{CO_2} in arterial plasma is maintained constant by respiratory activity. At the same time, the P_{CO_2} in the alveoli is maintained at a constant value (→ p. 92f.), to which that of the plasma equalizes at each passage through the lungs (i.e. the P_{CO_2} in the arterial blood is also constant). **Neutralization of H^+** adds CO_2 to the plasma ($HCO_3^- + H^+ \rightarrow CO_2 + H_2O$), but the excess CO_2 is eliminated each time the blood passes through the lungs, and, despite the addition of H^+, both the arterial P_{CO_2} and pH are kept relatively constant (open system; → **B**, left). Quantitatively, this compensation is trivial. For example, if net H^+ production should double, the excess CO_2 (assuming all other buffers were inoperative) would be only 60 mmol/d or 0.3 % of the normal daily load. On the other hand, the HCO_3^- consumed for buffering these 60 mmol/d of fixed acid is quantitatively much more important. Per day, it constitutes about one sixth of the HCO_3^- contained in the ECF. Renal excretion of H^+ is therefore necessary. For each H^+ ion excreted, one HCO_3^- ion is generated (→ p. 147, C1).

In principle, **addition of base** operates similarly. The reaction is $OH^- + CO_2 \rightarrow HCO_3^-$. The concentration of HCO_3^- increases, and initially the concentration of CO_2 is relatively lower than normal. However, the rate of elimination of CO_2 declines (lowered alveolar CO_2 gradient) and therefore arterial P_{CO_2} remains normal (→ **B**, right), so that the slight rise in arterial pH value is due only to the elevated HCO_3^- concentration.

At pH 7.4, the HCO_3^--CO_2 buffer (open system, i.e. P_{CO_2} is constant at 40 mmHg) accounts for about two thirds of the buffering capacity of blood. **Nonbicarbonate buffers (NBBs)** are predominantly intracellular and make up the remaining third of the buffering capacity of the blood.

The *buffering capacity* at a certain pH is the ratio: (H^+ or OH^- added per volume): (pH change), that is, the slope of the titration curve of this buffer (→ p. 335, D). The unit is $mol \cdot l^{-1} \cdot (\Delta pH)^{-1}$ (→ p. 110).

The NBBs operate as in a **closed system**; the total concentration of the buffer pair remains constant when pH changes occur. In the blood, **hemoglobin** is the chief NBB (→ p. 118). The buffering capacity of the blood may be reduced in anemia. NBBs not only cooperate with the HCO_3^--CO_2 buffer in metabolic acid-base disturbances (→ p. 114), but they become the **only** effective buffers in *respiratory* acid-base disturbances where prolonged changes of plasma CO_2 are the primary events (→ p. 116).

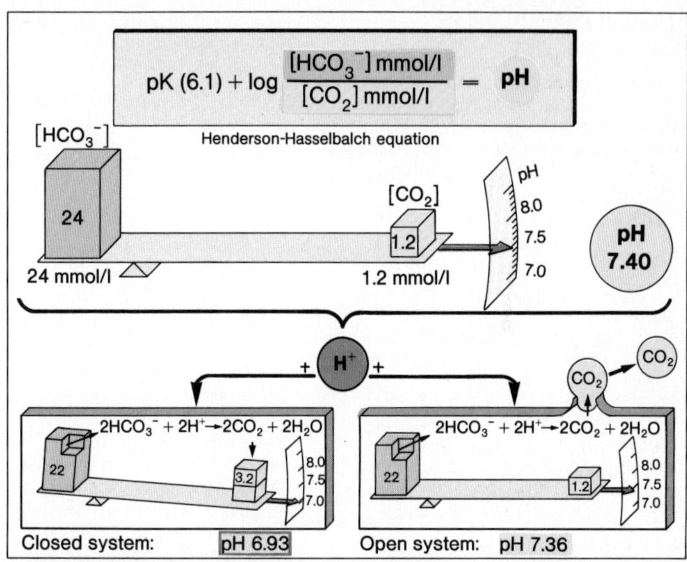

A. Bicarbonate buffer: open and closed systems

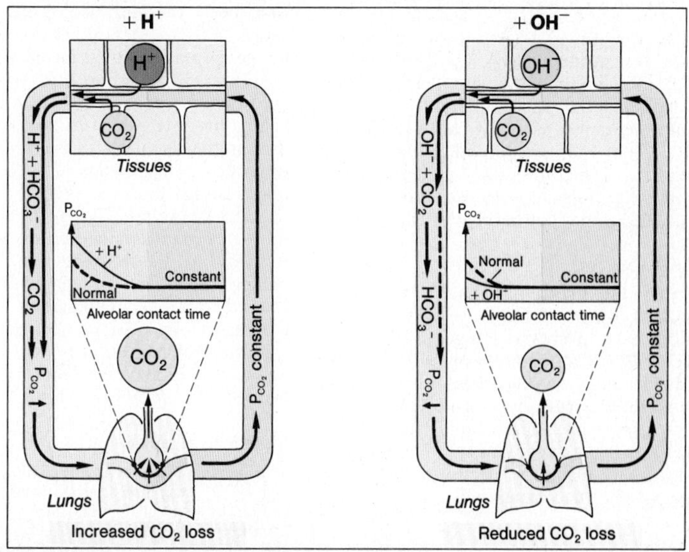

B. Bicarbonate as a blood buffer: open system

Acid-Base Balance and its Disturbances

Acid-base balance refers to the homeostatic mechanisms that maintain a constant pH. **Normal acid-base parameters of arterial plasma** (although *measured* in arterialized capillary blood) are as follows (for RBC values see table on p. 96):

	Females	Males
$[H^+]$	39.8 ± 1.4 nmol/l	40.7 ± 1.4 nmol/l
pH	7.40 ± 0.015	7.39 ± 0.015
P_{CO_2}	5.07 ± 0.3 kPa ($= 38 \pm 2$ mmHg)	5.47 ± 0.3 kPa ($= 41 \pm 2$ mmHg)
$[HCO_3^-]$	24 ± 2.5 mmol/l	24 ± 2.5 mmol/l

Acid-base equilibrium is achieved in the organism when the following equations are balanced:

1. H^+ production and uptake − HCO_3^- production and uptake = H^+ excretion − HCO_3^- excretion ≈ 60 mmol/d (dependent on nutrition);

2. CO_2 production = CO_2 excretion ≈ 15000 to 20000 mmol/d

For equation 1, the primary contributions are H^+ production as hydrochloric and sulfuric acids (end products of protein metabolism) and as phosphoric and lactic acid, and, for compensation, renal excretion of H^+ (→ p. 144). On the other hand, a diet high in vegetable matter generates large amounts of HCO_3^- and accounts for the alkaline urine that herbivores excrete.

Disturbances

When the pH falls below 7.35, acidosis occurs; when it rises above 7.45, alkalosis occurs. A primary change in P_{CO_2} is a **respiratory** disturbance; a primary change in HCO_3^- is a **nonrespiratory** or **metabolic** disturbance. Metabolic disturbances most often represent a discrepancy between production and excretion of H^+. The acid-base disturbance may be partly or fully **compensated** or **uncompensated**.

Metabolic acidosis

Metabolic acidosis (low pH, low HCO_3^-) may occur from (1) renal failure to excrete the normal acid load; (2) ingestion of acid; (3) excess endogenous production of acid as occurs in diabetes mellitus and starvation (incomplete metabolism of fats, ketoacids); (4) anaerobic production of lactic acid (incomplete oxidation of carbohydrate); (5) increased production of hydrochloric and sulfuric acid (high protein intake); (6) loss of HCO_3^- as in diarrhea or by the kidneys (renal tubular acidosis, or use of carbonic anhydrase inhibitors; → p. 142).

All of these events are followed, first, by **buffering** (→ **A**) of the excess H^+ ions (any loss of HCO_3^- by kidneys or intestine is equivalent to an increase in H^+ ions). HCO_3^- and the nonbicarbonate buffer bases (NBB-) contribute to this buffering in the ratio of 2 : 3 to 1 : 3, and the CO_2 arising from the HCO_3^- leaves the organism via the lungs (open system; → p. 112). The **buffer base (BB) concentration drops** (negative excess of BB; → p. 118). The second step is the **compensation of the metabolic acidosis:** the lowered pH value leads (via central chemoreceptors; → p. 104) to a rise in the ventilation rate, which in turn results in a drop in the alveolar and arterial P_{CO_2}. This **respiratory compensation** (→ **A, bottom**) not only restores the ratio $[HCO_3^-] : [CO_2]$ to near-normal values (20 : 1), but also (due to the increasing pH value) reconverts NBB-H to NBB^-. This consumes HCO_3^-, which makes necessary, as compensation, increased respiratory elimination of CO_2. If the reason for acidosis persists, then respiratory compensation is inadequate and an increase in **H^+ excretion by the kidneys** becomes necessary (→ p. 144 ff.).

Metabolic alkalosis

A metabolic (or nonspiratory) alkalosis (high pH, high HCO_3^-) may arise from (1) intake of alkaline substances (e.g. HCO_3^- infusion); (2) increased metabolism of organic anions such as lactate and citrate to CO_2 and H_2O; (3) loss of H^+ from vomiting, or from increased renal excretion of H^+ in K^+ deficiency (→ p. 148). **Buffering** of this disturbance is principally similar to that in metabo-

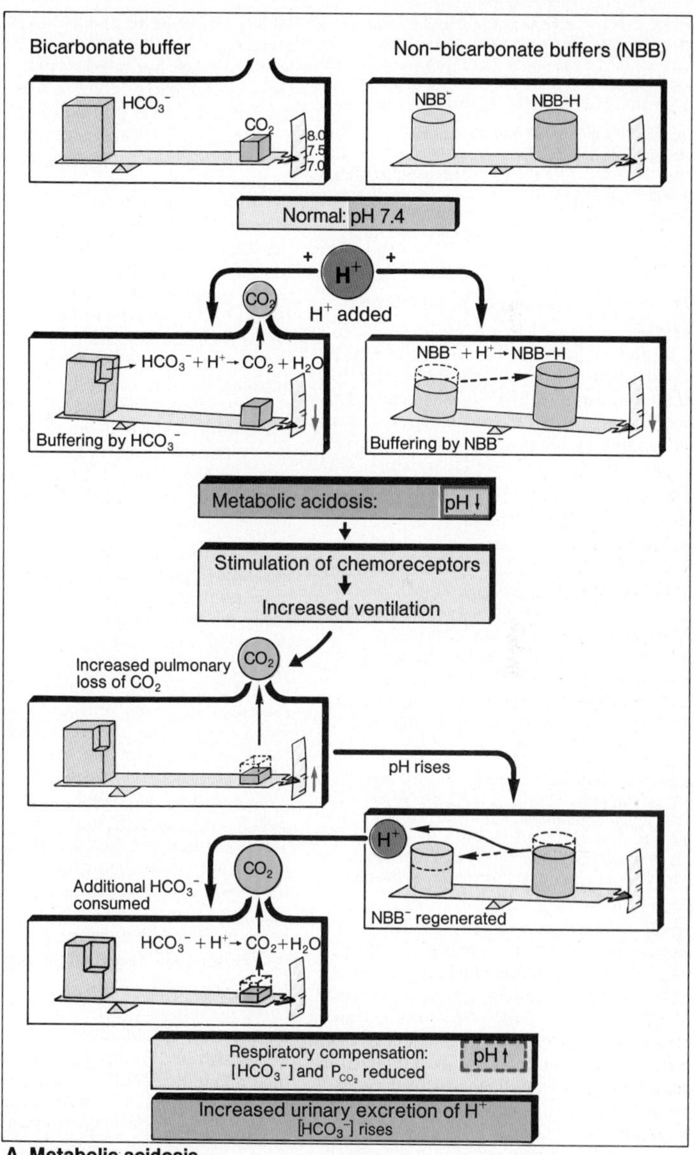

A. Metabolic acidosis

lic acidosis (positive excess of BB), but *respiratory compensation* by hypoventilation is limited on account of the accompanying O_2-*deficiency*. If the alkalosis is not of renal origin, it can be normalized by means of **increased HCO_3^- excretion in the urine**.

Metabolic alkalosis is common in the clinic. The elevated pH reduces free serum calcium and may lead to *tetany*. Irritability of the nervous system increases and may produce convulsions. Electrocardiogram (ECG) changes resemble hypokalemia (→ p. 172). Alkalosis is especially hazardous to patients receiving digitalis.

Respiratory Disturbances in Acid-Base Balance

If more CO_2 is expired than is produced by metabolism, as in hyperventilation, the P_{CO_2} of the plasma drops (*hypocapnia*). This condition increases blood pH and is known as **respiratory alkalosis** (→ **A**). If, conversely, relatively too little CO_2 is expired, the plasma P_{CO_2} rises (*hypercapnia*), causing **respiratory acidosis** (→ **A**).

In metabolic disturbances of acid-base balance, the concentrations of bicarbonate and of the bases of the NBBs change in parallel (→ p. 114); in respiratory disturbances, changes in these concentrations are dissociated. The HCO_3^--CO_2 buffer is no longer effective because in respiratory disturbances the change in P_{CO_2} is the *cause* and not the *result* as it is in metabolic disturbances (→ p. 114).

Causes of respiratory acidosis (low pH, high P_{CO_2}) are an impaired ability of the lungs to eliminate CO_2 (**hypoventilation**), as in (1) reduction of pulmonary tissue (tuberculosis, pneumonia), (2) reduced ventilation (poliomyelitis, barbiturate overdosage), and (3) abnormalities of the rib cage (scoliosis).

In these states, decreased respiratory exchange elevates the P_{CO_2} and lowers the pH. The H^+ excess is buffered by the NBB at the expense of the base form of the buffer pairs ($NBB^- + H^+ \rightarrow NBB - H$). As a result, HCO_3^- is formed to a small extent (→ **A**), but the combined total of bicarbonate concentration and NBB bases (**buffer base, BB**; → p. 110) remains **unchanged**, unlike the situation in metabolic disturbances. (A slight drop in BB results because some of the HCO_3^-, but not the Hb, diffuses into the interstitial space.) Nevertheless, the rise in HCO_3^- is too small to account fully for the high P_{CO_2}, so that the ratio HCO_3^-/CO_2 remains below normal (low pH, acidosis). If the P_{CO_2} remains high, a compensatory mechanism comes into action (→ **A**) after 1–2 days: H^+ ions are increasingly excreted by the kidney (as titratable acid and NH_4^+; → p. 144 ff.). For every H^+ ion secreted by the tubule cell, one HCO_3^- ion is released into the blood. As a result, the $[HCO_3^-]$ in the blood increases until the pH value, despite elevated P_{CO_2}, returns to normal (**renal compensation**). A portion of the retained HCO_3^- is used to neutralize the H^+, which is generated during the pH rise, from the reaction $NBB\text{-}H \rightarrow NBB^- + H^+$ (→ **A, bottom**). Because of the 1 to 2 day delay in renal metabolic response to acidosis, **acute respiratory acidosis** is poorly compensated in comparison to chronic respiratory acidosis. In the latter, HCO_3^- may increase 1 mmol for each 10 mmHg rise in P_{CO_2} above normal.

Hyperventilation is the cause of respiratory alkalosis (high pH; low P_{CO_2}). It is elicited by respiration during O_2 deficiency (→ p. 108) or by psychic changes. The plasma P_{CO_2} is depressed. This causes a slight drop in $[HCO_3^-]$ since part of the HCO_3^- converts to CO_2 ($H^+ + HCO_3^- \rightarrow CO_2 + H_2O$); for this reaction, H^+ ions are continuously provided by the NBB ($NBB\text{-}H \rightarrow NBB^- + H^+$). For the same reason, the $[HCO_3^-]$ also drops further in respiratory compensation of metabolic acidosis (→ p. 115; → **A, bottom**, and p. 118). To bring about normalization of the pH value (compensation), a further drop in $[HCO_3^-]$ is necessary. This is achieved by an increase in the renal excretion of HCO_3^-, which is brought about by a decrease in the tubular secretion of H^+ (**renal compensation**).

CO_2 enters the cerebrospinal fluid (**CSF**) from the blood much more rapidly than HCO_3^- and H^+. Additionally, the protein concentration in the CSF is low, which means that little NBB is available; therefore, an acute respiratory acidosis or alkalosis causes relatively large fluctuations in the pH of the CSF. These constitute an adequate stimulus for the central chemoreceptors (→ pp. 98 and 104).

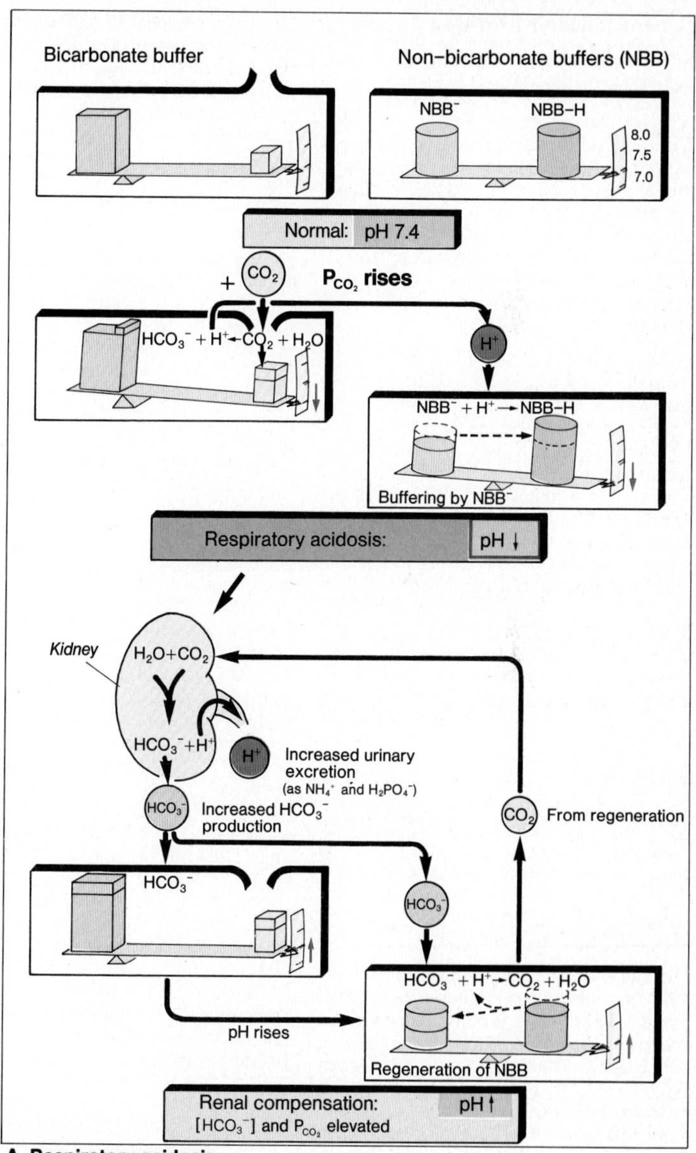

A. Respiratory acidosis

Determination of Acid-Base Status

In the Henderson-Hasselbalch equation:

$pH = pK_a + \log[HCO_3^-]/[CO_2]$,

$[CO_2]$ can be replaced by $\alpha \cdot P_{CO_2}$ (→ p. 98). The equation therefore contains two constants (plasma; 37 °C): The pK_a (= 6.1) and the solubility coefficient, α (= 0.225 mmol · $l^{-1} \cdot kPa^{-1}$ = 0.03 mmol · $l^{-1} \cdot mmHg^{-1}$). The remaining values, **pH**, **$[HCO_3^-]$**, and **P_{CO_2}** vary interdependently; when one is held constant the other two vary proportionally. Plotting log P_{CO_2} against pH yields a straight line (→ **A–C**).

The blue line in **A** and **B** shows data from a solution of HCO_3^- at 24 mmol/l, with *no other buffers*. Changing P_{CO_2} changes the pH without significantly influencing $[HCO_3^-]$. A family of parallel lines can be drawn to describe the relationships for other $[HCO_3^-]$ (→ **A** and **B**, orange lines). The scale of the coordinates is adjusted so that the slope of these lines is 45°.

In **blood**, the presence of nonbicarbonate buffers, NBB, influences the titration curve; when P_{CO_2} is changed, the pH changes much less than when HCO_3^- is the only buffer. The slope of the titration curve is therefore steeper than 45° (→ **B**, green and red lines). This means that a change in blood P_{CO_2}, for example from 40 (→ **B**, d) to 20 mmHg (→ **B**, c), is followed by a change in the **actual $[HCO_3^-]$** in the same direction (→ **B**; c now lies on the interrupted line that represents a lower $[HCO_3^-]$ than the yellow line). The **"standard" $[HCO_3^-]$**, however, which is always measured after equilibration with a P_{CO_2} of 5.33 kPa or 40 mmHg, does not change. It allows the evaluation $[HCO_3^-]$ independent of P_{CO_2} changes.

The **Siggaard-Andersen nomogram** (→ **C**) is used for clinical evaluation of acid-base balance. The log P_{CO_2} is plotted on the y axis and pH on the x axis. A horizontal reference scale is drawn from the y axis at the normal P_{CO_2} (= 5.33 kPa = 40 mmHg). On this scale are values for $[HCO_3^-]$, which represent the family of HCO_3^- titration curves in **A**. To any point on this scale, the 45° line can be drawn, from which the actual $[HCO_3^-]$ value can be read off.

On the upper part of the nomogram, the curve has two scales: one for Hb and one for **buffer bases (BB)**. The BB are composed chiefly of HCO_3^-, Hb, and other proteins (→ p. 110). A line drawn from a given value for Hb to the point P_{CO_2} = 40; pH = 7.4 describes the normal HCO_3^- titration curve in the presence of NBB. When Hb is low, as in anemia, the slope of this line falls to some extent.

To **use the nomogram**, arterial blood is drawn and measured for pH. It is then equilibrated with CO_2 at two different pressures, and the two corresponding pH values are plotted. *Example:* **C**, point A: pH at P_{CO_2} of 10 kPa (= 75 mmHg) and point B: pH at P_{CO_2} of 2.67 kPa (= 20 mmHg). The pH of the blood before equilibration is plotted along this line, and the P_{CO_2} corresponding to that pH is read from the y axis (→ **C**, green line for normal acid-base values).

In **C**, the green capitals show normal values whereas the red letters represent the respective values in an abnormal state; a and b are the equilibrated values. The original pH of 7.2 indicates acidosis. The *standard* $[HCO_3^-]$ of 13 mmol/l for this curve (d) is low. The P_{CO_2} of the original blood (4 kPa = 30 mmHg) is also low and represents a partial respiratory compensation by hyperventilation. A 45° line from c to e shows the *actual* $[HCO_3^-]$ in the original blood to be lower (11 mmol/l) than the standard $[HCO_3^-]$. This difference reflects the decreased P_{CO_2}. The BB can be read from the upper scale at g (normal BB = 48 meq/l when Hb = 150 g/l). **Base excess (BE)** is shown where the titration line crosses the lower curve on the nomogram. BE is the amount of acid or base needed to titrate 1 liter of blood to normal at a P_{CO_2} of 5.33 kPa (= 40 mmHg). The BE is (+) in metabolic alkalosis and (−) in metabolic acidosis. In respiratory acid-base disturbances, BB is virtually unchanged (→ p. 116) and BE ≈ 0.

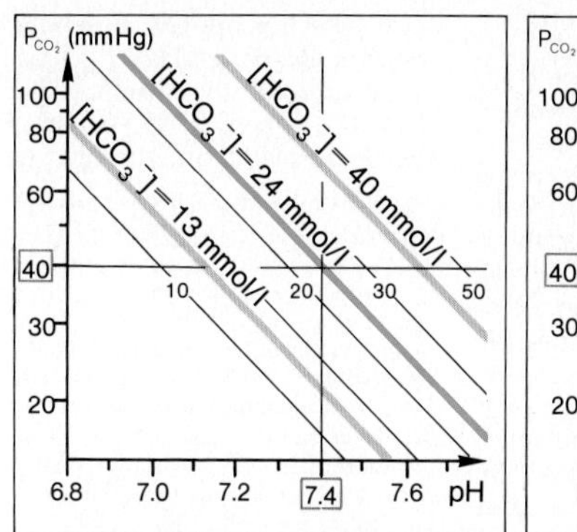

A. P_{CO_2}/pH nomogram (without NBB)

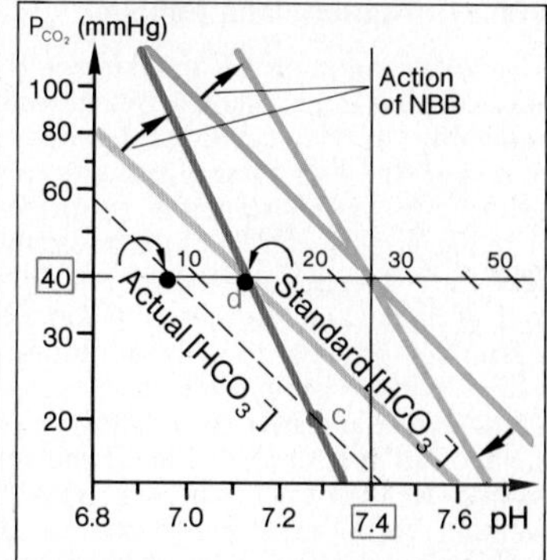

B. P_{CO_2}/pH nomogram (with NBB)

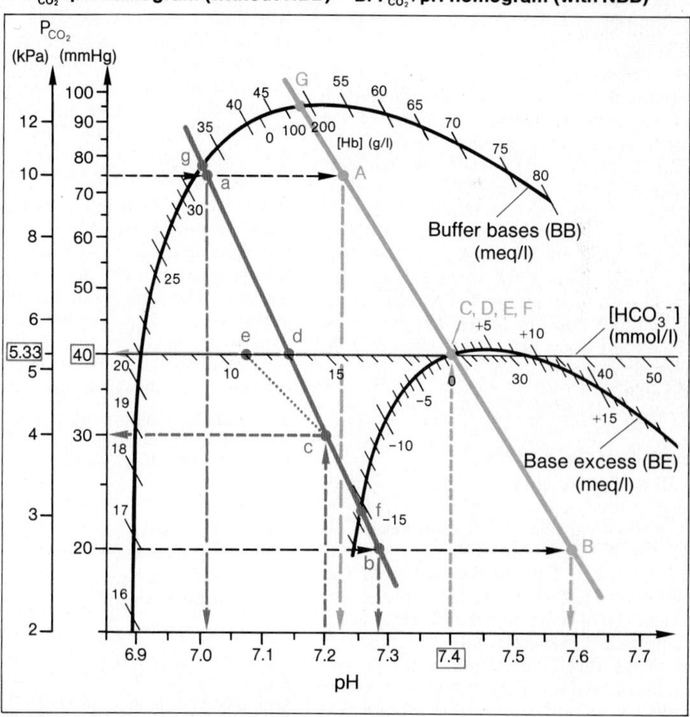

C. Siggaard-Andersen nomogram

Kidney: Anatomy and Function

A primary function of the kidney is to maintain constancy of **ECF volume** and of **osmolality** by balancing intake and excretion of **Na**$^+$ and **water**. Furthermore, the kidney achieves constancy of extracellular **K**$^+$ concentration and of blood and cellular **pH** by adjusting excretion of H^+ and HCO_3^- to their intake and to respiration and metabolism (→ pp. 110–116). In addition, the kidney conserves nutrients (e.g. glucose, amino acids) and **excretes** end products of metabolism (e.g. urea, uric acid) and xenobiotics. It also has numerous metabolic functions (e.g. arginine formation, gluconeogenesis, peptide hydrolysis) and is a source of hormones (e.g., angiotensin II, erythropoietin, D-hormone, prostaglandins).

The functional unit of the kidney is the **nephron**; 1.2 million nephrons make up each human kidney. At the beginning of the nephron, in the **glomerulus**, the blood is *filtered*: proteins and cells are retained, whereas water and all smaller dissolved substances pass into the **tubule**, whence the greater part of this *ultrafiltrate* is transported across the tubule wall and reenters the blood (**resorption, reabsorption**). The fraction that is not resorbed remains in the tubule and appears in the terminal **urine** (**excretion**). Some urinary solvents enter the nephron lumen from tubule cells by **secretion**.

Subunits of the Nephron

1. The **renal corpuscle** (→ **A 1, 2**) is situated in the renal cortex and is made up of Bowman's capsule and the **glomerulus** (→ **B**), which contains the filtering surface of the nephron. The **afferent** arteriole (*vas afferens*) brings blood to the nephron and breaks up into several capillaries, which comprise the **glomerular tuft**. The capillaries rejoin to form the **efferent** arteriole (*vas efferens*), which carries blood to a second capillary network around the tubular cells (→ p. 122). The glomerular capillaries invaginate Bowman's capsule like fingers pushed into an inflated balloon, forming two layers within the capsule: a visceral layer, which is applied closely to the capillaries, and a parietal layer, which forms the outer surface of the corpuscle. The capsular space (→ **B**) between the two layers collects the glomerular filtrate (**primary urine**).

The **glomerular filter** has several layers (→ **B**): the visceral layer of the capsule contains the **podocytes**. Their appendages, the *pedicels*, interdigitate closely. The slitlike spaces between the pedicels are convered by the *slit membrane* with pores of 5 nm diameter. The other face is made up of the capillary **endothelium**; spaces between its cells are 50 to 100 nm. A **basal membrane** lies between the podocyte layer and the capillary layer. The gaps in the filtration surface allow all components of blood to pass except blood cells and solutes above a certain molecular size (→ p. 126).

2. The **proximal tubule** has a convoluted (→ **A 3**) and a straight (*pars recta*; → **A 4**) segment. Cellular characteristics include: a tall **brush border** on the luminal surface and elaborate infoldings of the basal and lateral cell membrane to form a complex **basolateral labyrinth** in close contact with numerous mitochondria at the cellular side of the baso-lateral membrane (→ **A** and p. 5, C).

3. The **loop of Henle** has a thick descending limb (*pars recta*; see above) followed by a thin descending limb (→ **A 5**) and a thin (in long loops only) and a thick ascending limb (→ **A 6**). The long loops of Henle of the *juxtamedullary nephrons* dip into the inner zone of the medulla and represent about 20% of all nephrons; the remaining *cortical nephrons* have shorter loops.

4. The **distal tubule** begins with a straight segment (= thick ascending limb of Henle's loop; → **A 6**) followed by a convoluted segment (→ **A 7**). It has no brush border and fewer mitochondria than the proximal tubule. The early convoluted distal tubule contacts its own glomerulus. Here, the tubule wall contains specialized cells, the **macula densa** (→ p. 152).

Several distal tubules connect with a **collecting duct** (→ **A 8**) consisting of anatomically and functionally distinct cortical and (outer and inner) medullary segments, which make final modifications to the urine and conduct it to the **renal papillae** and pelvis for excretion.

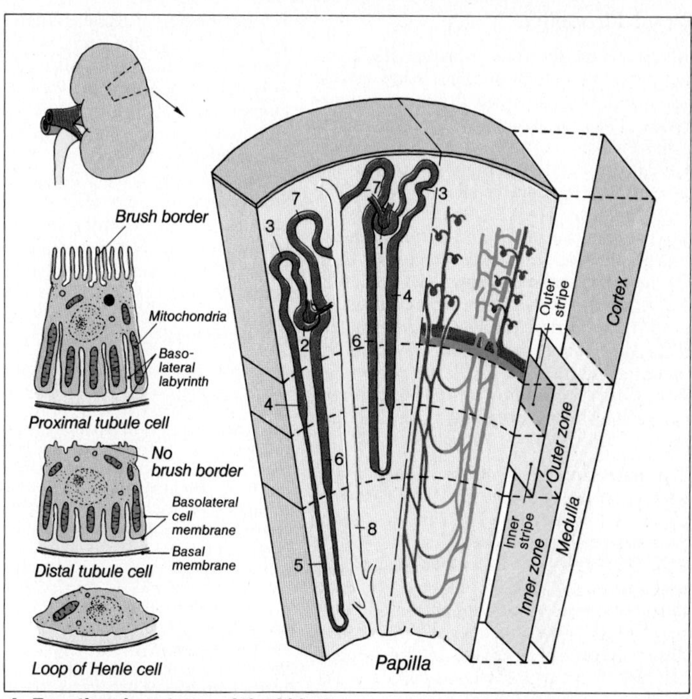

A. Functional anatomy of the kidney

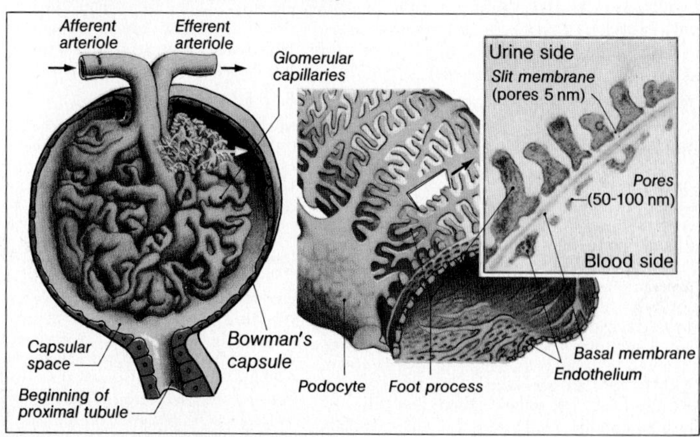

B. Glomerulus and Bowman's capsule

Renal Circulation

Blood reaches the kidney through the short, large-diameter, high-pressure **renal artery**, which branches to form the **interlobar arteries**. Their branches are the **arcuate arteries**, which, in turn, give off the **interlobular arteries** that supply the cortex and give off the **afferent arterioles (→ A2c)**. The arcuate arteries (→ **A1a, 2a**) form a boundary between the **cortex** and **medulla**. Distribution of blood to these two regions differs. Of the total **renal blood flow (RBF)** of about 1.2 l/min (20–25% of the cardiac output; → p. 154) the cortex receives approximately 90%, the outer medulla about 10%, and the inner medulla 1–2%. The latter percentage represents a blood flow of 0.4–0.7 ml/min per g tissue; this value is still higher than that in most other organs.

The **microcirculation of the kidney** is unusual; it consists of two capillary networks in series (→ **A2; B**). The first is a high-pressure net between the afferent (→ **A2c**) and efferent (→ **A2d**) arterioles of the glomerulus (→ p. 120). Changes in resistance of these vessels influence not only renal blood flow but also the glomerular filtration pressure. Blood in the efferent arteriole flows to the second capillary network, a low-pressure net surrounding the tubules (→ **B**). It supplies the tubular cells with blood, as well as effecting the exchange of substances with the tubular lumen (resorption, secretion; → p. 126ff.).

The kidney contains two classes of nephrons; the peritubular capillary net is adapted differently to each of these classes (→ **A2**):

cortical nephrons have short loops of Henle; around their tubules, the capillaries form a network as in any other tissue; (2) **juxtamedullary nephrons** have long loops of Henle, and their glomeruli are situated in the cortex near the medullary border; around these nephrons, the capillaries break up into long (40 mm) loops or slings that parallel the loops of Henle. The capillary loops, the **vasa recta**, are the only blood supply to the medulla. Changes in distribution of blood to these two networks influence the extent of salt excretion. Medullary blood flow also influences osmolal concentration of the medullary ECF; increased flow reduces the concentration and limits the capacity of the kidney to conserve water (→ p. 136).

Autoregulation of RBF. When the mean systemic blood pressure is varied from about 11 to 27 kPa ($\approx$ 80 to 200 mmHg), only trivial changes are observed in the glomerular filtration rate (GFR; → p. 124) and RBF (→ **C**). This phenomenon is **autoregulation** and occurs even in a denervated kidney. It is influenced largely by changes in the resistance of interlobular arteries and of afferent and efferent arterioles of the cortical nephrons (→ **B**). When the mean systemic pressure falls below about 11 kPa ($\approx$ 80 mmHg), autoregulation is no longer effective and both GFR and RBF fall off sharply (→ **C**).

Measurement of renal plasma flow (RPF) and RBF can be accomplished by employing the **Fick principle**: the rate (amount/time) of removal of a test substance from the plasma of an organ equals the rate entering the organ in the arterial plasma (RPF · arterial concentration) minus the rate leaving it in venous plasma (RPF · venous concentration). In other words, RPF = rate of removal/(arterial minus venous concentration). It is applicable to any organ, provided an appropriate test substance is used. Concentration of the test substance in the plasma entering and leaving the organ must be determined. For the kidney, the concentration of the test substance, p-aminohippurate (**PAH**), in venous plasma leaving the kidney is very small (at low arterial concentrations, renal venous concentration of PAH is roughly one tenth of the arterial value), and therefore only the arterial concentration is measured. The rate of removal from the plasma (or excretion in the urine) is urinary flow rate · urinary concentration of PAH ($\dot{V}_u \cdot U_{PAH}$). Thus

$$RPF = U_{PAH} \cdot \dot{V}_u/(0.9 \cdot P_{PAH}),$$

where P_{PAH} is the arterial plasma concentration. In other words, **RPF = PAH-clearance/0.9**. If P_{PAH} is too high PAH secretion is saturated, and $U_{PAH} \cdot \dot{V}_u/(0.9 \cdot P_{PAH})$, becomes much lower than RPF. **RBF** is calculated from RPF/(1-hematocrit) (→ p. 60).

Renal O_2 consumption in humans is about 18 ml/min. Because of the large RBF, the arteriovenous difference for O_2 is small (14 ml/l blood). O_2 consumption of the cortex (active transport!) is more than twentyfold that of the inner medulla. In the cortex, **energy metabolism** is oxidative and uses fatty acids and other substances as substrates. In the medulla, energy metabolism is anaerobic, O_2 utilization is low, and glucose is the chief substrate.

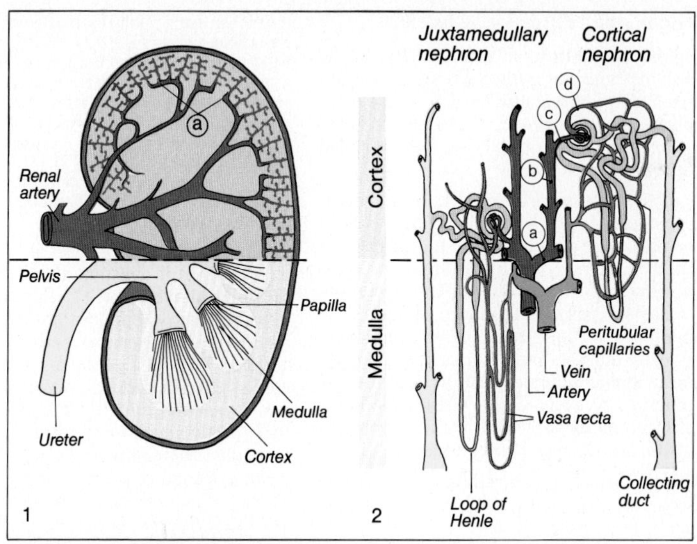

A. Vascular arrangement in kidney

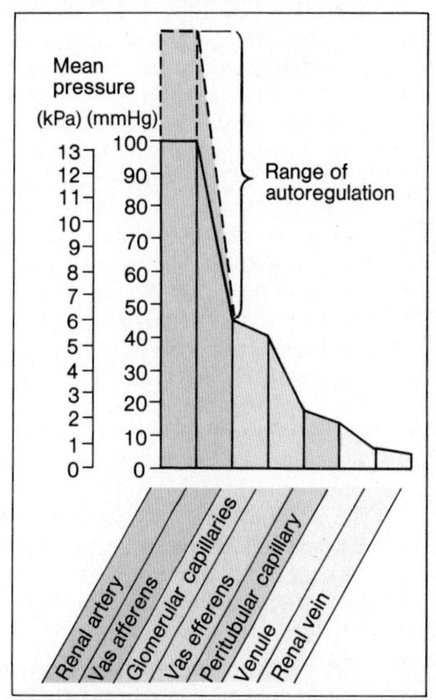

B. Blood pressure gradients in renal vessels

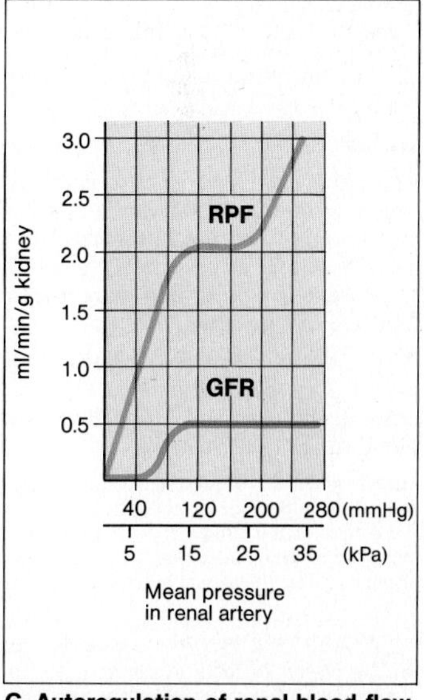

C. Autoregulation of renal blood flow and glomerular filtration rate

Glomerular Filtration, Clearance

Glomerular filtration rate (GFR) is the volume/time that is filtered by all glomeruli. An average of one fifth or 20% of the renal plasma flow (RPF, → p. 122) is filtered at the glomerulus. This ratio, GFR/RPF, is called **filtration fraction (FF)**.

For the **measurement of GFR**, an indicator substance (In) with specific properties must be present in the blood. The indicator must enter the tubule only by filtration and must not undergo resorption, tubular excretion, or metabolism. Furthermore, it should be inert and without influence on renal function. A suitable indicator is the carbohydrate **inulin**, which can be infused into the blood to measure GFR. With some limitations, **creatinine**, which is already present in the blood, may also be used.

The GFR may be determined (→ **A**) if urine flow rate ($\dot{V}_u$) and plasma and urine concentration of the indicator (P_{in} and U_{in}) are known. The rate at which the indicator is filtered is GFR (l/min) · P_{in} (g/l). Since all of the indicator must appear in the final urine (see indicator properties above), the rate of filtration = rate of excretion or $GFR \cdot P_{in} = \dot{V}_u \cdot U_{in}$. Thus,

$GFR \doteq \dot{V}_u \cdot U_{in}/P_{in}$ (→ **A**).

The GFR is therefore the same as the **clearance** (see below) of **inulin** or **creatinine** and amounts to about **120 ml/min** at a body surface of 1.73 m^2. Some 99% of GFR is resorbed (→ p. 136), and only about **1 to 2 l/d** are lost as **urine**.

Although the **plasma concentration of creatinine** (P_{cr}) rises with a drop in GFR, the P_{cr} alone is only a very *inexact* indicator of the GFR value.

Aspects of GFR. About 180 l of fluid are filtered daily. Since about 25% of the body weight of a 70 kg man, or 17 l, are ECF (→ p. 138), his ECF is reworked in the kidney more than 10 times/d, his plasma volume (3.2 l) even once every 25 minutes.

GFR varies with capillary hydrostatic pressure (P_{cap}; → p. 158), Bowman's capsular pressure (P_{BOW}), plasma oncotic pressure (π_{cap}; → p. 158), glomerular surface area (also dependent upon the number of intact glomeruli), and glomerular water permeability. P_{cap} (→ p. 123, B) at the afferent end of the glomerular capillary is 6 kPa (45 mmHg); at the efferent end, 5.7 kPa (43 mmHg). In contrast to other tissues where FF is only 0.005 (→ p. 158), the FF in the glomerulus is 0.2. π_{cap} therefore increases from roughly 2.7 kPa (20 mmHg) at the afferent end to roughly 4.4 kPa (33 mmHg) at the efferent end. P_{BOW} is 1.6 kPa (12 mmHg). The effective filtration pressure (= $P_{cap} - P_{BOW} - \pi_{cap}$) is, therefore, initially around 1.7 kPa (13 mmHg) and drops to 0 probably near the end of the glomerular capillary. Beyond this point filtration cannot occur (**filtration equilibrium**).

Paralleling the autoregulation of RBF (→ p. 122), GFR is constant at systemic blood pressures above about 11 kPa. Thus, filtration pressure is kept constant in this range by altering prearteriolar and arteriolar resistance. The signals for this autoregulation are (a) the arteriolar blood pressure (*myogenic response*) and (b) a feedback signal (only partly understood) transmitted by the *juxtaglomerular apparatus* (→ p. 152). If the mean systemic blood pressure falls below 8 kPa (= 60 mmHg), filtration ceases.

The expression $U \cdot \dot{V}_u/P$ is the **clearance** (C). The clearance of inulin (C_{in}) is the same as the GFR (see above). The clearance of any other substance X is compared to the GFR by its **clearance ratio** or **fractional excretion** (C_x/GFR). A substance X, which is removed from the tubule by **resorption** (→ **B1**) has a clearance ratio of less than 1.0 (e.g. Na^+, glucose). Other substances are removed from the blood by tubular cells and are added to the urine by **transepithelial secretion** (→ **B2**). In the case of **PAH** (→ p. 122), this secretion is so effective that after only a single passage through the kidney 90% of the PAH is removed from the blood (**extraction rate**). The PAH clearance thus provides an approximate measure for the renal plasma flow (RPF, → p. 122) and is roughly five times the value of C_{in} or the GFR. Still other substances are formed in cellular metabolism and are **secreted across the luminal cell membrane** (e.g. NH_3/NH_4^+). In both of these cases, the amount appearing in the urine is greater than the amount filtered, and the clearance ratio is greater than 1.0.

The absolute **rate of resorption** or **of net secretion** can be calculated as the difference between the rate of filtration (P_x · GFR) and the rate of urinary excretion ($U_x \cdot \dot{V}_u$).

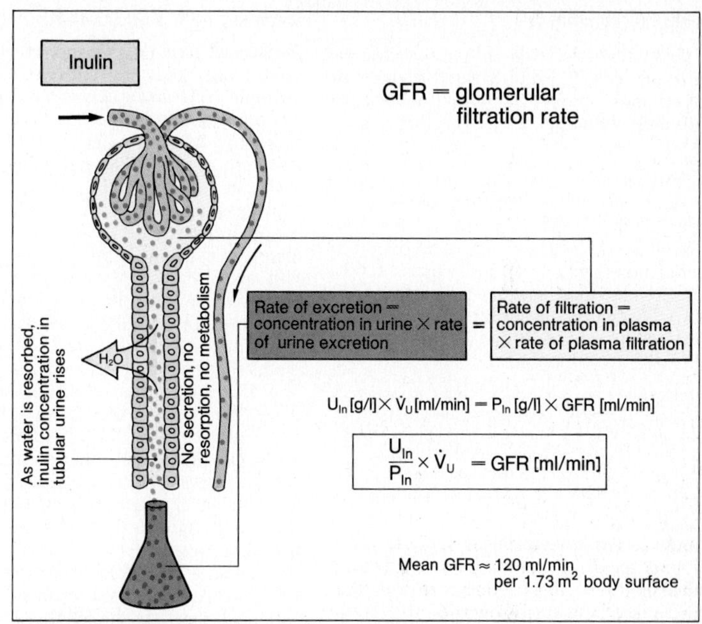

A. Glomerular filtration rate = inulin clearance

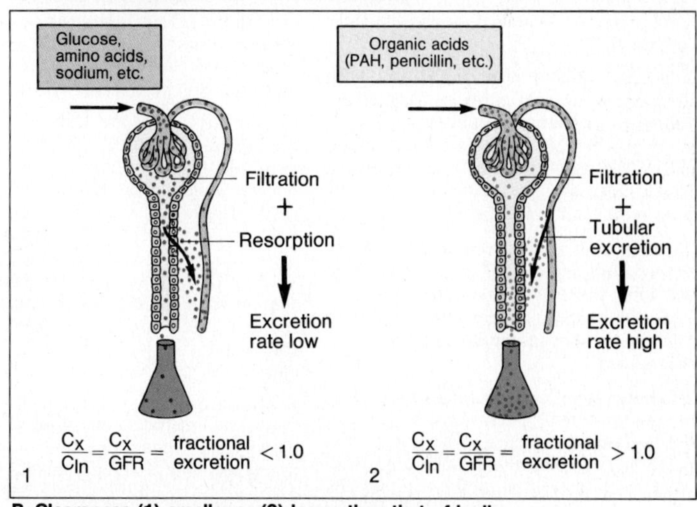

B. Clearances (1) smaller or (2) larger than that of inulin

Tubular Transport I

Substances dissolved in the plasma enter the urine either by filtration at the glomerulus or by secretion across the tubule wall. They can leave the tubule again by re(ab)sorption (→ **A**).

1. **Filtration:** At the glomerulus approximately one fifth of the plasma water is filtered off. The glomerular filter (→ p. 120) also allows the free passage of all dissolved substances with a molecular radius r < 1.8 nm (molecular mass ≈ 15000 Dalton). Normally, substances with r > 4.4 nm (molecular mass ≈ 80000 Dalton) are not filterable (e.g. globulins).

Molecules with 1.8 nm < r < 4.4 nm are only partially filterable. Negatively charged particles (e.g. *albumin*, r = 3.5 nm) pass through less readily (albumin to 0.05%) than neutral substances. The probable reason is that negative charges on the walls of the glomerular filter (→ p. 120) exert a repellent effect on anions.

Some of the low-molecular substances are bound to larger plasma proteins (**protein binding**; → p. 10), and are virtually nonfilterable in this form (→ **B**: T).

The glomerular filter is probably freed of "debris" by the phagocytic (→ p. 66) mesangial cells of the glomerulus.

2. **Re(ab)sorption** (→ **A**, w, x, y): After filtration, water and many substances are resorbed. These are electrolytes (Na^+, Cl^-, K^+, Ca^{2+} HCO_3^-, phosphate, etc.), amino acids, uric acid, lactate, urea, peptides, proteins, ascorbic acid, glucose, and many more (→ **C** and p. 128ff.).

3. **Active transcellular secretion:** (→ **A**, z): *Endogenous products of metabolism* (e.g. uric acid, glucuronides, sulfates) and *exogenous substances* (penicillin, diuretics, PAH, and other xenobiotics) are excreted in this manner (→ **C**).

4. **Cellular secretion:** Some substances (e.g. ammonium = NH_4^+, H^+-ions, hippurate, and mercapturates) are products of the metabolism within the tubule cell, from which they enter the tubular lumen by cellular secretion. Whereas NH_4^+ diffuses passively in its nonionic form (NH_3) into the tubular lumen (→ **A**, v), H^+ ions are actively secreted (→ **A**, u and p. 144ff.).

Active and passive transport processes (→ p. 8ff.) are frequently closely linked. For example, $\mathbf{H_2O}$ can be *passively resorbed* if an osmotic gradient (→ p. 335f.) is created by the active resorption of a dissolved substance (e.g. Na^+). The **resorption of water** leads, on the one hand, to the so-called "solvent drag" (→ p. 10 and p. 132), and on the other hand to the *concentration* of other dissolved substances present in the tubule (e.g. **urea**) that can then, in turn, be passively resorbed into the blood down a concentration gradient. In the case of ions, there are additional *electrical effects*: when Na^+ is reabsorbed, either an anion has to follow it (e.g. $\mathbf{Cl^-}$ in the proximal tubule; → p. 132), or a cation has to be secreted (e.g. $\mathbf{K^+}$ in the distal tubule; → p. 148ff.). **Glucose, amino acids, phosphate**, and other substances are *actively transported* in most cases, the energy for this type of transport being obtained by close coupling with the Na^+ transport into the cell (**co-transport**); this type of **secondary active transport** is described on pp. 11 and 128. **PAH** is transported across the basolateral membrane into the cell by a "tertiary"-active process. First of all, dicarboxylates (e.g. glutarate) enter the cell by a secondary-active process (→ p. 129, B, c), after which, in exchange for PAH (antiport, → p. 11), they leave the cell and enter the blood again.

Certain substances are reabsorbed by *passive diffusion* (e.g. urea). The permeability of such substances depends in part upon their *lipid solubility*. The nonionized form of weak electrolytes is more soluble in lipids than the ionized form and can thus cross the membrane more easily (**nonionic diffusion**, → **A**: "$Y^- \rightleftarrows Y^0$"). The *pH value* of the tubular urine therefore also influences passive resorption. Molecular size also plays a role in diffusion since the smaller the molecule the better it diffuses (→ p. 8).

Diffusion pathways through the tubule epithelium include both **transcellular** and **paracellular** routes. Especially the "tight" junctions of the proximal tubule (→ p. 8) are relatively leaky and allow paracellular diffusional transport of smaller molecules and ions to some extent.

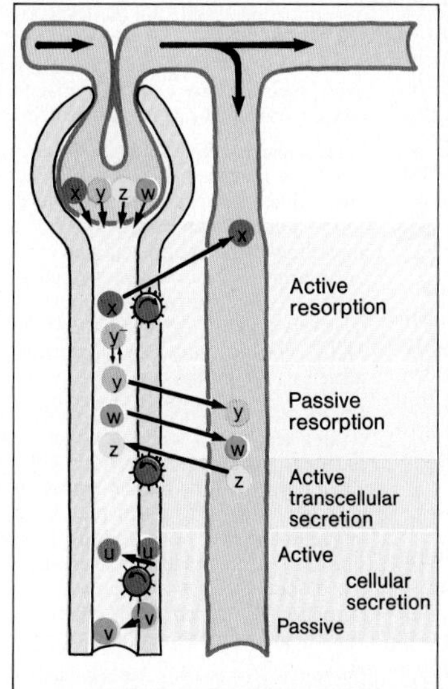

A. Filtration and tubular transport

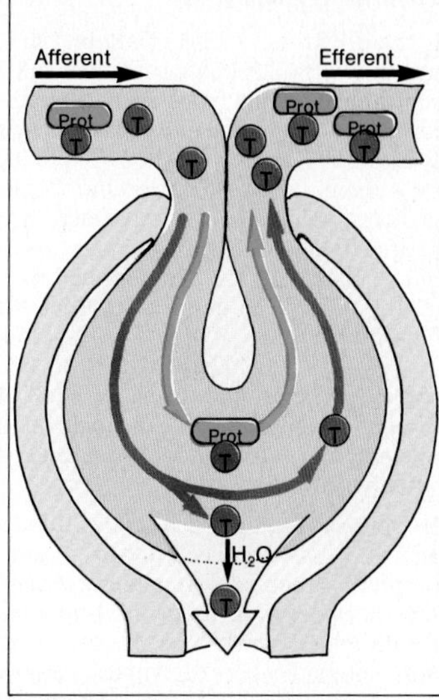

B. Protein binding and filtration

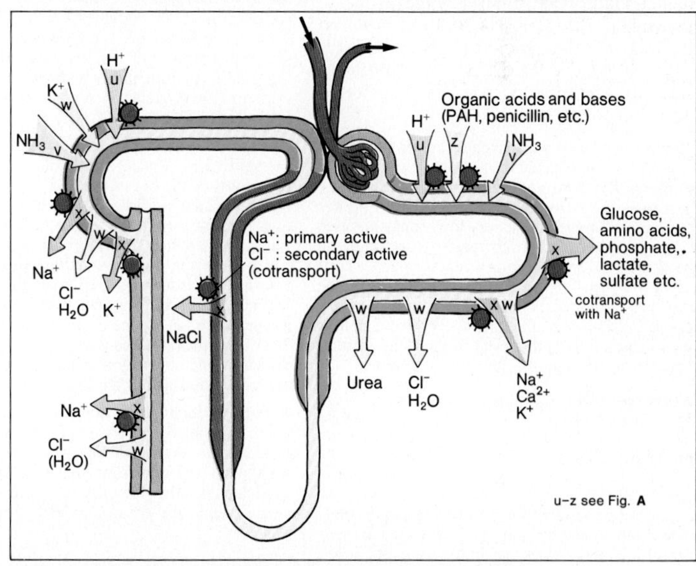

C. Locus of transport processes in the nephron (simplified)

Tubular Transport II

If the GFR (→ p. 124) of about 180 l/d in humans is multiplied by the plasma concentration of the substance filtered, the resulting **"load" (quantity filtered/time)** is 160 g/d for glucose and about 70 g/d for the total amino acids. If not bound to plasma proteins, solutes of this molecular size are filtered freely. Their glomerular "permeability" thus is 100%. Despite the low "permeability" of the glomerular filter for albumin (about 0.01–0.05%) roughly 1–3 g/d (180 l/d · 42 g/l · 0.0001 to 0.0005) still enter the primary urine. The job of the resorption systems of the nephron is to prevent the excretion of these and other valuable substances.

D-Glucose is resorbed into the tubule cell against a concentration gradient ("uphill" = actively). It makes use of one of the transport systems in the brush border membrane (*"carrier"* → **B, a**) via which Na^+ enters the cell ("downhill"): *secondary active cotransport with* Na^+. The Na^+ gradient is maintained, in turn, by the (primary) active Na^+ "pump" (**Na^+-K^+-ATPase**) on the basolateral membrane (→ **B**). Glucose is believed to leave the cell on the antiluminal side by a process of so-called facilitated diffusion (→ **B, b** and p. 11).

Glucose is normally resorbed to about 99%. The concentration profile in the proximal tubule (→ **A, a**) shows that resorption takes place rapidly at the beginning of the tubule, but towards its end the concentration settles to a very low, constant value (C_∞), at which active resorption from, and the passive "leak" into the lumen exactly balance each other. Further "downstream", glucose can still be resorbed when its concentration rises slightly above C_∞ due to the resorption of water. The **leak** mentioned above is the result of passive transport between or through the cells into the lumen.

Active resorption, J_{act}, of glucose can be saturated by increased glucose concentrations and can be described by the Michaelis-Menten equation (→ pp. 11 and 333):

$J_{act} = J_{max} \cdot C(K_m + C)$, where

J_{max} is the maximum transport rate, K_m is the half-saturation constant, and C is the actual concentration of glucose in the tubule lumen. If C rises above 10 mmol/l in the plasma and primary urine (e.g. in *diabetes mellitus*), the resorption system becomes saturated (J_{act} approaches J_{max}) and the concentration profile (→ **A**) levels off to such an extent that glucose appears in the final urine (**glucosuria**). Besides this *prerenal* glucosuria, a *renal* glucosuria has also been observed. Here, the tubular transport mechanism is defective.

In principle, **L-amino acid** resorption is very similar to that of glucose (→ **A** and **B**). There are about seven different transport systems (mostly Na^+ cotransport), which means that J_{max} and K_m vary according to the amino acid and carrier. L-valine, for example, is resorbed very rapidly (→ **A, b**); L-glutamine, like glucose (→ **A, a**); and L-histidine, relatively slowly (→ **A, c**). Of the quantities filtered, 99.9%, 99.2%, and 94.4%, respectively, are returned to the blood.

An increased excretion of amino acids (**hyperaminoaciduria**) can originate either *prerenally*, with elevated plasma concentrations (saturation of reabsorption, see above), or *renally* as a result of defective transport, which may be either specific (e.g. *cystinuria*) or nonspecific (*Fanconi syndrome*). The tubule cells can also take up amino acids from the blood, (→ **B** and **C**), probably to serve as nutrients for the cell itself, especially in the more distal regions of the nephron.

Phosphate (→ p. 144), **lactate, citrate**, and many other substances are reabsorbed in the proximal tubule by secondary active Na^+ cotransport. Na^+ cotransport of $\mathbf{Cl^-}$ (and K^+) takes place in the distal nephron (→ p. 132). **Uric acid** and **oxalate** are both reabsorbed and secreted concurrently: uric acid is predominantly reabsorbed (only 10% excretion), and oxalate is predominantly secreted (excretion > filtered load).

Oligopeptides such as glutathione and angiotensin are split so rapidly in the tubular lumen by the peptidases of the brush border (γ-glutamyl transferase, aminopeptidase M, etc) that they can be totally reabsorbed in the form of free amino acids. The same applies to the disaccharide *maltose*, which is broken down by maltase to glucose and resorbed in this form (→ **C**). Dipeptides resistant to luminal hydrolysis (e.g. the normal plasma constituent *carnosine*) are resorbed as intact molecules. They enter the tubule cells by a cotransport system driven by the inwardly directed H^+ gradient and are hydrolyzed *within* the cell.

Proteins (albumin, lysozyme, β-microglobulin, etc) are reabsorbed by *endocytosis* and undergo mainly intracellular lysosomal "digestion" (→ **D**). Normally, this type of resorption is saturated, so that any increase in the glomerular protein permeability (e.g. in nephrotic syndrome) leads to **proteinuria**.

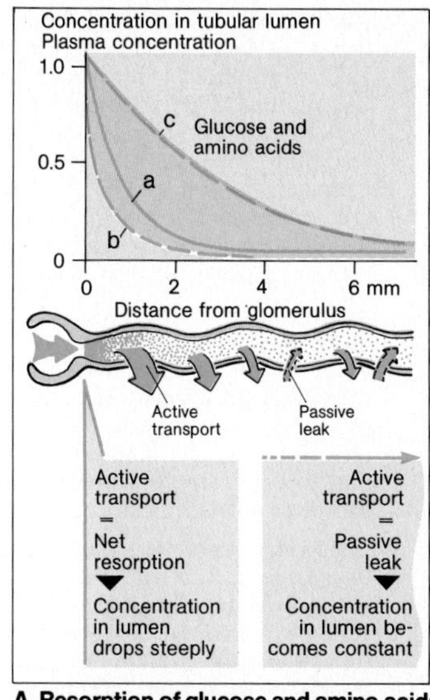

A. Resorption of glucose and amino acids

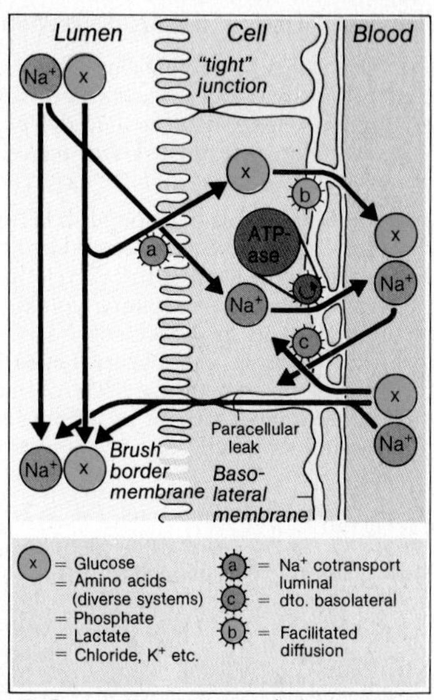

B. Secondary active Na^+ cotransport

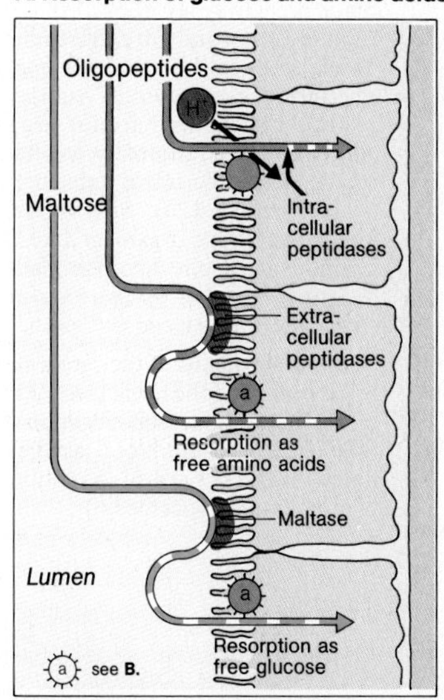

C. Resorption of oligopeptides and maltose

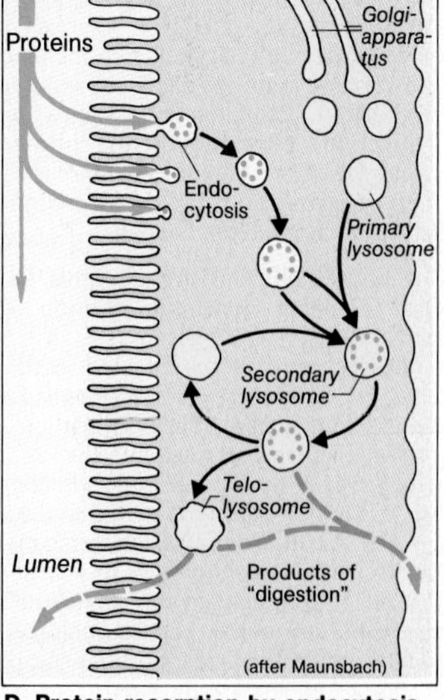

D. Protein resorption by endocytosis

Selection, Detoxification, Excretion

The **diet** (→ **A1**) contains not only substances that are useful to the organism for its growth and energy balance, but also *physiologically inert* or even *injurious materials*. They may also be present in the inspired air.

The organism is able to distinguish between the useful and the undesirable substances and can reject the latter at three levels: in the gastrointestinal tract, in cellular uptake and metabolism, and in the kidneys. The first distinction is made in the **process of eating**: the *smell* and *taste* (→ p. 296) of many *noxious substances* prevent their being consumed at all, or quickly lead to *vomiting* (→ p. 204).

In the **gastrointestinal tract** (→ **A2**) the **specificity of** the digestive **enzymes and mechanism of absorption** ensure that predominantly useful substances are broken down and absorbed (→ p. 218 f.), whereas physiologically inert substances cross the intestinal epithelium usually to only a very small extent and are mainly evacuated in the feces. From the intestine, the absorbed substances pass to the **liver**, where as much as 95% may be extracted from the blood in the portal vein during the first passage (*first pass effect*). The liver cell makes a further distinction between useful substances, which are either stored or metabolized, and inert or harmful substances, the majority of which undergo **detoxification**.

Various mechanisms are available to this end: following enzymatic addition of an OH^-- or a COO^--group (→ **A3**), the substance may be conjugated with *glucuronic acid, sulfate, acetate*, or *amino acids* (→ **A4** and p. 214). The conjugation products are actively secreted into the **bile**, with which they enter the intestine, whence they are to a major extent evacuated in the **stool** (→ **A5**). Another conjugation mechanism in the liver involves *glutathione* to which, with the aid of specific enzymes, toxic and carcinogenic substances such as chloroform, methyl iodide, epoxides, naphthalene, phenanthrene, to name only a few, are conjugated, and subsequently excreted by the **kidney** in the form of *mercapturates* (→ p. 128).

The **lungs** also play the role of an *excretory organ* (→ **A6**). They receive the blood that has passed through the liver en route from the intestine. The lungs are especially efficient in trapping *fat-soluble substances* (e.g. serotonin, methadon), which are then inactivated and to some extent *excreted* in the *bronchial mucus*. In this way, the *brain*, which is particularly sensitive to such substances, is protected from damage.

Some of the compounds taken up in the intestine enter the systemic circulation in their original form and become available to the **cells of other organs**. Here, too, a process of selection takes place. Brain and muscle cells, for example, take up the substances that are important for their own metabolism by means of specific transport mechanisms.

The **kidney**, whose blood flow is equivalent to a large part of the cardiac output (→ p. 154), exerts an efficient *regulating effect on the composition of the blood*. The renal tubules behave similarly to the intestinal tract: the unwanted substances from the blood are barely resorbed at all after glomerular filtration and are excreted in the urine. This applies to the metabolic end products of nitrogen-containing substances (→ p. 146). Useless or harmful organic acids and bases are additionally secreted into the tubule by specific active transport processes (→ **A7** and p. 126). Substances that are important for the organism (like D-glucose, L-amino acids, and so on) are, just as in the intestine, resorbed *by specific transport systems* and thus escape excretion.

The **choroid plexus** of the cérebral ventricles in the brain and the uveal tract of the **eye** can transport the same organic anions out of the tissue and into the blood as the liver and kidney transport out of the blood and into bile or urine, respectively.

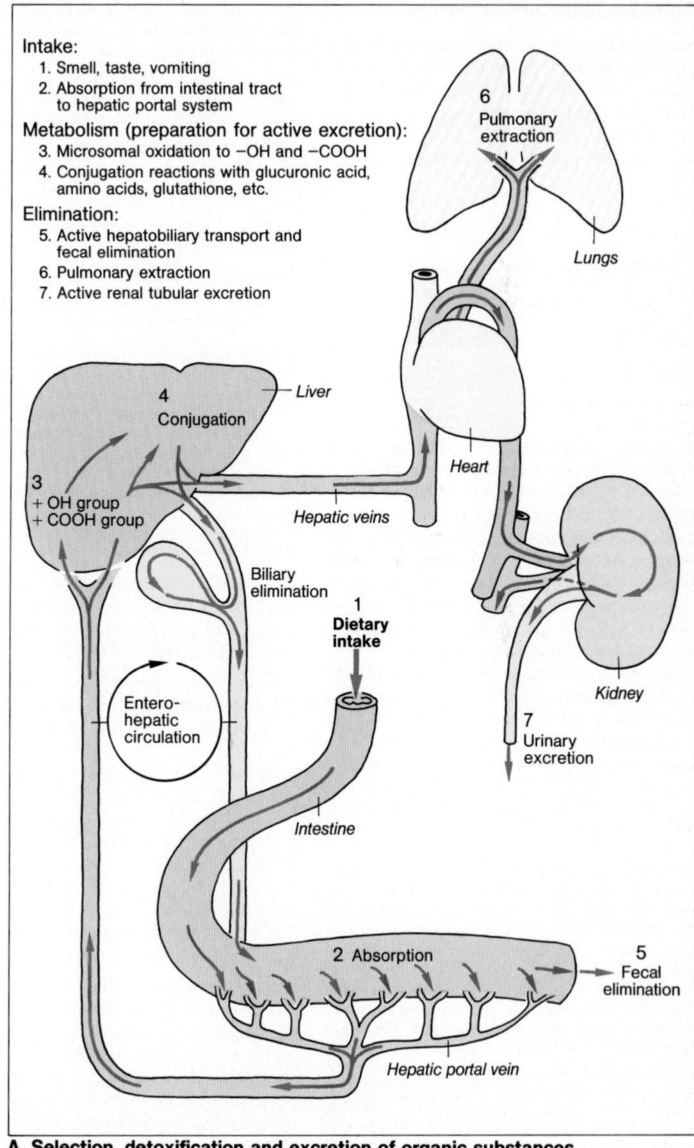

A. Selection, detoxification and excretion of organic substances

Renal Handling of Na^+ and Cl^-

NaCl is taken in the diet in amounts that vary, according to personal taste, from 8 to 15 g/d. Many hypertensive subjects consume NaCl at the upper end of this range. The 180 l of plasma water that are filtered daily at the glomerulus contain roughly 1.5 kg NaCl (*load*; → p. 128). Of this, less than 1 % is normally excreted, the remaining 99 % being **resorbed** from the tubules. The quantity of Na^+ excreted is exactly adjusted by the kidney to match the salt intake in such a way that the Na^+ concentration and thus the extracellular volume (→ p. 138) of the body remain constant.

Whereas metabolic energy is required for part of the tubular Na^+ transport, Cl^- is resorbed passively or by secondary active transport. NaCl resorbed over the entire length of the **proximal tubule** is passively followed by water, whereby the fluid resorbed, as well as that remaining in the tubule, is at all times virtually *isotonic* with the plasma. At the end of the proximal tubule about 60 % to 70 % of the filtered Na^+ and water have already been returned to the blood (→ **B**). In the early proximal tubule, H^+ ions are secreted in exchange for Na^+. Thus, HCO_3^- is reasorbed rapidly (about 85 % to 90 % are resorbed by the time the end of the proximal tubule is reached). The resorption of Cl^- lags behind, establishing a lumen to blood chemical gradient for Cl^- (less than 60 % of the Cl^- filtered are resorbed in the proximal convoluted tubule).

The **mechanism** of **"proximal" NaCl resorption**. The active transport mechanism for Na^+, by means of which it is "pumped" out of the cell and into the interstitium, is the Na^+-K^+-ATPase localized in the basolateral membrane (*basal labyrinth*; → p. 5 f.). This keeps the Na^+ concentration within the cell low, with the result that further Na^+ can enter the cell passively via the brush border membrane. The high electrochemical Na^+ gradient between lumen and cell maintains this influx and drives "carriers", with the help of which $\mathbf{H^+}$ ions are secreted into the lumen (→ p. 144) and *glucose, amino acids*, and so on (→ p. 128) are resorbed from the primary urine. A considerable portion of the Na^+ in the proximal tubule is *passively* resorbed and, in contrast to its active transport, this occurs chiefly via the spaces *between* the tubule cells (*paracellularly*). Two mechanism are responsible for the passive, transepithelial resorption of Na^+ and Cl^-:

(a) Cl^- *diffuses* from the lumen into the interstitium along its chemical gradient, and Na^+ follows the Cl^- to maintain electroneutrality.

(b) all resorbed solutes (including Na^+ and Cl^-) are followed trans- and paracellularly by *water*. The latter drags Na^+ and Cl^- ions (and urea for example) out of the lumen because their reflexion coefficient in the proximal tubule < 1.0 (**solvent drag** → p. 10).

In the *thick ascending limb* (TAL) of **the loop of Henle** a further 15–20 % of the filtered NaCl are actively resorbed.

Here, too, the *primary active transport of* Na^+ is mediated by the basolateral Na^+-K^+-ATPase; on the luminal side there is a common carrier for Na^+, K^+, and $2Cl^-$ (*secondary active cotransport*; → p. 149, B2). This carrier can be inhibited by so-called **loop diuretics**, for example, furosemide (→ p. 142). The epithelium of the TAL has a very low permeability for water, in addition to which the passive permeability for Na^+ and Cl^- is also low, so that these ions cannot diffuse back into the tubular lumen.

The Na-Cl cotransport in the TAL therefore produces a hypotonic urine in the tubule lumen and a hypertonic milieu in the surrounding interstitium (→ **B** and p. 134 ff.).

The remaining 10–20 % of the Na^+ reaches the **distal convoluted tubule** and finally the **collecting duct**. In both of these segments, Na^+ (followed by Cl^-) is again actively resorbed. Chiefly in the *cortical collecting ducts*, Na^+ resorption (through Na^+ channels or Na^+/H^+ antiporters in the luminal membrane) is influenced by **aldosterone** (→ p. 140 and p. 150), which thus controls the excretion of NaCl in the urine. Depending upon the salt and water intake, fractional **NaCl excretion** fluctuates between 5 % and 0.5 % of the quantity filtered.

Apart from aldosterone, *GFR, renal medullary blood flow, sympathetic renal nerves, atriopeptin* (→ p. 140), and probably a *natriuretic hormone* are involved in the regulation of NaCl excretion. The sum of these, part stimulatory, part inhibitory (some in completely different segments of the tubule), determine the final amount of Na^+ (and Cl^-) excreted.

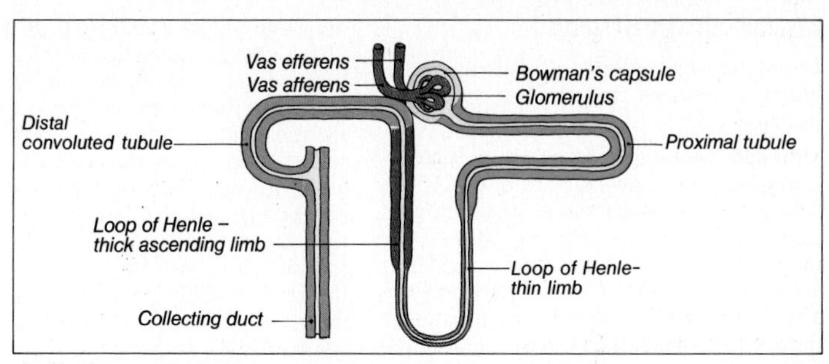

A. Nephron segments

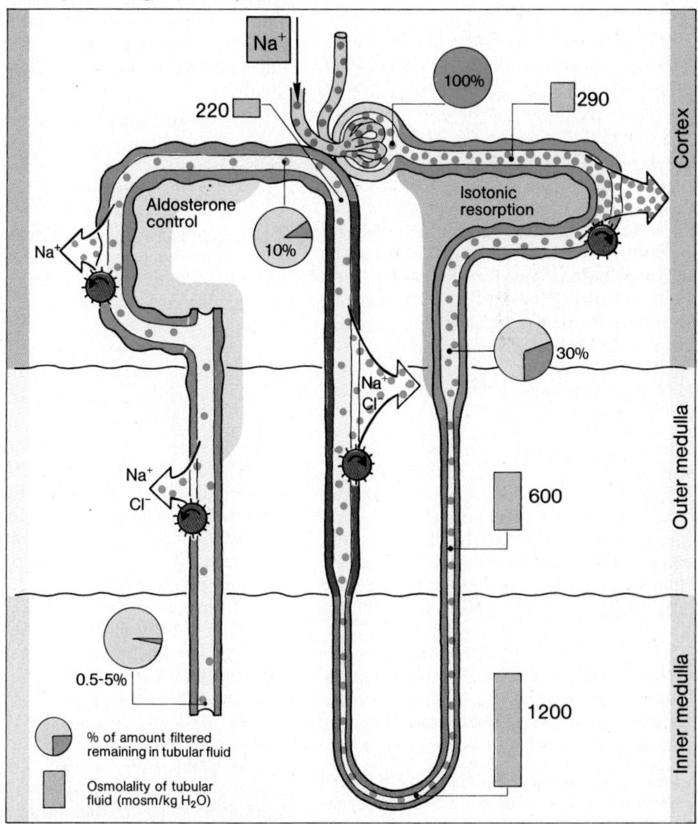

B. Na^+ resorption in the nephron

Countercurrent Systems

Countercurrent systems are widely distributed in animals and serve a variety of functions.

1. **Simple exchanger.** Consider a **heat** exchanger in which two streams of water, one cold (0 °C), the other hot (100 °C), flow in parallel at the same rate (→ **A1**). At the input end, the heat gradient is steep and heat exchange is rapid. The gradient decreases along the length of the system until, at the output end, temperature has equilibrated; 50 °C is the limiting temperature for both the cold and the hot stream.

2. **Countercurrent exchanger**. If the direction of flow in one stream is reversed, there will be a temperature gradient at all points, allowing heat to be exchanged along the whole length of the two streams (→ **A2**).

In the same way, **water** may be exchanged instead of heat. Changes in rate of flow may alter the magnitude of the exchange and of the gradient, but the general principle remains. Instead of heat, **solutes** may also be exchanged along a concentration gradient. Such an exchange takes place, for example, in the **portal triads of the liver** where the bile flows in the opposite direction to the arterial and portal blood; in this way, some of the substances released with the bile may reenter the blood.

The same conditions apply when a single stream reverses itself as in a **hairpin loop** (→ **A3**). If the tip of the loop is in contact with a heat sink (ice), heat will be lost. A temperature gradient will be established, however, which will allow countercurrent heat exchange, and the returning stream will be only slightly cooler than the entering stream. This mechanism enables ducks, storks, and other birds to stand on ice without losing much heat. In human beings, too, a similar exchange of heat between arteries and veins in the extremities plays a role in heat balance (→ also p. 194f.).

In the **vasa recta** of the **kidneys** (→ p. 122) a comparable *countercurrent exchange* exists for the plasma water and substances dissolved in the blood. A prerequisite is an increasing hypertonicity of the renal medulla (see below). For osmotic reasons some of the plasma water leaves the descending limb of the vasa recta and reenters the ascending limb, thus, in effect, leaving the renal medulla unaffected (→ **A4**). Conversely, solutes from the ascending, venous limb, in which they are conveyed from the hypertonic medulla, are continuously returned to the descending limb. This applies especially to substances that are formed in the renal medulla (e. g. CO_2) or resorbed from the collecting duct there, such as urea (→ p. 137, B).

Consequently, despite the essential blood supply, the high osmolality and tonicity (→ p. 335ff.) of the renal medulla (see below) is, under normal conditions, subject to little alteration.

3. In a **countercurrent multiplier system**, a concentration gradient between the two limbs of a loop is established by means of the **expenditure of energy** (→ **A5**). The relatively small gradient between the two streams (single step) is magnified by the countercurrent to a relatively large gradient along the limb of the loop involved. The gradient is directly proportional to the length of the hairpin loop and to the concentration gradient between the two limbs, and inversely porportional to (the square of) the flow rate.

From the **thick ascending limb of the loop of Henle** (→ **A5** and **A6**), continuous **active transport** (i.e. with the expenditure of energy) of Na^+ and Cl^- takes place into the surrounding interstitium (→ p. 132). In contrast to the descending loop of Henle, whose tonicity (→ **A5**) is in equilibrium with that of the interstitium, the ascending loop has a **very low permeability to water**. Thus the active transport of NaCl produces the *single step gradient* between the ascending loop of Henle, on the one hand, and the descending loop and the interstitium of the renal medulla on the other. Since it is the high tonicity of the interstitium that is responsible for the withdrawal of water from the collecting duct (→ p. 136), active NaCl transport can be looked upon as the **driving event in the concentration mechanism of the kidney.**

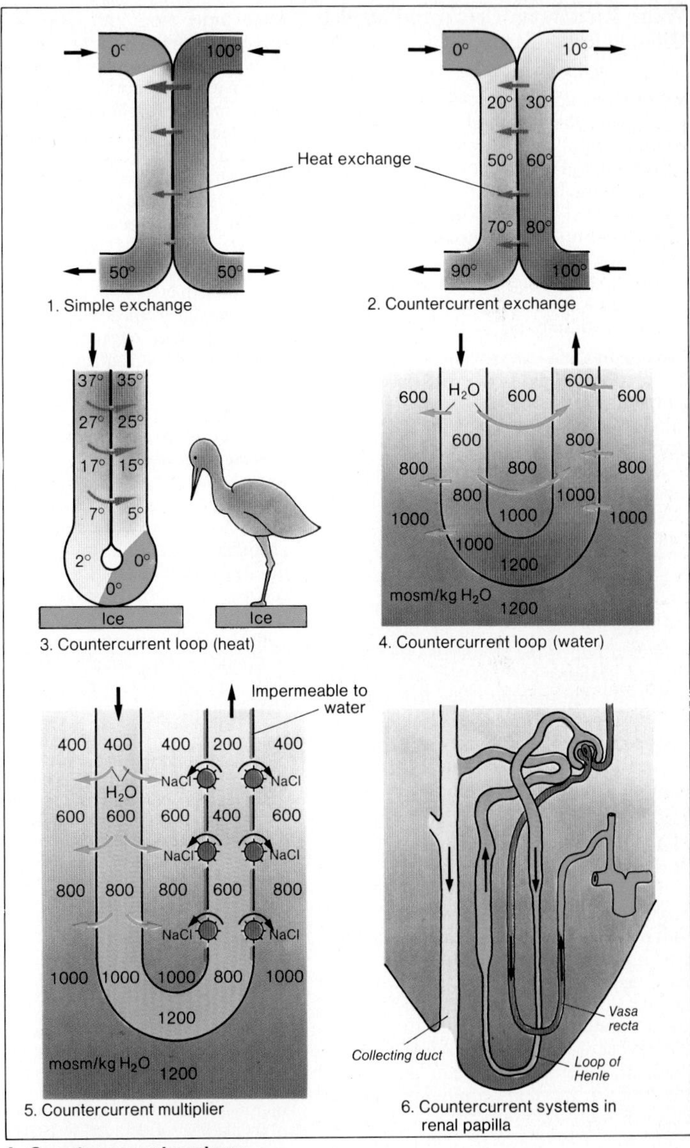

A. Countercurrent systems

Water Resorption. Concentration and Dilution in the Kidney

Plasma water is filtered in the kidney at about 120 ml/min or 180 l/d (GFR; → p. 124). At the glomerulus, the filtrate is isotonic with the plasma (osmolality is 290 mosm/kg H_2O). In contrast, the volume of the terminal urine is about 1.5 l/d and its osmolality (→ p. 335 ff.) may vary from 50 to 1200 mosm/kg H_2O, according to water intake; the urine may be hypotonic with a flow of up to 18 ml/min (**diuresis**) or hypertonic with a flow of only a few tenths of a ml/min (**antidiuresis**).

Proximal tubule. Approximately **two thirds** of the tubular fluid is **resorbed** between the glomerulus and the end of the proximal tubule (→ **A**). Resorption of Na^+ is the primary driving event in water resorption; it is followed by resorption of Cl^-, HCO_3^-, etc. and establishes a (very small) osmolal gradient along which an osmotically equivalent volume of water diffuses passively out of the lumen (*isotonic resorption*). Oncotic pressure (π; → pp. 124, 158, 335 ff.) in the peritubular capillaries provides an additional force for water movement. When water is filtered out of the glomerular capillaries, plasma proteins become concentrated (→ p. 127, B); the oncotic drive for water resorption is therefore especially high in the early sections of peritubular capillaries. The greater the fraction of water filtered, the higher the oncotic pressure.

Loop of Henle. The **thick ascending limb** (TAL) of the loop of Henle actively transports NaCl into the medullary ECF (→ **A** and p. 134). Since cells of the TAL are relatively impermeable to water, the fluid that remains in the tubule is made **hypotonic**. NaCl, which is delivered into the medullary ECF, establishes an osmotic gradient that is greater towards the papillary tip. **Urea** and other solutes also contribute to this gradient. The loop of Henle dips into this gradient. Its tubular fluid equilibrates with the ECF; in the thin descending limb, net efflux of water occurs. This water is mainly returned to the vasa recta, that is, there is a net resorption of water (→ **A**).

The convoluted **distal tubule** receives hypotonic urine from the loop of Henle. In presence of **antidiuretic hormone (ADH)**, i.e. in *antidiuresis*, water is resorbed along this section until the tubular fluid becomes isotonic (osmotic equilibration with the isotonic environment in the renal cortex). Thus, isotonic fluid is delivered to the **collecting duct**, where the final adjustment of urinary volume and concentration takes place (→ p. 132). In the absence of ADH, i.e. in *water diuresis*, (→ p. 140), water permeability of the "late" distal tubule as well as of the collecting duct is low. Therefore, water is not removed from the urine, and since the NaCl transport continues, the osmolality even sinks below that of the "early" distal tubule.

However, when the need for water conservation develops, **ADH** is secreted and **increases the permeability of the collecting ducts to water**. Water (→ **A**) diffuses out of the medullary collecting ducts into the hypertonic environment of the medullary ECF. The absorbed water in the medullary ECF equilibrates with blood in the vasa recta and is carried away. In this way, urine in the collecting ducts is concentrated. Water is removed from the urine until the osmolality of the urine (U_{osm}) is more than four times that of the plasma (P_{osm}), that is, $U/P_{osm} = 4.5$ (antidiuresis).

Urea plays an important role in the concentration of the urine; the details of the mechanism, however, are not yet completely understood. The distal tubule and the cortical collecting ducts are only slightly permeable to urea, whose concentration therefore increases steadily in these sections (→ **B**). The parts of the collecting ducts near the papilla, on the other hand, are readily permeable to urea, some of which therefore diffuses back into the interstitium (where it helps maintain a high osmolality), and the rest is excreted (→ **B** and p. 146). Excretion of urea is increased in diuresis.

Medullary osmolality can be decreased by a number of factors: increased medullary blood flow can wash out solutes (NaCl and urea) from the medulla; osmotic diuresis abolishes the medullary osmotic gradient; water diuresis reduces the medullary gradient, limiting it to the inner medulla; loop diuretics block the NaCl transport in the thick ascending limb of the loop of Henle; ADH deficiency (see above).

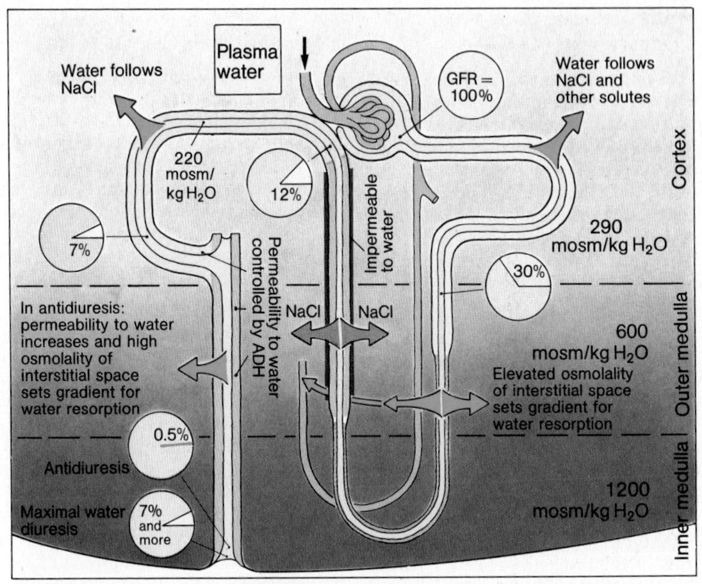

A. Fluxes of water in the nephron

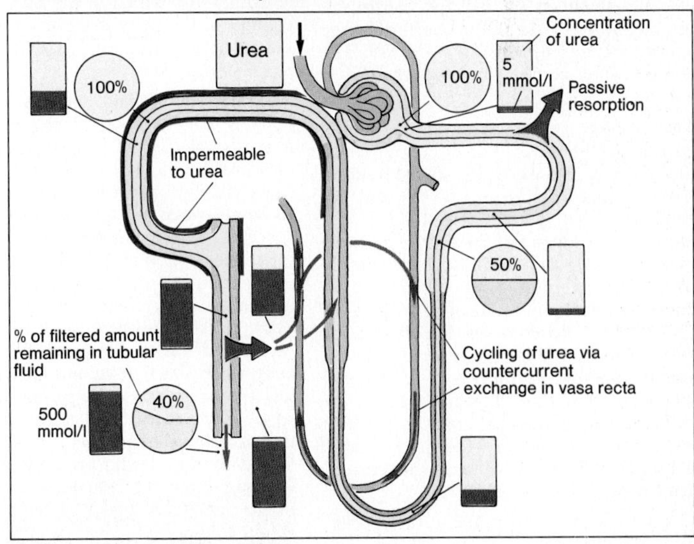

B. Fluxes of urea in the nephron

Water Balance, Body Fluid Compartments

Without water, life on earth could not exist. The cells of all multicellular animals not only contain water (intracellular fluid, **ICF**) but are also surrounded by fluid, the **internal environment** (→ p. 1). This fluid environment, the extracellular fluid or **ECF**, is the connecting link between the external world and the inner space of the cell; it carries nutrients, disposes of wastes, and carries humoral messages.

The water content of the body is normally very exactly regulated (regulation → p. 21) and is the result of the **water balance**: that is, *water losses must continuously be balanced to match water intake and production* (→ **A**). This is achieved via the ECF (with the exception of the water formed in cellular metabolism). The average **water intake** (2.5 l/d) consists of (→ **A**): (a) *drinks* (roughly 1.3 l/d); (b) *water in solid food* (0.9 l/d); and (c) *water of oxidation* arising from metabolism (→ p. 199, C). In the mean this is exactly balanced by the **water losses** (→ **A**), which on an average consist of: (a) the *urine* (1.5 l/d); (b) water lost in the *expired air* and from the *skin* (→ p. 193, B3); and (c) water contained in the *feces* (→ **A** and p. 231, C).

The average **water turnover** in adults amounts to about 1/30 of the body weight (2.4 l/70 kg), whereas in babies the fraction is much higher (1/10; 0.7 l/7 kg) and accounts for their greater sensitivity to disturbances in the water balance. The turnover of water may deviate considerably from the quantities given above, but at intervals the essential state of balance has to be restored. A march in very hot weather, or work in an iron foundry, for example, may lead to copious **sweating** (→ p. 192; several liters per hour), which has to be made up for by the intake of an equal quantity of water (and salt). Conversely, if too much fluid is imbibed this has to be balanced by increased urine excretion (→ p. 140).

A water deficit results in **thirst**; this mechanism is under the control of the *thirst center in the hypothalamus*. Thirst is elicited by an elevated osmolality of the body fluids and by an increased angiotensin II concentration in the CSF (→ p. 140 and p. 290).

Depending upon age and sex, the contribution of water to the body weight (1.0) lies between 0.46 and 0.75 (→ **B**). In *babies*, the proportion of water is 0.75; in a *young male*, 0.64 (*female*: 0.53); in an *elderly male*, 0.53 (*female*: 0.46). The differences between the sexes (as well as individual differences) are mainly due to the *proportion of fat* in the body: most tissues have an average water content of 0.73 (young adults) whereas in fat it is only about 0.2 (→ **B**).

With an average total water content of about 0.6 (body weight = 1), roughly 3/5 (= 0.35 of body weight) of this water is in the cells (**ICF**) and 2/5 (= 0.25 of body weight) in the extracellular space, the latter containing **interstitial water** (0.19), **plasma water** (0.045), and the so-called *transcellular fluid* (CSF, intestinal tract, etc: 0.015) (→ **C**). Plasma differs from the rest of the ECF chiefly on account of its *protein content*; the ICF differs considerably in its *ionic composition* from the ECF (→ p. 65, B). Since Na^+ is mainly distributed in the ECF, the Na^+ content of the body is the most important factor determining the ECF volume (→ p. 132 and p. 140).

Measurement of fluid volumes within the body is usually based on the principle of *indicator dilution*. Provided that the indicator substance (which is injected into the blood circulation) confines itself to the space to be measured (→ **C**), the volume of this space (l) = the quantity of indicator injected (g)/concentration of the indicator (g/l) after its distribution in the space in question (measured in a blood sample). *Inulin*, for example, is an indicator for the greater part of the **ECF volume**; *antipyrin*, for the **total body water**. The ICF volume is thus more or less the antipyrin space minus the inulin space. *Evans Blue* is used as an indicator for the **plasma volume** since it is completely bound by the plasma proteins. The **blood volume** can then be calculated with the help of the hematocrit (→ p. 65, A) as follows: Blood volume =
(plasma volume)/(1 – hematocrit).

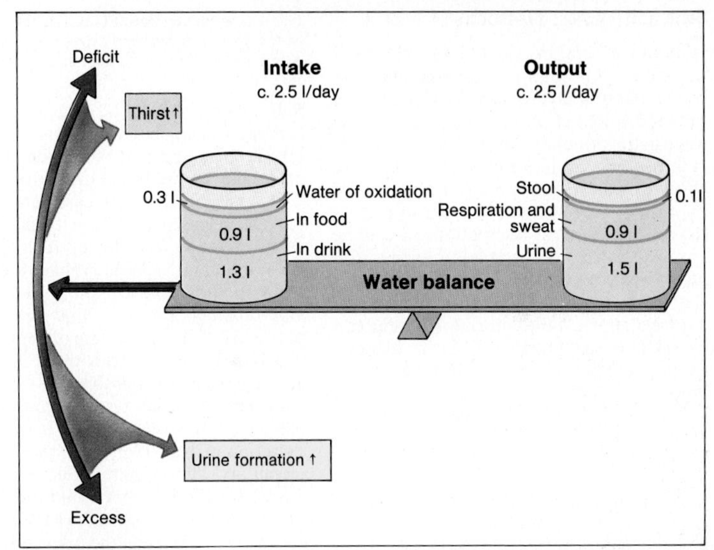

A. Water balance

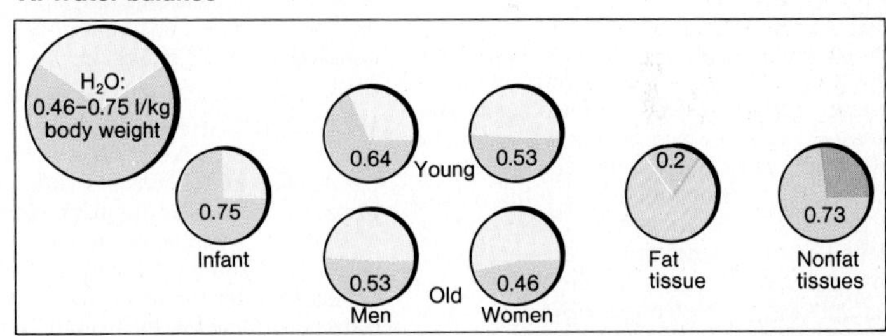

B. Total body water

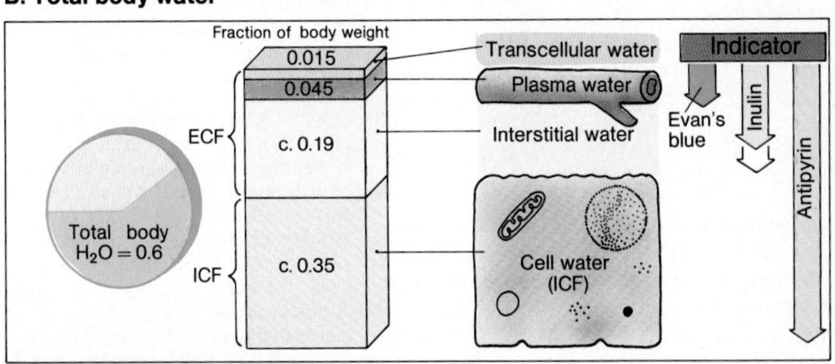

C. Body fluid compartments

Salt and Water: Hormonal Control

With rare exceptions, the ICFs and ECFs of the body (→ p. 138) have an **osmolality** (→ p. 335) of around **290 mosm/kg H_2O**. Intake of NaCl or loss of water, for example, cause a rise in the osmolality of the **ECF**. Since this is in osmotic equilibrium with the **ICF**, an efflux of water from the latter ought to be the result (→ p. 143, A). In order to protect the cells from larger fluctuations in volume and osmolality, the osmolality of the ECF has to be strictly controlled. This function is shared by **osmoreceptors** (mainly in the hypothalamus), the hormone **adiuretin** (= **ADH** = vasopressin), and the **kidneys**, the target organs (→ p. 136). If, for example, intake of NaCl is too high, the control system normalizes the osmolality by means of water retention. The price paid is an increased ECF volume **(ECV)**. Thus the *NaCl content of the body determines the size of the ECV*, and since the hormone **aldosterone** controls the excretion of NaCl (→ p. 132 and p. 150) it consequently also regulates the ECV.

Water deficit (→ **A 1**): If the water lost by the body (in sweat, expired air, etc) is inadequately replaced, the ECF becomes hypertonic. A *rise in osmolality* of only 3 mosm/kg H_2O suffices to increase the **release** of **ADH** from the posterior lobe of the pituitary (→ **A 1** and p. 240). In addition, the volume deficit causes a reduction of pressure in the low-pressure system of the circulation. This is registered by stretch receptors in the atria of the heart and reported directly via nervous pathways to the hypothalamus, with a resultant release of ADH (*Henry-Gauer reflex*). ADH reaches the kidney via the blood and brings about a decrease in water excretion (→ p. 136). The simultaneous sensation of **thirst** signals that water is required by the body (→ p. 138).

Water Excess (→ **A 2**): The intake of hypotonic fluid lowers the osmolality of the ECF. This signal **inhibits** the release of **ADH**, with a subsequent increase in the excretion of hyposmolal urine (→ p. 136). Within about 1 hour the excess water has been eliminated.

If too much water is taken up *too quickly*, **water intoxication** may occur (nausea, vomiting, shock). The reason for this reaction is that the osmolality of the plasma has already dropped considerably before the inhibition of ADH release can become effective.

Salt deficit (→ **A 4**): If, at a normal water content, too much NaCl is lost or too little is taken up in the diet, the decrease in the osmolality of the blood (and other factors) leads to a transitory decrease in the amount of ADH released and hence to an increase in water excretion (see above). As a result, the **ECV** and thus the **plasma volume are reduced** (→ **A 4**). The decrease in the plasma volume, and possibly also in blood pressure (→ p. 153, B), elicits the release of **angiotensin II** (via **renin**), which causes thirst and stimulates the release of **aldosterone** (→ p. 150 ff.). Aldosterone promotes the resorption of Na^+ (→ p. 153, B), that is, it inhibits the excretion of Na^+ (*Na^+ retention*), with the secondary effect that water is retained. Together with the water intake due to thirst, this is sufficient to return the ECV to normal (→ **A 4**). Here, too, the reduction in the plasma volume elicits the Henry-Gauer reflex (see above), and the amount of ADH released quickly rises again.

Salt excess (→ **A 3**) elevates plasma osmolality and subsequently leads to an increase in the release of ADH (*H_2O retention and* thirst). The ECV increases, and with it the plasma volume. The situation is the reverse of that seen in connection with salt deficit: via an inhibition of the renin-angiotensin II-aldosterone mechanism, and increased release of *atriopeptin* (see below) and of other mechanisms (→ p. 132 and p. 152) more NaCl is excreted, followed by more water, and the ECV returns to its normal volume.

The **a**trial **n**atriuretic **p**eptide or **f**actor (**ANP** = **ANF**) or **atriopeptin** is stored in vesicles in cells of the cardiac atrial wall. An increased stretching of the atrium (high ECV!) elicits release of atriopeptin, one of whose effects is to increase Na^+ excretion by the kidney to some extent. As atriopeptin also has direct effects on the cardiovascular system, on other hormones, and on the CNS, the name atriopeptin should be preferred. Although atriopeptin seems to act in an integrative manner on a number of target organs to modulate cardiovascular function and fluid balance, the physiological role of this hormone is not fully understood.

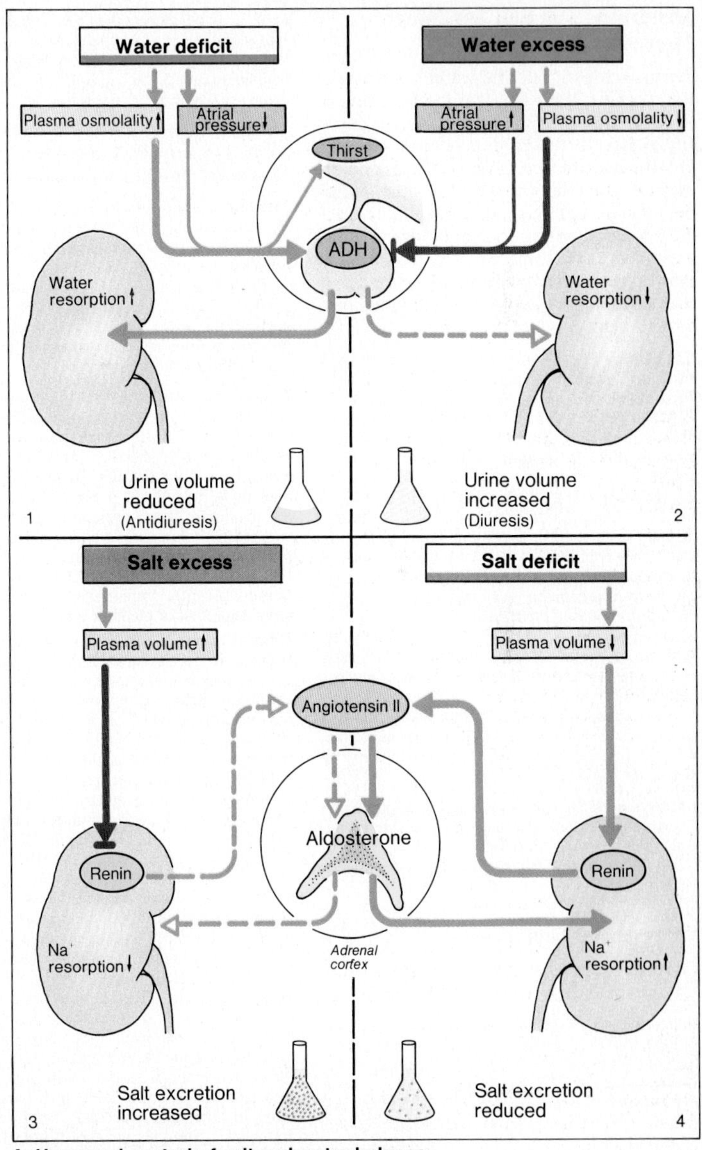

A. Hormonal control of salt and water balance

Disturbances in Salt and Water Homeostasis

Disturbances in salt and water homeostasis (→ **A** and p. 140) may be due to errors in (a) the water and salt balance (i.e. an imbalance between intake and losses); (b) the distribution between plasma, interstitium (together constituting the extracellular space, ECS), and intracellular space ICS; and (c) hormonal regulation. The disturbances and some of their *causes* and *consequences* are briefly listed below (↓ = decrease; ↑ = increase; Ex = examples; osm = osmolality):

1. **Isosmotic loss** (→ **A 1**): ECS ↓; ICS no change; osm no change. Ex: vomiting, diarrhea, diuretic therapy, blood loss, burns, drainage of ascites.
2. **Water deficit** (→ **A 2**): ECS ↓; osm ↑; fluid shifts from ICS to ECS. Ex: sweating, hyperventilation, osmotic diuresis, chronic renal disease, diabetes insipidus.
3. **Salt deficit** (→ **A 3**): osm ↓; fluid shifts from ECS to ICS; ECS ↓. Ex: vomiting, diarrhea, sweating, adrenal insufficiency, hypokalemia, CNS lesions, salt-losing nephritis.
4. **Isosmotic excess** (→ **A 4**): ECS ↑; osm no change. Ex: heart failure, nephrosis, acute glomerulonephritis, decompensated cirrhosis.
5. **Water excess** (→ **A 5**): ECS ↑; osm ↓; fluid shifts to ICS. Ex: water drinking, excessive ADH secretion; intensive gastric lavage, infusion of glucose solutions.
6. **Salt excess** (→ **A 6**): osm ↑; fluid shifts from ICS to ECS; ECS ↑. Ex: infusion of hypertonic saline, adrenal hyperactivity, steroid therapy, drinking sea water, CNS lesions.

Consequences: 1), 2), and 3) result in **hypovolemia** (→ p. 186); 3) and 5) lead to **intracellular edema** (including *cerebral swelling*); and 4), 5), and 6) result in **extracellular edema** (including *pulmonary edema*).

Diuresis and Diuretics

Diuresis is an elevated excretion of urine (> approximately 1 ml/min). Types:

Water diuresis. Water drinking dilutes plasma and reduces ADH secretion (→ p. 140). As a result, hyposmolal urine (minimum ≈ 40 mosm/kg H_2O) is excreted or, in other words, there is an *excretion of free water*. The same events occur when there is a failure of ADH secretion, as in *diabetes insipidus*.

Free water is the quantity of water that would have to be withdrawn from this kind of urine in order for it to attain the osmolality of plasma (= P_{osm} = 290 mosm/kg H_2O). The proportion of free water can be calculated from $1 - (U_{osm}/P_{osm})$, where U_{osm} = urine osmolality. Hence in 1 liter of urine containing, for example, 58 mosm/kg H_2O, there are 0.8 liters of free water.

Osmotic diuresis. When a nonresorbable solute is excreted, it must be accompanied by a corresponding volume of water.

Mannitol and even excesses of salt or bicarbonate may act in this manner to drag fluid into the urine. If the tubular resorption capacity (→ p. 128) for a normally resorbable substance (e.g. glucose) is exceeded due to its very high concentration in the plasma, water is excreted with the nonresorbed part of the substance. Thus, the glycosuria of **diabetes mellitus** is accompanied by diuresis and, secondarily, by increased thirst.

Pressure diuresis. When the blood pressure rises, autoregulation prevents an increased RPF in the cortex (→ p. 122). In the medulla, however, autoregulation is less effective; medullary blood flow increases and washes out the concentration gradient in the medulla (→ p. 136). This action reduces the maximum urine osmolality and leads to diuresis. Since a high ECF volume elevates the blood pressure (→ p. 180), causing a pressure diuresis that in turn can bring the ECF volume back to normal, pressure diuresis may play a role in **long-term regulation of blood pressure**.

Diuretic agents (→ **B**) act on the renal tubule to suppress resorption of solutes. Because an osmotically equivalent volume of water must accompany the solute, urine volume increases; the extra urinary water is derived from the ECF. A reduction of ECF volume is a therapeutic objective in edema or hypertension. However, depletion of ECF activates the renin-aldosterone system, which curbs salt and water loss (*secondary hyperaldosteronism*). Some agents (e.g. acetazolamide) inhibit carbonic anhydrase (CA; → p. 144 ff.) and produce a modest diuresis (high pH, increased $NaHCO_3$, and decreased NH_4^+ in urine) but they are no longer used as diuretics. **"Loop diuretics"** (e.g. furosemide) inhibit the Na^+-K^+-$2Cl^-$ co-transport system in the thick ascending limb of the loop of Henle (→ p. 132) and produce a copious diuresis and decrease free water clearance as well as the medullary osmotic gradient. **Thiazide** diuretics act as inhibitors of Ca^{2+} and Na^+ transport, acting primarily in the distal nephron; they may increase K^+ excretion. **Amiloride** and similar diuretics inhibit the exchange of Na^+ for H^+ in the proximal tubule to a small extent and decrease K^+ excretion by blocking Na^+-channels and thereby the transepithelial potential in the distal tubule (→ p. 148 ff.). A decreased K^+ excretion is also observed with **aldosterone antagonists** (→ p. 152) which cause a slight loss of Na^+.

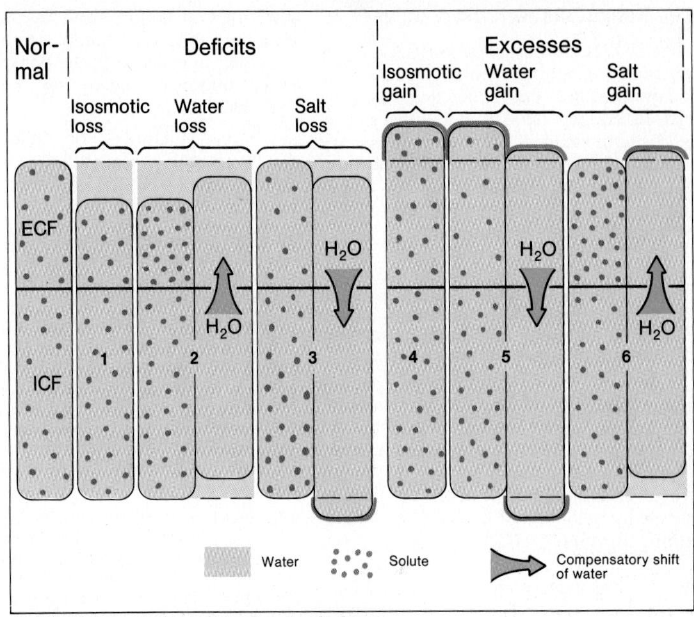

A. Disturbances in salt and water balance

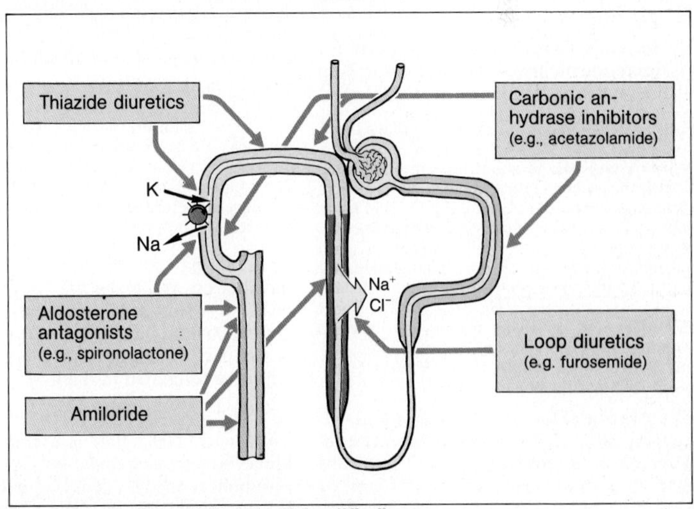

B. Locus of action of diuretics (simplified)

The Kidney and Acid-Base Balance

The enzyme **carbonic anhydrase (CA)** plays a key role at all points in the body where a H^+ ion gradient (i.e. a pH difference) has to be established: the renal tubules, the gastric mucosa, the small intestine, and the salivary glands are only a few examples. In addition, CA plays an important part in CO_2 transport in the erythrocytes (→ p. 96). CA catalyzes the *overall reaction*

$$H_2O + CO_2 \rightleftarrows H^+ + HCO_3^-.$$

Usually, carbonic acid (H_2CO_3) is said to be the intermediate compound in this reaction (→ **A**). However, it has been proposed more recently that OH^-, instead of H_2O, also binds to the enzyme (→ **A**).

The **H^+ secretion** into the lumen of the renal tubule is performed by two mechanisms: (1) exchange of H^+ for Na^+ (common carrier; → p. 132), (2) primary active H^+ pump (H^+-ATPase). The bulk of the H^+ ions is secreted in the proximal tubule, where, during titration by HCO_3^-, phosphate, etc., the pH drops from 7.4 to approximately 6.7. In the collecting duct, H^+ are secreted primarily by the H^+ pump of the *intercalated cells* (→ p. 148). There, the luminal pH can drop to about 4.5.

H^+ secretion serves two primary functions: (1) excretion of "fixed" acid; (2) resorption of the filtered bicarbonate (HCO_3^-).

Sources and Excretion of "fixed" acid in the urine: An average diet containing roughly 70–100 g protein/d gives rise to about 190 mmol H^+ ions. HCl (from arginine, lysine, and histidine), H_2SO_4 (from methionine and cystine), H_3PO_4, and lactic acid are the most important acids involved (*"fixed" acids*). About 130 mmol H^+ per day go into metabolism of anionic amino acids (glutamate$^-$ and aspartate$^-$; from protein) and of other dietary organic anions (lactate$^-$, etc.); this results in a **net H^+ production** of about **60** (40–80) **mmol/d.** Although these H^+ ions are buffered where they are set free, they still have to be excreted for the regeneration of the buffers.

The **pH value of the urine** drops as low as 4 in extreme cases, that is, its H^+ concentration is at most 0.1 mmol/l (→ p. 334). This means that in a daily urine volume of 1.5 l a maximum of only 0.15 mmol or less than 1 % of the daily production of H^+ ions can be excreted in the *free form*. However, the low urinary pH established, especially in the collecting duct (H^+-pump; see above), is necessary in order to titrate phosphate and other substances and to trap the secreted NH_3 as NH_4^+ (see below).

A large quantity of fixed acids, **10–30 mmol/d**, is excreted as so-called **titratable acid** (→ **B**) (80 % phosphate, 20 % urate, citrate, etc.).

The acid excreted in this way is termed *titratable* because its quantity can be determined by titrating the urine with NaOH to the pH value of the plasma (normally pH 7.4).

Phosphate is present *in the blood* (pH 7.4) to 80 % as HPO_4^{2-}, and in the acidic *urine* almost entirely as $H_2PO_4^-$ (→ p. 335), that is, the secreted H^+ ions are buffered by HPO_4^{2-} (→ **B**). Approximately 150–250 mmol phosphate are filtered daily, of which 80–95 % are resorbed (see also p. 154 f.); the rest is excreted. Of this, 80 % have taken up an equimolar amount of H^+ ions during passage through the tubules.

In *acidosis* (→ p. 114 ff.) the excretion of phosphate rises to some extent. The moderate increase in H^+ ion excretion achieved in this way, which is followed by an increasing NH_4^+ excretion (see below), is mainly the result of phosphate mobilization from the bones due to acidosis (→ p. 254 ff.).

The urinary **excretion of NH_4^+** is *not per se* but indirectly a form of H^+ excretion (normally about **25–30 mmol/d**).

NH_3, traditionally, has been said to be produced by the tubule cell and subsequently secreted by nonionic diffusion into the tubule lumen, where it is "titrated" by (secreted) H^+ and excreted in the form of NH_4^+. However, glutamine deamidation in the tubule cell yields **NH_4^+** (not NH_3). As a consequence, a revised view of NH_4^+ excretion has been recently proposed:

At a normal protein intake, amino acid metabolism yields HCO_3^- and NH_4^+ in nearly equimolar amounts (about 700–1000 mmol/d). The majority of these compounds is used for the (energy dependent) **urea formation** in the **liver**:

$$2HCO_3^- + 2NH_4^+ \rightleftarrows \text{urea} + CO_2 + 3H_2O.$$

Thus, each NH_4^+ that is excreted by the kidney saves one hepatic HCO_3^-, which in turn buffers one H^+. This is called "indirect H^+ excretion" above. However, the liver exports only about 15–30 % of the NH_4^+ to the kidney in an unaltered form. Most of it is

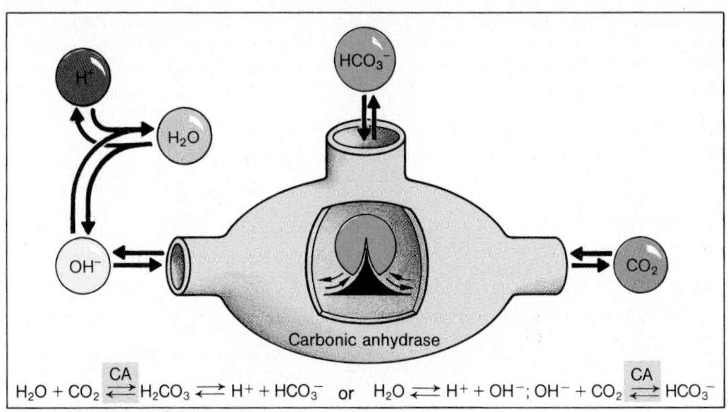

A. Carbonic anhydrase (CA)

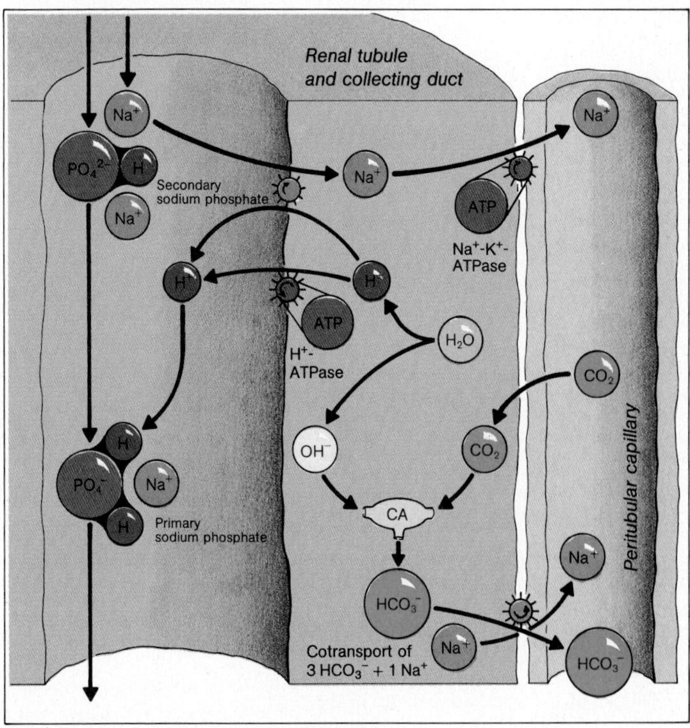

B. H^+ excretion as $H_2PO_4^-$ (titratable acid)

first incorporated into **glutamine** (= Glu-NH_2) in the hepatocytes, the required glutamate$^-$ being temporarily withdrawn from its H^+-consuming metabolism (see above).

It has been discussed that the liver may be involved in pH homeostasis by *regulating* glutamine export to the kidney (see also below).

In the **kidney**, glutamine is split by renal **glutaminases** (→ **C2, a**), yielding NH_4^+ and *glutamate* (= Glu^-). Glu^- may be further metabolized to 2-oxoglutarate^{2-} and a second NH_4^+ by renal *glutamate dehydrogenase* (→ **C2, b**). Only when the bivalent metabolite of glutamine, 2-oxoglutarate^{2-}, is finally converted to an uncharged compound, such as glucose or CO_2 (usually taking place in the kidney), are two H^+ neutralized, and therefore finally "indirectly excreted" by NH_4^+ excretion.

NH_4^+ is dissociated into $NH_3 + H^+$ to a small extent in the proximal tubule cell and is secreted into the lumen via nonionic diffusion (NH_3) and via H^+ secretion (in exchange for Na^+ or pumped by the H^+-ATPase) into the lumen where NH_4^+ is formed again (→ **C2**).

Part of this NH_4^+ is apparently resorbed (in its charged form!) by using the cotransport mechanisms for K^+ in the thick ascending limb (→ p. 149, B2). By nonionic diffusion (i.e. as NH_3) it reenters the urinary space in the collecting duct where, due to the very low pH in this segment, it is ultimately trapped in the urine as NH_4^+.

In **chronic metabolic acidosis** (→ p. 114), NH_4^+ excretion increases. *Hepatic glutamine formation* rises at the expense of urea formation and is accompanied by an increased flux through *renal glutaminase*. As a result, NH_4^+ excretion may rise to threefold the normal amount. It takes 1–2 days before this adaptation is fully developed. The regulatory signal(s) of this process is not completely understood.

Resorption of bicarbonate (HCO_3^-; → C1). Approximately 4300 mmol HCO_3^- per day are filtered, a quantity about 40 times that present in the blood. A highly efficient resorption of HCO_3^- is therefore of the utmost importance in order to prevent the breakdown of the acid-base balance in the body (→ p. 110ff.).

The H^+ ions secreted in the tubule react on the spot with HCO_3^- to give CO_2 and H_2O (→ **C1**), with the participation of a luminal CA (at the brush border). The CO_2 readily diffuses into the cell, H^+ and HCO_3^- are again formed (cytoplasmic CA!), and the H^+ ions are secreted into the tubule while the HCO_3^- enters the blood (→ **C1**). Thus HCO_3^- traverses the luminal membrane in the form of CO_2. About 85–90% of the filtered HCO_3^- is resorbed in the early proximal tubule, while the rest is mainly resorbed in the collecting tubule.

An elevated (or lowered) P_{CO_2} in the plasma leads to an increase (or drop) in the H^+ secretion and HCO_3^- resorption, which is important in the *compensation of respiratory acid-base disturbances* (→ p. 116).

Nitrogen Metabolism and Excretion

Whereas carbohydrates and fats are almost entirely broken down to water and CO_2 by the organism (→ p. 198), the nitrogen of nitrogen-containing substances such as **proteins, amino acids, nucleotides**, and so on is excreted by the kidney in the form of other nitrogen-containing substances, largely as **urea**, and to a smaller extent as $\mathbf{NH_4^+}$ (see above), **uric acid, creatinine**, and the like.

Compared with the excretion of 300–500 mmol urea daily (depends on protein intake), the 0.5–6 mmol of **uric acid** (from nucleotide metabolism) excreted daily are insignificant. Nevertheless, the excretion of uric acid is of clinical interest since it is a substance with a very low solubility and can form kidney stones. A high blood level of uric acid can cause *gout*.

As far as the energy balance of the organism is concerned, the best way of excreting nitrogen would be in the form of NH_3, which is what aquatic animals do. Terrestrial animals, however, would be unable to wash out the toxic NH_3 quickly enough. Urea, on the other hand, is nontoxic, readily soluble in water, and carries two N atoms per molecule, even though its synthesis requires ATP. Snakes and birds excrete nitrogen mainly in the form of uric acid crystals, that is, nitrogen excretion is combined with H^+ excretion (uric *acid*), without the necessity of water as a solute. This mechanism is therefore well suited to desert animals.

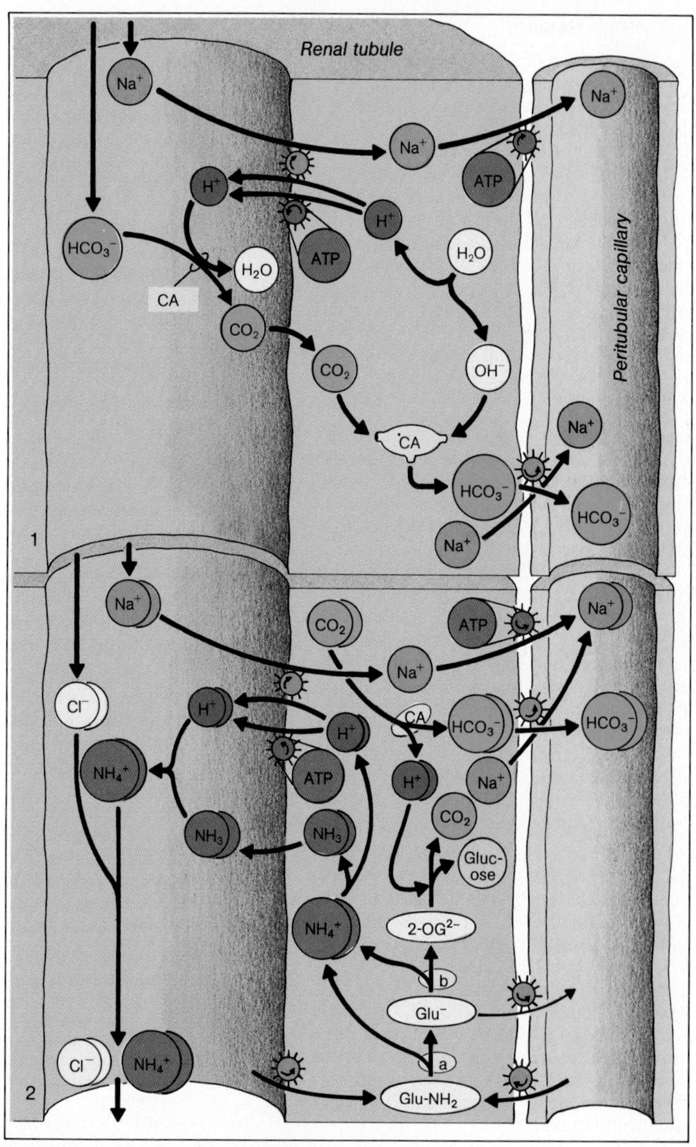

C. Bicarbonate resorption (1) and NH_4^+ excretion (2)

Potassium Balance

About 50–150 mmol K^+ are taken up daily (*minimum requirement* 25 mmol), of which about 90% are excreted in the urine and 10% leave the body in the feces. The **plasma K^+ concentration** amounts to **3.4–5.2 mmol/l** while a 20–30-fold concentration (→ p. 65, B) prevails in the cells, that is, about 98–99% of the roughly 4500 mmol K^+ ions in the body are present in the cells (3000 mmol in muscle cells, 200 mmol each in liver and erythrocytes). Although extracellular K^+ accounts for only 1–2% of the total, it is of great importance since one of its main functions is that of mediating the regulation of the entire K^+ balance of the organism.

Acute regulation of the extracellular K^+ concentration is achieved by **alterations in the distribution of K^+** between the ECF and the ICF. This is a relatively rapid way of preventing or lessening, for instance, a dangerous rise in the K^+ concentration in the ECF following either the intake of larger quantities of K^+ in the food or the release of K^+ within the body (e.g. by hemolysis). The movement of K^+ is largely under *hormonal control*: an acute rise in the K^+ in the ECF results in the release of **insulin**, which promotes the uptake of K^+ by the cells and consequently lowers the K^+ concentration in the ECF. *Epinephrine, aldosterone*, and *alkalosis* also promote the cellular uptake of K^+.

Chronic regulation of the K^+ balance of the body is primarily achieved by the **kidney** (see below), but also to some extent by the **colon**. K^+ excretion is primarily influenced by the K^+- and H^+-concentration in the ECF, by aldosterone, and by the excretion of Na^+ (→ **C**). If the K^+ intake is chronically elevated, then the efficiency of the K^+ excretory mechanism becomes enhanced (**K^+ adaptation**). In cases where renal function is much reduced, this adaptation of the remaining, functioning tubular apparatus ensures that the K^+ balance is for the most part maintained.

In the **kidney**, K^+ is filtered at the glomerulus and most of it is normally resorbed (net *resorption*); nevertheless, the quantity excreted may sometimes exceed the quantity filtered (net *secretion*, see below). Regardless of the K^+ intake, about 70% of the quantity filtered is *resorbed* (→ **A**) before the end of the **proximal tubule**. In all probability, this transport takes place against a small electrochemical gradient. Some K^+ ions must therefore be *actively* removed from the tubular lumen, although the majority of the K^+ leaves the proximal tubule *passively* (→ **B 1**). About 10–20% of the filtered K^+ leaves the tubular fluid in the **loop of Henle** (secretion in the descending limb is less than the resorption in the ascending limb; → **B 2**), so that only 10% of the quantity filtered appears in the **distal tubule** (→ **A**).

A **high intake of K^+** results in increased excretion of K^+ in the urine (in extreme cases as much as 150% of the quantity filtered), whereas in **K^+ deficiency** the urine is low in K^+ (minimum ≈ 3% of quantity filtered). This adaptation to the prevailing need is achieved almost entirely by a drastic rise in, or the absence of, K^+ secretion in the "late" **distal tubule** and in the beginning of the **collecting duct** (→ **B**); in addition, K^+ can be (actively) resorbed at these sites.

The cellular mechanisms of the **K^+ transport** in the distal tubule and in the initial part of the collecting duct have not yet been completely elucidated. *K^+ is actively transported* into the cell (see above, → **B**) from the blood *in exchange for Na^+* and probably also from the lumen of the tubule. This results in the *high intracellular K^+ concentration* that constitutes the main driving force for the *passive efflux of K^+* from the cell (→ **B**). Changes in these fluxes thus influence K^+ secretion via the intracellular K^+ concentration (sum of "pump and leak" fluxes).

There are two types of cells in the "late" distal tubule and in the collecting duct: the **principal cells**, which secrete K^+ (and resorb Na^+), and the **intercalated cells**, which are thought to be responsible for active K^+ resorption during K^+ deficiency and, by the way, for H^+ secretion (cell type A) or HCO_3^- secretion (cell type B) in these segments.

The **mechanism of K^+ secretion** by the principal cells (→ **B 3**): As with all other tubule cells, the **Na^+-K^+-ATPase** of the basolateral cell membrane lowers the intracellular Na^+ concentration and at the same time raises that of K^+. K^+ can leave the

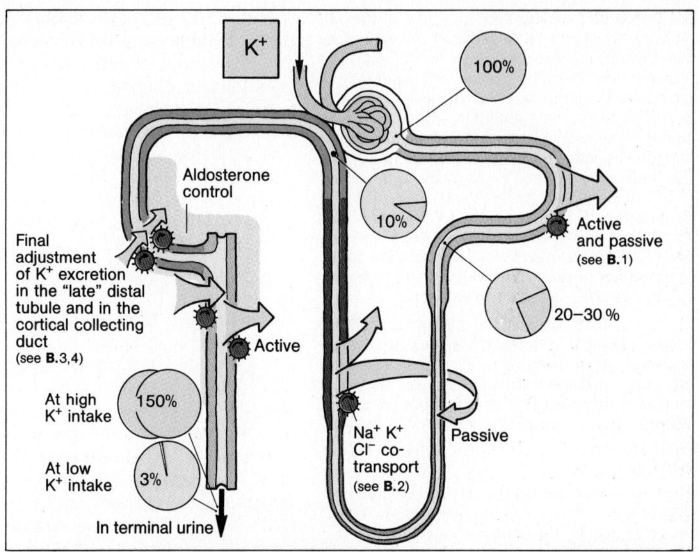

A. Resorption, secretion and excretion of K^+

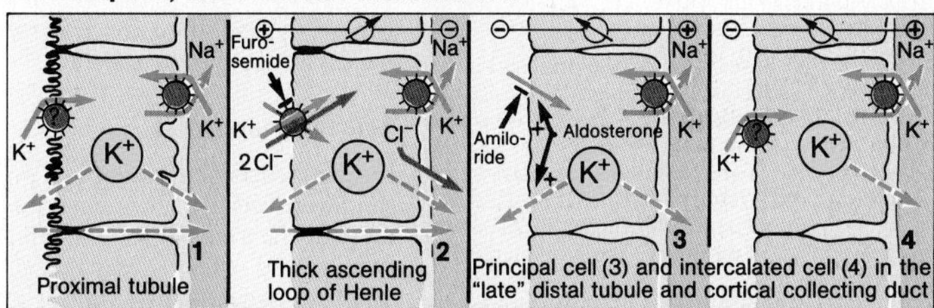

B. K^+ resorption (1, 2, 4) and K^+ secretion (3) in the renal tubule

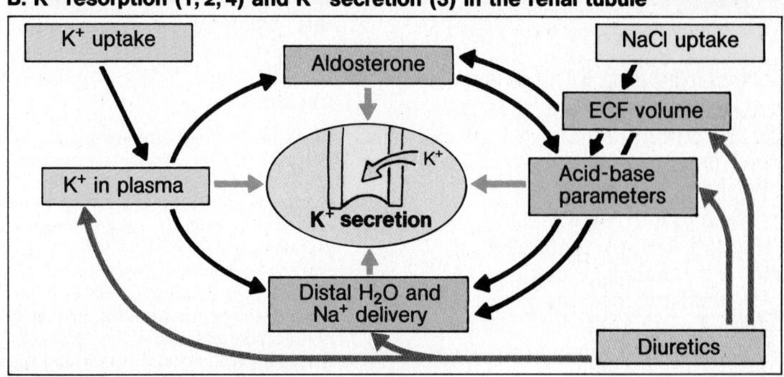

C. Influences on K^+ secretion and excretion

(after Wright and coworkers)

cell through **K^+ channels** on both sides of the cell, whereby the electrochemical gradient across the respective membrane determines the K^+ diffusion rate. In addition, on the luminal cell membrane of the principal cell there are **Na^+ channels** (which can be inhibited by the diuretic agent *Amiloride*), through which the Na^+ flows from the lumen into the cell. This influx is electrogenic (→ p. 15), so that the luminal membrane is depolarized to about 30 mV (lumen: +), whereas the basolateral membrane retains its normal potential of approximately 70 mV (exterior: +). There is thus a greater driving force for K^+ efflux on the luminal side than on the opposite side of the cell, which favors K^+ efflux in the direction of the lumen: **secretion**.

The result of the asymmetrical membrane potential of the principal cell is a **lumen-negative transepithelial potential** of roughly 40 mV. Among other uses, this might be a driving force for the paracellular resorption of Cl^-, but it is not yet certain whether this is the main route or even the only route for Cl^- resorption in this part of the nephron.

The *intercalated cells* (→ **B 4**) have no luminal Na^+ channels, and the K^+ conductivity is low, so that scarcely any K^+ is secreted here. *Active K^+ resorption* and *H^+ secretion* by type A of these cells are possibly accomplished by a **H^+-K^+-ATPase** in their luminal membrane.

The more Na^+ is resorbed by the principal cells, the more K^+ is secreted. There are two probable reasons for this **coupling of Na^+- and K^+- transport** in the "late" distal tubule and in the cortical collecting duct: (1) greater luminal depolarization (see above) as Na^+ resorption increases; and (2) the rise in Na^+ concentration inside the cell. This slows down Na^+-Ca^{2+} exchange at the basolateral cell membrane, resulting in the rise of the intracellular Ca^{2+} concentration. This is a signal for the luminal K^+ channels to open.

Factors influencing K^+ excretion (→ C):

1. An increase in **K^+ uptake** causes a rise in the K^+ concentration in plasma and cells, which in turn increases the chemical driving force for K^+ secretion.

2. **Blood pH.** Alkalosis raises, and acute acidosis lowers, the intracellular concentration and hence the excretion of K^+; however, the latter rises again in chronic acidosis. The reasons for this are (a) an increased distal urinary flow (see point 3); and (b) the resulting hyperkalemia that causes the release of aldosterone (see point 4).

3. If there is a rise in the **urinary flow rate** in the distal tubule (e. g. due to increased NaCl intake), osmotic diuresis, or any other form of inhibition of upstream Na^+ resorption, more K^+ is excreted (see, for example, K^+ losses caused by certain diuretics; → p. 142). The probable explanation for this is that secretion of K^+ is limited by a certain luminal K^+ concentration. This means that a greater volume per time can take with it more K^+ per time.

4. **Aldosterone** (see also below) increases the incorporation, formation, or the "bringing into operation" of Na·- and K^+-channels at the luminal membrane, which directly or indirectly (depolarization) leads to an increased K^+ secretion (and Na^+ resorption). Aldosterone enhances the activity of the Na^+-K^+-ATPase and also brings about, over a longer period of time (e. g. in the case of K^+ adaptation; see above), important morphological changes in the target cells. (Although K^+ excretion remains high in cases of chronically elevated aldosterone levels, Na^+ resorption, for reasons unknown, subsides again after about two weeks: *escape phenomenon.*)

Mineralocorticoids

Mineralocortico(stero)ids are produced in the **adrenal cortex**. Their chief function is to regulate the transport of Na^+ and K^+ in the kidney and other organs (gall bladder, intestine, sweat glands, salivary glands, and so on). **Aldosterone** is the most important of this group of hormones, although corticosterone, desoxycorticosterone, and even the glucocorticoids (→ p. 260) affect salt transport.

Biochemistry: Aldosterone is a C_{21} steroide (i.e. possesses 21 C atoms) and is produced in the *zona glomerulosa* (→ p. 261) of the adrenal cortex. The starting point for the *biosynthesis* of aldosterone is *cholesterol*, which is largely taken up from the plasma but can also be formed in the adrenal cortex. The *rate of formation* of aldosterone is 80–240 µg/d; its *plasma concentration* amounts to 0.10–0.15 µg/l. These values fluctuate according to NaCl intake and time of day. The rate of secretion is at its highest in the early morning and is lowest in the late evening. Aldosterone is coupled to glucuronic acid in the liver (→ p. 214) and in this form is *excreted* in the bile and urine.

Control of the release of aldosterone. Aldosterone produces Na^{+} retention throughout the body and increases K^{+} excretion. Due to the resultant secondary retention of water, one of the effects of aldosterone release is an *increase in the extracellular volume (ECV)* (→ p. 140ff.). It is natural, therefore, that the release of aldosterone is stimulated by conditions that are accompanied by (a) *reduction in blood volume*, (b) *hyponatremia*, and (c) *hyperkalemia*. *Angiotensin II* (→ p. 152) plays an important role in aldosterone release, but what constitutes the primary stimulus is still uncertain. The formation of aldosterone is also stimulated by ACTH (→ p. 261, A); it is inhibited by atriopeptin from the atria of the heart (→ p. 140).

Effects of Aldosterone: Aldosterone stimulates Na^{+} resorption and loss of K^{+} by salt-transporting cells. Its effects are first observed ½ to 1 hour after its administration (or release) and attain a maximum after several hours.

This delayed onset of the effect reflects the time required for the following intracellular reaction steps necessary (?) for steroid hormones to take effect (see also p. 244): (1) diffusion of aldosterone (= A) across the cell membrane; (2) specific binding to cytoplasmic receptor proteins (= R); (3) "activation" and changes in conformation of the activated A-R on chromatin receptors (DNA) in the cell nucleus; (5) RNA induction; and (6) production of **a**ldosterone-**i**nduced **p**roteins (**AIPs**) that mediate (at least most of) the cellular effects of A.

Hyperaldosteronism occurs when the adrenals secrete excessive amounts of aldosterone.

Primary hyperaldosteronism (due to adrenal cortical tumors producing aldosterone; Conn syndrome), does not respond to normal feedback controls: Na^{+} retention leads to an increase in ECV and hyptertension; K^{+} loss results in a hypokalemia that is accompanied by a hypokalemic alkalosis.

Secondary hyperaldosteronism is much more common and occurs when there is a reduction of the *effective* plasma volume, as occurs in pregnancy, heart failure, chronic diuretic therapy, dietary salt restriction, nephrosis, and cirrhosis of the liver with ascites. In each of these conditions, activation of the renin-angiotensin II mechanism (→ p. 152) leads to enhanced release of aldosterone.

In **adrenocortical insufficiency** (Addison's disease) the deficiency of aldosterone results in a greatly enhanced Na^{+} excretion and K^{+} retention which, added to the lack of glucocorticoids (→ p. 260), produce an extremely dangerous situation.

Excretion of Ca^{2+} and Phosphate

The kidney is an important organ contributing to the Ca^{2+} **balance** due to its excretory function (→ p. 254ff.). The *total* Ca^{2+} *plasma concentration* (free and bound Ca^{2+}) amounts to 2.3–2.7 mmol/l (4.6–5.4 meq/l). Roughly 1.3 mmol/l of this are *ionized* Ca^{2+}, 0.2 mmol/l are *compexed* (by phosphate, citrate, etc.), the remaining 0.8–1.2 mmol/l are *bound to plasma proteins* and therefore cannot be filtered at the glomerulus (→ p. 10 and p. 127, B). The daily filtered load of Ca^{2+}, therefore, amounts to about 270 mmol (180 l/d · 1.5 mmol/l). Only about **0.5–3%** of this load are **excreted** in the urine. The **site of resorption** of Ca^{2+} is the entire nephron with the exception of the thin parts of Henle's loop. Resorption of Ca^{2+} and Na^{+} often go in parallel (→ p. 132). This is true for the *effects of diuretic drugs* (→ p. 142) and bulk transport in the proximal tubule and in the thick ascending limb of Henle's loop. The site of the **fine adjustment of** Ca^{2+} **excretion** is the late distal nephron. **Parathyrin** (→ p. 254) and, to a small extent, **1,25-**$(OH)_2$**-calciferol** or **calcitriol** (a derivative of vitamin D; → p. 256) decrease Ca^{2+} excretion; **calcitonin (CT)** (→ p. 256) probably also influences Ca^{2+} excretion.

Of the **phosphate** filtered at the glomerulus (→ p. 144ff.), **80–95% are resorbed**, mainly in the proximal tubule. In contrast to Ca^{2+} excretion, the excretion of phosphate is increased by parathyrin. Phosphate excretion is also increased by calcitonin to some extent (→ p. 256).

Renin-Angiotensin Mechanism

The **juxtaglomerular apparatus (JGA; → A)** consists anatomically of (a) the *macula densa cells* of the distal tubule; (b) the closely adjoining parts of the *afferent* and *efferent arteriole* of the **same** nephron; and (c) the extraglomerular mesangial region or "polkissen" (polar cushion) with two types of cells: (1) agranular cells = lacis cells = Goormaghtigh's cells, and (2) granular mesangial cells = myoepitheloid cells. The latter are localized in the arteriolar wall (→ **A**).

The JGA is ideally situated for receiving signals regarding the renal arterial pressure (afferent arteriole) and the composition of the urine in the early distal tubule (macula densa), and for evaluating them for the regulation of renal blood flow, filtration pressure, and hence the GFR (→ p. 124). In addition, renal sympathetic nerves influence the JGA.

Biochemistry → **B**). Beside other organs (e.g. brain, heart, adrenal gland), the kidney contains (in the granular cells of the JGA) the proteolytic enzyme **renin**, which can be set free and enters the blood. Renin acts on *renin substrate*, **angiotensinogen** (from the liver), splitting off a decapeptide (a peptide with 10 amino acid groups), the so-called **angiotensin I. Converting enzyme**, a peptidase occuring in the lung, the kidney, and other tissues, splits two amino acids off the decapeptide angiotensin I to produce the highly active octapeptide **angiotensin II**. The latter is broken down in the liver and kidney.

Control of the renin-angiotensin mechanism (→ **B**) is not yet thoroughly understood. An acute low blood pressure (or acute reduction in the effective plasma volume) leads to release of renin (baroreception in the renal arterial vessels?), which results in restoration of the blood pressure or plasma volume, and in a reduced renin release (*negative feedback* → **B**). A decreased mean arterial pressure in only one kidney (e.g. stenosis of one renal artery) also increases renin release from that kidney, which leads in this case to systemic *hypertension* (→ p. 180).

Renin release is influenced by β-adrenergic stimulation (→ p. 56) and responds to circulating epinephrine. Angiotensin II and the aldosterone that is released by it (see below) also have an inhibitory effect on the release of renin (→ **B**).

Target Organs and Effects of Angiotensin II

1. *Cardiovascular system:* angiotensin II is the most potent *vasoconstrictor substance* in the organism and acts directly on the arterioles. The result is a rise in blood pressure (→ **B**). It is uncertain whether or not this effect is involved in the physiological regulation of blood pressure.

2. *CNS:* angiotensin II stimulates the circulation "center" with resultant vasoconstriction, which enhances the direct effect on the arterioles. Further, angiotensin II sets off the *thirst mechanism* in the hypothalamus and plays a stimulatory role in regulating the *appetite for NaCl* (→ **B**). This angiotensin II is probably released by the CNS itself.

3. *Kidney:* Here, too, angiotensin II exerts a vasoconstrictor effect, by means of which both renal blood flow and GFR (→ p. 122ff.) are influenced. Furthermore, there is some evidence that the JGA can act *locally* as part of a (tubuloglomerular) *feedback system in the individual nephron*. This would mean that a rise in the GFR would lead to increased NaCl concentration (or resorption) at the macula densa. Via an unknown signal, constriction of the vas afferens would then lower GFR and the Na^+ load, and the stimulus to the macula densa would be removed.

However, release of renin into the systemic circulation and subsequent systemic angiotensin II formation *decreases* if the NaCl concentration (or resorption) at the macula densa rises. Thus, plasma angiotensin II cannot be the signal for the feedback constriction of the vas afferens. Whether intracellular angiotensin II or other signals (prostaglandins?) are involved in this feedback mechanism remains to be elucidated.

4. *Adrenal cortex:* angiotensin II stimulates the release of aldosterone in the adrenal cortex (→ p. 150). Aldosterone increases Na^+ resorption in the distal tubule and thus enhances the Na^+- and H_2O-saving effect of the diminished GFR (→ **B**).

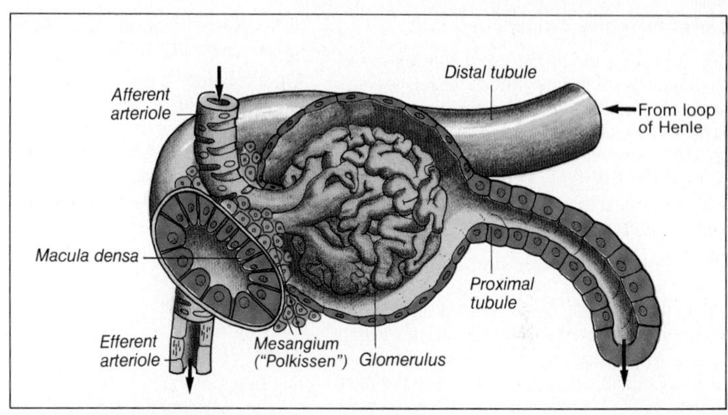

A. Juxtaglomerular apparatus

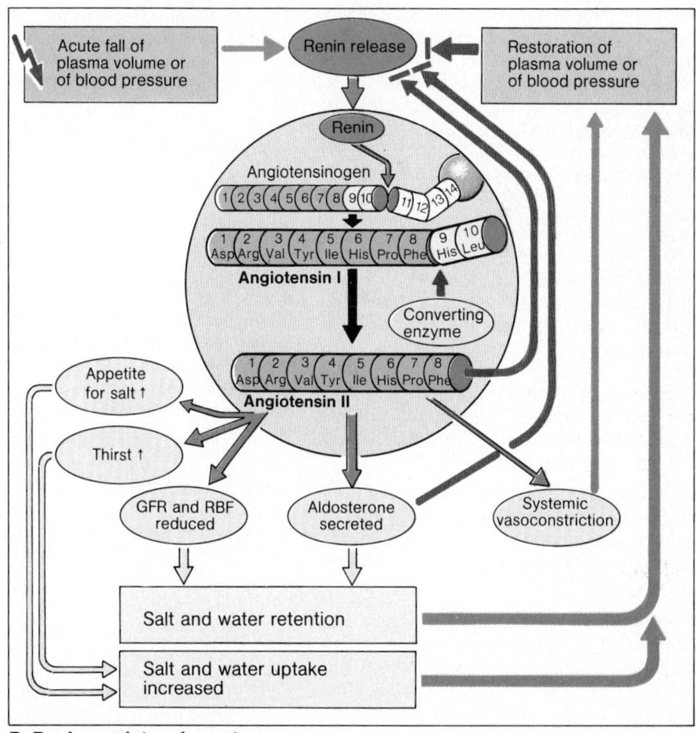

B. Renin-angiotensin system

Cardiovascular System

The left ventricle of the heart pumps blood through the arterial blood vessels of the **systemic circulation** to the peripheral capillaries. The blood is returned to the right side of the heart via the systemic veins and is pumped by the right ventricle into the lungs, whence it is returned to the left side of the heart (**pulmonary circulation**; → **A**).

The **total blood volume** amounts to about 4.5–5.5 l (approx. 6–8 % of the body weight), of which roughly 80 % are in the systemic veins, the right side of the heart, and in the vessels of the pulmonary circulation collectively known as the **low-pressure system** (→ **A**) on account of the relatively low pressure prevailing within it (mean 2 kPa = 15 mmHg). Due to its *large capacity* and its *high compliance*, the low-pressure system can also act as a *blood reservoir*. If the normal blood volume is increased (e.g. by a blood transfusion), more than 98 % of the transfused volume goes to the low-pressure system and less than 2 % to the arterial **high-pressure system**. Conversely, it is almost exclusively the low-pressure system that is reduced if the blood volume is lowered.

Cardiac output is the volume of blood ejected by each ventricle per unit of time and can be calculated from the *heart rate* × *stroke volume*: at rest it amounts to ($70\ \text{min}^{-1} \times 0.07\ \text{l} =$) about **5 l/min** (females, 4.5; males, 5.5). The cardiac output can be increased to many times this value by increased heart rate and stroke volume (→ p. 48).

One method of determining cardiac output employs **Fick's principle**: Cardiac output = oxygen consumption ($\dot{V}_{O_2}$) divided by the arteriovenous O_2 concentration difference, AVD_{O_2} (→ p. 92).

The cardiac output of the left ventricle is distributed in the **systemic circulation**, in which the organs are connected "in parallel" (brain, cardiac muscle, gastrointestinal tract, skeletal muscles, kidneys, skin, etc.) according to the momentary need. The **pulmonary circulation**, on the other hand, receives the entire cardiac output (of the right ventricle) since it is connected "in series" to the systemic circulation (→ **A**).

An adequate blood supply to the **brain** (about 13 % of cardiac output) is of primary importance since, besides being essential for life, it is an organ particularly sensitive to O_2 deficiency (hypoxia; → p. 102), and nerve cells once destroyed usually cannot be replaced.

The blood flow to the **cardiac muscle** must also be maintained at all costs (at rest, about 4 % of the cardiac output, → p. 188) since heart failure affects the entire circulatory system.

The **lungs** receive blood via two routes: (1) via the *pulmonary artery* (pulmonary circulation; see above), which conveys "venous" (low O_2 content) blood to the lungs where it is loaded with O_2 or " arterialized" (100 % of right ventricular output); (2) via the *bronchial arteries*, which provide the lung tissue itself with arterialized blood. Drainage in each case takes place via the *pulmonary veins*.

The **kidneys** receive 20 to 25 % of the cardiac output (→ p. 122). This high blood flow in relation to their weight (only 0.5 % of the body weight) is mainly connected with the *regulatory and excretory functions* of the organ. A relatively small proportion of the blood flow suffices for the basic needs the kidney tissue itself. In circulatory shock (→ p. 186), therefore, a part of the blood flow can be shunted from the kidneys to the high-priority brain and heart.

During heavy muscular work, up to two thirds of the cardiac output flows through the **skeletal musculature** (→ p. 48). A similar proportion of the cardiac output flows through the **gastrointestinal tract** during digestion; it is therefore logical that these two groups of organs do not simultaneously obtain their maximum blood flow (→ **A**).

Blood flow through the **skin** (at rest, about 10 % of cardiac output) serves primarily for the *loss of heat* (→ p. 192), which is why it is especially well supplied with blood in situations (such as heavy work) in which much heat is produced, or when the surrounding temperature is high, or both.

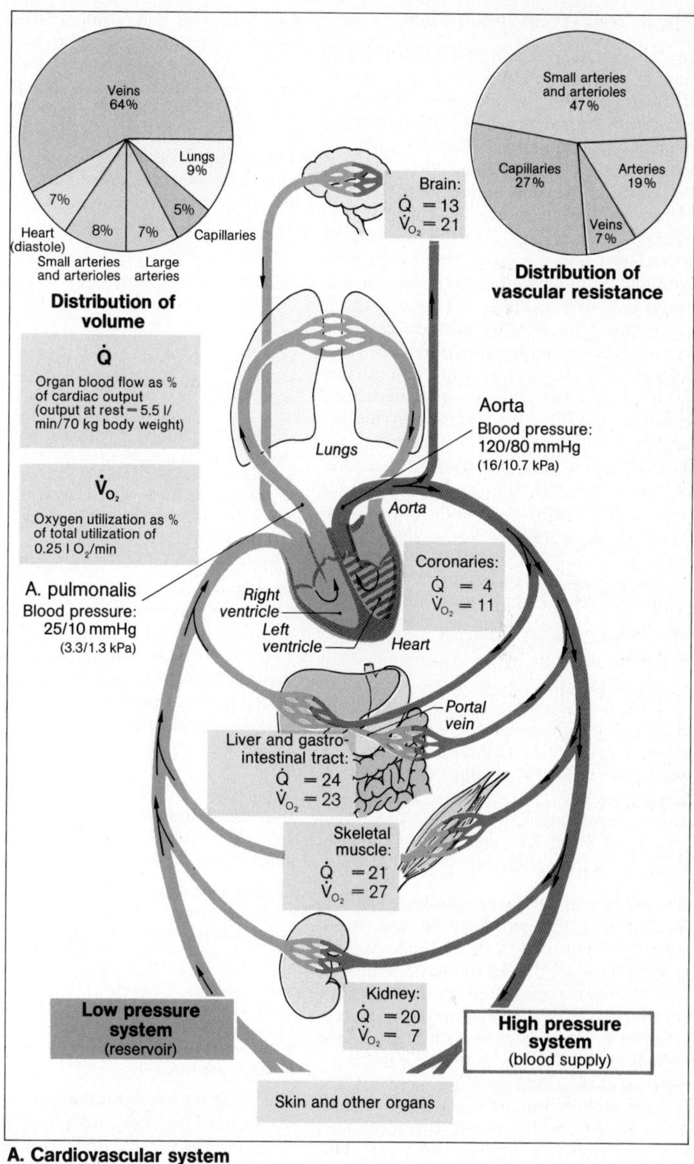

A. Cardiovascular system

Blood Vessels and Blood Flow

In the **systemic circulation** the blood leaves the left ventricle in the aorta and flows into the arteries, which divide and subdivide to form the arterioles via which the blood reaches the capillaries. These reunite to form venules, from which the blood is passed on to the veins and reenters the right atrium of the heart via the superior and inferior venae cavae (→ **A**). On the way, the **mean blood pressure** (→ p. 160) drops from about 13.3 kPa (100 mmHg) in the aorta to about 0.25 to 0.5 kPa (2 to 4 mmHg) in the venae cavae (→ p. 94 for pulmonary circulation). This **mean pressure difference** (**ΔP**) between the aorta and the right atrium (roughly 13 kPa) and the **total peripheral resistance (TPR)** in the systemic circulation (about $2.4\ \text{kPa} \cdot \text{min} \cdot \text{l}^{-1}$) determine the **flow rate** of the blood (**$\dot{Q}$**), that is, the cardiac output (→ p. 154). Besides being valid for the entire systemic circulation, *Ohm's Law*, **$\Delta P = \dot{Q} \cdot R$**, applies to the individual sections of the circulation: in sections with a high flow resistance (R) there is a large drop in pressure and a small blood flow.

The **volume flow $\dot{Q}$ (m^3/s)** is equal in the successive sections of the vascular system, or, in other words, the same volume of blood flows through the aorta, per unit of time, as through all arterioles or all capillaries of the systemic circulation. On the other hand, the **velocity of flow (m/s)**, which determines the contact time, is inversely proportional to the common cross-sectional area (→ **A 2**) of the vessels (fast flow in aorta, slow in capillaries).

The **aorta** and the **large arteries** distribute the blood to the periphery (at rest, mean velocity of flow 0.2 or 0.05–0.1 m/s, respectively); they also, thanks to their compliance (decreases with age), convert the intermittent blood flow at the beginning of the aorta (systole: 0.7 m/s) to a smooth type of flow (**Windkessel effect**; → p. 163). As the heart contracts and blood pressure increases, the arteries stretch and store potential energy; when it relaxes, blood pressure decreases and the stretched arteries rebound. This keeps the blood flowing during diastole, even though the aortic valve is closed.

The **arterioles** and **small arteries** together account for about 50% of the TPR (→ p. 155, A), so that here the blood pressure undergoes a steep drop (**resistance vessels**). Changes in the arteriolar resistance will thus considerably affect the TPR (→ p. 176ff.). The width of the individual arterioles, and especially of their **precapillary sphincters**, also determines the *blood flow* in the capillary network (i.e. the size of the *capillary exchange area*).

According to the **Hagen-Poiseuille law**, the resistance (R) to flow is calculated from

$$R = 8 \cdot l \cdot \eta \cdot \pi^{-1} \cdot r^{-4}.$$

Thus, R in tubes of *length* (l) depends on the viscosity (η) of the fluid and is inversely proportional to the *fourth power* of the *radius* (r^{-4}). Thus a reduction in radius of only 16% (e.g. in the arterioles) suffices to double the resistance. The **viscosity, η, of the blood** rises with increasing *hematocrit value* (→ p. 65, A) and, since blood is a heterogeneous fluid, with decreasing *velocity of flow* (piling up of the slowly moving erythrocytes; → p. 64 and p. 186).

Although the radius of the **capillaries** is still smaller (→ **A**) than that of the arterioles, their total number is so large (about $5 \cdot 10^9$) that they contribute only 27% to the TPR. The resulting drop in blood pressure along the capillaries is largely responsible for the bidirectional **exchange of fluid** between blood and interstitial space (→ p. 158). On account of the *low velocity of flow* in the capillaries (0.3 mm/s), their *very large total surface area* (about 300 m^2), and extremely thin and therefore *permeable walls*, they are especially suitable for the exchange of solutes and fluids.

Laplace relationship: the wall tension (T) in a cyclindric vessel = the *transmural pressure* P_t (= pressure in the vessel minus surrounding pressure) times the *vessel radius* (*r*).

$$T = P_t \cdot r.$$

Since *r* is very small in the capillaries (3000 times smaller than in the aorta; → **A**), the wall tension is low, so that the thin walls of the capillaries suffice to resist pressure.

The **veins** collect the blood and return it to the heart. They make up a large part of the low-pressure system and serve as a **blood reservoir:** *capacitance vessels* (→ p. 184).

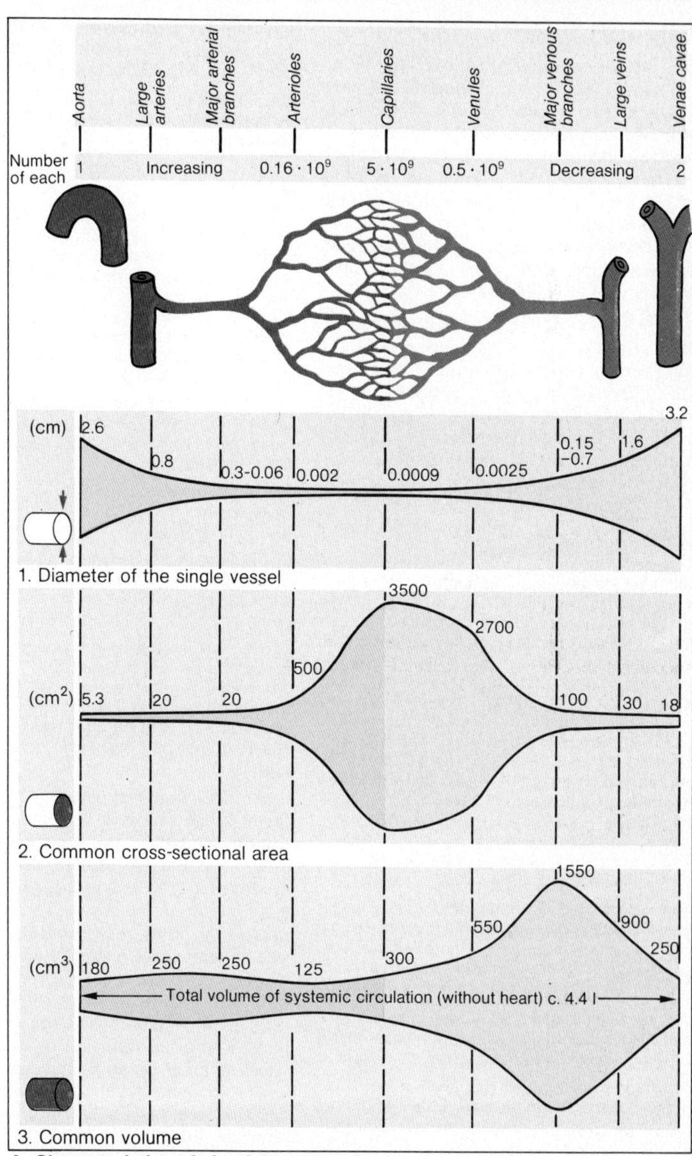

A. Characteristics of circulatory vessels

Capillary Exchange

Cell nutrition is effected via the blood capillaries. Water and substances dissolved in the plasma (except blood cells and large proteins) can freely cross the *thin capillary walls* through pores (8 nm diameter). Each day, a total of roughly 20 l of fluid (i. e. 0.5 % of plasma flow) are **filtered** by the nonrenal capillaries into the interstitial space. About 18 l/d reenter the capillaries by **resorption**; the remaining 2 l/d return to the bloodstream via the **lymph** (→ **A**).

The **driving forces for filtration and resorption** at the capillary wall are (→ **A**) the *difference in hydrostatic pressure* (**Δ P**) and the difference in oncotic pressure × the reflexion coefficient (**$\Delta\pi \cdot \sigma$**; → p. 336) between the inside and outside of the capillary (*Starling hypothesis*). The difference between Δ P and $\Delta\pi$, together with the *permeability* (hydraulic conductance) and the *exchange area* of the capillary wall determine the extent of fluid exchange.

Δ P is the driving force for filtration and at *heart level* amounts to about 3.9 kPa (29 mm Hg) at the *arterial end* of the capillaries, dropping to roughly 1.9 kPa (14 mm Hg) at the *venous end* (→ **A** and **B**, red and mauve lines). This is opposed by the oncotic pressure difference, ($\Delta\pi$), of about 2.7 kPa (20 mm Hg) (→ **A** and **B**, green line), whereby $\sigma \approx 1$ in most organs (i.e. the capillary wall is nearly impermeable to proteins). The difference between Δ P and $\Delta\pi$ at the arterial end of the capillaries is 3.9 minus 2.7 = 1.2 kPa (i. e. *filtration* occurs here), whereas at the venous end it is 1.9 minus 2.7 = − 0.8 kPa (i. e. *resorption* takes place). Where the protein permeability of the capillary wall is higher (e. g. in the liver) the oncotic forces ($\Delta\pi \cdot \sigma$) are reduced, since $\sigma < 1$ (→ p. 336).

The venous limb of the capillaries has a larger diameter and is more permeable than the arterial portion, which means that resorption can take place over a shorter distance and at a lower pressure difference than the preceding filtration.

The above data represent *average* values; in fact, under certain circumstances, it is possible that a capillary filters but does not resorb, or vice versa.

The above pressure values only apply at the level of the heart; lower down, for example the "weight" of the blood column (**hydrostatic pressure**) considerably increases the pressure in the capillary lumen (feet, standing: + 12 kPa!). Especially when standing still, the high Δ P values in the foot capillaries (arterial, about 16 kPa = 120 mmHg) lead to high local filtration rates, which, as in the kidney, are compensated because $\Delta\pi$ also rises (disproportionately; → p. 336, E) due to this efflux of water.

The following factors can influence **transcapillary exchange of fluid**, and in some cases may lead to **edema** (see also p. 142).

(a) *A change in blood pressure at the arterial end* of the capillary.

(b) *A change in the venous pressure* at the end of the capillary as a result of variations in the contraction of the venules. In the case of *venous stagnation* (e. g. due to cardiac insufficiency), the venous pressure may be pathologically high with the result that filtration predominates (→ **B 1**) and fluid collects in the interstitium (*edema*). At a high portal venous pressure (e. g. in liver cirrhosis), fluid is filtered into the peritoneal space: *ascites*.

(c) *Changes in the protein content of the plasma affect* $\Delta\pi$ (→ **B 2**) The alteration in $\Delta\pi$ is much greater than would be expected from van't Hoff's law (→ p. 336, E).

(d) *If protein permeability is increased* ($\sigma < 1$), for example, due to histamine (→ p. 72), filtration predominates (*edema*).

(e) *A reduction in lymphatic drainage* (due to blockage of the lymphatics) leads to an accumulation of fluid in the interstitium and, again, to edema.

Any increase in interstitial volume results in a local rise in pressure and consequently in a decrease in Δ P. Edema therefore only increases the interstitial space to the point at which a new equilibrium is established between filtration, on the one hand, and resorption and lymphatic drainage on the other.

Although dissolved particles are dragged through the capillary wall with the water during filtration and resorption (*solvent drag*; → p. 10), the **exchange of solutes by diffusion** (→ p. 8 f.) plays a quantitatively far greater role. If the substance concerned is present in the same concentration on each side of the capillary wall, inward and outward diffusion are equal, with a resulting net diffusion of 0. If, however, there is a concentration difference between plasma and interstitium, net diffusion of the solute in question takes place. In this way nutrients and O_2 leave the circulating blood, and CO_2 and products of metabolism diffuse in the opposite direction.

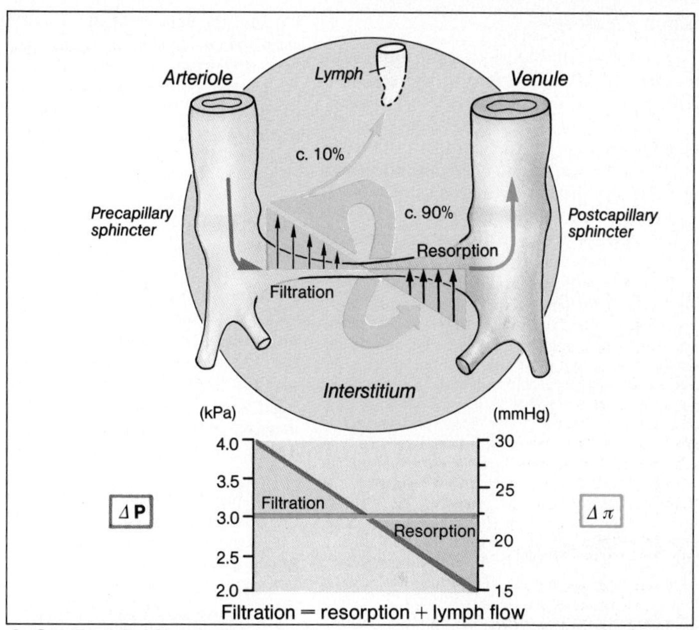

A. Capillary fluid exchange

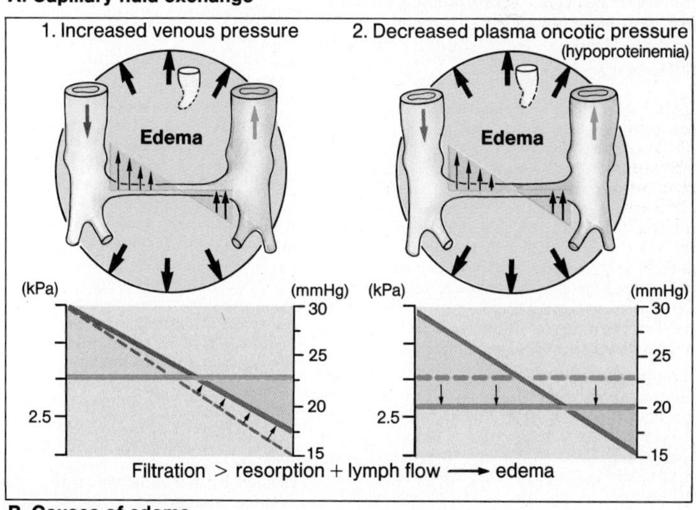

B. Causes of edema

Blood Pressure

The general term blood pressure applies to the **arterial blood pressure** in the systemic circulation. It fluctuates with each heart beat between a maximum value (**systolic blood pressure** P_s) during cardiac systole (→ p. 162) and a minimum value (**diastolic blood pressure** P_d) during cardiac diastole (→ **A** and **B 1**). Their geometric mean (see below) is the **mean pressure** ($\bar{P}$); their difference is known as **pulse pressure** (PP).

PP is principally a function of stroke volume (**SV**; → p. 154) and arterial compliance or capacitance (= volume change/pressure change). At a given SV and a *falling compliance* (vessels become more rigid), P_s increases more than P_d and, as a consequence, PP increases (common in old age). A *rising SV* at a given compliance also causes a greater increase in P_s than in P_d (PP increases). At a given heart rate and SV, $\bar{P}$ is proportional to TPR (→ p. 156). If *TPR rises* and the SV is ejected as fast as before, both P_s and P_d increase to the same extent, thus leaving PP unchanged. However, the TPR rise may prolong SV ejection. In this case, the ratio arterial volume increase/peripheral runoff during ejection becomes smaller. As a consequence, P_s rises less than P_d, and PP is lowered.

At rest (sitting or lying), the systolic blood pressure $\mathbf{P_s}$ (measured on the upper arm) is normally about **16 kPa (120 mmHg)** and the diastolic pressure $\mathbf{P_d}$ is roughly **10.7 kPa (80 mmHg)**.

Blood pressure can be **measured** *directly* by means of a needle situated in the blood stream (→ **A** and **B 1**; blood pressure curve) or *indirectly* using an inflatable *cuff* (Riva-Rocci). The cuff is applied to the upper arm and inflated until, with certainty, the cuff pressure exceeds the anticipated P_s. The cuff pressure is gradually released, with the operator at the same time listening through a stethoscope applied to the inner angle of the elbow. The pressure indicated on the dial (in mmHg or in kPa) at the moment when a hissing noise is heard is the systolic pressure. At a lower pressure, the diastolic pressure, the sounds become quieter.

It is the **mean blood pressure** (→ **A**) that is decisive for the blood flow through an organ. It can be determined graphically by drawing a line through the directly measured blood pressure curve in such a way that equal areas above and below this line are enclosed by the blood pressure curve. The position of this line indicates the magnitude of the mean blood pressure. Although it drops from the aorta to the femoral artery, the *systolic* pressure in the latter may exceed that in the aorta (→ **A 1** and **A 2**).

Optimal regulation of the blood pressure is essential (→ p. 176 ff.). If the (mean) blood pressure is *too low*, the result may be **shock** (→ p. 186), *anoxia* (→ p. 102), or even death of the tissue. Chronically elevated blood pressure (**hypertension**, → p. 180) is injurious because the blood vessels (especially those of the heart, brain, kidneys, and retina) also suffer damage.

The **blood pressure in the pulmonary artery** is lower than that in the aorta (→ **B 2**), the systolic pressure amounting to approx. 3.3 kPa (25 mm Hg) and the diastolic pressure to about 1.3 kPa (10 mmHg). The pulmonary circulation therefore belongs to the *low-pressure system* (→ p. 154). The surroundings of the pulmonary vessels (air-filled lung tissue) are highly compliant. An increase in the pulmonary volume flow (e. g. during exercise) thus results not so much in a rise in pulmonary blood pressure as in dilatation of the pulmonary vessels (lowering of resistance). At the same time they act as a kind of short-term **reservoir** (→ p. 94 and p. 184).

Whereas the arterial blood pressure primarily depends upon cardiac output and TPR (→ p. 156), the **blood pressure in the veins** is chiefly determined by blood volume and capacitance, and amounts to only 0.2–0.5 kPa (1.5–4 mm Hg) in the veins near the heart.

Such low pressure inside the vessels means that their width, which is determined by the transmural pressure (→ p. 156 and p. 184), depends to a very large extent on the *pressure of the surroundings*. Within the thorax this (*intrathoracic*) *pressure varies* with respiration (→ **B 4** and p. 80) so that the width of the venae cavae fluctuates, and respiration thus has a *pump-like action* on the venous return to the heart (→ see also p. 184). At inspiration the intrathoracic pressure drops (→ **B 4**) more than the mean pressure in the venae cavae (→ **B 3**), which leads to an increase in transmural pressure (→ **B 5**), in dilatation of the veins, and thus to an increased venous return to the right side of the heart. By means of the Frank-Starling mechanism (→ p. 182 ff.), the stroke volume of the right ventricle (→ **B 7**) and the flow in the pulmonary artery (→ **B 6**) are then temporarily increased. At the same time, the stroke volume of the left ventricle drops slightly, due to the dilatation of the pulmonary vein resulting from inspiration, a temporary reduction in the amount of blood entering the left side of the heart also occurs.

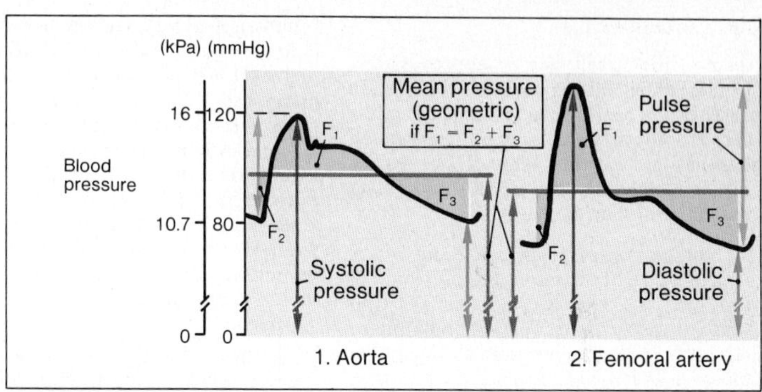

A. Arterial pulse waves

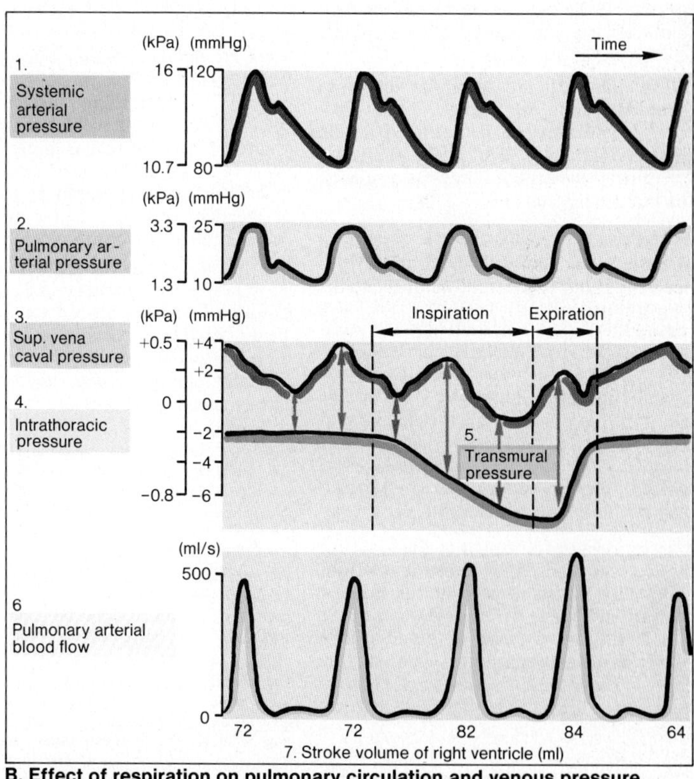

B. Effect of respiration on pulmonary circulation and venous pressure (schematic)

The Cardiac Cycle

At rest, the **heart rate** is approximately 70/min, which means that the **four action phases of the heart** are accomplished in less than a second. These are the **contraction phase (I)** and **ejection phase (II) of systole** and the **relaxation phase (III)** and **filling phase (IV) of diastole** (→ A).

The **cardiac valves** ensure that the blood flows in the right direction, that is, from the atria into the ventricles (phase **IV**) and thence into the aorta or the pulmonary artery (phase **II**). Opening and closing of the valves (→ **A 10**) is determined by the pressure on either side of the valve: if, e.g., the pressure in the left ventricle is higher than that in the aorta, the aortal valve is open; if lower, the valve is shut.

The mechanical phases of cardiac action are preceded, within the heart itself, by electrical events that can be recorded as the **ECG** (→ **A 1** and p. 168 ff.). The events are not exactly synchronous in the two sides of the heart (→ **A 1 a**).

The following sequence of events constitutes a cardiac cycle: **atrial systole** (phase **IV c**): while the ventricle is still in diastole, depolarization of the pacemaker of the heart (sinoatrial node; → p. 164) initiates excitation of the atrial musculature (**P wave** in ECG; → **A 1**), which then contracts (elevated atrial pressure in phase **IV c**; → **A 4**). The blood thus expelled completes the filling of the ventricle (see below). This is the end of diastole. Normally, the **end-diastolic** (ventricular) **volume** is about 125 ml (→ **A 6**), but may rise to as much as 250 ml.

The electrical excitation within the heart has by this time reached the ventricle (**QRS** in **A 1**), which then contracts. In this **phase of contraction** (phase **I**) all four valves are closed, i.e., the volume of blood in the ventricles remains constant (→ **A 6**), but the pressure rises rapidly (**isovolumetric contraction**). When the pressure in the left ventricle exceeds that in the aorta (10.7 kPa or 80 mmHg; → **A 2** and **A 3**) the semilunar valves open. This marks the beginning of the **ejection phase** (phase **II**), during which the pressure in the left ventricle and in the aorta briefly rises to a maximum of about 16 kPa (120 mmHg; → **A 2** in phase **II b**). (Systolic and diastolic pressures in the pulmonary artery are 3.3/1.1 kPa or 25/8 mmHg; → p. 94). Following ejection of the blood (→ **A 6** and **A 7**) the ventricles relax (isovolumetric **relaxation**, phase **III**), and their pressure immediately sinks below that of the aorta and the pulmonary artery, respectively (→ **A 3**). This causes closure of the semilunar valves and marks the beginning of the **filling phase, IV**, the first part of **diastole**.

In the meantime the atria have refilled. A decisive role is played by the suction effect resulting from the downward displacement of the level of the valves during the ejection phase. This lowers atrial pressure and furthers atrial filling, thus ensuring that the ventricles fill rapidly at the very beginning of phase **IV a** (→ **A 6**). Active **atrial contraction** (phase **IV c**, see above) contributes only about 15% to ventricular filling at normal heart rates. At higher heart rates, the cardiac cycle becomes shorter at the expense of the diastole. Here, active atrial filling becomes more important.

Blood is supplied to the cardiac musculature via the *coronary arteries* (→ **A 8** and p. 188). **Coronary blood flow** takes place almost *only during diastole* (especially in the left ventricle) since the vessels are compressed by the contracting cardiac muscle during systole (transmural pressure < 0).

Normal cardiac function can also be detected acoustically. The first of the **heart sounds** (→ **A 9**) is produced in the phase of isovolumetric contraction and the second at closure of the semilunar valves. Pathological **heart murmurs** result from too tight (*stenosis*) or leaky (*insufficiency*) valves, or from other abnormalities.

The cardiac cycle is also reflected in the waves of the **venous pulse** (→ **A 5, a, c, x, v, y**). The positive *a wave*, e.g., is caused by atrial contraction; the negative *x wave*, by depression of the level of the valves.

The intermittent pumping of the heart produces a **pulse wave** (→ **A**, "Windkessel"effect), which spreads through the arterial vessels at the **pulse wave velocity**. This is much higher than the flow velocity (→ p. 156) and is greater the thicker and more rigid the vessel walls (rises in hypertension and with age) and the less the diameter of the vessel (aorta: about 6 m/s; a. radialis: about 10 m/s).

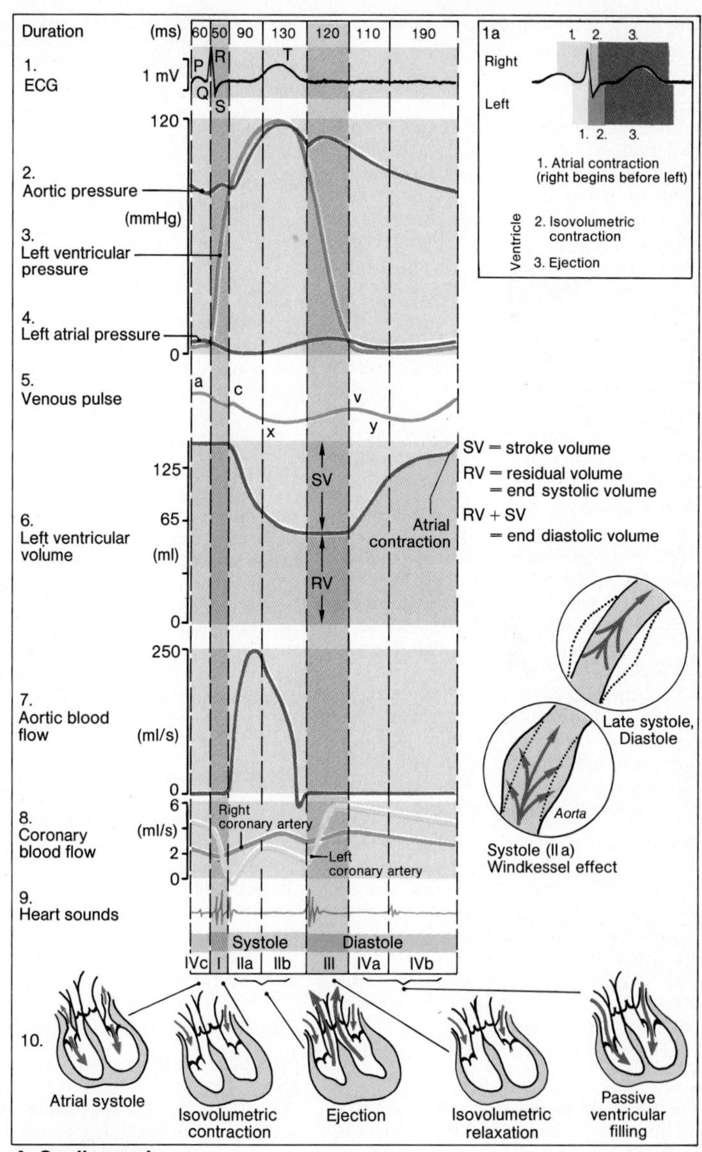

A. Cardiac cycle

Excitation and Conduction in the Heart

The cardiac musculature consists of *two types of muscle cells* (or fibers): (1) cells that initiate and conduct impulses; and (2) cells that, besides conducting, respond to stimuli by contracting. The latter constitute the working musculature of the heart (*myocardium*). Contrary to the situation in skeletal muscle (→ p. 32), excitation is generated within the organ itself (in cells of type 1, see above): **autorhythmicity** or autonomy of the heart.

The myocardium of the ventricles is a *functional syncytium*, that is, the cells are not insulated from one another; a stimulus arising at any point in the ventricle leads to complete contraction of both chambers (**all-or-none contraction**). The same applies to the atria.

Excitation of the heart normally proceeds from the **sinoatrial (SA) node** (→ **B** and **C**), which is the **physiological pacemaker** of the heart (→ **A**). From the SA node, excitation spreads through both *atria* to the **atrioventricular (AV) node** (→ **A** and **B**), whence, via the *bundle of His* and its two *branches* (Tawara), it reaches the *Purkinje network*, which carries the impulse to the ventricular muscle. Activation of the ventricular muscle takes place from endocardium to epicardium and from the apex of the ventricles to the base. This can be followed on the **ECG** of the intact organism (→ **B** and p. 168).

The SA node and the conducting system do *not* have a constant resting potential; instead, the cell slowly depolarizes immediately after each repolarization (the most negative value of which is known as the **maximum diastolic potential (MDP)**: This slow depolarization (**pacemaker potential, prepotential**) continues until the **threshold potential (TP)** is reached again and the next **action potential** is set off (**AP**; → pp. 26 and 45 and **D**).

The **AP of the pacemaker cells** results from the following changes in conductances (**g**) and in ionic currents (**I**): beginning with the **MDP** (SA node: about − 70 mV), g_k decreases continuously. Although g_{Ca} and g_{Na} are low at this point, it is I_{Ca} and I_{Na} that lead to the slow depolarization or *prepotential.* Slowly, g_{Ca} also begins to rise (and to some extent g_{Na} as well) so that a higher I_{Ca} contributes to the late prepotential. At the **TP** (about − 40 mV in the SA node) g_{Ca} rises relatively rapidly, but then is inactivated again whereas g_k now rises steeply. Following a moderately steep rise and a "rounded off" maximum of the AP the cell is again repolarized to the MDP (→ **D**).

In the working **myocardium**, the rapid rise in the AP (→ p. 31, A) is due to the brief but fast influx of Na^+ (→ p. 42). In contrast, in the sinus and AV nodes, where the density of the Na^+ channels is much lower, the rise in the AP (mainly caused by Ca^{2+} influx) is relatively slow.

Every AP in the SA node causes a heartbeat. In other words, the frequency of impulses from this pacemaker determines the heart rate, which can therefore be modified (e.g. lowered) by the following **changes in pacemaker potential** (in the SA node): (1) If the *TP becomes less negative*, so that the threshold is reached later (→ **D 1**); (2) If the *slope* of the prepotential becomes *less steep*, so that the (unaltered) threshold is attained later (→ **D 2**); (3) If the *MDP* becomes *more negative*, thus, taking more time to attain the TP (→ **D 3**); (4) If *repolarization* following an AP is slower (flatter).

The dominant role of the SA node in normal cardiac excitation is due to the fact that the lower portions of the system responsible for excitation and conduction in the heart have a lower inherent pacemaker frequency than the SA node (reasons (2) and (4) above). This means that excitation from the SA node arrives before the spontaneous depolarization of the lower portions has reached its own TP.

Control of Excitation of the Heart

Although, due to its autonomy, the heart is able to beat even without the participation of external nerves, **adaptation of cardiac activity** to the changing needs of the organism (→ p. 48) is to a large degree dependent upon intact cardiac nerves (see below). The following aspects of cardiac activity can be

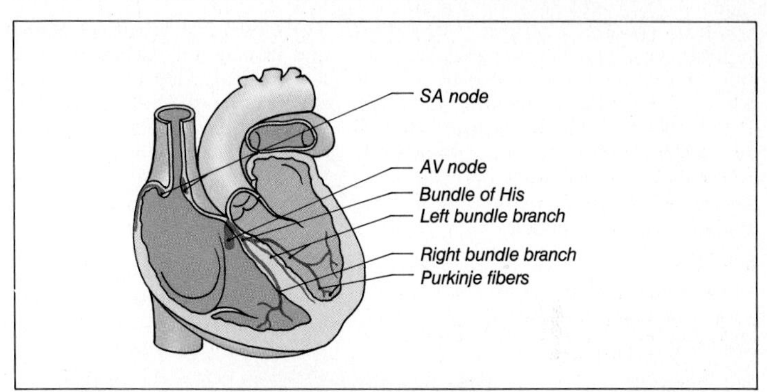

A. Excitation and conducting systems of the heart

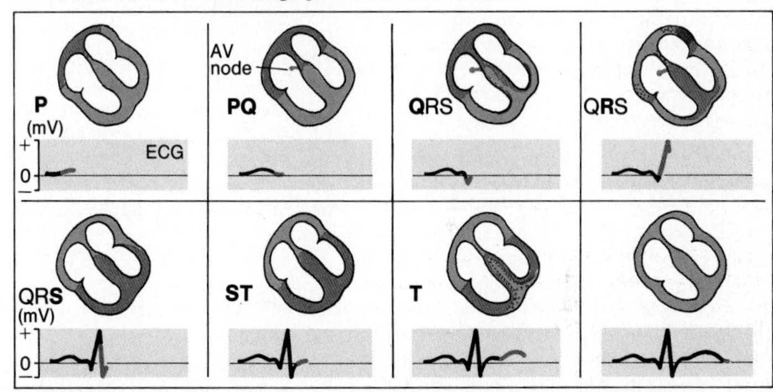

B. Correlation of ECG with de- and repolarization wave in the heart

Event		Time (ms)	ECG	Conduction velocity (m/s)	Intrinsic auto-maticity (min^{-1})
SA node					
Impulse generated		0	P wave	0.05	70–80
Atrial depolarization	right	50		0.8–1.0	
	left	85			
AV node					40–60
Arrival of impulse		50	P-Q interval (delay in excitation)	0.05	
Departure of impulse		125			
Bundle of His activated		130		1.0–1.5	20–40
Bundle branches activated		145		1.0–1.5	
Purkinje fibers activated		150		3.0–3.5	
Endocardium depolarized	right ventricle	175	QRS complex	1.0 in myocardium	
	left ventricle	190			
Myocardium depolarized	right ventricle	205			
	left ventricle	225			

C. Excitation of the heart: Time course, ECG and conduction velocity

modified: (1) The *frequency* with which impulses are initiated in the pacemaker, and thus the heart rate (**chronotropism**); (2) The speed of conduction of excitation, especially in the AV node (**dromotropism**); (3) The *force* of contraction, that is, the *contractility* of the heart (**inotropism**); (4) The *excitability*, via alteration in the threshold of excitation (**bathmotropism**).

The **efferent cardiac nerves**, via which the function of the heart can be influenced (→ p. 51 ff.), are **parasympathetic** fibers of the **vagus** (cholinergic) and **symphathetic nerves**. The heart rate is lowered by the vagal fibers that terminate at the SA node (**negative chronotropic effect**), and raised by those of the sympathetic nerves via β_1-adrenoceptors (**positive chronotropic effect**). The effects result from an alteration in the slope of the prepotential (→ **D 2** and **E 1**) and a change in the MDP in the SA node (→ **D 3**).

The flattening off of the prepotential and the increase in negativity of the MDP under vagus influence result from an *increased* K^+ *conductance* (g_K; → p. 26); the *increased slope* of the prepotential under the influence of the sympathetic nerve, or epinephrine, is due to an *elevated* g_{Ca} and in some cases to a *decreased* g_K. In the lower-lying portions of the conducting system, only the sympathetic nerve has a chronotropic effect, so that it plays a decisive role in the event of the pacemaker function being taken over by this part of the conducting system (see below).

The left branch of the vagus nerve retards, and the symphathetic nerves accelerate the **transmission** of the stimulus in the **AV node** (→ p. 164): **negative and positive dromotropic effects**, respectively. The main influence is on the steepness of the prepotential (→ **E 2**) and on the MDP (→ **D 3**). Here too, changes in g_K and g_{Ca} play important roles.

Whereas in chronotropism and dromotropism it is the conducting system that is influenced by the sympathetic and parasympathetic nerves, contractility can be enhanced by direct sympathetic stimulation of the working myocardium: **positive inotropic** effect. The underlying effect is a rise in the intracellular concentration of $\mathbf{Ca^{2+}}$ (= $[\mathbf{Ca^{2+}}]_i$).

The **myocardial action potential** (→ pp. 42 and 45) causes the **release of** $\mathbf{Ca^{2+}}$ from the longitudinal system but in smaller quantities than in the skeletal muscle (→ p. 34 ff.). During the AP there is also a $\mathbf{Ca^{2+}}$ **influx from the ECS** into the myocardial fibers via Ca^{2+} channels, which is probably what triggers the release of the intracellular stores of Ca^{2+}. The influx of Ca^{2+} from outside the cell increases when the extracellular $[Ca^{2+}]$ is raised; it can be enhanced β_1-adrenergically (direct positive inotropic sympathetic effect) and can be inhibited pharmacologically by blockers of the Ca^{2+} channels, also known as *Ca^{2+} antagonists*. The $[Ca^{2+}]_i$, which is crucial for electromechanical coupling (→ p. 36 ff.), depends not only on the influx but also on the $\mathbf{Ca^{2+}}$ **efflux** from the sarcoplasm. It proceeds via primary-active "pumps" (Ca^{2+}-ATPase) into the longitudinal system and into the ECS. Ca^{2+} also enters the latter in exchange for Na^+ (*Ca^{2+}/Na^+ antiport*). This secondary-active outward transport of Ca^{2+} is driven by the Na^+ gradient established by the Na^+-K^+-ATPase (→ p. 11).

Disturbances in Cardiac Excitation

(see also p. 174)

Changes in **serum electrolyte concentrations** influence heart excitation. One of the results of a *slight* degree of **hyperkalemia** is an elevation of the MDP in the sinoatrial node, and therefore possibly a positive chronotropic effect. In *severe hyperkalemia* the more positive MDP leads to inactivation of the Na^+ channels (→ p. 26), that is, to a decrease in the slope and amplitude of the AP in the atrioventricular node (negative dromotropic effect). In addition, there is a rise in g_K so that the prepotential becomes less steep (negative chronotropic effect, → **D 2** and **E 1**) and repolarization of the myocardium is accelerated, thus reducing the $[Ca^{2+}]_i$ in the cell. The result is a negative inotropic effect and reentry phenomena in the myocardium (see below). In extreme cases the pacemaker activity even comes to a stop (*cardiac arrest*). **Hypokalemia** (moderate) has positive chronotropic and inotropic effects (→ **E**). **Hypercalcemia** probably raises g_K, thus reducing the duration of the AP.

Temperature also influences excitation of the heart. *Fever*, for example, has a positive-chronotropic (→ **E 1**) and a negative-inotropic effect, whereas the effect of *cooling* is negative-chronotropic, negative-dromotropic (→ **E 1**), and positive-inotropic.

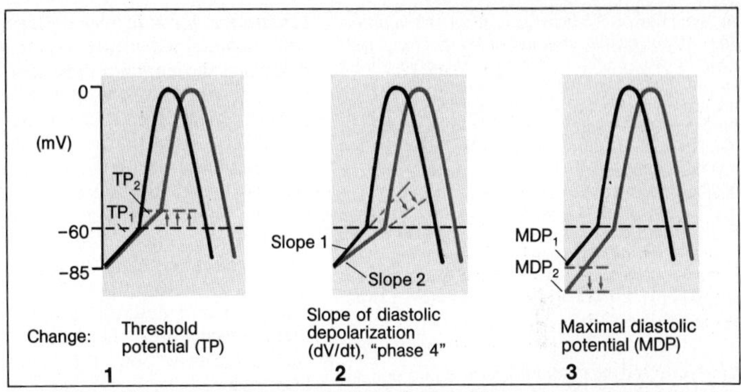

D. Changes in the pacemaker potential influence (here: decrease) heart rate

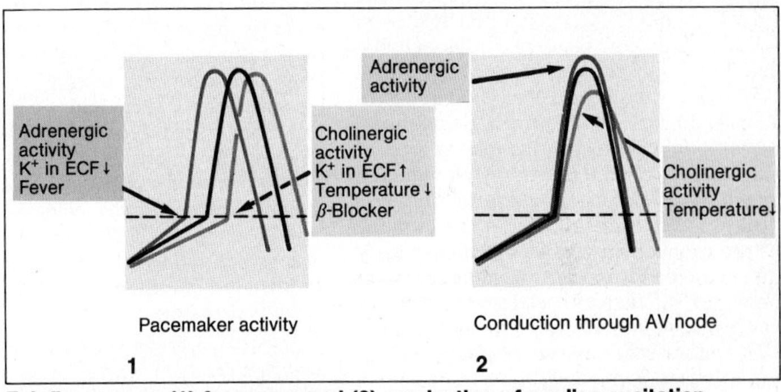

E. Influences on (1) frequency and (2) conduction of cardiac excitation

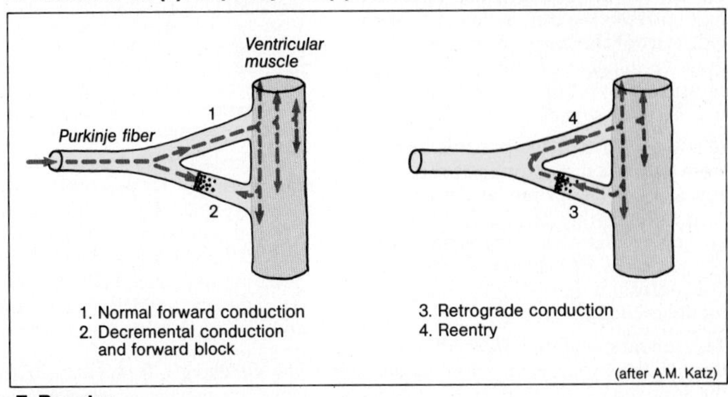

F. Reentry

Arrhythmias of the heart can affect the *initiation* and the *conduction of impulses* (→ p. 174). Probably the most common cause of ventricular flutter and fibrillation is a circus movement (**reentry**) of the impulse (→ p. 167, F). One possible reason for this is that the forward conduction of the impulse is blocked at the junction between Purkinje fiber and myocardium (→ p. 167, F 2), but the impulse can reach the myocardium from the other direction (→ p. 167, F 1 and p. 167, F 3), reenters the Purkinje fiber, and takes the same pathway again (→ p. 167, F 4). In the healthy myocardium a sharp reduction in the duration of the AP (e. g. in hyperkalemia, see above) and consequently in the refractory period, can result in reentry.

Electrocardiogram (ECG)

The ECG is a graphic recording of the changes occurring in the electrical **potentials** (in **mV**; → p. 329) between different sites on the skin (*leads*) as a result of cardiac activity. The ECG thus reflects the *electrical* events connected with cardiac excitation and provides information about the anatomical orientation of the heart, the relative sizes of the heart chambers, heart rate, rhythm and origin of excitation, spread of the impulse, decay of excitation, and disturbances in the above events, irrespective of whether they are due to anatomical, mechanical, metabolic, or circulatory defects. Changes in electrolyte concentration (→ pp. 166 and 172) and the effects of certain pharmacological agents (e. g. digitalis) can be detected on the ECG. The ECG gives **no** direct information about the contraction and pumping efficiency of the heart; these properties can only be judged on the basis of blood pressure (→ p. 160), cardiac output (→ p. 154), heart sounds (→ p. 162), and so on.

It is assumed that the potentials recorded from the body surface originate between the stimulated and nonstimulated regions of the myocardium, that is, the ECG curve describes the **migration** of the **excitatory front**. No potentials are detectable on the ECG of a myocardium that is, either nonexcited or totally excited (→ p. 170).

The migration of the wave of excitation through the myocardium is accompanied by the development of innumerable potentials, varying in their amplitude and direction. Parameters that also have a direction are in general known as *vectors* and can be graphically represented as arrows. In this particular case the length of the arrow represents the amplitude of the potential, and its direction indicates that of the potential (arrow tip +). Just as a parallelogram of forces can be constructed, so can an **integral vector** (→ **A**) be constructed from any desired number of single vectors. In the heart, the size and direction of the integral vector changes during the progress of excitation, that is, the tip of the arrow of the integral vector describes loops (→ **C**). In a **vector cardiogram**, these vector loops are shown directly on the screen of an oscilloscope.

Recordings from the extremities and thoracic wall, as usually performed in clinical practice (indirect leads: electrodes located on the skin), can also be used to follow the time course of the integral vector by a recorder. Here, the vector is projected to scale (scalar ECG). Each ECG lead gives a unidimensional picture of the integral vector, so that with *two leads* in one plane the integral vector *in this plane* (usually the frontal plane) is determined. For a *three-dimensional estimate* of the integral vector, at least one additional lead in a different plane is necessary (→ **F**).

It should be remembered that the magnitude of the potentials measured via the individual leads can only by compared with each other if the electrical resistance (distance, type of tissue) between the heart and the recording electrodes is equal. This condition is largely fulfilled in the case of the leads from the three extremities. In recording a vector diagram (see above), inequalities in resistance can be compensated by the insertion of electrical resistances (corrected, orthogonal *leads of Frank*).

The standard limb leads, I, II, and III of Einthoven (→ **D**) are *bipolar leads* in the frontal plane. Recordings are made from electrodes situated on the skin of both arms and of the left leg, and the changes in the potential difference between the two arms (I), right arm and left leg (II), and left arm and left leg (III) are measured (→ **D**).

The **Goldberger leads**, also in the frontal plane, are *augmented unipolar leads* from the extremities, again from both arms and the left leg; but in this case the electrodes from

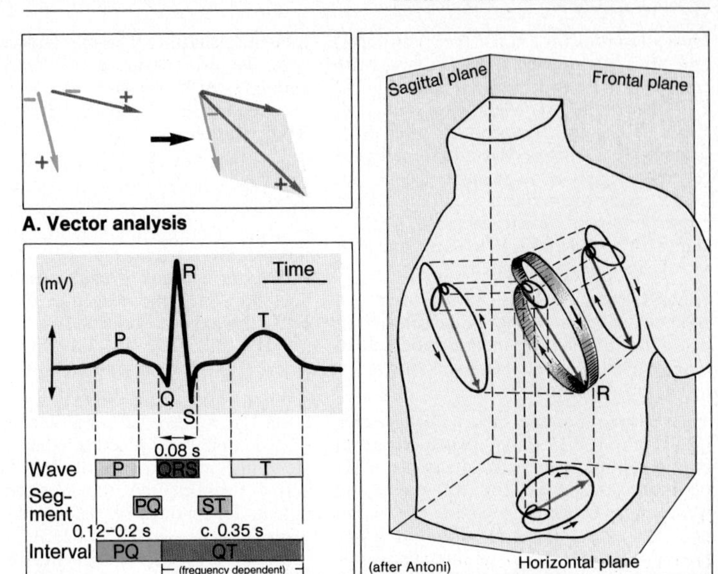

A. Vector analysis

B. ECG

C. Spatial vector loops

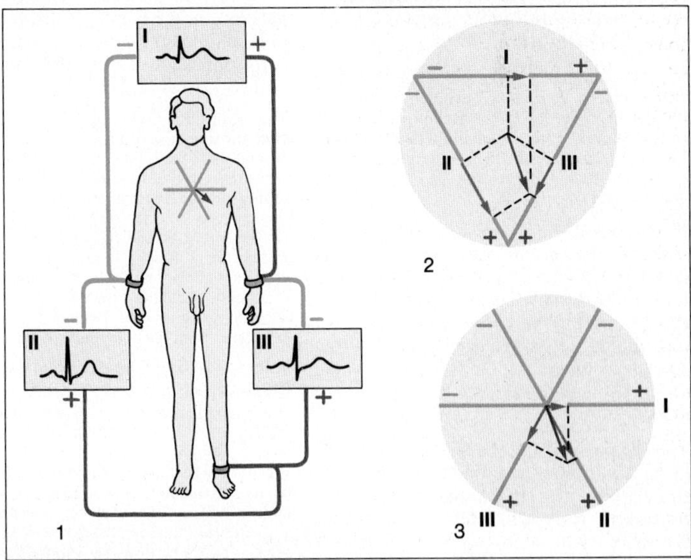

D. Bipolar leads (I, II, III) of Einthoven

two extremities are coupled (via resistances) and serve as an *indifferent electrode* with respect to the third, *different* electrode (→ **E**). The excitatory events are not only "seen" from another aspect than when using leads I–III (see below) but, additionally, the recorded signal is augmented. The Goldberger leads are named according to the corresponding different electrode: aVR = right arm (→ **E 1**), aVL = left arm, and aVF = left leg (a = augmented).

An **ECG curve** (→ **B**) is characterized by a series of deflections or **waves**, the convention being that a positive potential produces an *upward* deflection, a negative potential a *downward* deflection.

The **P wave** represents *atrial depolarization* (≤ 0.2 mV, ≤ 0.11 s). The wave for atrial repolarization is *not* visible on the ECG because it is masked by the succeeding waves. The **Q wave** (≤ 0.04 s; mV < 1/4 of R), the **R** and the **S waves** (R + S = 0.6–2.6 mV), together constituting the **QRS complex** (≤ 0.11 s) (it is given the same name even if one of the three components is missing), represent *ventricular depolarization.*

In the QRS complex, the convention is that every initial negative deflection is denoted by Q, every positive deflection (with or without preceding Q) by R, and every negative deflection that follows R, by S. This may result in a situation in which deflections of the QRS bearing the same name are not synchronous in all leads.

Next comes the **T wave**, which reflects *repolarization of the ventricles*. Although depolarization and repolarization are contrary processes, the T wave and the R wave usually point in the same direction (in most recordings +), which indicates that the spread of activation and its decay take different pathways through the myocardium.

The **PQ segment** and the **ST segment** (→ **B**) normally lie on or near the 0 mV line. Fully stimulated atria (**PQ segment**) and ventricles (**ST segment**) therefore produce *no* externally detectable potential. The **PQ interval** (0.12–0.21 s; → **B** and p. 165, C) is the time elapsing between the beginning of atrial excitation and the beginning of ventricular excitation. The **QT interval** varies with the heart rate: at 75 beats/min it amounts to 0.35–0.40 s; this is the time required for ventricular depolarization and repolarization.

The various ECG recordings "observe", for any one moment, the integral vector of cardiac activity from different angles. A recording (lead) of the potential parallel to the integral vector thus shows the full size of deflection, whereas a lead at right angles does not "see" the deflection. With leads I–III, the vector is observed from three sides (→ **D 2** and **D 3**); the Goldberger leads record from three additional aspects (→ **E**).

If the potentials from two of these leads (e. g. I and II) are recorded simultaneously (e. g. QRS complex) it is possible to construct the mean integral vector (in the frontal plane) and, at the same time, the amplitude of the potential from the other leads in the frontal plane (e. g. III; → **D 2** and **D 3**). A similar procedure is used for the practical determination of the electrical axis of the heart, or, in other words, the **mean QRS vector**, whose positions in the frontal plane, if the spread of excitation takes a normal course, roughly corresponds with the anatomical longitudinal axis of the heart.

The mean QRS vector can be calculated exactly from the sum of the area as occupied by the Q, the R, and the S waves (taking into account whether they are + or –). In practice, however, it suffices to measure the height of the waves and to subtract the negative from the positive. If this is done for two leads (e. g. in lead I: 0.5 mV–0.5 mV = 0 mV; in lead II: + 1.1 mV [→ **G 1**]) it is possible to construct the *electrical axis of the heart*. (The evaluation of the third lead, in this case lead III, is superfluous since the potential in III can be calculated from the potential in II minus that in I.)

Axis (→ G, H). The electrical axis or mean QRS vector normally points to the left, downward and posteriorly (normal limits, → **H**).

When it points outside the normal limits, **right** or **left axis deviation** (RAD or LAD) is noted. Body build influences the axis; stocky, squat people have a horizontal axis, sometimes with LAD; thin, tall subjects have a vertical axis. Infants have a *relative* right ventricular hypertrophy that is seen on the ECG as RAD. As the individual ages, the axis shifts toward the left. Increased pulmonary artery

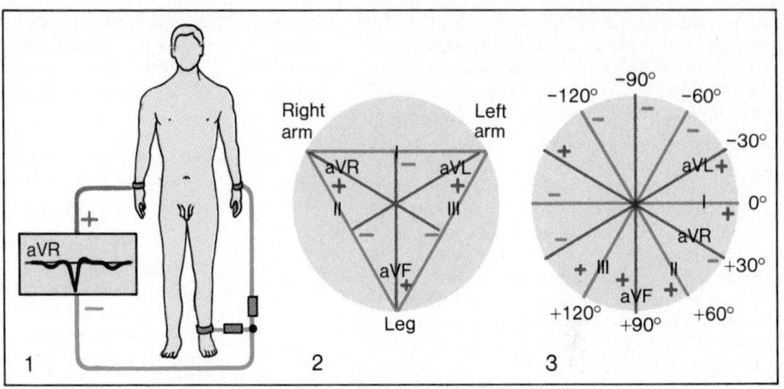

E. Unipolar augmented leads in frontal plane (Goldberger)

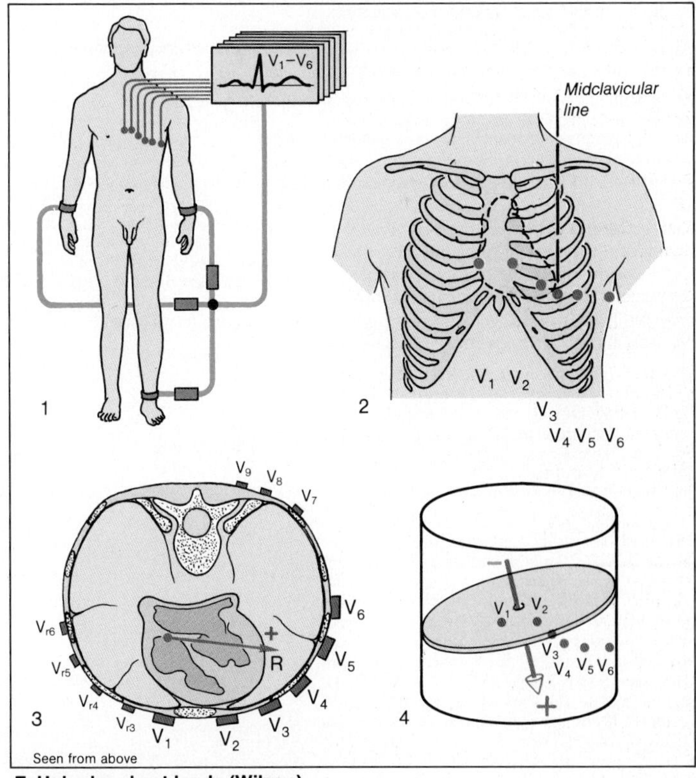

F. Unipolar chest leads (Wilson)

resistance means increased work for the right ventricle and causes hypertrophy of the right ventricle and RAD. In left ventricular hypertrophy (e.g. in hypertension) there is commonly LAD. Large myocardial infarcts also may shift the axis.

If the six *precardial leads* according to *Wilson*, $\mathbf{V_1-V_6}$, are combined with the above leads in the frontal plane, a *three-dimensional* view of the integral vector is obtained. V_1-V_6 are chest leads, the individual exploratory electrodes being situated approximately in the *horizointal plane* (→ **F**). The indifferent electrode (**central terminal**) used for V_1-V_6 is made by connecting the three limb leads by high (5 kΩ) resistances (→ **F 1**). Chest leads are useful in recording vectors pointing backwards since they produce very small potentials, or none at all, in the frontal plane. Since the mean QRS vector usually points to the left and somewhat posteriorly, the chest (represented as a cylinder for simplification) is divided into + and – halves (→ **F 4**), with respect to this vector. Therefore the QRS vector is usually – in V_1 and V_2 and + in V_5 and V_6.

For specific purposes the above 12 leads can be supplemented by (a) an electrode situated behind the heart (swallowed, recordings via esophagus), and with additional chest electrodes (b) on the left back (V_7-V_9) or (c) on the right thorax ($V_{r3}-V_{r6}$) (→ **F 3**).

Repolarization of the ventricle produces its own vector loop (→ **C**), which is recorded as the **T wave**. It is therefore possible to construct a spatial *T vector* from the various recordings. It normally forms an angle of 60° (maximum) with the mean QRS vector.

This angle increases in older individuals, possibly indicating O_2 deficiency in the heart. A *QRS-T angle* of 180° usually points to a pathological condition that may be (1) excessive ventricular pressure, (2) bundle block, (3) a digitalis effect. The QRS duration and the QT interval can be used to distinguish between these possibilities (→ **J**).

In **myocardial infarction**, the blood supply to certain areas of the myocardium is obstructed. The muscle tissue at the *center* of the infarct dies off (*necrosis*), which means that *no depolarization* can take place at this site. During the first 0.04 seconds of ventricular activity an "0.04 vector" is produced, "pointing away" from the infarct. Since the infarct is usually in the left ventricle, and the mean QRS vector also points to the left, the "0.04, vector" will be the opposite of the mean QRS vector (→ **K**); that is, for example, if there is a high positive R, there is also a *larger negative Q-wave* (→ **K 2**). Between the necrotic region of the myocardium and the normal surrounding tissue is a zone in which the *blood supply is reduced*, with consequent disturbances in excitatory properties.

The abnormal repolarization frequently results in *reversal* of the *T wave* (becomes negative in most recordings; T vector "points away" from the ischemic zone of the infarct; → **K**). Additionally, in the acute stage of the infarct the ST segment is usually displaced above or below the 0 line ("injury potential" of the zone of injury). The injury potential of the damaged area of the myocardium deforms the QRST region of the ECG in the direction of the monophasic AP of the myocardium (→ p. 31, A 3). This is known as the *monophasic deformation* of the ECG in the fresh infarct (→ **K 1**). The displaced ST segment is the first to return to normal (→ **K 2**), whereas the abnormal T wave can still be seen months later (→ **K 2**). The 0.04 vector (*enlarged Q*) occurs, at the earliest, after several hours (→ **K 2**), but may still be clearly visible after many years have elapsed (→ **K 3**).

ECG and serum electrolytes

Changes in the concentrations of $\mathbf{K^+}$ and $\mathbf{Ca^{2+}}$ in the serum can lead to disturbances in the excitation of the myocardium and hence to **alterations in the ECG**. At $[K^+] > 6.5$ mmol/l the T wave is higher and pointed, and conduction is defective, leading to an extended PQ and a wider QRS. In extreme cases, cardiac arrest occurs (→ p. 166). At $[K^+] < 2.5$ mmol/l ST is lowered, T is biphasic (first +, then –), and an additional, positive *U wave* follows on the T wave. At $[Ca^{2+}] > 2.75$ mmol/l (5.5 meq/l), the duration of QT is reduced at the expense of the ST segment; at $[Ca^{2+}] < 2.25$ mmol/l (4.5 meq/l), QT duration is increased.

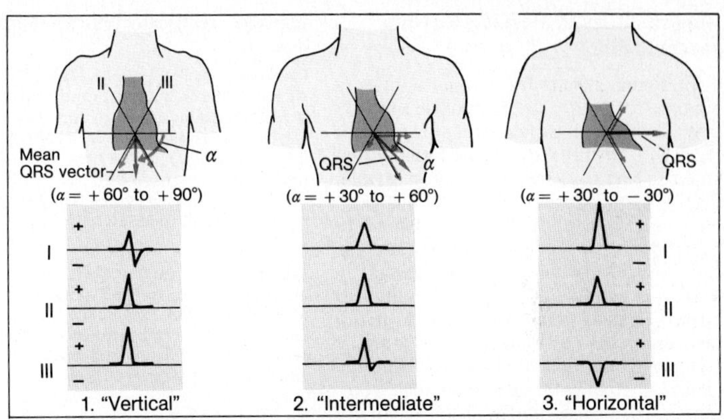

G. Normal mean QRS vectors determined from limb leads I–III

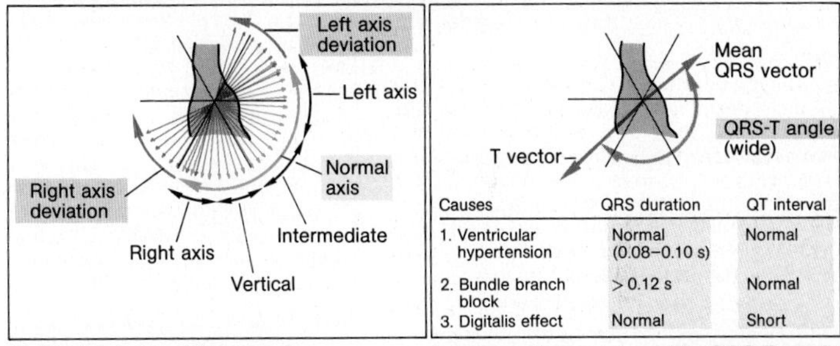

Causes	QRS duration	QT interval
1. Ventricular hypertension	Normal (0.08–0.10 s)	Normal
2. Bundle branch block	> 0.12 s	Normal
3. Digitalis effect	Normal	Short

H. Electric axis of the heart

J. Interpretation of a wide QRS-T angle

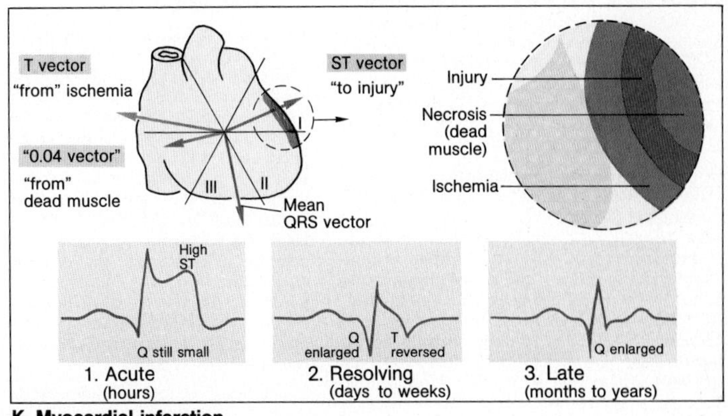

K. Myocardial infarction

Abnormalities in Cardiac Rhythm (Arrhythmias)

Arrhythmias results from abnormalities in impulse formation or in impulse conduction. Their diagnosis lies within the scope of the ECG. **Disturbances in the formation of impulses** lead to *change in the sinus rhythm.* If sinus frequency rises above 100/min (exercise, psychic excitation, fever [rise of 10 beats/min per 1 °C], hyperthyreosis, etc) the condition is known as **sinus tachycardia** (→ **A 2**); if it dropw below 50–60/min it is known as **sinus bradycardia** (e.g. in hypothyreosis). In both cases the rhythm is regular, in contrast to **sinus arrhythmia**, a condition frequently encountered in adolescents. In this case, the frequency fluctuates and is dependent upon *respiration*; it is accelerated by inspiration, retarded by expiration.

Even if impulse formation in the SA node proceeds normally (nomotopic formation of stimulus; → **A**), abnormal or *ectopic* (*heterotopic*) impulses may arise in the atria (*atrial*), in the AV node (*nodal*), or in the ventricle (*ventricular*). The impulses from an atrial (or nodal) ectopic focus are transmitted to the ventricle, which is thus thrown out of its sinus rhythm: **supraventricular arrhythmia** due to **atrial** or **nodal extrasystole (ES)**.

In atrial ES the *P wave is deformed* but the QRS complex is normal. In nodal extrasystole, stimulation of the atria is retrograde; the *P wave* is thus *negative* and is either masked by the QRS wave or appears shortly after it (→ **B 1, right**). Because in supraventricular extrasystole the sinus nodes often also depolarize, the interval between the R wave of the ES (= R_{ES}) and the next normal R is lengthened by the time required for the impulse to travel from the focus to the sinus node: **postextrasystolic interval**. Here $R_{ES}R > RR$ and $(RR_{ES} + R_{ES}R) < 2RR$ (→ **B 1**).

In **atrial tachycardia** (depolarization of focus > 180/min; P wave replaced by jagged "base line"), the ventricle adheres to the rhythm of excitation up to a frequency of about 200/min. At higher frequencies, usually only every second or third impulse is transmitted, the intervening impulses falling within the refractory phase (→ p. 26) of the AV node. Such high atrial frequencies (up to 350/min) are termed **atrial flutter**. In **atrial fibrillation** the focus may depolarize as often as 500/min, but only an occasional impulse is transmitted. Ventricular excitation is thus totally irregular (**absolute arrhythmia**).

An ectopic stimulus can also arise in the ventricle: **ventricular extrasystole** (→ **B 2, B 3**).

In this case the *QRS complex of the ES is deformed.* At a low sinus frequency, the next sinus excitation is transmitted normally to the ventricles: **interpolated extrasystole** (→ **B 2**). At higher sinus frequencies, however, the next sinus stimulus arrives while the myocardium is still refractory, so that only the next but one sinus impulse is effective: **compensatory interval**. Here $(RR_{ES} + R_{ES}R) = 2RR$.

Ventricular tachycardia results from a rapid sequence of (ectopic) ventricular impulses, beginning with an ES (→ **B 3**). Ventricular filling and cardiac output decrease and **ventricular fibrillation** can even ensue, that is, a high-frequency uncoordinated twitching of the myocardium (→ **B 4**). Unless treated, the failure to eject blood can be just as dangerous as cardiac arrest. Ventricular fibrillation mostly occurs if the extrasystole falls within the **vulnerable phase** of the preceding impulse (*relative refractory phase*, synchronous with T wave in the ECG). APs elicited during this phase show (a) as less steep rise and are therefore transmitted more slowly, and (b) are of shorter duration. Both properties contribute to the chances that the myocardial excitation repeatedly encounters areas that are already reexcitable ("re-entry"; → p. 167, F). Ventricular fibrillation can also be elicited by accidental contact with high voltages, but can usually be reversed by application of high DC voltage (*defibrillator*; → p. 30).

Arrhythmias also result from **disturbances in** conduction in the AV nodes (**AV block**), or in the branches of the bundle of His (right or left **bundle block**).

First degree **AV block**: AV conduction is abnormally slow (PQ interval > 0.25); in *second degree AV block every* 2nd or 3rd impulse is conducted, and in *third degree AV block* there is a total block (→ **B 5**), which results in temporary cardiac standstill (*Adam-Stokes attack*). In this case, ventricular pacemakers with a slow frequency usually take over (*ventricular bradycardia* despite normal frequency of atrial excitation). Partial or total independence of the QRS complex from the P wave results (→ **B 5**). Although the SA frequency at rest amounts to 60–80/min, the heart rate drops to 40–60/min when the AV node takes over the pacemaker function (→ **B 5**). The impulse frequency of the so-called tertiary (ventricular) pacemakers is 20–40/min. In such cases an **artifical pacemaker** is invaluable.

Bundle block produces considerable deformities in the ECG since the affected side of the myocardium is stimulated by the healthy side via aberrant routes.

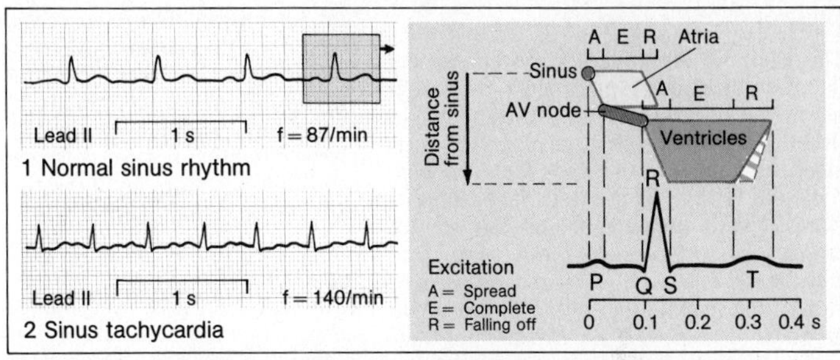

A. Nomotopic production of excitation with normal conduction (after Trautwein)

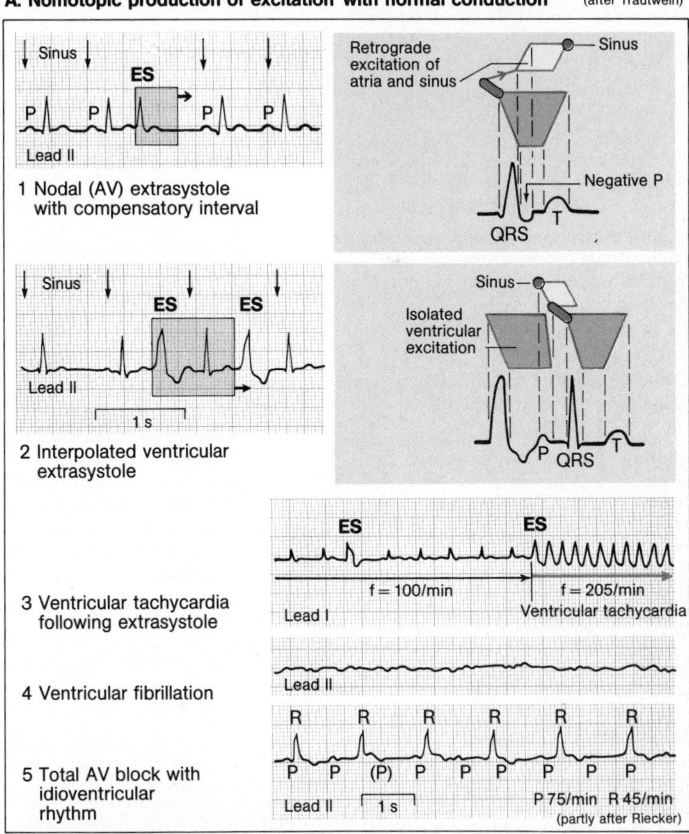

B. Heterotopic production of excitation (1-5) **and disturbances in conduction** (5)

Control of the Circulation

The regulation of the vascular system ensures that the entire body is provided with enough blood, not only when it is at rest but also in situations involving changes in the environment or changes in the work load (see also p. 48 and p. 154). This entails (a) the maintenance of a **minimum blood flow to all organs**, (b) *optimum regulation* of cardiac activity and of blood pressure (**homeostasis**), and (c) **redistribution** of blood to the active organs (e.g. muscle) at the expense of other, resting organs (e.g. gastrointestinal tract; → p. 48), since a maximum blood supply to all organs simultaneously (→ **A**) would overtax the heart.

Regulation of blood flow in the various organs is mainly achieved by alterations in the *diameter* of the vessels. The state of tension (**tone**) of the vascular musculature is influenced by (**1**) **local effects**, (**2**) **neural activity**, and (**3**) **hormonal signals**.

At rest, most blood vessels are in an intermediate state of tension (**resting tone**; → **C**). Denervation leads to partial dilatation of the vessels (**basal tone**; → **C**). Basal tone is the result of spontaneous depolarization in the vascular musculature (see also p. 44).

1. Local control of the circulation or **autoregulation** has two functions:

(A) when the metabolism within an organ is constant, autoregulation serves, in many cases, to keep the *blood flow constant* in the face of *changing blood pressure* (e.g. vasoconstriction in the kidney when the blood pressure rises; → p. 122);

(B) to adapt the blood flow in the organ to local changes in activity, i.e., in metabolism (*metabolic autoregulation*). This may involve a manifold increase in blood flow over and above the resting value (e.g. in cardiac and skeletal muscle; → **A** and **B**).

Mechanisms of autoregulation:

(a) **myogenic effects** (originating in the vascular musculature), consisting of vasoconstriction (e.g. in kidney, brain etc. but *not* in skin and lungs) in response to vasodilatation due to a rise in blood pressure.

(b) $\mathbf{O_2}$ **deficiency** usually causes vasodilatation, i.e. blood flow and hence the provision of O_2 rise with increasing O_2 consumption. In the **lungs**, however, a low P_{O_2} in the vicinity of the vessels results in vascular constriction. In this way, the blood flow is diverted to other portions of the lung where conditions are more favorable for reloading with O_2 (→ p. 94).

(c) A rise in the **local concentration of metabolites** (CO_2 and H^+ ions, ADP, AMP, adenosine, etc) and of osmotically active substances in general (K^+) results in an increased blood flow, which promotes their removal. The accumulation of such substances and a lack of O_2 explain why, following the release of an occlusion (*tourniquet*), the blood flow may rise by as much as fivefold (**reactive hyperemia**).

(d) **Vasoactive substances** ("tissue hormones") such as kallidin, bradykinin, histamine (vasodilator), and angiotensin II (vasoconstrictor; → p. 152) can probably be released in response to local as well as nervous stimuli. In addition, like the catecholamines (see below), they are carried as hormonal signals in the blood to the vessel that has to be regulated.

Cerebral and coronary blood flow are almost exclusively under *local metabolic control* (point 1 b and 1 c).

2. Nervous control of the blood vessels (principally the arterioles) is mediated, with only a few exceptions, by the **symphathetic nerves** (→ p. 53): postganglionic transmission involves **α-receptors** (α_1; → p. 56), the stimulation of which leads to *constriction* (→ **C**), and $\boldsymbol{\beta_2}$**-receptors** (*vasodilatation*).

In **kidneys** and **skin**, for example, α-receptors are predominant, but in **skeletal muscle** there are more β_2- than α-receptors; the stomach, intestines, and coronary vessels have about equal numbers of α- and β-receptors. In *skin* (and *kidneys?*) practically only constriction takes place (→ **C**, right) beyond the basal tone of the vessels, whereas in muscle and intestines the blood vessels can be constricted and dilated beyond the basal tone (→ **C**, left). The β-receptors in the vessels of skeletal muscle respond more to the **epinephrine** circulating in the blood than to nervous stimuli (→ p. 56ff.).

Nervous coordination of local blood flow is mainly effected in one of two ways:

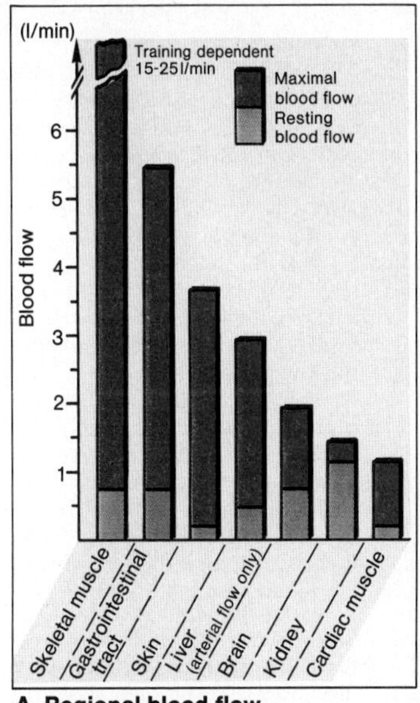

A. Regional blood flow

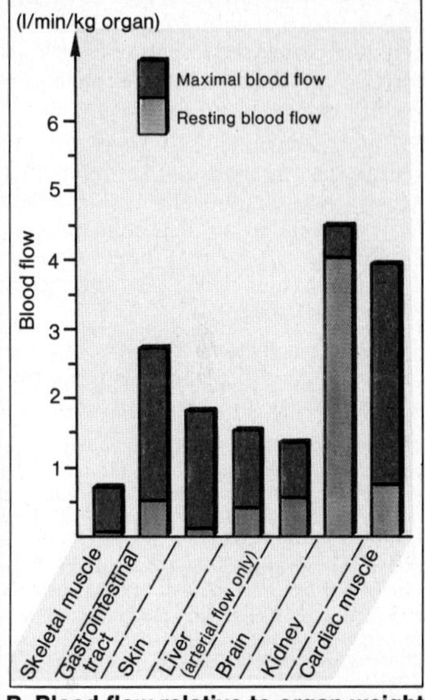

B. Blood flow relative to organ weight

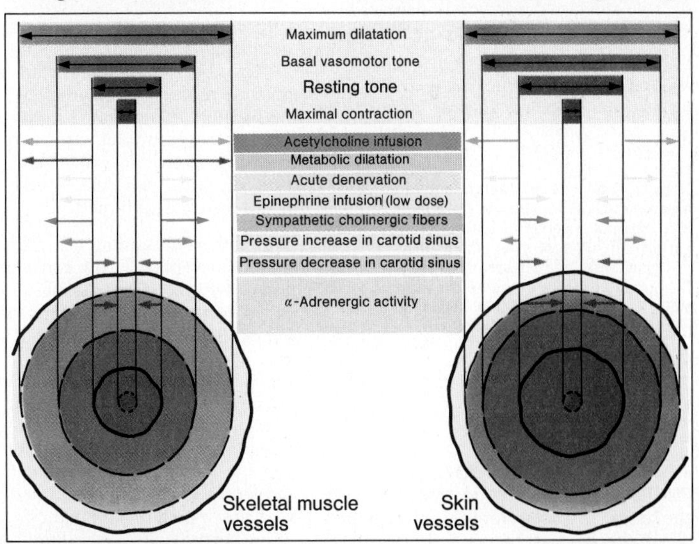

C. Vasomotor influences

(after Koepchen)

(a) via *central co-innervation*: for example, simultaneously with the activation of a group of muscles an impulse is sent from the cerebral cortex to the cardiovascular centers (→ **D**); or

(b) by reflexes elicited by nervous signals sent to the centers from the organ in which activity and metabolism have changed. If nervous and local metabolic effects are conflicting, e.g. sympathetic stimulation during muscular work, then the metabolic effects have priority.

The **blood supply to the skin** is largely subject to *central control* and serves primarily for temperature regulation (→ p. 194).

The vasoconstriction in the skin caused by severe cold is interrupted periodically (*Lewis's reaction*) to prevent tissue damage. Direct temperature effects on the vessels as well as so-called "**axon reflexes**" are thought to be involved in this reaction; that is, an afferent impulse travelling from the skin is transferred to efferent vascular nerves while still in the periphery. The reddening of the skin caused by scratching (*dermographism*) also arises in this way.

In volume deficiency, the skin can assume the role of a blood reservoir (paleness of the skin as a result of circulatory centralization; → p. 186).

Via the α- and β-receptors in the **veins**, their *volume* (**capacitance vessels**) and consequently the *venous return* to the heart can be regulated (→ p. 184).

In many mammals the arteriovenous anastomoses of the blood vessels in the skeletal muscle have their own special vasodilator nerve fibers that run peripherally with the sympathetic fibers, although postganglionically they are cholinergic (→ p. 53). In this way the blood flow to the muscles can be elevated while the **muscular action** is still being planned (*start- or anticipatory-reaction*). Whether or not this pathway exists in humans is uncertain.

Parasympathetic vasodilatation occurs in the *genital organs* (*erection*), in some vessels in the meninges, and (indirectly via *kinins*) in the *salivary* and *sweat glands*.

3. A **humoral hormonal effect** is exerted on the diameter of the blood vessels by catecholamines released from the adrenal medulla. **Epinephrine** in *low* concentrations has a *vasodilatory* effect (*β_2-receptors*), but in *high* concentrations it *constricts* (*α-receptors*) (→ also p. 58). **Norepinephrine** (via *α-receptors*) has only a *constrictor* effect (→ **C**).

The **central control of the circulation** is effected by the *medullopontine regions* of the brain (→ p. 272). Pathways from the **receptors** in the *high-pressure system* (stretch or pressure receptors in aorta and carotid artery; R_D in **D**), in the *low-pressure system* (stretch receptors in the vena cava and atria; R_A and R_B in **D**), and in the *left ventricle* (R_V) terminate in the region. The receptors measure the arterial **blood pressure**, the **pulse rate** (R_D and R_V), and the **filling pressure** in the low-pressure system (and thus, indirectly, the *blood volume*), whereby the A receptors (R_A) respond chiefly to atrial contraction and the B receptors (R_B) to passive filling (→ **D**). Alterations in these two values elicit corrective responses from the centers in the CNS (**cardiovascular "center"**) in the form of efferent impulses to the heart and to the blood vessels (→ **E**).

Situated laterally in the cardiovascular center is an area of continuous sympathetic discharge (*pressor zone*; → **D**) to the heart and blood vessels, i.e. *cardioacceleratory* (rate and strength of contraction) and (predominantly) *vasoconstrictory* (**resting tone**; → **C**). The pressor regions are intimately connected with neurons situated medially in the cardiovascular center (*depressor zone*; → **D**); both zones are connected with the *vagus nuclei*, the activity of which leads to reduction of heart rate and of the speed of transmission within the heart (→ **D**).

The pathways running centrally from the pressure receptors in the aorta and carotid sinus carry the afferent impulses of the so-called **homeostatic circulatory reflexes** (→ **E**), which are primarily aimed at stabilizing the blood pressure (*depressant*). An **acute rise in blood pressure** increases the rate of afferent impulses and activates the depressor zone, thus setting off the reflex (depressor) response by means of which (a) via the vagus nerve, cardiac activity is reduced, and (b) via inhibition of the sympathetic vascular innervation, vasodilatation occurs, with a resultant decrease in peripheral resistance. These two responses bring about a lowering of the elevated blood

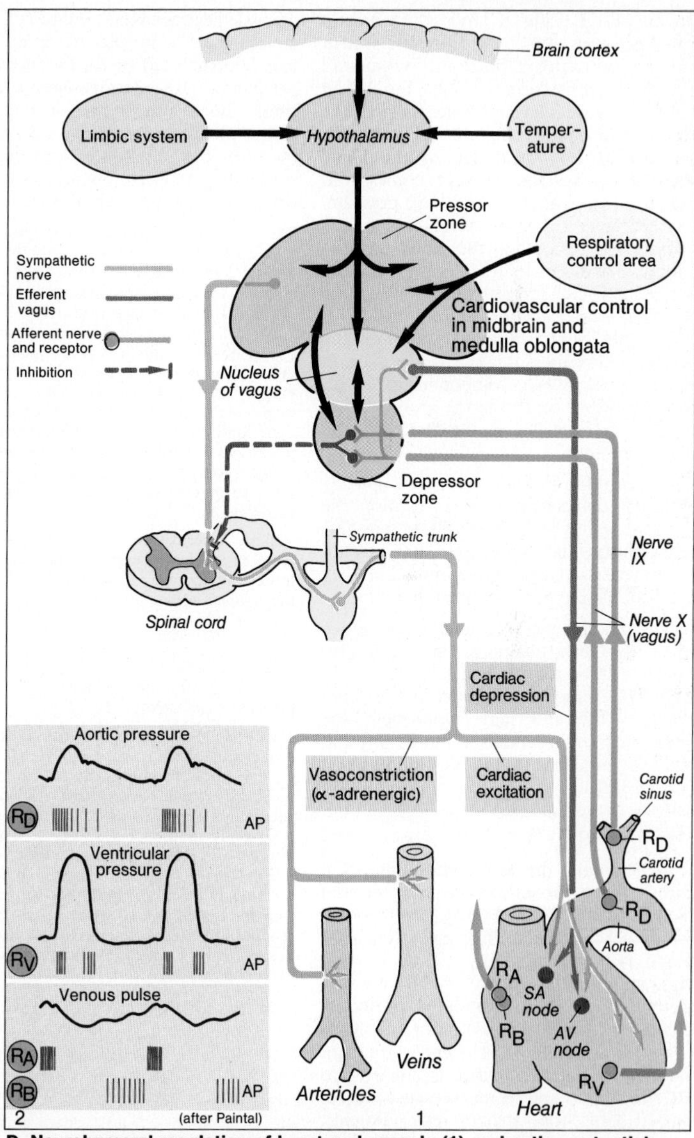

D. Neurohumoral regulation of heart and vessels (1) and action potentials (AP) originating from circulatory receptors

pressure (→ **E**, right). Conversely, a **drop in blood pressure** activates the pressor system and results in a rise in the cardiac output and peripheral resistance, so that the blood pressure is again elevated. Since the pressure receptors also possess differential properties (→ p. 276), the **regulation of the blood pressure** as described above comes into action in *acute* changes in blood pressure. For example, if the position of the body is altered (lying/standing) the blood is redistributed, and without the help of the homeostatic circulatory reflexes (*orthostatic reaction*; → p. 184), the altered venous return would lead to severe fluctuations in the arterial blood pressure. Further, a drop in P_{O_2} or a rise in P_{CO_2} (via connections with the respiratory center) in the blood elicit a pressor reaction, that is, the desired rise in blood pressure.

However, if the blood pressure is *chronically* elevated (**hypertension**; see below) the elevated value is also stabilized via this reflex, that is, the "depressants" not only fail to prevent the high blood pressure, they even contribute to its stabilization.

A momentary increase in venous return (e.g. due to intravenous infusion) also causes accelerated cardiac activity (→ **E**, left). This response is known as the *Bainbridge reflex* and may supplement the Frank-Starling mechanism, but its physiological significance is not fully understood (→ p. 182ff.).

Hypertension

Hypertension is the term used to describe a chronic elevation of the systemic arterial blood pressure. As a rule, the decisive criterion is the repeated finding of a diastolic blood pressure exceeding 12 kPa (90 mm Hg), measured at rest by the Riva-Rocci method (→ p. 160). Untreated or inadequately treated hypertension not only leads to overloading (→ p 184f.) and permanent damage to the left ventricle, but also to arteriosclerosis and its consequences (myocardial infarction, stroke, renal damage, etc.), thus considerably shortening the lives of a large part of the population.

Hypertension may be caused by (a) an increased extracellular volume (ECV) and a consequent rise in the cardiac output (**volume hypertension**) or (b) by an increase in peripheral resistance (**resistance hypertension**). Since every type of hypertension causes alterations in the blood vessels that increase their resistance, type (a) in time turns into type (b), which, whatever its origin, is the beginning of a vicious circle.

The ECV becomes greater if the uptake of NaCl (and water) is increased. To restore the balance, a normal kidney excretes more Na^+ and water, although the blood pressure is slightly raised in the process. Thus the widespread habit of *high NaCl intake* in the diet is held partly responsible for the, by far, most common form of hypertension, that is, **essential** or **primary hypertension** (apart from a primary increased vasoconstriction of unknown origin and other factors). Increasing the NaCl excretion by a decrease in aldosterone release (→ p. 141, A 3) is hardly possible in this case since the plasma concentration of the hormone already sinks to practically zero in response to even a moderate NaCl intake.

Volume hypertension also arises if, e.g., due to renal insufficiency, even a relatively low NaCl intake can no longer be balanced or if a tumor of the adrenal cortex produces uncontrolled quantities of aldosterone and thus causes Na^+ retention.

One of the known causes of resistance hypertension is, e.g., a catecholamine-producing tumor (*pheochromocytoma*).

If the **renal blood flow** is lowered in one kidney (e.g. stenosis of the renal artery) or in both (e.g. stenosis of the aortic isthmus), the release of **renin** is increased in the affected kidney(s). This leads to the production of greater quantities of angiotensin II (→ p. 152) and hence to resistance hypertension. Angiotensin II also increases aldosterone release, which causes Na^+ retention (→ p. 150f.) and, consequently, volume hypertension. In untreated unilateral renal arterial stenosis, this **renal hypertension** also causes damage to the vessels of the healthy kidney, and thus further aggravates the general hypertension.

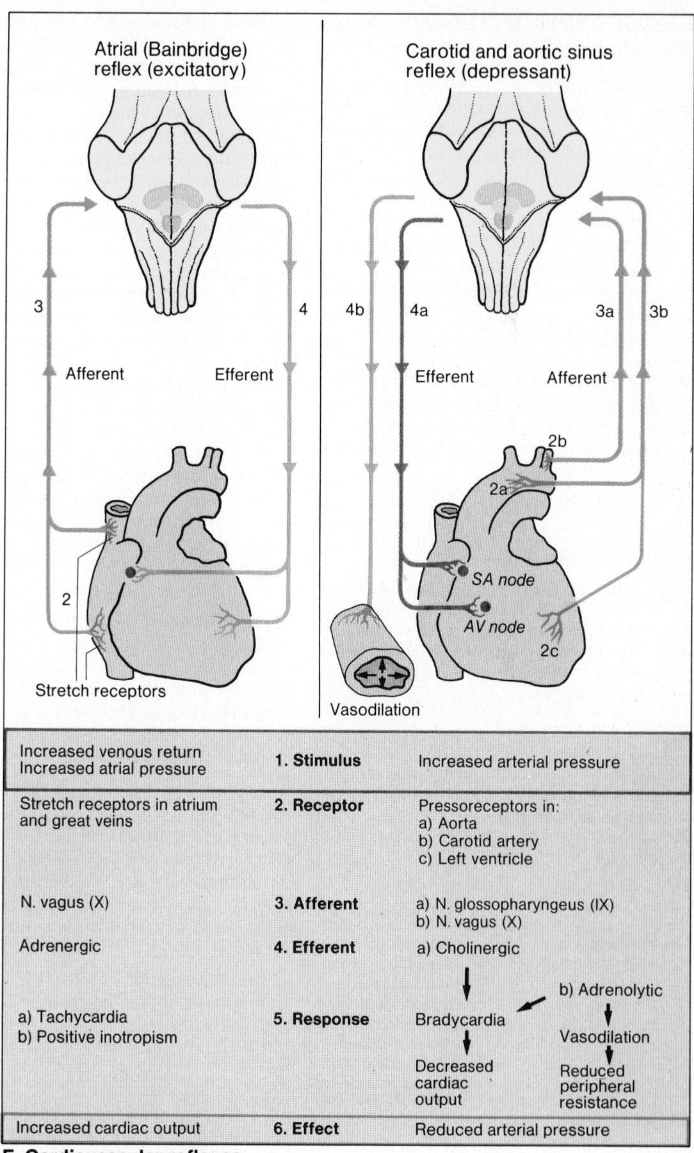

Atrial (Bainbridge) reflex		Carotid and aortic sinus reflex
Increased venous return Increased atrial pressure	**1. Stimulus**	Increased arterial pressure
Stretch receptors in atrium and great veins	**2. Receptor**	Pressoreceptors in: a) Aorta b) Carotid artery c) Left ventricle
N. vagus (X)	**3. Afferent**	a) N. glossopharyngeus (IX) b) N. vagus (X)
Adrenergic	**4. Efferent**	a) Cholinergic b) Adrenolytic
a) Tachycardia b) Positive inotropism	**5. Response**	Bradycardia → Decreased cardiac output Vasodilation → Reduced peripheral resistance
Increased cardiac output	**6. Effect**	Reduced arterial pressure

E. Cardiovascular reflexes

Pressure–Volume Relationships in the Ventricle

Myocardial function is best expressed in terms of pressure developed and cardiac ejection; in clinical terms, these refer to ventricular end-diastolic pressure or volume (left or right), arterial pressure, and cardiac output. The relationship between the *length* and the *tension* of a muscle (→ p. 40ff.) is expressed in the heart by the relationship between **ventricular volume** (corresponds to muscle length) and **ventricular pressure** (tension in the muscle). By recording the changes in volume and pressure occurring during a complete cardiac cycle **(pressure–volume, PV, diagram)** it is possible to construct a *cardiac work diagram* (*Frank* 1895; → **A 2**, points A-D-S-V-A for the left ventricle). For its construction the following PV curves are required:

a) The **resting tension curve**, which indicates the pressure that develops passively (without muscular contraction) at different ventricular filling volumes (→ **A 1** and **A 2**, blue curve).

b) The **curve for the isovolumetric maxima** (→ **A 1** and **A 2**, green curve). This is obtained experimentally, starting from different filling volumes, by measuring *the maximal pressure developed* by the ventricle at a *constant ventricular volume* (isovolumetric, i.e. no ejection; → **A 1**, vertical arrows). This curve is elevated if the *contractility* is increased (→ p. 185, C 3).

c) The **curve for the isotonic maxima** (→ **A 1** and **A 2**, mauve curve). Again starting from different filling values, the volume ejected is experimentally adjusted so as to maintain a *constant pressure* during the decrease in volume (isotony; → **A 1**, horizontal arrows).

d) The **curve of the afterloaded maxima (AM curve)** for each filling volume (→ **A 1** and **A 2**, brown curves). Systole (→ p. 162) includes an isovolumetric contraction phase (→ **A 2**, points A-D), which is followed by the auxotonic expulsion phase (volume decreases but pressure continues to rise; → **A 2**, points D-S). Mixed contraction of this kind (→ **A 2**, points A-D-S) is termed *afterloaded contraction* (→ p. 40). At a certain filling volume (→ **A 2**, point A), its maximum changes (→ **A 2**, point S) depending on the end-diastolic aortic pressure (→ **A 2**, point D); however, all of these maxima lie on the AM curve. The AM curve is the line joining the isovolumetric maximum (→ **A 2**, point T) with the corresponding isotonic maximum (→ **A 2**, point M). Changes of filling volume (→ **A 1**) or contractility will displace the AM curve (→ p. 184).

Pressure–volume loop. If the actual values for a cardiac cycle are entered on a PV diagram, the result for the left ventricle is as follows (→ **A 2** and p. 162f.): if, for example, the *end diastolic volume EDV* is 130 ml (→ **A 2**, point A), the ventricular pressure rises during the *isometric contraction phase* (all valves closed), until the pressure in the aorta (in this case 10.7 kPa or 80 mmHg) is reached (→ **A 2**, point D). During the *ejection phase* the pressure continues to rise at first, the ventricular volume being at the same time reduced by the stroke volume: *auxotonic contraction*. When the maximum (systolic) pressure has been attained (→ **A 2**, point S) the volume remains practically constant, although the pressure drops until it is below that in the aorta (→ **A 2**, point K). In the *relaxation phase*, the pressure (at constant volume) rapidly drops to (almost) zero (→ **A 2**, point V). At this time only the *residual volume* is present in the ventricle (*endsystolic volume ESV*; in our example: 60 ml). During the *filling phase*, the pressure within the ventricle rises slightly again (along the resting tension curve).

The **adaptation of the heart** to alterations in ventricular filling (*preload*) and to aortal pressure (*afterload*) is achieved *autonomically* by changes in the **resting tension: Frank-Starling mechanism** (→ p. 184).

Cardiac work and power

Since work [N · m] = pressure [N · m^{-2}] × volume [m^3], the area contained by the points A-D-S-V-A in the PV diagram (→ **A 2**) corresponds to the **PV work** performed by the heart (in **A 2**, left chamber) during systole. The area below the resting tension curve (→ **A 2**) corresponds to the diastolic PV work (filling).

In addition to the systolic PV work of the two ventricles (at rest approximately 1.1 J) the heart has to perform a further 20 % (0.22 J) *work for the pulse wave* (distension of the vascular walls). The work required to *accelerate the blood flow* is very small at rest (1 % of the PV work) but rises at high heart rates.

The overall **power** of the heart amounts to about 1.5 W at rest.

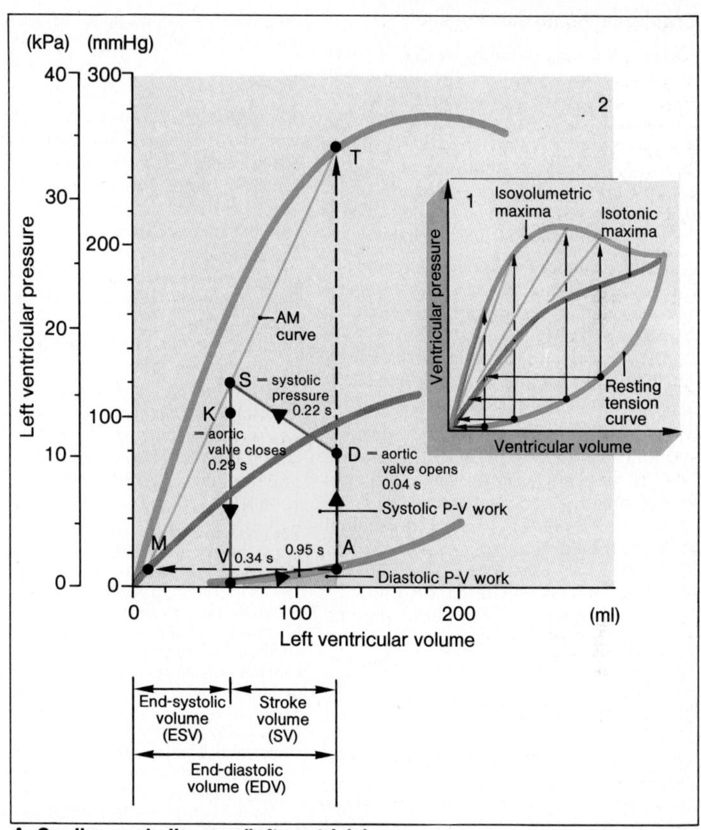

A. Cardiac work diagram (left ventricle)

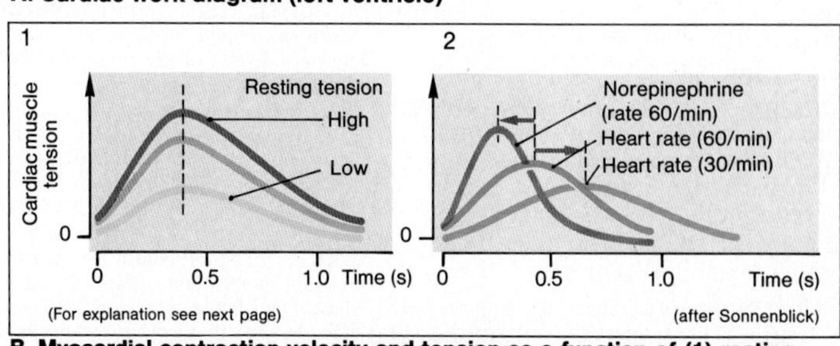

B. Myocardial contraction velocity and tension as a function of (1) resting tension, (2) heart rate and adrenergic activity

Regulation of Stroke Volume

Whereas the adaptation of cardiac activity to the O_2 requirements of the organism is under exogenous control (cardiac nerves; → pp. 164ff. and 178ff.), adaptation of the stroke volume (SV) to changes in filling (position of body, respiration) and aortic pressure takes place *autonomously*, by alterations in the end-diastolic resting tension (→ p. 41 ff.): **Frank-Starling mechanism**.

If the **filling volume (preload)** is increased (→ **C 1**), the beginning of the phase of rising tension on the resting tension curve is displaced to the right (→ **C 1**, point A_1). As a consequence, the SV and the work done by the heart are raised and there is also a slight rise in the endsystolic volume, ESV.

If the **aortic pressure (afterload)** is elevated due to an increase in peripheral resistance (→ **C 2**), the aortic valve does not open until a corresponding ventricular pressure is attained (→ **C 2**, point D′). In the transitional phase the SV is decreased (SV′), as a result of which the ESV increases to ESV′ (→ **C 2**). The next diastolic filling, therefore, displaces the beginning of the phase of rising tension to the right (→ **C 2**, point A_2). In this way the SV normalizes (SV_2) despite an elevated aortic pressure (→ **D 2**). The ESV has undergone a relatively large increase in the process (ESV_2).

An important **function of the Frank-Starling mechanism** is to match the SVs of the two ventricles so exactly that neither a stagnation in (pulmonary edema) not total emptying of the pulmonary circulation can occur, either of which would be fatal.

Contractility of the cardiac muscle can be increased *exogenously* (**positive inotropic effect**; → p. 166). One of the effects is to *displace the isovolumetric maxima* (→ **C 3**). This enables the heart to work against an elevated pressure (→ **C 3**, point D 3) or (at the cost of the ESV) to raise its stroke volume (→ **C 3**, SV_4) or both.

Whereas differences in preload affect only the **force of contraction** of the heart (→ **B 1**), the *speed* of contraction is altered by the *inotropic effect* (→ **B 2**) exerted by the catecholamines (β_1-receptors) or by an increased heart rate (frequency inotropism; → p. 166). The maximum isovolumetric rise in pressure (max dP/dt) is, therefore, also employed clinically as a measure of contractility.

Veins

Blood from the capillaries is returned by the veins to the heart. The **driving forces for venous return** are: (a) The *blood pressure* remaining after passage through the capillaries (*vis a tergo*; about 2 kPa or 15 mmHg); (b) The suction force created in systole by the *lowering of the level* of the *plane of the cardiac valves*; (c) The pressure exerted by the contracting skeletal muscle on the blood vessels („*muscular pump*"); the *valves in the veins* prevent the blood from being pumped in the wrong direction; (d) The *excess pressure in the abdominal cavity* and a simultaneous *negative pressure in the chest cavity* caused by *inspiration* (P_{pl}; → p. 80), which lead to vasodilatation in the thorax and thus to suction within the veins (see also p. 160).

The pressure in the blood vessels remote from the heart varies considerably in connection with changes in posture. For instance, in getting up from a horizontal position (**orthostasis**) the blood vessels in the legs have to support a column of blood, or, in other words, are subjected to an additional *hydrostatic pressure*. The result is a *dilatation* of the (in comparison to the arteries) highly distensible veins and the accumulation of about 0.4 l of blood, which is drawn from the *central blood volume* (primarily the pulmonary circulation). This leads to a reduction in the venous return to the left heart and thus also to a decreased SV and cardiac output. In order to prevent too drastic a drop in blood pressure (or even *orthostatic collapse*), the heart rate and the peripheral resistance are reflexly raised (**orthostatic reflex**). Much less venous blood accumulates in the legs during walking (muscle pump; see above) than in standing.

When the body is in an upright position a negative pressure prevails in the veins of the head. Somewhat below the level of the diaphragm is the point where there is no change in venous pressure connected with a change in posture (*indifferent point*).

What is termed the **central venous pressure** (mean pressure in the right atrium, normally 0–12 cm H_2O or 0–1.2 kPa is primarily *dependent upon the blood volume*. By its measurement the blood volume can be monitored (e.g. during infusions). Central venous pressure is elevated, for example, in cardiac insufficiency, or, physiologically, in pregnancy.

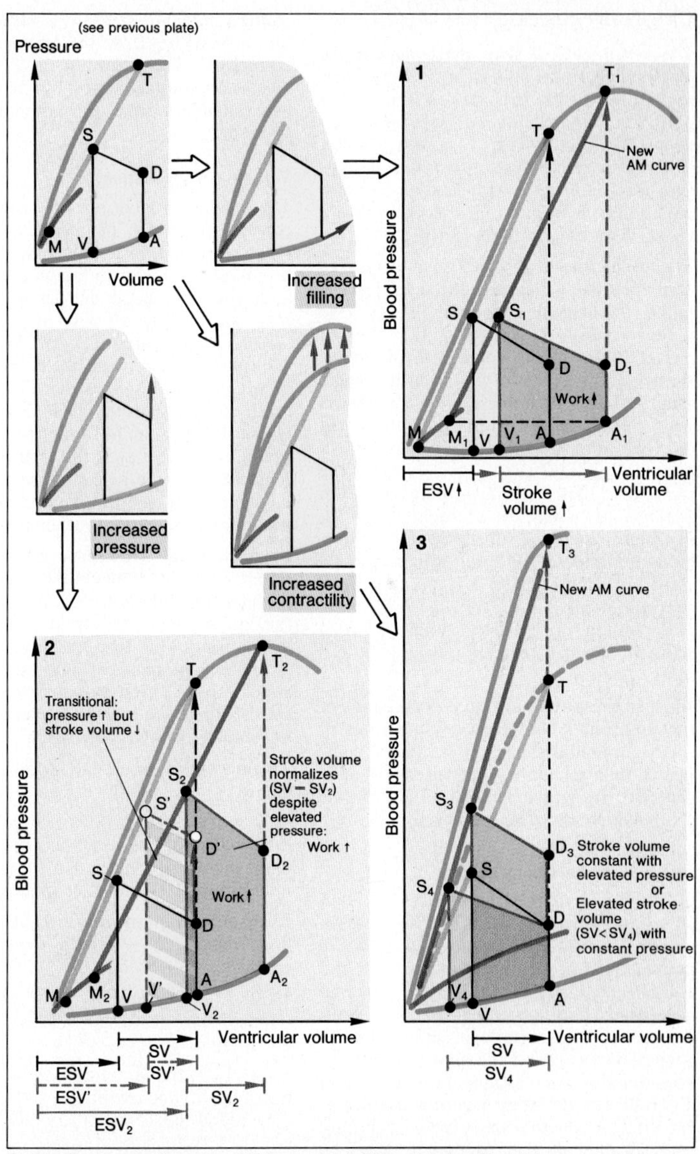

C. Cardiac work diagram: (1) increased filling (preload), (2) increased blood pressure (afterload), and (3) increased contractility

Circulatory Shock

Shock is a state of **acute inadequacy of the blood supply** to the vital organs of the body, leading to acute **hypoxia** or **anoxia** (→ p. 102) and *accumulation of metabolic products* (CO_2, lactic acid). In a wider sense, shock also includes *disturbances in O_2 release and utilization* with (initially) unimpaired blood supply, as, for example, in *septic shock* caused by bacterial toxins.

The usual **cause of shock** is a **decreased cardiac output**, which may be due to (1) *heart failure* (**cardiogenic shock**) or (2) *too little venous return*. Reasons for (2): are (a) reduction in blood volume (**hypovolemic shock**) following loss of blood (**hemorrhagic shock**) or loss of fluid (e.g. in connection with burns, severe vomiting, prolonged diarrhea, and so on); (b) *peripheral vasodilatation*, and hence too much blood in the periphery. This mechanism plays a role in *anaphylactic shock*, for example (→ p. 72), in which vasoactive substances (histamine, etc.) are released. Among other factors, shock is accompanied by *reduced blood pressure* (weak pulse), *elevated heart rate*, *pallor* (not in shock due to vasodilatation), a reduction in urine excretion (*oliguria*), and severe *thirst*.

Most of these symptoms are expressions of the organism's compensatory measures at the beginning of shock (→ **A**). Rapid compensatory mechanisms that **raise blood pressure (BP)** are supplemented by a slower type that **counteracts volume deficiency.**

Blood Pressure Compensation

The falling blood pressure reduces the activity of pressure receptors in the arterial system (→ p. 176ff.), which leads to the activation of pressor areas in the CNS and to *elevated sympathetic tonus*. **Arterial vasoconstriction** accompanied by withdrawal of blood (not in shock due to vasodilatation) from the skin (pallor), the abdominal space, the kidneys (oliguria), and so on directs the smaller cardiac output to the vital organs (coronary arteries, brain): **centralization of the circulation**. *Vasoconstriction of the venous capacity vessels* increases cardiac filling. The accompanying *tachycardia* helps raise the lowered cardiac output resulting from the decreased stroke volume, SV. These neurally mediated mechanisms are supplemented by the release of **catecholamines** from the adrenal medulla (→ p. 58).

Volume Compensation

The reduced blood pressure and arterial constriction *lower the capillary filtration pressure* (→ p. 158) so that there is an influx of interstitial fluid into the bloodstream. In addition, the volume deficiency and lower blood pressure set off the *renin-angiotensin-aldosterone mechanism* (→ p. 152). This results in **thirst** and lowers the renal excretion of salt and thus also of water. The *decreased atrial pressure* leads to ADH release (*Henry-Gauer reflex*; → p. 140) and thus also to a saving of water. Later, the erythrocytes are replaced with the help of greater *erythropoietin release* (→ p. 60ff.) and the *plasma proteins* by increased synthesis in the liver.

In **shock, in the strict sense of the word**, the organism is no longer able, *without help from outside* (**infusion**, etc.), to recover with the aid of its homeostatic compensatory mechanisms. In this case, certain *self-reinforcing mechanisms* may develop that aggravate the shock to such a degree that it can no longer be relieved even by therapy: **irreversible** or **refractory shock**.

This may develop via the following vicious circles:

1. Blood volume ↓ → vasoconstriction → disturbance of tissue metabolism → vessel damage → vasodilatation and filtration into interstitium ↑ → blood volume ↓↓...

2. Vasoconstriction and low BP → blood flow velocity ↓ → blood viscosity ↑ → resistance ↑ → blood flow ↓↓...

3. BP ↓ → O_2 deficiency and acidosis → damage of myocardium → cardiac output ↓ → BP ↓↓...

4. BP ↓ → tissue metabolism ↓ → vessel damage → blood clotting → clotting factors ↓ → blood loss → BP ↓↓...

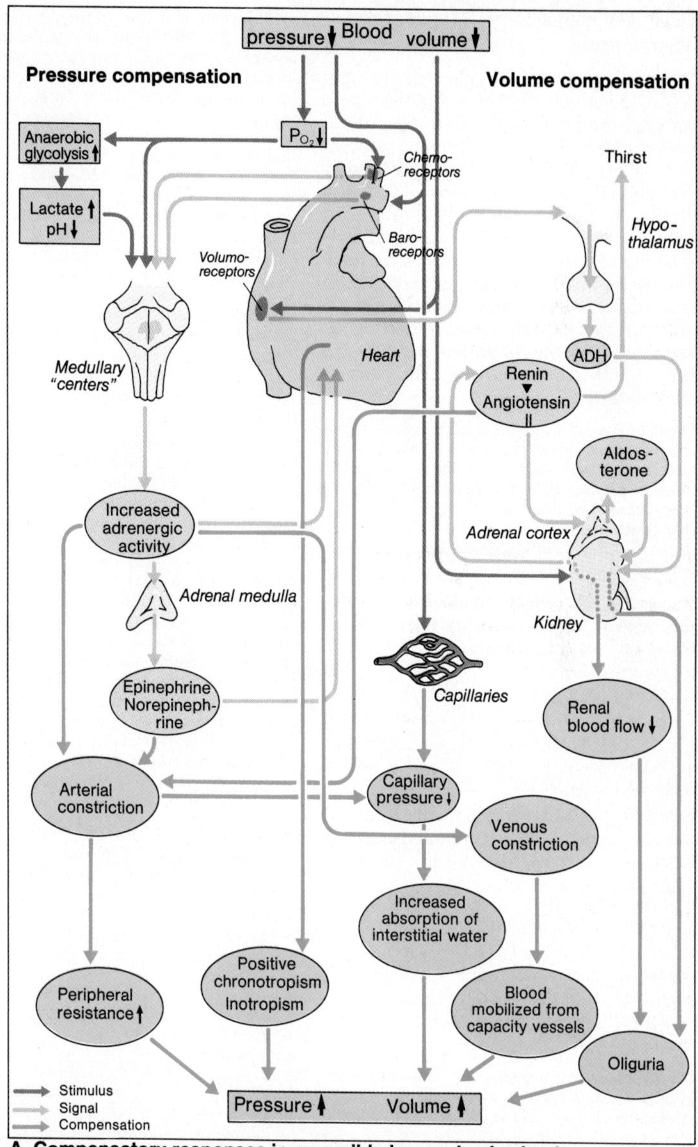

A. Compensatory responses in reversible hypovolemic shock

Coronary Blood Flow, Myocardial Metabolism

The cardiac muscle is supplied via the right (1/7 of the blood) and left (6/7 of the blood) **coronary arteries** from the aorta. The *venous return* is effected (about 2/3) via the *coronary sinus* and (roughly 1/3) via the small coronary veins into the right atrium (→ **A**). **Blood flow** in the **cardiac muscle** ($\dot{Q}_{cor}$) of a heart weighing 300 g *at rest* is approximately 250 ml/min, but can rise threefold to fourfold *during work* (→ **A**). $\dot{Q}_{cor}$ depends (1) on the arterial *blood pressure*, and (2) on the *width of the coronary vessels*, which is influenced (a) by the *transmural pressure* (P_t) and (b) by the *tonus* of the vascular musculature.

The $\mathbf{P_t}$ of a vessel is the difference between internal pressure (P_i = blood pressure) and surrounding pressure (P_o). In the subendocardial sections of the coronary arteries, P_o is so high during systole due to myocardial contraction that P_t drops appreciably (right ventricle), or is $\leqslant 0$ (left ventricle). Hence, blood flows in the coronary arteries practically only during *diastole* (→ p. 163, A 8).

Regulation of coronary blood flow. At rest, the venous coronary sinus blood has a fractional O_2 concentration of about 0.06 (l/l blood). Since the fractional O_2 concentration of the arterial blood is roughly 0.20, the resting heart has an **arteriovenous O_2 difference** (AVD_{O_2}) of approximately 0.14, which can only rise slightly (to approximately 0.16) during work (→ **A**).

From $\dot{Q}_{cor} \cdot AVD_{O_2}$ it is possible to calculate the increase in **O_2 consumption of the heart** ($\dot{V}_{O_2}$) from about 35 (rest) to 120–160 ml/min. The rise in O_2 consumption is thus mainly achieved by an *increased blood flow*. This is elicited, above all, by **local** chemical factors such as O_2 lack and the release of adenosine and K^+, although it seems that endothelial factors and stimulation by autonomic nerves contribute to the vasodilatation.

The $\dot{V}_{O_2}$ of the myocardium rises, on the one hand, with the maximal speed of shortening (V_{max}) of the cardiac muscle (→ p. 42 f.) and, on the other hand, with the product of *myocardial tension and duration of systole* (**tension-time index**). For a small SV and high blood pressure (myocardial tension ↑) $\dot{V}_{O_2}$ is greater, despite the same amount of work (→ p. 182), than at a low pressure and high stroke volume, SV. This means that when the heart is performing the same amount of work (P · V), **cardiac efficiency**, too, is considerably higher in the latter case than in the former. Therefore, in a heart with a poor blood supply (**coronary insufficiency**) an *elevated ventricular pressure* (e.g. in hypertension) further reduces the O_2 supply. On the other hand, these patients are also endangered by a *drop in blood pressure* (e.g. in the early morning), since this also entails a lower $\dot{Q}_{cor}$. In coronary insufficiency, even a drop in the P_{O_2} of the air, and thus of the AVD_{O_2} (*rapid rise in altitude* in cable car or airplane), can result in hypoxia and anoxia (→ p. 102) of the cardiac muscle (**angina pectoris, coronary infarct**). Myocardial hypoxia can be alleviated, e.g. by lowering the peripheral resistance and, hence, also the blood pressure (e.g. with nitroglycerol), or by a direct reduction of cardiac work by β-receptor blocking agents (→ p. 59).

Depending on availability, the myocardium can utilize as nutrients (**substrates**) *glucose, free fatty acids, lactate*, etc. for the production of ATP. *At rest*, the proportion of the O_2 consumption ("*O_2 extraction coefficient*") expended on each of these three substrates is usually about one third of the total. During physical *work*, more of the **lactate** formed in the skeletal muscle is oxidized by the myocardium (→ p. 46, p. 247, and **A**). In O_2 deficiency, ATP is also produced anaerobically (in this case: lactate *formation* in the myocardium). *On its own*, anaerobic production of energy from glycogen (e.g. in anoxia) suffices to maintain cardiac activity for only about eight minutes (*functional maintenance time*). The heart can be resuscitated (at 37 °C) up to 30 minutes after anoxia. A *temperature drop* of 10 °C doubles this resuscitation time, since the energy requirements also drop (cooling of organs for transplantation).

Measurement of Blood Flow

1. **Plethysmography** (→ **B**): An extremity is placed in a chamber and the venous outflow is occluded. Blood flow can be measured as a rate of volume change within the chamber.
2. **Magnetic flow meter** (→ **C**): Flow of the electricity-conducting blood through a magnetic coil can be measured by the magnitude of the induced current.
3. **Indicator gas method**: The gas (e.g. *argon*) is inspired for about 10 minutes. From repeated samples of blood, the gas concentrations in the artery (C_{art}) and vein (C_{ven}) can be measured and for each a *mean value per time* calculated, whose difference (AVD_{indic}, → **D**), together with the equilibrium gas constant C_e (in which $C_{art} = C_{ven} = C_{tissue}$) and the equilibrium time (t), can be used to calculate the blood flow ($\dot{Q}$) through the particular tissue (e.g. brain, myocardium).

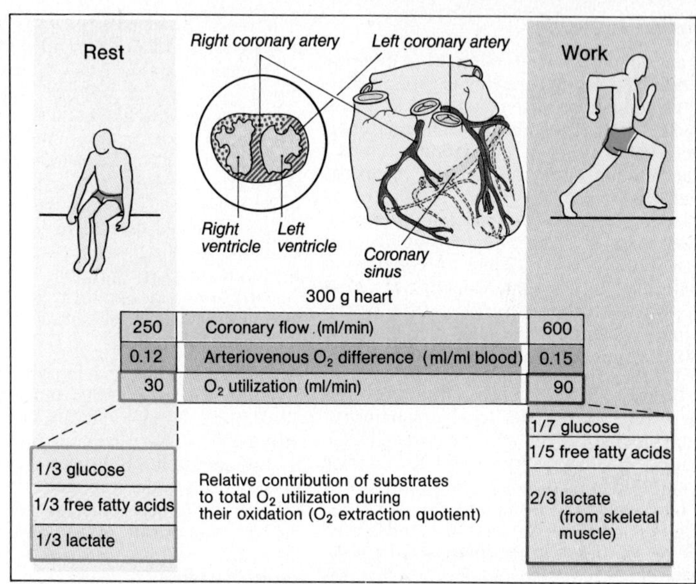

A. Blood flow, substrate, and O_2 utilization in heart muscle

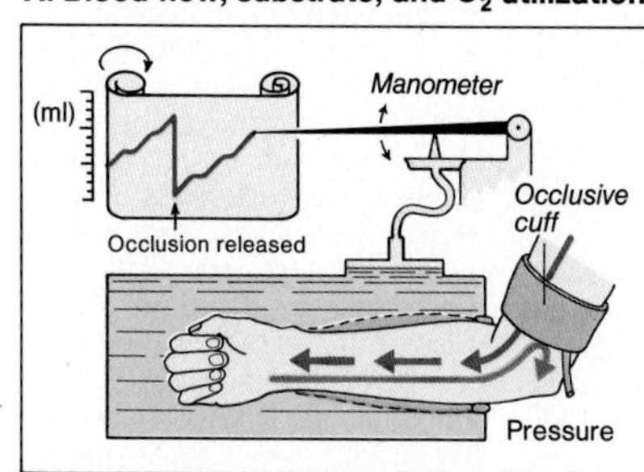

B. Plethysmography to measure blood flow

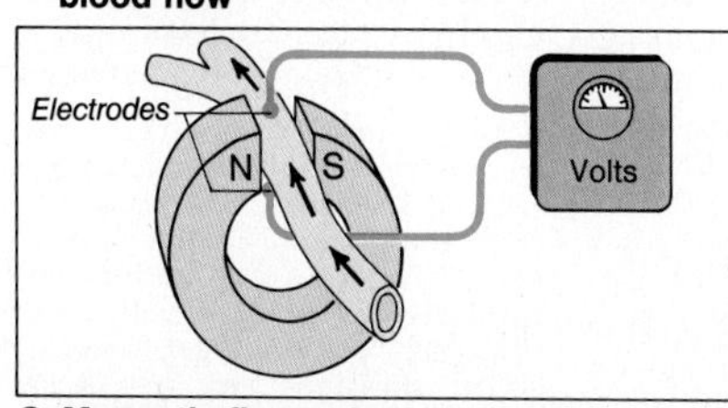

C. Magnetic flowmeter to measure blood flow

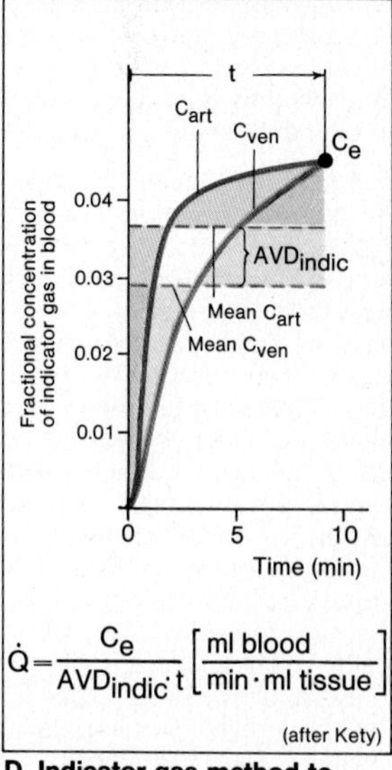

D. Indicator gas method to measure blood flow

Fetal Circulation

The fetal circulation is designed to meet the needs of a rapidly growing organism existing in a state of relative hypoxia. The only connection between fetal and external environment is the maternal **placenta**: it serves the fetus in the role of "intestine" (**uptake of nutrients**, partly by active transport), as "kidney" (**removal of breakdown products**), and also as "lung", (**uptake of O_2** and **removal of CO_2**). Despite (in comparison with the adult) *a displacement to the left in the O_2 dissociation curve* (→ p. 101, C) of the fetal hemoglobin, the fetal blood is only 80% saturated with O_2 in the placenta.

The blood is distributed within the fetus according to its own local requirements. Organs that are not yet active, such as the lungs, are largely bypassed. The **cardiac output** amounts to roughly 0.25 l/min per kg body weight and the **heart rate** to 130–160 beats/min. About 50% of the blood ejected from the fetal heart flows through the placenta, while the remaining 50% supplies the body (35%) and the lungs (15%) of the fetus. The left and right sides of the heart function predominantly *in parallel*, since a fully developed pulmonary circulation in series with the systemic circulation, present in the adult, is not yet necessary.

The fetal blood takes the following course (→ **A**): after *arterialization in the placenta* (saturation = 0.8; → **A**) the blood enters the fetus in the **umbilical vein** and, via the **ductus venosus**, largely bypassing the liver, joins the inferior vena cava where it mixes with the venous blood returning from the lower torso. Guided by folds in the vena cava, this blood is so directed in the right atrium that it passes through the **foramen ovale** (a hole in the interatrial septum) into the left atrium, and thence into the left ventricle. In the right atrium the pathway of this mixed blood is crossed by the stream of venous blood from the superior vena cava, which (with only slight mixing) enters the right ventricle. Only about one third of this blood goes to the lungs (high resistance: the lungs not yet inflated). The remaining two thirds reach the aorta via the **ductus arteriosus** in which, on account of the *low peripheral resistance* (placenta), the **blood pressure** is relatively low: about 8.7 kPa or 65 mmHg at the end of pregnancy. The partially arterialized blood from the left ventricle supplies the *arteries of the head* (the brain is particularly sensitive to O_2 deficiency; → p. 102) and the *upper torso* (→ **A**). The venous blood from the ductus arteriosus enters the aorta at a point beyond the origin of the vessels supplying these sensitive regions. The blood available for the lower torso is thus relatively low in O_2 (saturation = 0.6; → **A**). The major portion is returned in the **umbilical artery** to the placenta, where it is replenished with O_2.

At birth, nutrition and removal of waste products via the placenta cease abruptly. The *blood CO_2 rises* and (via the chemoreceptors; → p. 104) provides a potent stimulus to respiration. The **inspiratory movement** thus elicited produces a negative pressure in the thoracic cavity, which, as well as drawing the blood out of the placenta and umbilical vein (*placental transfusion*), also causes the lungs to inflate.

Inflation of the lungs lowers the resistance in the pulmonary circulation, whereas the resistance in the systemic circulation rises due to the constriction of the umbilical cord. This causes a *change in direction of the flow in the* **ductus arteriosus**. The pulmonary circulation receives aortal blood in this way for several days after birth. Filling of the right atrium is reduced (no placental blood), whereas that of the left atrium rises (pulmonary blood flow increases). This gives rise to a pressure gradient from the left to the right atrium that *closes*, by means of folds, the **foramen ovale**, which, like the ductus arteriosus and ductus venosus, closes up completely in time. Systemic and pulmonary circulation are now functioning in series.

If the foramen ovale or the ductus arteriosus remains *open* (together accounting for 20% of the congenital heart diseases) the blood takes short cuts (**shunts**), which constitute an extra load for the heart. In the case of an *open foramen ovale* the circulation is as follows: left atrium – right atrium – right ventricle (*overloaded right heart*) – lungs – left atrium; circulation with *open ductus arteriosus*: aorta – pulmonary artery – lungs (*excessive pressure load*) – left heart (*excessive volume load*) – aorta.

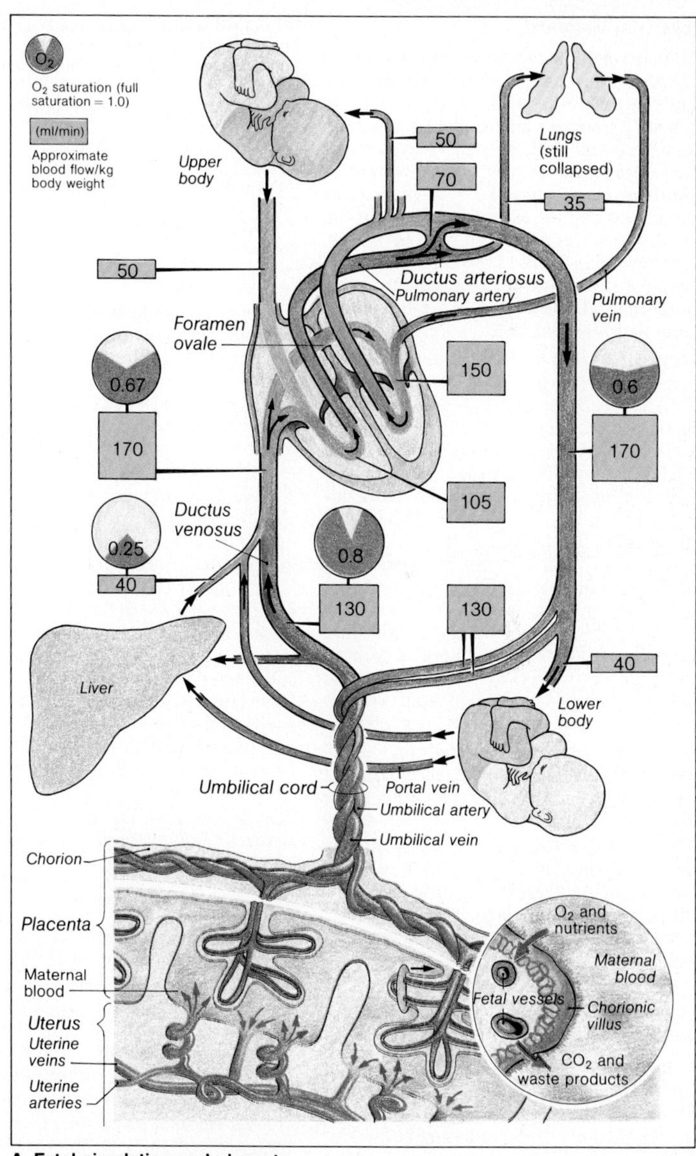

A. Fetal circulation and placenta

Thermal Balance

Humans belong to the group of animals termed **homeothermic** that maintain their body temperature within a narrow range despite a changing environmental temperature. In actual fact, only the depths of the body cavities have a truly constant temperature (**core temperature** ≈ **37** °C), whereas the temperature of the extremities and skin can vary over several °C and are, in a way, *poikilothermic* (→ p. 195, A). The core temperature can only be held constant if **heat production** and **heat uptake** are in equilibrium with the **heat loss**.

Heat production is a function of energy metabolism (→ p. 196 ff.). *At rest*, about 56% of the basal heat is generated by the internal organs and 18% by musculature and skin (→ **A 2, top**). During *exertion*, there is a considerable rise in heat production; the contribution made by muscular work increases both absolutely and relatively, and can account for as much as 90% of the heat produced (→ **A 2, bottom**). To *maintain the body temperature* additional muscular activity may become necessary, either voluntarily or due to shivering, or both. An infant can generate additional heat, if needed, without shivering (→ p. 194). This kind of "cold defense" can only take effect if enough *thyroid hormone* is available (→ p. 250).

Heat uptake (by irradiation and conduction, see below) assumes a significance when the surrounding temperature exceeds that of the skin.

The heat generated within the body is taken up and conveyed by the blood to the surface, but this **internal heat flow** is only possible if the skin temperature is lower than the core temperature. The transport of heat to the skin depends primarily upon the *blood supply to the skin* (→ p. 194).

Heat loss (external heat flow) involves several mechanisms (→ **B**):

1. **Radiation** (→ **B 1** and **C**). The quantity of heat transferred from one body to another by radiation is a fourth power function of the temperature of the radiating body. This applies to the skin, on the one hand, and to nearby human beings or objects, on the other.

If an object is hotter than the skin, the body takes up radiation from it, but if the object is colder (or is not a radiating body) the skin may lose heat by radiation. No vehicle is required for the transfer of heat by radiation. The temperature of the air through which the heat radiates has little influence on heat transfer, and is hardly influenced by the temperature of the air (air itself is a poor radiator); for example, despite intervening warmer air, heat can be lost to a cold wall of a room, but on the other hand, despite the intervening "air-free" outer space, irradiation from the sun can be absorbed. Similarly, despite cold air, radiation can be taken up from an infrared lamp.

2. **Conduction of heat** from the skin to the surrounding air. This requires that the air is cooler than the skin, that is, a *temperature gradient* must exist. This kind of heat loss is greatly enhanced when the air layer warmed by the skin is removed (e.g. by a breeze) and replaced by a cooler, drier layer (**convection**; → **B 2** and **C**).

3. Radiation and conduction are insufficient to prevent warming up of the body during heavy exertion or at high surrounding temperatures. Under these circumstances, heat loss is enhanced by the **evaporation** of water (→ **B 3** and **C**). At environmental temperatures *above about 36 °C* (→ **C**) heat is lost *exclusively by evaporation*. At still higher temperatures, heat is *taken up* from the environment by radiation and conduction (+ convection). Perspiration becomes profuse in order to maintain the balance between heat uptake and heat loss by evaporation.

Heat can only be lost by evaporation if the surrounding air is relatively dry (desert, sauna), in which case, as long as the water and salt lost in the perspiration are restored, air temperatures much higher than the body temperature can be tolerated (> 100 °C!). At very high humidities, on the other hand (tropical rainforest), the body cannot tolerate air temperatures above about 33 °C, even at rest.

The water required for evaporation reaches the surface of the skin by *diffusion* (*perspiratio insensibilis*) and via the neurally activated **sweat glands** (→ **B 3**; p. 53; and p. 195, D). The evaporation of one liter of fluid causes a heat loss of 2428 kJ (580 kcal).

Acclimatization to continued exposure to high temperatures, as in the tropics, often takes years: (1) the rate of sweat secretion increases; (2) the salt content of the sweat decreases; and (3) thirst and water intake are increased.

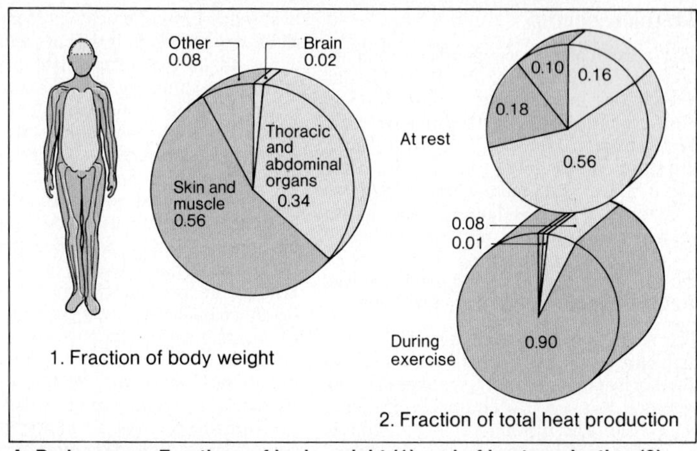

A. Body organs: Fractions of body weight (1) and of heat production (2)

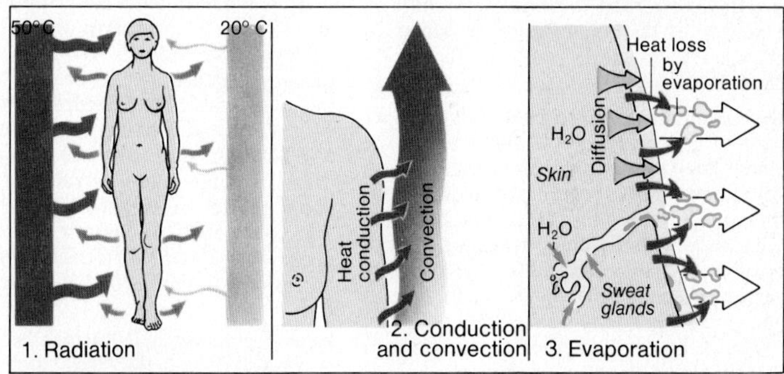

B. Mechanisms of heat loss

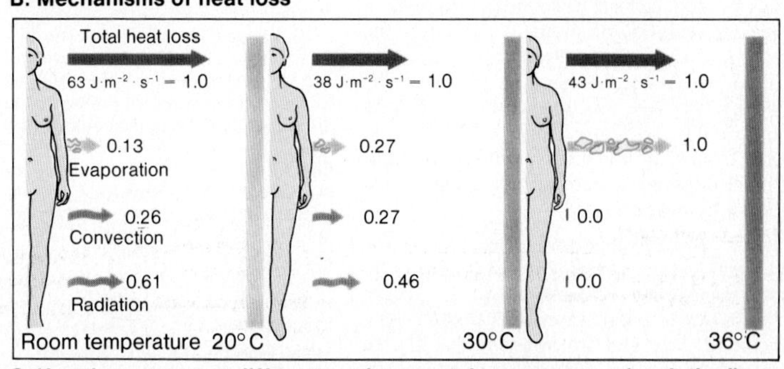

C. Heat loss at rest at different environmental temperatures (unclothed)

Thermoregulation

The **function of thermoregulation** is to maintain the body core temperature, despite fluctuations in heat uptake, heat production, and heat loss (→ p. 192), at a **set value**. The average set value is 37 °C, with *diurnal variations* of roughly ± 0.5 °C (minimum about 3 AM, maximum about 6 PM; → p. 331, A). These **deviations from the set value** are controlled by an "*internal clock*" (→ p. 292). A longer-lasting deviation from the set value occurs during the *menstrual cycle* (→ p. 263) and in *fever* (see below).

Thermoregulation is centered in the **hypothalamus** (→ p. 290), where heat-sensitive **thermoreceptors** respond to changes in core temperature (→ **A**). Additional information is received from *thermoreceptors in the skin* (→ p. 276) and *spinal cord.* The hypothalamus integrates these data and initiates a variety of **thermoregulatory responses** to counteract deviations from normal body temperature (→ **D**).

If the **core temperature rises** above the set value (e. g. during physical work), firstly, the **blood flow to the skin** is increased, and consequently also the transport of heat from the core to the skin. Not only does the increased volume/time transport more heat/time, it also diminishes the countercurrent exchange of heat between arteries and the accompanying veins (→ **B** and p. 134 f.). In addition, the venous return is diverted from the deep-lying veins to those nearer the surface. Secondly, there is a greater **secretion of sweat**, which cools the skin and thus creates the temperature gradient necessary for heat to be lost. The signal for this reaction comes from the *central heat receptors*. Those of the skin, in this case, cannot report warming-up since their surroundings are in fact being cooled down by evaporation.

If the **body temperature drops** below the normal value, the loss of heat is cut back and **heat production** is raised (up to four times the basic metabolic rate) by *voluntary muscular activity* and by *shivering* (→ **D**). In *infants*, the ratio of surface to volume is very high; therefore, cooling can occur very easily. They possess an additional source of heat in the form of *brown fat*, located in the shoulders and back. If the core temperature drops, the rate of metabolism is raised by adrenergic stimulation, and more heat is generated. These counterregulations are elicited by cool surroundings via the *cold receptors of the skin* (→ p. 276) *before* the core temperature drops.

An environmental temperature is "comfortable" if cutaneous blood flow is at a medium level and neither activation of sweat glands nor shivering is necessary for keeping the core temperature constant. This **thermic comfort** not only depends on environmental temperature but also on clothing, muscular work, wind, air humidity, and radiation. If, in a room, relative humidity is 50 % and wall temperature equals air temperature, the comfortable temperature for a sitting, lightly dressed person at rest is 25 °C, or 28 °C if naked, but is 22 °C for office work and 35 to 36 °C (at rest) in water. The latter value is lower (down to 31 °C) in obese persons, because skin fat insulates. Beyond the comfortable temperature not only heat production and sweating come into play but also **behavior**, such as choice of appropriate dress, seeking shade in summer, heating living quarters in winter. Below about 0 °C and above 50 °C (in very dry air, as in the sauna, even above 100 °C), these are the only means of maintaining body temperature (→ **C**).

Fever is produced by circulating exogenous or endogenous *pyrogens* that disturb the thermoregulatory mechanism in the hypothalamus; the body's "thermostat" is set at a higher value, i.e. the *set value is raised in fever*. Relative to this value, the body is too cold, as the fever develops, and *chills* and *shivering* accompany the rising temperature. Conversely, as the fever subsides and the set value is returned to normal, the body is too warm. This results in *vasodilatation* and *sweating*. In connection with infection, inflammation, and necrosis, *macrophages become active* (→ p. 66 ff.). They release *interleukin 1 and 6* (→ p. 70), which promote the synthesis of certain proteins in the liver, brain, and other organs. These, in turn, act as *endogenous pyrogens* (partly mediated by prostaglandin metabolism) on the thermoregulatory center in the hypothalamus, and thus elicit fever.

Measurement of the core temperature can be fairly accurately made either *rectally* or *orally*. Reliable *axillary* measurement (in the closed armpit) takes up to half an hour.

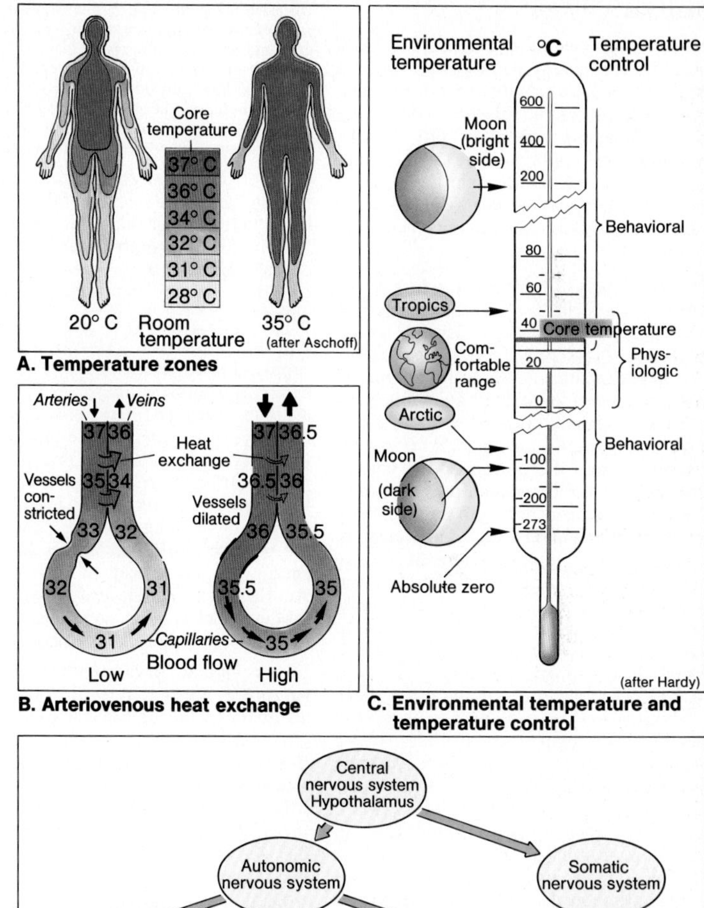

A. Temperature zones

B. Arteriovenous heat exchange

C. Environmental temperature and temperature control

Central nervous system Hypothalamus

Autonomic nervous system

Somatic nervous system

Sweat glands

Blood vessels

Brown fat

Skeletal muscle

Evaporative heat loss

Heat flow (core→skin)

Heat production (non-shivering)

(in newborns)

Heat production (shivering)

D. Neural influences on heat metabolism

Nutrition

An **adequate diet** must provide the body with sufficient energy, a minimum of protein (with all essential amino acids) and carbohydrates, inorganic substances (including trace elements), essential fatty acids, vitamins, and enough water. To ensure normal passage times through the colon, the diet must contain *roughage* ("fiber"), that is, nondigestible plant constitutents (cellulose, lignin, etc.)

The **daily energy requirement**, which is the equivalent of metabolic rate and power (1 J/s = 1 W = 86.4 kJ/d) depends on a number of factors and varies considerably even at rest (resting metabolism). Therefore, it was necessary to establish a definition for a **basal metabolic rate** (BMR): This is measured (1) in the morning; (2) fasting; (3) resting, supine; (4) at normal body temperature; and (5) at thermic comfort (→ p. 194). The BMR varies according to sex, age, body weight, and height; in adults it amounts, on a very rough average, to something more than 7 MJ/d (= approx. 80 W). Physical work raises the energy requirements (**working metabolic rate**): for office work to approximately 11 MJ/d (127 W); for women doing heavy work to about 15 MJ/d (= 175 W); and for men doing heavy work to about 20 MJ/d (= 230 W) per 70 kg body weight. These figures are averages over a number of years, but on any one day a male doing the heaviest type of work may perform up to a maximum of 50 MJ/d (= 600 W). In competitive sports (e.g. in marathon running), as much as about 1600 W may be maintained for over two hours, although the daily metabolic rate is much lower.

Energy needs are covered primarily by three basic nutrients: proteins, carbohydrates, and fats (→ **A**, **B**). The *minimum* **protein** intake is about 0.5 g protein/kg weight. This intake is necessary to balance the output of endogenous protein lost as result of degradative reactions. However, for normal activity, about twice this value is needed (*functional minimum*), of which half should be in the form of animal proteins to furnish the **essential amino acids**.

An essential amino acid is one that can only be synthesized by the organism in insufficient amounts or not at all. Many plant proteins are deficient in one or more of the essential amino acids, and their nutritive value is correspondingly less. In humans, the essential amino acids are: histidine, isoleucine, leucine, lysine, methionine, phenylalanine, threonine, tryptophan, and valine.

Most of the energy requirements are satisfied by carbohydrates and fats, which are largely interchangeable. The energy contribution of **carbohydrates** (starch, sugar, glycogen), normally 60% of the total, can fall to 10% before metabolic disturbances appear.

Fats are superfluous, provided there is a supply of essential fatty acids (e.g., linoleic) and of fat-soluble vitamins (A, D, E, K). On average, dietary fat provides about 25–30% of the energy (one third as essential fatty acids), although the proportion rises according to energy requirements (e.g. for a laborer: about 40%). For the physically undemanding working conditions in the western world, the diet is too rich in energy (i.e. fat instead of carbohydrates). Another factor here is *alcohol* consumption (about 30 kJ/g). The consequence of this excessive energy intake is *excess weight*.

Dietary intake of a number of **mineral** substances is essential for the body, e.g. an adequate intake of **calcium** (0.8 g/d), **iron** (10 mg/d; women 15 mg/d), and **iodine** (0.15 mg/d). A number of other "**trace elements**" (As, F, Cu, Si, V, Sn, Ni, Se, Mn, Mo, Cr, Co) are also vital, but they are provided in sufficient quantities in a normal diet. On the other hand, taken in excessive quantities, they may be toxic.

Vitamins (A, B_1, B_2, B_6, B_{12}, C, D_2, D_3, E, H, K_1, K_2, folic acid, niacinamide, pantothenic acid) are organic substances that play a vital role in metabolism (usually as coenzymes), but *cannot be synthesized* within the body, or only in insufficient amounts.

Although mostly required in very minute quantities a deficiency can lead to characteristic disorders (*avitaminoses*), for example, night blindness (A), scurvy (C), rickets (D), megaloblastic anemia (folic acid), pernicious anemia (B_{12} = cobalamine), beri-beri (B_2), pellagra (nicotinic acid), and disturbances in blood-clotting (K).

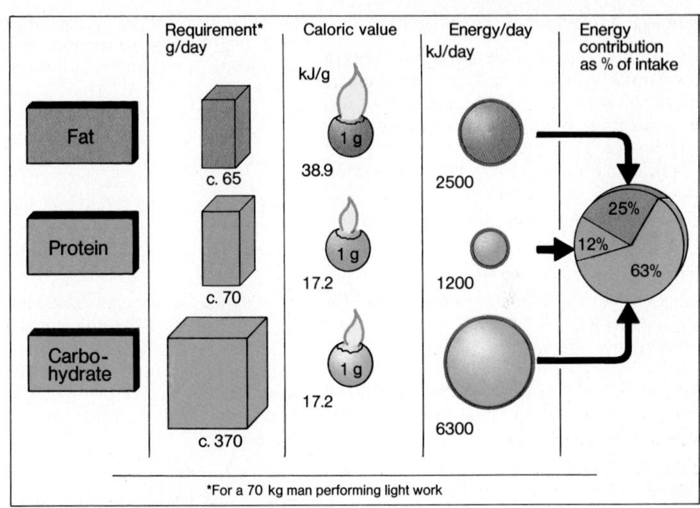

A. Energy requirements and sources

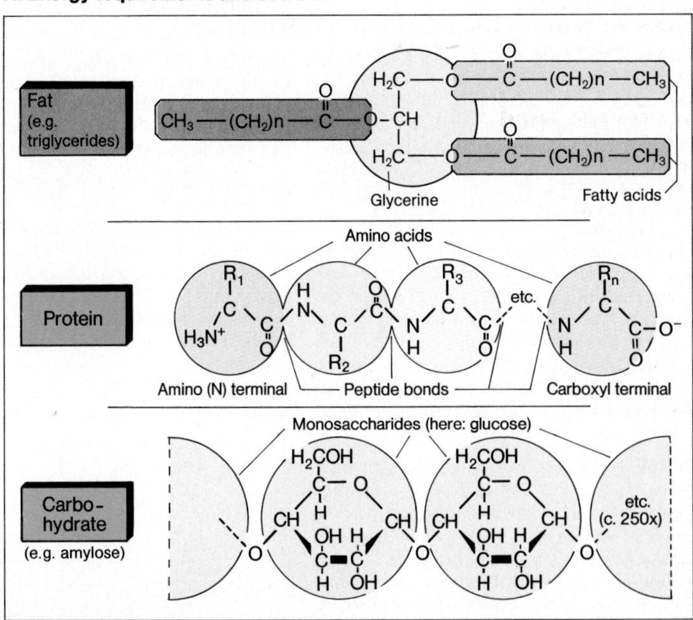

B. Chemical structure of basic nutrients

Energy Metabolism. Calorimetry

Metabolism converts the chemical energy of foodstuffs into heat and into mechanical work (muscles) (→ p. 18f.), but part of it is also used for the synthesis of endogenous substances. The utilizable energy content of the food made available by its total combustion (i.e. its breakdown to CO_2 and H_2O with the consumption of O_2) is its **physical caloric value** (C_{pc}).

The C_{pc} can be determined with the *bomb calorimeter* (→ **A**). A known quantity of the food substance in question is placed in a combustion chamber surrounded by water in an insulated container and then incinerated in O_2. The heat produced is taken up by the surrounding water, thus providing a measure of the C_{pc}.

Fats and carbohydrates are also completely oxidized to CO_2 and H_2O in the body. Their **physiological caloric value** (C_{pl}) is therefore identical with their C_{pc}. The *mean value* for fats is 38.9 kJ/g; for digestible carbohydrates, 17.2 kJ/g (→ p. 197, A). **Protein**, on the other hand, is only broken down in the organism as far as urea (and other substances), which, if its combustion were to be completed, would deliver additional energy. The C_{pc} of proteins (about 23 kJ/g) is thus greater than their C_{pl}, which only averages about 17.2 kJ/g (→ p. 197, A).

At rest, the energy received by the body in the form of food is for the most part converted into **heat energy** since almost no external mechanical work is being performed. In this situation, the heat lost (at constant body temperature) is equivalent to the metabolic rate within the organism (e.g. work done by cardiac and respiratory musculature; active transport; etc.).

In human beings, energy production may more easily be determined by **indirect calorimetry**. In this case, measurement is made of the utilization of O_2 ($\dot{V}_{O_2}$), which is a measure of energy liberation. For exact measurements by this technique, it is necessary to establish the **caloric equivalent CE** of oxygen for the various foodstuffs being oxidized. For the oxidation of 1 mol of glucose, 6 mol of O_2 (6×22.4 l) are required (→ **C**). The caloric value of glucose is 15.7 kJ/g. Thus the energy content of 180 g of glucose is 2827 kJ with an oxygen consumption of 134.4 l, that is, 21 kJ/l O_2, which is the CE of oxygen with glucose as the substrate under standard conditions. The average CE for carbohydrates is 21.15 (18.8) kJ/l O_2; for fats, 19.6 (17.6) kJ/l O_2; for proteins, 19.65 (16.8) kJ/l O_2, whereby in each case the first figure applies under standard conditions (0 °C), and the figure in parentheses at 37 °C.

The metabolic rate can be calculated from the CE, provided the nutrient that is being oxidized is known. This can be approximated by determining the **respiratory quotient RQ** ($= \dot{V}_{CO_2}/\dot{V}_{O_2}$, → p. 92). For carbohydrate, RQ = 1.00, as can be seen from the reaction:

$$C_6H_{12}O_6 + 6O_2 \rightleftarrows 6CO_2 + 6H_2O.$$

For the fat tripalmitin, the reaction is

$$2C_{51}H_{98}O_6 + 145O_2 \rightleftarrows 102CO_2 + 98H_2O.$$

The RQ = 102/145 = 0.7. Since the proportion of protein in the diet is fairly constant, it is possible to determine a CE for every RQ between 1 and 0.7 (→ **D**). **The metabolic rate** can then be calculated from **CE** · $\dot{V}_{O_2}$.

A protein diet raises the metabolic rate by about 15–20% (the so-called **specific dynamic action**). The higher metabolic rate is due to the fact that 89 kJ are required to produce 1 mol ATP from proteins (amino acids) but only 74 kJ if the source is carbohydrate. The utilization of the free energy of the amino acids is thus less than that of glucose.

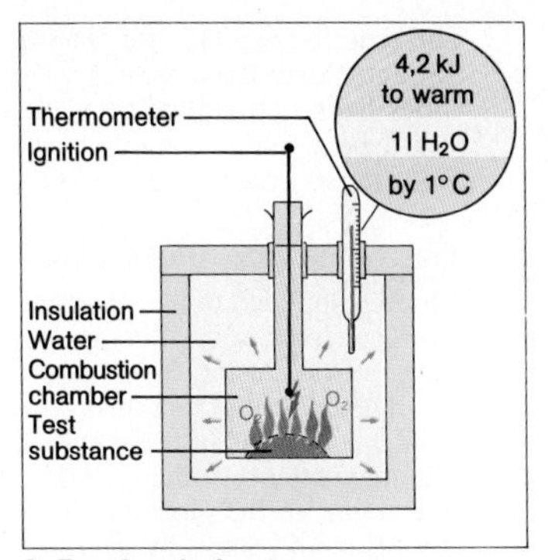

A. Bomb calorimeter

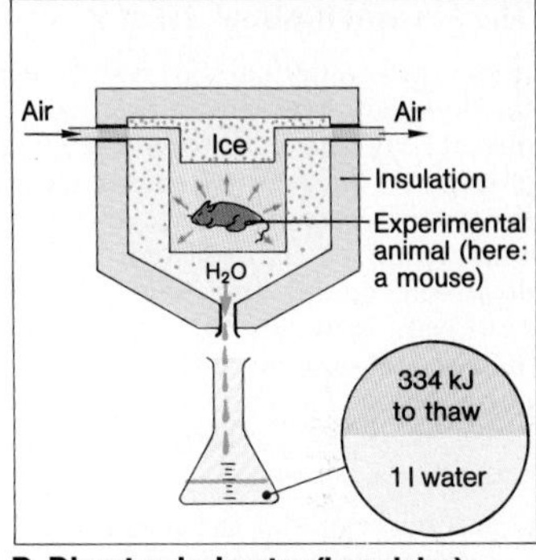

B. Direct calorimetry (Lavoisier)

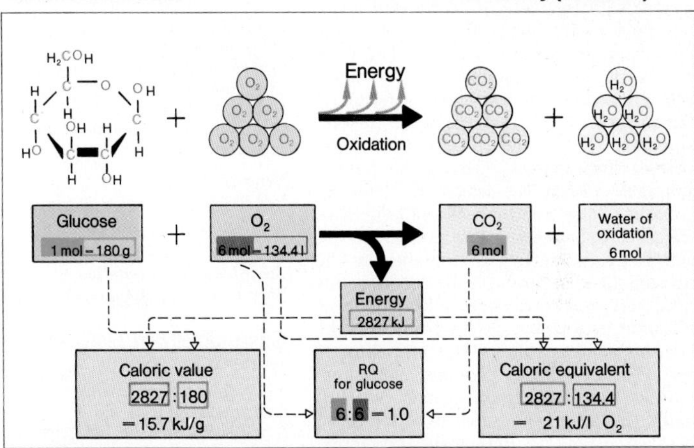

C. Oxidation of glucose: respiratory quotient, caloric value, and caloric equivalent

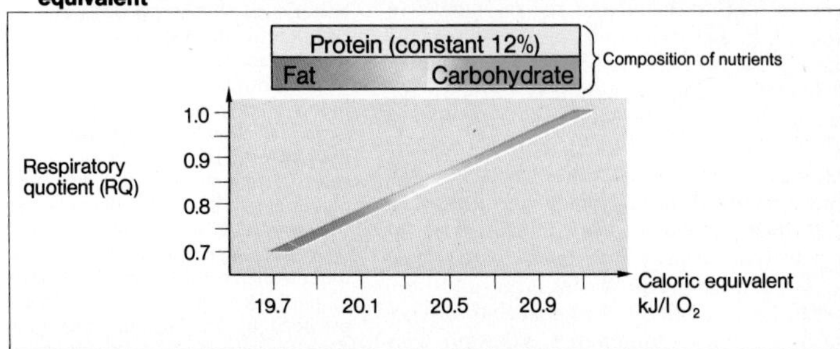

D. Caloric equivalent and RQ relative to nutrient composition

The Gastrointestinal Tract

In the gastrointestinal (GI) system, which is the boundary between the external and the internal environments, food is prepared for passage to the internal environment. Nutrients are propelled and mixed by the **musculature** of the GI tract and broken down into smaller units (**digestion**) that are absorbed through the intestinal tract mucosa (**absorption**) into the lymph or portal blood. The absorption process takes place by diffusion, carrier transport, or endocytosis.

Digestion begins in the **mouth**, where the large food particles are reduced in size, mixed with saliva, and converted to a semifluid mass. Swallowing transfers the chewed food to the **esophagus**. The food passes through the esophagus to the **stomach**, where it is mixed with gastric juice and liquefied by the contractions of the "distal" stomach. The liquefied food is now called **chyme** which passes through the pyloric sphincter into the **duodenum**. Exocrine secretions of the intestinal cells and digestive juices from the **pancreas** and **gall bladder** are added to the intestinal contents in the duodenum. Bile produced by the **liver** aids the digestion of fats as well as the removal of bilirubin, toxins, etc. Apart from this, the liver plays a key role in the metabolism of carbohydrates, fats, proteins, and hormones. The **pancreas** contributes bicarbonate (HCO_3^-) and digestive enzymes, besides being an important endocrine organ. A number of **gastrointestinal hormones** that contribute to the regulation of digestion are primarily produced in the upper part of the small intestine, the lower stomach, and in the pancreas. Most of the absorption of the digested food, as well as of the fluids secreted by the salivary glands, the stomach, etc., takes place in the **small intestine** (**duodenum, jejunum**, and **ileum**). In the **large intestine**, final absorption of electrolytes and water takes place. The contents of the end of the large intestine, **feces**, consist chiefly of unabsorbable vegetable matter, desquamated cells and bacteria, and a minimum of water. They are stored in the **rectum** until they are voluntarily eliminated (**defecation**).

The **time** required for the **passage** of food through the different parts of the GI tract varies from one individual to another and also depends upon the composition of the food (mean values → **A**; see also p. 206).

Gastrointestinal Blood Flow

Blood is supplied to the stomach, intestines, liver, pancreas, and spleen by three branches of the aorta. During the process of digestion the intestinal blood supply, which is independent of the systemic blood pressure (**autoregulation**), is increased by the pump-like action of the intestinal movements on the intestinal vessels, by the vagus nerve, by hormones (VIP, etc), and by local reflexes.

Paralysis or obstruction of the intestine (**ileus**) gradually reduces the blood supply, which is still further impaired by the rising intraluminal pressure caused by accumulating gases (CO_2, methane, H_2S, etc.). When the intraluminal pressure exceeds that in the intestinal blood vessels, the blood supply is totally cut off.

Venous blood containing substances absorbed from the intestine is drained to the liver via the hepatic **portal vein**. Most of the absorbed fatty components are taken up by the intestinal **lymphatics**, thus bypassing the liver to reach the general circulation.

Gastrointestinal Defense Systems

To cope with bacteria, viruses, and foreign macromolecules entering the body via the mouth, the GI tract possesses highly efficient defense systems (→ p. 66ff.). The first line of defense is formed by mucins, immunoglobulin **A** (**IgA**), and lysozyme (→ p. 202) present in the saliva. In the stomach a bactericidal effect is exerted by HCl and pepsins, in addition to which the intestinal tract is equipped with immune-competent tissue, the Peyer's patches. Probably via specialized "M" cells (**m**embranous cells) in the mucosa, antigens from the lumen can enter Peyer's patches, which respond by pouring out IgA (*oral immunisation* or, in some cases, oral allergy; → p. 72). In contrast to the other immunoglobulins, IgA is protected from the digestive enzymes of the intestinal lumen by a so-called *secretory component*, which is attached to IgA in the epithelial cells. Yet another barrier to the entry of parthogens from the gastrointestinal tract is provided by the large number of macrophages (Kupffer cells) in the ramifications of the hepatic portal vein. In the newborn infant the GI mucosa is primarily protected by the IgA from the mother's milk.

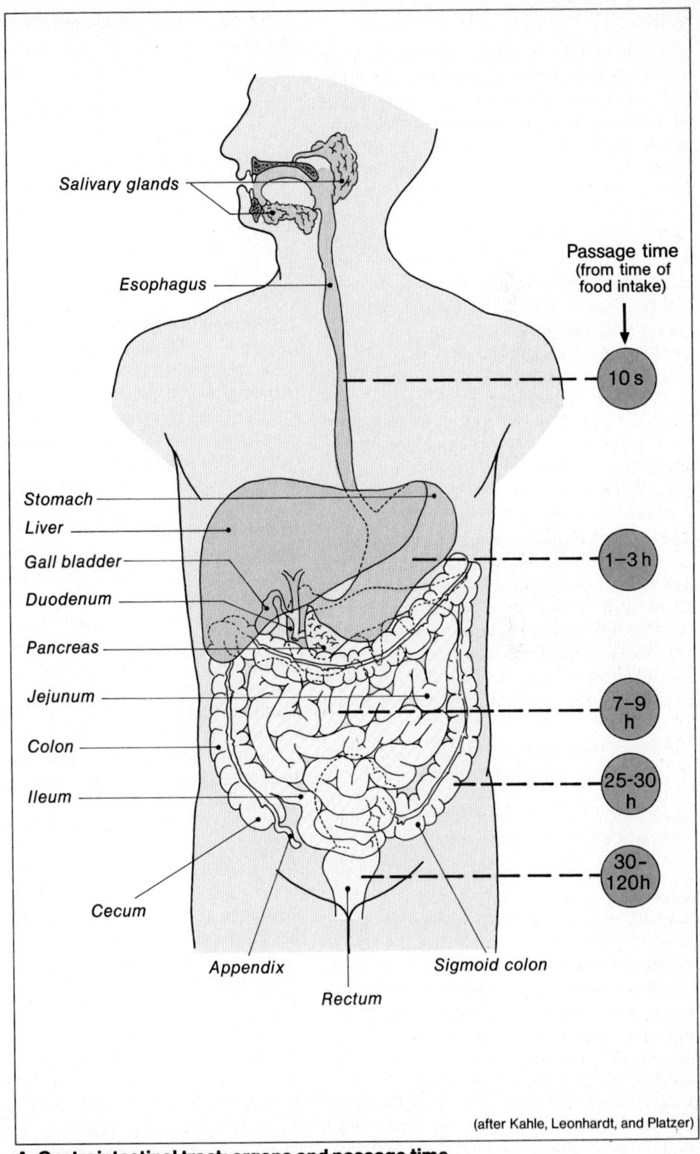

A. Gastrointestinal tract; organs and passage time

Saliva

The **role of saliva** is reflected in its composition: **mucus** (mucins) moistens and lubricates the food particles, making them easier to swallow, and also facilitates the movements involved in mastication and speech. Saliva is also important for **oral hygiene** (irrigation of mouth and teeth) and serves as a fluid seal in suckling of the infant. Some parts of the food are dissolved in the saliva, which not only makes possible the stimulation of chemoreceptors involved in taste (→ p. 296) but also starts off the action of the α-**amylase** contained in the saliva (*ptyalin*), which initiates the digestion of carbohydrates (starches). Immunoglobulin A (**IgA**), *lysozyme* (→ p. 65ff.) and *peroxidase* (→ p. 68), provide **defense** against pathogenic organisms. The high $\mathbf{HCO_3^-}$ content buffers the pH to a value of 7–8; an acid pH would inhibit the amylase and damage the dental enamel.

The **salivary output** amounts to **0.5–1.5 l/d**. Depending on the degree of stimulation, the **flow rate** my vary from 0.1 to 4 ml/min. At 0.5 ml/min about 95% of the saliva is secreted by the *parotid glands* (watery saliva) and the *submandibular glands* (mucin-rich saliva); the rest is secreted by the *sublingual glands* and the glands in the mucosa lining the mouth.

The site of **formation** of the saliva is the **acini** which secrete the *primary saliva* (→ **A**, **C**). It is similar in its electrolyte composition to blood plasma, and its formation is driven by the transcellular **transport of** $\mathbf{Cl^-}$.

Cl^- is taken up from the blood into the acinar cells by secondary-active $\mathbf{Na^+}$**-**$\mathbf{K^+}$**-2Cl-Cotransport** (reversed direction as compared to p. 149, B2), and reaches the acinar lumen through $\mathbf{Cl^-}$**-channels**. This causes a lumen-negative transcellular potential which also drives Na^+ into the lumen (paracellular diffusion); water follows for osmotic reasons. Neurotransmitters stimulating salivary secretion increase the intracellular concentration of $\mathbf{Ca^{2+}}$ (→ **C**) which not only opens the Cl^- channels (and thereby increases fluid secretion) but also stimulates **exocytosis** (→ p. 12) of the salivary proteins.

The primary saliva is **modified in the ducts** of the glands (→ **A**): Na^+ and Cl^- are resorbed, while K^+ and HCO_3^- are secreted. HCO_3^- secretion is accompanied by H^+ transport into the blood (Na^+/H^+ antiport). *Carbonic anhydrase* is involved in these processes (→ p. 145, A). As NaCl resorption outweighs $KHCO_3$ secretion the saliva becomes *hypoosmolal*, at rest down to 50 mosm/kg H_2O. The low NaCl concentration (→ **B**) improves the protein solubility and lowers the threshold of the taste receptors for salt (→ p. 296).

The **stimulation of salivary output** is a **reflex** response (→ **D**) set off by taste-, smell- and touch receptors in the mouth as well as by chewing. *Conditioned reflexes* may also be involved, as illustrated by the common experience that the mere rattling of plates can turn on salivary secretion. The production of watery saliva is elicited by cholinergic, α-adrenergic, and peptidergic (substance P) stimulation triggered (partly via IP_3; → p. 244) by an increase in the cytoplasmic concentration of $\mathbf{Ca^+}$ entering the cytoplasm from intracellular stores and from the ECS (→ **C** and p. 17). Cholinergic stimulation (→ p. 54) also releases enzymes (*kallikreins*) that form the potent vasodilator *bradykinin* from plasma kininogens. **V**asoactive **i**ntestinal **p**eptide (**VIP**) probably acts here as a cotransmitter. *Vasodilatation* is essential because the maximum salivary flow exceeds the resting blood supply. β-Adrenergic activation of the salivary glands leads (via cAMP; →**C** and p. 56ff.; p. 242) to a highly viscous saliva, rich in mucin. Viscous saliva is produced by dogs when eating meat, for example, whereas cholinergically elicited watery saliva predominates if the food consumed is dry. The biological significance of the dual secretomotor control in humans, and the reason why the two control systems evoke saliva of different compositions, is unknown.

The mean salivary output is closely dependent upon the water content of the body: if the latter is low the mouth and throat are dry, which not only represents a saving of water but also contributes to the sensation of **thirst** essential for the maintenance of the overall water balance (→ pp. 138 and 152).

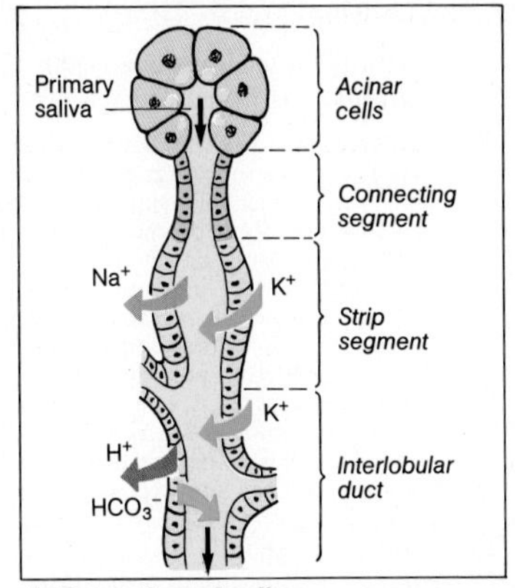

A. Formation of saliva

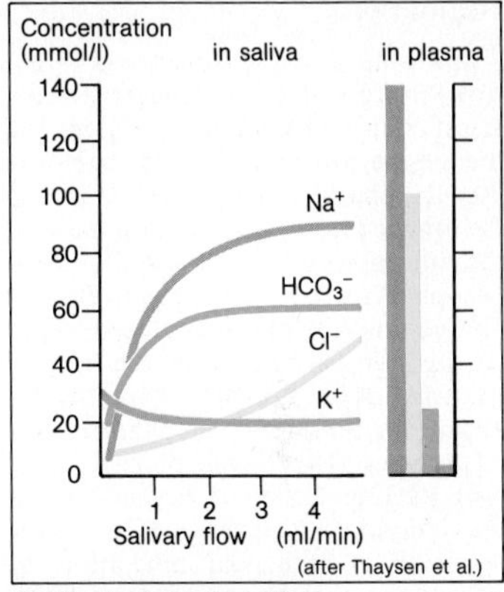

B. Electrolytes in saliva

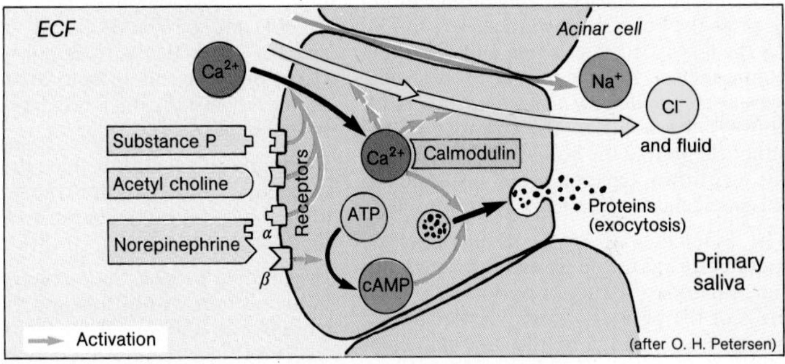

C. Control of formation of saliva in acinar cells

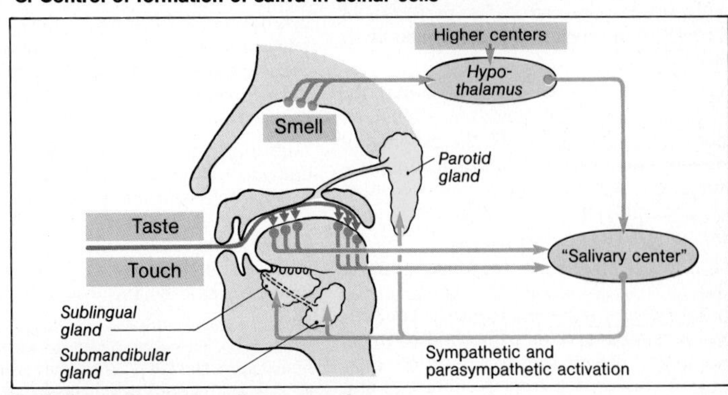

D. Reflex pathways in salivary secretion

Deglutition

Swallowing or deglutition (→ **A 1–10**) is a *reflex* process. It is initiated by the voluntary act of collecting food on the tongue, raising the tongue, and propelling the **bolus** of food into the pharynx. The mouth is closed and the tongue presses against the gums (→ **A 1**); the soft palate is raised (→ **A 2**) to close the nasopharynx (→ **A 3**); respiration is inhibited, and the glottis is closed to prevent passage of particles into the respiratory tract (→ **A 5**). Pressure from the tongue pushes the bolus into the pharynx. Sphincter muscles of the lower pharynx relax (→ **A 6**). Resting tension of the 3 cm segment at the junction of pharynx and esophagus normally is high, but relaxation occurs reflexly, so that the tongue can push the food into the *esophagus*, and then contract to press the bolus downwards (→ **A 7, A 8**). As the larynx subsides again and breathing continues (→ **A 9**), a (*primary*) *peristaltic wave in the esophageal musculature* (→ **A 10**) conveys the bolus to the entrance of the stomach. Should the bolus become stuck on the way down, the esophageal distension at this point elicits a *secondary peristaltic wave*.

The esophagus is 25 to 30 cm long; its musculature at the upper end is striated, but the remainder is smooth muscle. The progress of the peristaltic wave in the striated muscle is controlled by the *medulla oblongata*; afferent and efferent signals are conveyed in the *vagus nerve*. The peristalsis of the smooth musculature is controlled by its own *ganglia*.

At the entrance to the stomach (**cardia**), the esophagus is normally closed by a sphincter mechanism. A twisting of the esophageal musculature (wringing mechanism), pressure in the abdominal space, and a cushion of veins play a role in this mechanism (opening reflex, → p. 206).

Motility of the gastrointestinal smooth muscle is closely related to function of the *myenteric plexus* (→ p. 210); defects of the plexus result in disordered peristalsis. In the esophagus, defects lead to failure of relaxation of the cardia; food accumulates and the esophagus dilates to accommodate the mass (*achalasia*).

Vomiting

Vomiting is a **protective reflex** that is integrated in the medulla oblongata.

Many **stimuli** can induce vomiting; dilatation of the stomach from overeating; gastric mucosal irritation (e.g., by alcohol); unpleasant sights, smells, and ideas contribute psychic stimuli; touching the pharyngeal mucosa or stimulating the equilibration organs of the inner ear (*motion sickness*); pregnancy, in which normal "*morning sickness*" may progress to a threatening *hyperemesis gravidarum*; severe pain; poisons; toxins and medications (*apomorphine* is one of the most potent stimuli to vomiting and is used on occasion for the treatment of poisoning); irradiation effects (tumor therapy); elevated intracranial pressure (e.g. in brain edema; → p. 142); or hemorrhage or tumors in the brain region.

These stimuli activate the **vomiting "center"** in the medulla oblongata, between the olive (→ **B 1**) and the solitary tract (→ **B 2**) in the area of the reticular formation (→ **B 3**). Occasionally, chemoreceptors in the vicinity of the vomiting center (area postrema → **B 4**) also play a role.

Prodromes of vomiting include the sensation of nausea, hypersalivation, rapid heart rate, altered respiration, paleness, sweating, and dilatation of the pupils.

In vomiting proper, the diaphragm is fixed in the inspiratory position and the glottis is closed. The abdominal wall muscles contract and increase intra-abdominal pressure. The duodenum constricts; the cardia simultaneously relaxes, allowing the gastric contents to be passed into the esophagus. The esophagopharyngeal sphincter relaxes and the soft palate rises to permit the bolus to be expelled through the mouth (→ **B**).

Vomiting is a protective reflex to hinder damage to the stomach and to the total organism by poisoning. Prolonged vomiting, however, is associated with losses of acidic gastric juice and leads to nonrespiratory alkalosis (→ p. 114) and disturbances in fluid balance (→ p. 142).

Motion sickness, with accompanying nausea and vomiting, is elicited by stimuli to which the vestibular apparatus (→ p. 298) is unaccustomed (e.g. at sea); the sickness is enhanced if the head is additionally moved, thus producing a discrepancy between the optical and vestibular input.

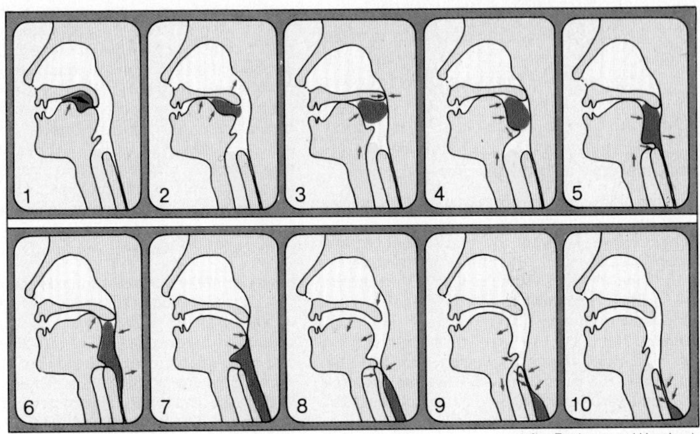

A. Swallowing

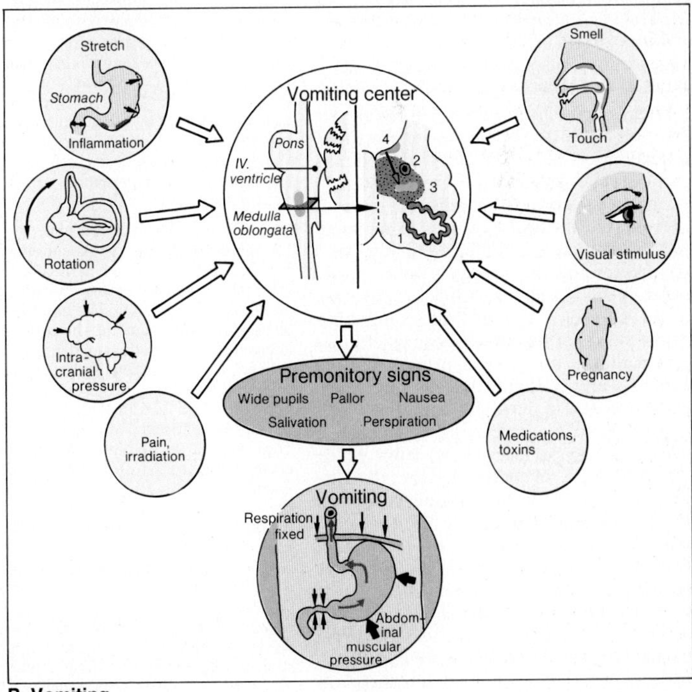

B. Vomiting

Stomach: Structure and Motility

The esophagus enters the upper third of the stomach, the *fundus*, at the *cardia*. The adjoining *corpus* of the stomach is followed by the *antrum*, which narrows to the *pylorus*. This leads to the *duodenum* (→ **A**). Functionally, the stomach can be divided into a **proximal** and a **distal portion** (→ **A**).

The **size** of the stomach depends upon how much it contains; it is chiefly the proximal portion that enlarges (→ **A, B**), while the intraluminal pressure undergoes only slight alterations. In **structure**, the wall of the stomach resembles that of the intestine (→ p. 211). The **tubular glands** of corpus and fundus contain *chief* (or peptic) *cells* (CC) and *parietal cells* (PC; → **A**), which produce the constituents of the gastric juice (→ p. 208). Additionally, the gastric mucosa contains *endocrine cells* and *mucus-producing cells*, i.e. mucous neck cells and superficial epithelial cells. The two *autonomic plexuses* (→ p. 208) of the stomach wall control gastric motility, which, via these plexuses, can also be influenced by the *autonomic nervous system* (→ p. 50ff. and **D**).

When food is swallowed, the cardiac sphincter opens reflexly and stimulation of inhibitory vagus fibers causes temporary relaxation of the proximal stomach (**receptive relaxation**; → **D 2**). The entry of food also causes reflex relaxation of the "proximal" stomach (**accommodation reflex**) in order to prevent the internal pressure from rising further with increased filling. Finally, local stimulation of the stomach wall (partly reflex, partly via gastrin) leads to activation of the stomach (→ **D 1**). Tonic contraction of the **proximal stomach** (whose chief function is that of a *reservoir*) ensures that the contents are gradually passed on to the **distal stomach**, at the upper border of which (middle third of the corpus) is situated a *pacemaker region* (see below) from which *peristaltic waves* of contraction spread to the pylorus. Contraction is particularly strong in the antrum: the gastric contents are *pressed* towards the pylorus (→ **C 5, C 6, C 1**), *compressed* (→ **C 2, C 3**), and *propelled back* again following closure of the pylorus (→ **C 3, C 4**). In this way the food is *liquefied*, thoroughly *mixed* with the gastric juice, partially digested and the *fats emulsified*.

In the **pacemaker cells** of the **distal stomach** (see above) oscillations in the membrane potentials can be registered every 20 s, proceeding with increasing speed (0.5–4 cm/s) and amplitude (0.5–4 mV) towards the pylorus. As in the heart, the pacemaker activity in the distal parts of the stomach is overridden by the more proximal pacemaker due to its higher frequency. Whether and how often contraction follows such a wave of stimulation depends upon the summated neural and humoral factors. **Gastrin** and **CCK** (= cholecystokinin = pancreozymin) increase the frequency of the pacemaker as well as of response. Other peptide hormones inhibit this motility either directly, like **GIP** (**g**astric **i**nhibitory **p**eptide), or indirectly, like somatostatin (**SIH**; → p. 246 and **D 1**).

Emptying of the stomach during digestion is primarily dependent upon the **tone** of the **proximal stomach** and the *pylorus*, which are under reflex and hormonal control (→ **D 2**). Cholinergic fibers of the **vagus nerve** increase the tone of the proximal stomach, whereas other vagus efferents (**ATP** and **VIP** [vasoactive intestinal peptide] as cotransmitters) and andrenergic **sympathetic fibers** inhibit it. **Motilin** promotes the emptying of the stomach (tone of proximal stomach increases; pylorus dilates), whereas **CCK**, **gastrin**, and other substances inhibit it by producing the opposite effects. Most of the time the pylorus is slightly open (free outflow of "finished" chyme). It contracts only (1) at the end of antral "systole" (see above), in order to retain solid food; and (2) during duodenal contraction, to prevent reflux (bile salts) into the stomach. If this should nevertheless occur, refluxed free amino acids, not normally present in the gastric lumen, elicit reflex closure of the pylorus.

Emptying rate. The length of time for which food remains in the stomach varies considerably: solids, e.g., can only flow into the duodenum as **chyme** after they have become suspended as particles of about 0.3 mm diameter. The time elapsing before 50 % of the water entering the stomach has left it again is 10–20 minutes, and depends largely upon the tonus of the proximal stomach. The value for solids lies between 1 and 4 hours, according to the ease with which they are liquefied and the intensity of peristalsis. The emptying rate for carbohydrates > proteins > fats. The decrease in this rate, which occurs when the pH of the expelled chyme drops and the osmolality rises is mediated by receptors in the duodenum, by enterogastric reflexes, and by peptide hormones (see below; → **D 2**). *Indigestible components* such as bone, fibers, and foreign bodies remain in the stomach during the *digestive phase*. Their evacuation together with digestive secretions and sloughed-off cells takes place during the ensuing **interdigestive phase**, with the help of the waves of contraction that spread through the stomach and intestine at 2-hourly intervals ("internal clock"): **migrating motor complex**. **Motilin** from the mucosa of the duodenum is involved in the regulation of this phase.

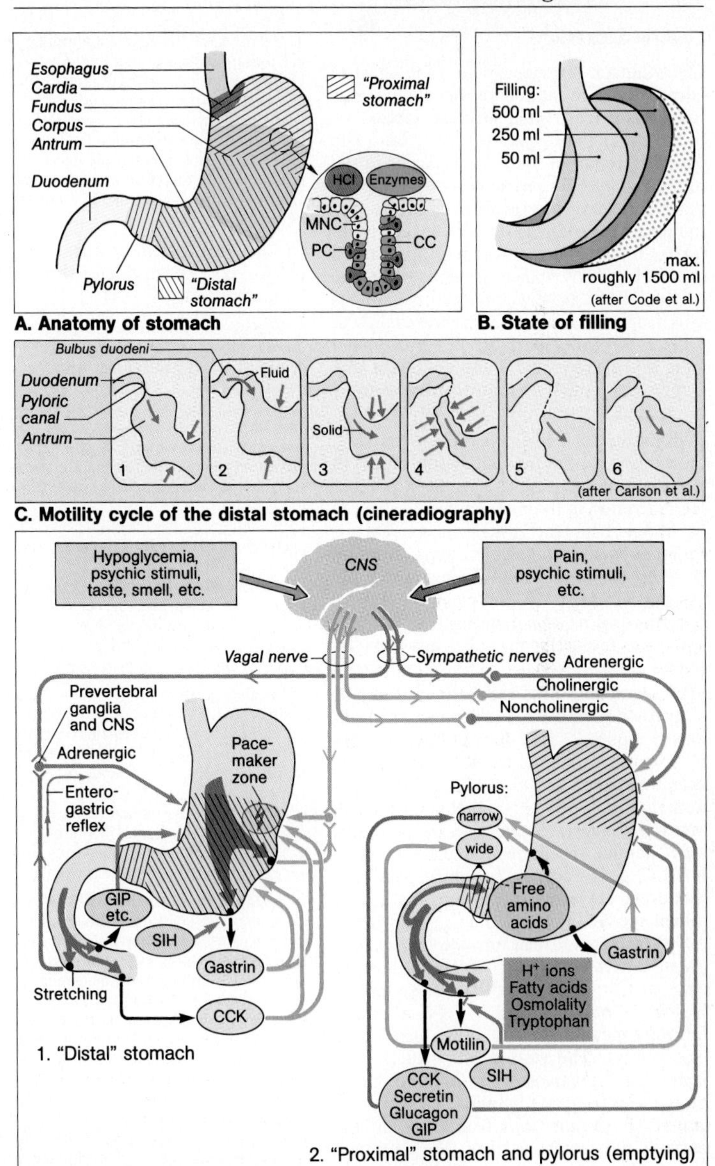

D. Influences on gastric motility

Gastric Secretion

The stomach secretes up to 3 l of **gastric juice** daily. The main components are **pepsinogens, mucus, HCl, intrinsic factor** (→ p. 226), and "*gastroferrin*" (→ p. 62). The **site of secretion** of the gastric juice is in the tubular glands or **gastric pits** of the gastric mucosa, where the constituents of the gastric juice are produced by different cell types (→ p. 207, A). The pepsinogens are formed in the *chief cells*, which are situated in the fundus of the stomach. Specialized mucus cells produce the **mucus** that serves mainly to *protect* the inner surface of the stomach from the digestive juices. Hydrochloric acid is formed by the *parietal* or *oxyntic cells* of the fundus and corpus.

Pepsins are formed from their precursors, the *pepsinogens*, by the splitting off of part of the molecule at a pH below 6. Maximum HCl secretion in the stomach results in a pH of about 1 in the gastric juice. This is buffered by the stomach contents to values between 1.8 and 4, which covers the pH optima of most of the pepsins. Besides contributing to the denaturation of proteins, the low pH value also has a bactericidal effect.

HCl secretion: With the help of *carbonic anhydrase* (CA; → p. 144 ff.) and an **ATP**-driven "pump" (**H^+-K^+-ATPase**; → **B**), H^+ ions (in exchange for K^+) are actively accumulated in the lumen to a concentration that is about 10^7-fold that in the cell (*active transport*). K^+ returns to the lumen by a passive mechanism (recirculation of K^+). Cl^- also enters the lumen passively. For every H^+ ion secreted into the gastric lumen one HCO_3^- ion (from CO_2 and OH^-; → **B**) leaves the cell on the blood side (passive exchange for Cl^-). In addition to this, as in all cells, there is an active Na^+/K^+-"pump" (*Na^+- K^+-ATPase*). During a meal the **parietal cells** are **activated** (see below). The resting H^+ secretion of 2 mmol/h can increase up to 20 mmol/h and more. This is made possible by the activation of invaginations (*canaliculi*) of the parietal cell which, in turn, are equipped with a dense brush border, offering a vastly increased cell surface for the secretion of H^+ into the gastric lumen.

The mucosa of the stomach is protected from the H^+ ions in the gastric juice by the active **secretion of HCO_3^-** ions which, without noticeably altering the pH of the gastric contents, adequately buffer any acid that may succeed in penetrating the mucus covering of the epithelium. The danger of *gastric ulcers* is therefore enhanced by substances that inhibit HCO_3^- secretion, whereas substances that enhance HCO_3^- secretion, like **prostaglandin E_2**, afford protection.

In the **stimulation** of the secretion of a physiological **gastric juice** three *phases* can be distinguished (→ **A**):

1. **Cephalic** or **psychic-neural phase**: Food in the mouth initiates **reflex** stimulation of the secretion of gastric juice. The afferent limbs of this, in part *conditioned* (→ p. 202), *reflex* are the nerves of taste, sight, and smell. The reflex can also be elicited by *glucose deficiency* in the brain. Aggressive emotions are capable of evoking an increase in secretion, whereas fear has an inhibitory action. In each case the efferent limb of the reflex is the **vagus** (X). (Following *vagotomy*, sometimes performed as ulcer therapy, all such influences disappear). The **acetyl choline** released by the vagus and intramural nerves activates (via IP_3 and **Ca^{2+}** influx), besides the chief cells, the parietal cells, the adjacent **H**(istamine)-cells, and the **G**(astrin)-cells of the antrum. Thus, indirectly, the vagus has both a paracrine (**histamine**) and an endocrine (**gastrin**) effect on the secretion of the gastric juice (→ **C**).

2. **Local** or **gastric phase**: When the gastric contents come into contact with the lower portion of the stomach (antrum), **gastrin** is released as a result of *mechanical* (stretching) and *chemical* stimulation (peptides, amino acids, Ca^{2+}, products of roasting, alcohol, etc.). Gastrin is conveyed to the upper parts of the stomach via the *blood* (endocrine activation, see above) and increases the secretion of gastric acid. The release of gastrin is inhibited by a very low luminal pH value (negative feedback).

3. **Intestinal phase**: As the chyme enters the duodenum, stretching of the intestinal wall has a positive influence on the secretion of gastric juice, mediated by endocrine pathways (*enterooxyntin? gastrin?*). Absorbed amino acids have a similar effect. A low pH and fat in the duodenal chyme inhibit the secretion of gastric acid via the release of various hormones (*secretin, GIP, SIH*). In this way not only the quantity but also the composition of the chyme leaving the stomach are adapted by the duodenum to the requirements of the small intestine. In general, SIH has a retarding effect on the absorption of food. The secretion of SIH and insulin by the pancreas is probably coordinated in some way (see also p. 246).

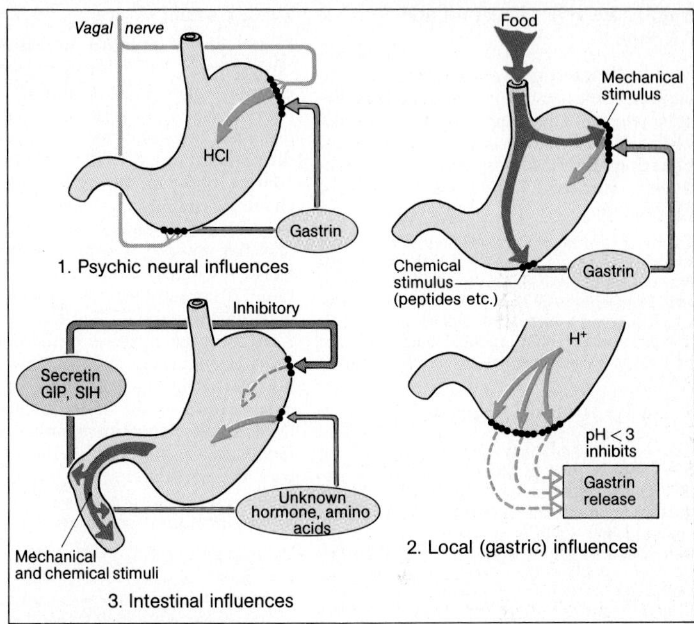

A. Secretion of gastric juice

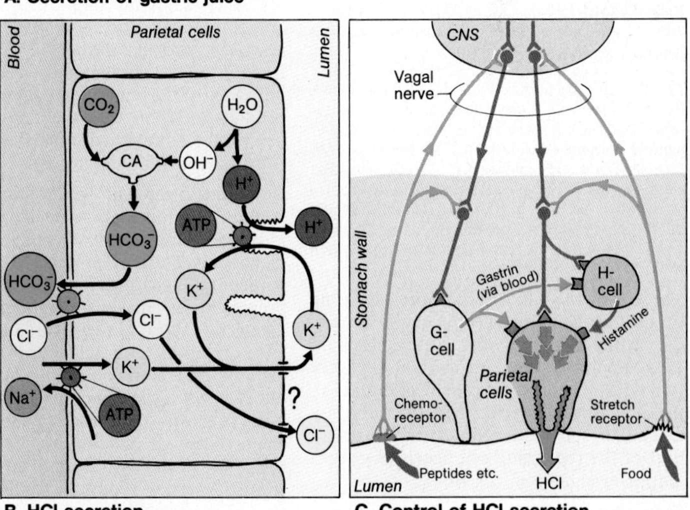

B. HCl secretion

C. Control of HCl secretion

Small Intestine: Structure and Motility

The small intestine (in vivo about 2 m in length) comprises three segments: **duodenum, jejunum**, and **ileum**. The **main function** of the small intestine is the completion of digestion and the absorption of the breakdown products, of water, and of electrolytes.

The outer covering of the small intestine consists of the peritoneum and **serosa** (→ **A 1**); followed by a *longitudinal layer of muscle* (→ **A 2**); a layer of *circular muscle* (→ **A 3**); and the *mucosa* (→ **A 4**), which contains a further muscular layer (→ **A 5**). The inner border to the intestinal lumen is formed by a layer of various types of epithelial cells (→ **A 6–A 8**).

The surface area of the epithelium-lumen boundary is increased to about 300-fold to 1600-fold (more than 100 m^2) that of a smooth cylindrical tube: about 3-fold by the annular folds in the mucosa and submucosa (→ **A**), 7-14-fold by the fingerlike projections known as *villi*, which are approximately 1 mm long and 0.1 mm thick (→ **A 9**), and 15-40-fold by the microvilli of the brush border (→ **A 10**) on the luminal membrane of the epithelial cells responsible for *absorption* (→ **A 7**). Scattered between the resorbing cells of the villi are *mucus-producing cells* (→ **A 6**). The **crypts of Lieberkühn** at the base of the villi (→ **A 8**) consist of various cells: (a) cells producing the **mucus** that forms a protective and lubricant layer in the intestinal lumen; (b) *undifferentiated* and *mitotic cells* that give rise to the cells of the villi (see below); (c) *endocrine cells*, which probably possess receptors on the luminal side, but which release their **peptide hormones** (*secretin*, CCK, motilin, *SIH, GIP*, etc.) into the blood; (d) *Paneth cells* that release proteins (*enzymes, immunoglobulin*); and (e) *membranous cells* (→ p. 200). Situated more deeply in the duodenal wall are the *glands of Brunner*, which release into the lumen a secretion that is rich in HCO_3^- and glycoproteins.

The tips of the villi are continuously being sloughed-off and replaced by new cells from the crypts. In this way the entire epithelium of the small intestine is renewed about every two days. The cast-off cells disintegrate in the lumen, setting free digestive enzymes and other substances (e.g. iron; → p. 62); part of the remaining cell debris (e.g. protein) is evacuated in the stool.

The mesentery (→ **A 11**) supplies the small intestine with blood and lymph vessels (→ **A 12–14**) and with sympathetic and parasympathetic nerves (→ **A 15** and p. 50 ff.).

Four types of **intestinal movements** can be distinguished, all of them independent of external innervation (autonomous). The *backward and forward movements of the individual villi*, effected by contraction of their smooth muscle fibers, ensure intimate contact between the epithelium and the chyme. *Pendular movements* (longitudinal muscle; → **C 1**) and *rhythmic segmentation* (circular muscle; → **C 2**) are **mixing movements**, whereas the *peristaltic waves* (30–120 cm/min) serve to **propel the intestinal contents** (about 1 cm/min; → **C 3**) toward the large intestine. In the small intestine the frequency of the slow potential fluctuations in the smooth musculature decreases towards the anus. In this way the orally situated portions assume a *pacemaker function* (→ p. 44), which is the main reason why the peristaltic waves (= continuous repetition of the *peristaltic reflex*) only run in the direction of the anus.

By stimulating stretch receptors, the bolus (→ **B**) elicits a **peristaltic reflex** that narrows the lumen in its wake and widens it in its path. So-called *type 2 cholinergic motoneurones* (with long-lasting discharge) are stimulated by *serotoninergic interneurones*, and simultaneously cause contraction of the circular muscle behind the bolus and of the longitudinal muscle in front of it. At the same time the circular muscle behind the bolus is deinhibited and that in front of it is inhibited (→ **B**).

Sympathetic efferent nerves are vasoconstrictor and bring about relaxation of the intestinal musculature via *inhibition* of the plexus myentericus (→ **A 16**). **Parasympathetic efferent fibers** are switched from preganglionic to postganglionic in the plexus myentericus; they exert a *stimulatory* effect on all three muscle layers and on the exocrine and endocrine glands of the intestine. The *plexus submucosus* (→ **A 17**) contains primarily the sensory neurons of the chemoreceptors and mechanoreceptors of the mucosa. Messages from these nerves and from the stretch receptors of the musculature elicit peripheral and, via **visceral afferent fibers**, central *reflexes*.

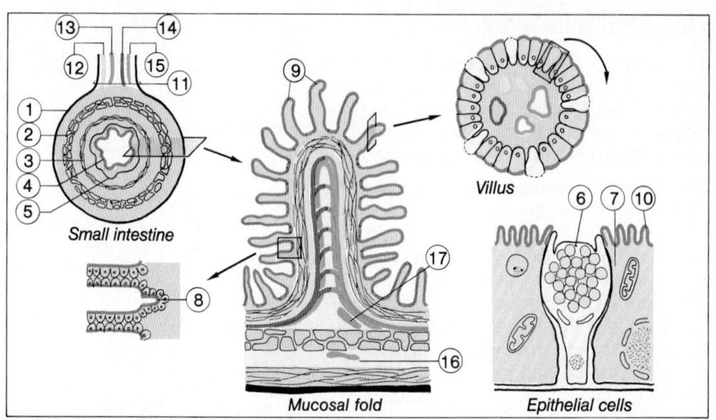

A. Structure of the intestine (schematic)

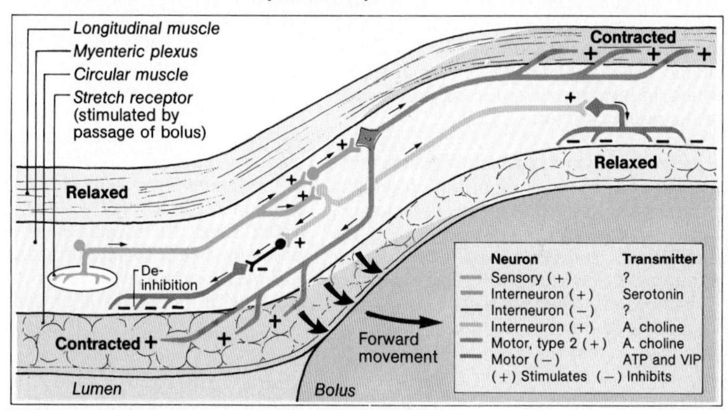

B. Neuronal control of peristalsis (see also C3) (after J. D. Wood)

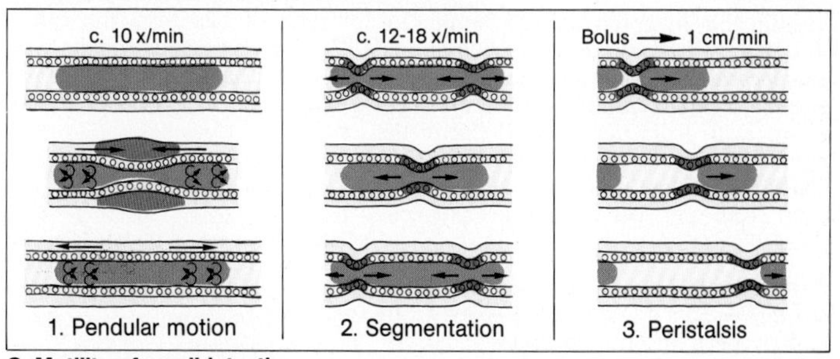

C. Motility of small intestine

Pancreatic Juice – Bile

The **pancreas** releases approximately 2 l of juice daily into the duodenum. The juice is rich in $\mathbf{HCO_3^-}$ and in **digestive enzymes** necessary for the breakdown of proteins, fats, and carbohydrates in the chyme.

Secretion of the pancreatic juice is under the control of the **vagus** nerve and two hormones produced by the duodenal epithelium: **secretin** and **CCK** (= **cholecystokinin**, which is identical to **pancreozymin**; → **A**). The stimulus for release of secretin is provided by fats and low pH of the duodenal contents. Secretin reaches the pancreas via the bloodstream and increases the volume flow of pancreatic juice and raises its HCO_3^- content. (The greater the rise in HCO_3^- the lower is the concentration of Cl^-; → **B**.) $\mathbf{HCO_3^-}$ is required for the neutralization of the acidic chyme (gastric acid!). Fat in the chyme also stimulates the release of CCK, which, in turn, increases the *enzyme content* of the pancreatic juice.

Pancreatic enzymes in protein digestion (proteases): The two most important proteolytic enzymes are secreted as inactive precursors, *trypsinogen* and *chymotrypsinogen*. In the duodenum, *enteropeptidase* (the former name, enterokinase, is obsolete) activates trypsinogen to **trypsin**, which, in turn, activates, among other proenzymes, chymotrypsinogen to **chymotrypsin** (→ **A**). If the activation occurs within the pancreas itself, autodigestion takes place and produces *acute pancreatic necrosis*.

Trypsin and chymotrypsin split certain peptide bonds *within* the protein molecule (*endopeptidases*) while another pancreatic enzyme, a **carboxypeptidase**, splits off single amino acids from the carboxyl end of the peptide chains (→ p. 197, B and p. 224). Carboxypeptidase is released (also by trypsin) from a precursor (procarboxypeptidase).

Carbohydrate digestion: pancreatic **α-amylase** breaks down starch (→ p. 197, B) and glycogen into trisaccharides and disaccharides (maltose, maltotriose, α-limit dextrin), intestinal *1,6-glucosidase* splits the branched dextrins, and *maltase, lactase*, and *saccharase* split the corresponding disaccharides, maltose, lactose, and saccharose (sucrose), into monosaccharides (→ p. 197, B and p. 224).

Fat digestion: *Pancreatic lipase* is the most important enzyme involved in fat digestion. It breaks down triglycerides into 2-monoglycerides and free fatty acids (→ p. 197, B). To be effective it requires another enzyme, *co-lipase*, which is formed (also by the action of trypsin) from the pro-colipase of the pancreatic juice (p. 218 ff.).

Bile is necessary for the normal digestion of fats (→ also p. 214 ff.). It is continuously produced in the **liver** (about 0.7 l/d), but is not always emptied at once into the intestine. When, as is the case between meals, the sphincter of Oddi (the muscle shutting off the bile duct from the duodenum) is closed, the bile enters the **gall bladder** where it is concentrated to 1/5–1/10th of its original volume by withdrawal of water that passively follows the actively transported Na^+ and Cl^- out of the lumen of the gall bladder (→ p. 215, D). The resulting concentrate contains large quantities of specific bile components in a small volume, a situation which favors the formation of *gall stones*. When the bile is required for digestion, the *gall bladder contracts* and its contents are mixed with the duodenal chyme.

The **contraction of the gall bladder** is partly under reflex and partly under hormonal control (*CCK*; see above and **A**). In addition to fat, egg yolk and $MgSO_4$ (so-called *cholagogues*) very effectively stimulate the release of bile from the gall bladder via CCK. Secretin and bile salts in the blood, on the other hand, stimulate the production of bile in the liver (so-called *choleretics*; → p. 214).

In addition to endogenous substances and inert or toxic xenobiotics (→ p. 214 ff.), the bile excretes compounds such as iodine-containing drugs administered to render the bile ducts visible on the X-ray screen (*cholangiography and cholecystography*).

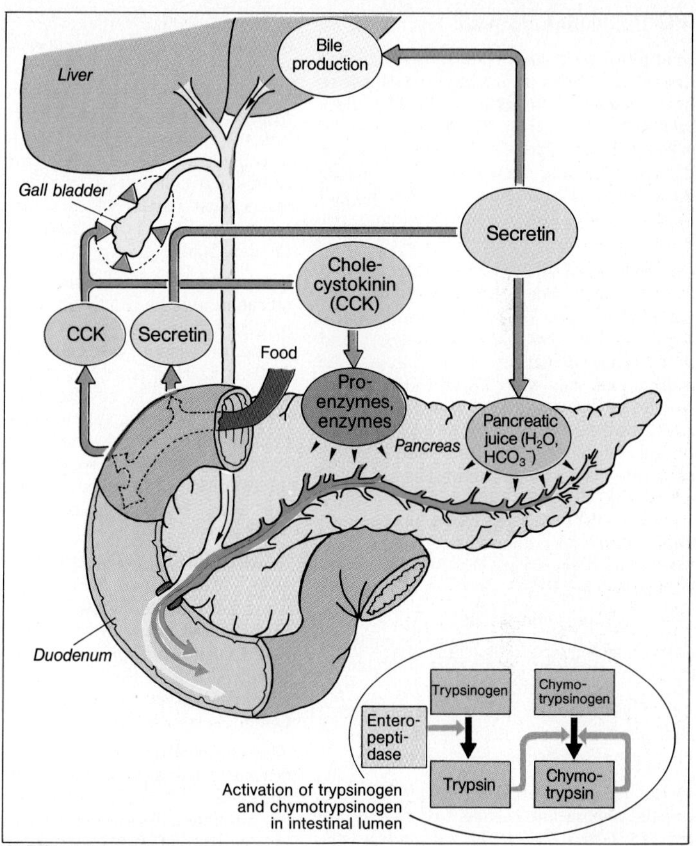

A. Bile and pancreatic juice

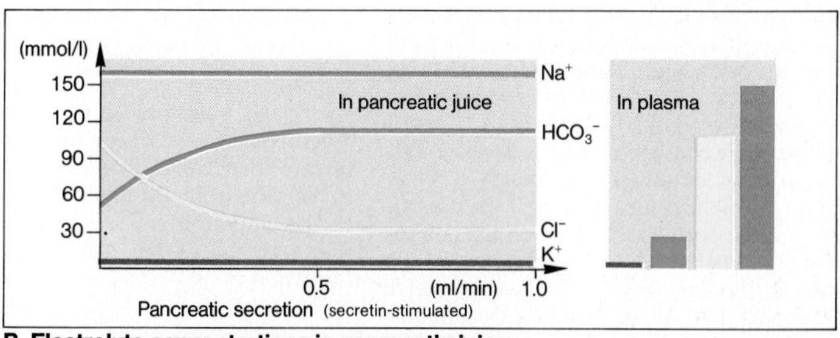

B. Electrolyte concentrations in pancreatic juice

Bile Production, Hepatic Excretion

In addition to its important role in *metabolism* (→ textbooks of biochemistry) the **liver** has an **excretory function** (→ **A**). The **bile** is secreted from the liver cells directly into the *bile canaliculi* that lie between adjacent hepatic cells, and then flows into the *bile ducts* (→ **A**). From here, it reaches the gall bladder (→ p. 212) or the duodenum, or both.

Composition of bile: Apart from *water* and *electrolytes* the bile contains *bilirubin, steroid hormones*, the *salts of bile acids, cholesterol, lecithin (phosphatidyl choline)*, and many other substances. *Drugs* may also be excreted via the liver. Certain of the above-listed substances (e.g. *bilirubin*) are not readily soluble in water and are therefore bound to albumin in the blood. After separation from the albumin they are taken up by the liver cells, and together with a transport protein they reach the smooth endoplasmic reticulum where they are *conjugated* with *glucuronic acid* and other substances (→ p. 216f.).

Bile secretion is **stimulated** by (a) increased hepatic blood flow; (b) vagus stimulation; (c) an elevated level of bile salts in the blood; and (d) secretin, etc. Hepatic bile is produced continuously and is stored and concentrated in the *gall bladder*, where it becomes more viscous (→ also p. 212).

In the liver, the so-called *primary* **bile acids** or **salts**, *cholate* and *chenodesoxycholate*, are formed from *cholesterol*. *Secondary bile acids* (e.g. desoxycholate, litocholate, and so on) are formed from primary bile acids in the intestine by bacterial action. They are then absorbed (like the primary bile salts) and returned to the liver: **enterohepatic circulation** (→ p. 220). Here both types of bile salts are conjugated with *taurine, ornithine, glycine*, and other substances (→ **A**), in which form they enter the bile. If the bile salt level in the portal vein rises (via the enterohepatic circulation), the production of bile salts in the liver is slowed down (negative feedback) and their secretion is simultaneously increased; this in turn, accelerates the flow of bile. The explanation for this so-called **bile salt-dependent choleresis** (see also p. 212) is probably that when the plasma concentration of bile salts is elevated, more accumulate (actively) in the hepatic cells, thus increasing the driving force for the *transport of bile salts* out of the hepatocytes into the canaliculi. Water then follows due to osmosis (→ **B**). A second component of canalicular choleresis is the **bile salt-independent choleresis** driven by active transport of NaCl from the hepatic cell into the canaliculi.

Independently of this, active secretion of **bilirubin** into the canaliculi also occurs (→ p. 216). Other endogenous substances like porphyrins, and exogenous compounds such as PAH, phenol red, sulfobromphthalein, penicillin, glycosides, etc. are transported by the same system (whereby competitive inhibition may occur; → p. 11). This secretory mechanism serves primarily for the **excretion** of these substances since they are less subjected to enterohepatic circulation.

As with bilirubin (→ p. 216), thyroxine, and many of the steroid hormones, exogenous substances are **conjugated** before transport (e.g. chloramphenicol with *glucuronic acid*, naphthalene and phenanthrene with *glutathione*), but in this case the conjugation primarily serves the purpose of **detoxification** (→ also p. 130).

The composition of the bile is modified during its passage through the **bile ductules and ducts** (→ **B**). This is where **secretin** exerts its choleretic effect (*ductular choleresis*); the modification of HCO_3^- secretion plays a role here, too. (cf. pancreas, → p. 212). Further modification of the composition of bile takes place in the **gallbladder** (→ **D** and p. 212).

As in the intestinal lumen, **cholesterol** is "dissolved" in the bile in *micelles*, which it forms with *lecithin* and *bile salts*. Alterations in the ratio of the three substances can lead to precipitation of *cholesterol crystals*, which is one of the causes of *gall stone formation* (→ **C**).

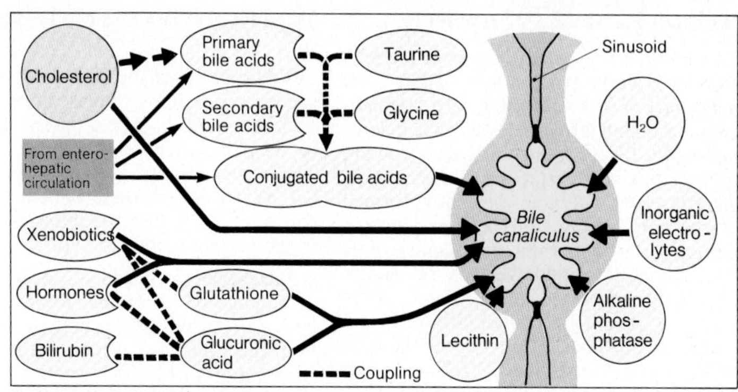

A. Excretory functions of liver

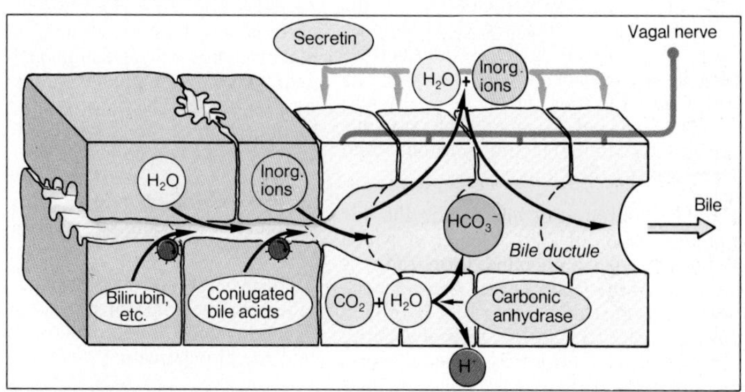

B. Transport in bile formation

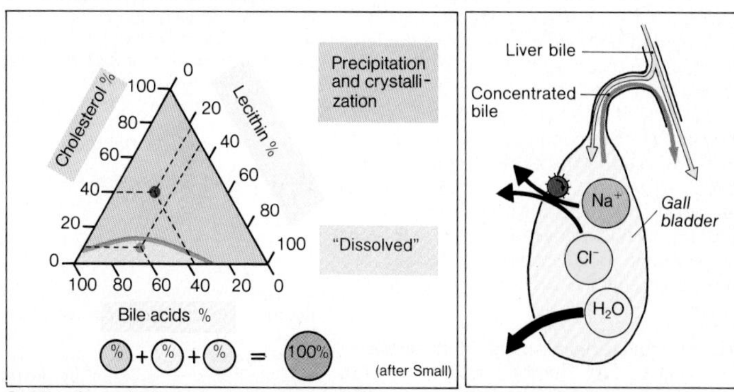

C. Micellar solution of cholesterol in bile

D. Concentration of bile

Bilirubin Excretion – Jaundice

Bilirubin is a major component of the bile. About 85% arise from the *breakdown of old red blood cells* (→ **A, B** and p. 60ff.). Hemoglobin is mainly broken down in the macrophages (→ p. 66ff.), the globin and iron are split off and, via intermediary stages including *biliverdin, bilirubin* is finally formed (35 mg bilirubin per 1 g Hb). Free bilirubin is poorly soluble in water, and because of its fat solubility it would have a toxic effect if it were not transported in the blood tightly *bound to albumin* (2 mol bilirubin/1 mol albumin). Bilirubin is taken up by the liver cells in its free form (→ **A**). *UDP glucuronic acid*, formed in the liver from glucose, ATP, and UTP with the participation of the enzyme *glucuronyl transferase*, conjugates with the bilirubin to give the water-soluble **bilirubin glucuronide**, which is actively *secreted* into the bile canaliculi (→ **A** and p. 214). Some of it enters the systemic circulation ("direct-reacting bilirubin") and is excreted by the kidneys.

The daily excretion of bilirubin in the bile amounts to about 200–250 mg, of which about 15% (unconjugated form only) is resorbed from the intestine: *enterohepatic circulation*. Part of the bilirubin is broken down in the liver and bile to *urobilinogen*, part of it in the intestine by bacterial action with the formation of *stercobilinogen*, both of which substances are colorless. Their partial oxidation leads to the formation of *urobilin* and *stercobilin*, which give the feces their brown color. Part of the urobilinogen is absorbed in the small intestine and reaches the liver, where it undergoes further breakdown. Stercobilinogen is to some extent absorbed in the rectum (bypassing the liver; → p. 230), and is excreted by the kidneys (2 mg/day), together with traces of urobilinogen. Damage to the liver cells leads to increased renal excretion of these two substances, which are thus of diagnostic value.

The normal **bilirubin content of the plasma** is 3–10 mg/l. At concentrations above 18 mg/l, **jaundice (icterus)** develops; first of all the scleras, later on the skin, take on a yellow color.

Three types of jaundice can be distinguished:

1. **Prehepatic jaundice**: bilirubin production is increased (e.g. by increased hemolysis) to such an extent that the capacity of the liver to form bilirubin conjugates is exceeded. The bottleneck is at glucuronyl transferase (→ **A**), so that in these patients the non-conjugated ("indirect-reacting") bilirubin in the plasma increases.

2. **Intrahepatic jaundice** arises from: a) damage to liver cells; for example, by poisons (poisonous agaric), or inflammation (hepatitis), in which transport and conjugation of bilirubin are impaired; b) total absence (*Crigler-Najjar syndrome*) or deficiency (*Gilbert's syndrome*) of glucuronyl transferase, or immaturity of the glucuronylisation system at birth, which is aggravated by the possible jaundice caused by a high rate of hemolysis in the newborn; c) inhibition of the glucuronyl transferase, for example, by steroids; d) an inborn defect (*Dubin-Johnson-syndrome*) or inhibition (e.g. by drugs or steroids) of bilirubin secretion into the bile canaliculi.

3. **Posthepatic jaundice**: obstruction of the bile ducts by stones or tumor causes reflux of conjugated ("direct-reacting") bilirubin into the bloodstream.

Alkaline phosphatase, which is a normal constituent of the bile, is then also present in the blood in abnormally large quantities, and is an important diagnostic aid. These bile constituents partly reach the blood by leaking through the "tight" junctions between the canaliculi and the sinusoids (→ p. 215, A), partly by diffusional countercurrent exchange (→ p. 134) at the *portal triads* where bile flow is opposed by nearby portal and arterial blood flow.

In type 2a) and d), and in 3), the amount of water-soluble, conjugated bilirubin in the urine is elevated (brown coloration). In 3), the stool is clay-colored because no bilirubin enters the intestine to serve as a source of stercobilirubin.

In prehepatic jaundice, unconjugated bilirubin, if not bound to albumin, can enter the brain where it exerts a toxic effect (*Kernicterus*). The severity of the situation is enhanced in the newborn, whose albumin levels are low, or by medication with substances that compete for the available albumin (organic anions, e.g. sulfonamides).

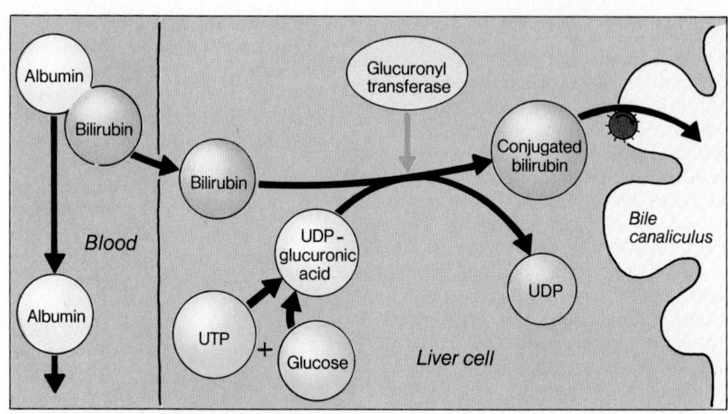

A. Conjugation of bilirubin in the liver

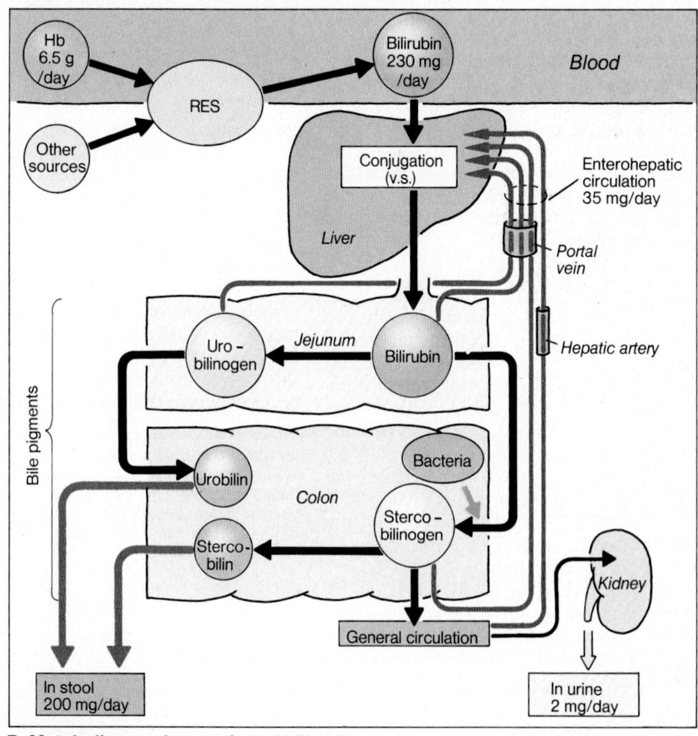

B. Metabolism and excretion of bilirubin

Digestion of Lipids

The **daily intake** of fats (butter, oil, margarine, milk, meat, eggs, etc.) varies considerably (10–250 g/d) and averages 60–100 g/d. The major portion of the dietary lipids (90%) consists of neutral fats or *triglycerides*. The remainder is made up of *phospholipids, cholesterol* and its *esters*, and the *fat-soluble vitamins* A, D, E, K. Normally, more than 95% of these lipids are absorbed in the small intestine. The fact that lipids are *poorly soluble in water* poses special problems in connection with their digestion and absorption in the aqueous milieu of the gastrointestinal tract and their further transport in the plasma (→ p. 220ff.).

Although small quantities of triglycerides can be absorbed without being broken down, absorption is normally preceded by enzymatic degradation. **Lipases**, the enzymes responsible for the breakdown of fats, are present in the secretion of *glands*, situated at the *base of tongue*, and in the *pancreatic juice*. 10–30% of the fats are broken down in the stomach (acid pH optimum of the lipases of the tongue glands), 70–90% in the duodenum and upper jejunum.

Lipases become active mainly at the *fat-water interface* (→ **B 1**), provided that mechanical **emulsification** of the fats (chiefly as a result of the "systolic" contraction of the distal stomach; → p. 206) has taken place; the relatively small drops of an emulsion (1–2 μm) present the lipases with a large surface area for attack. The *pancreatic lipase* (with a maximum turnover rate of 140 g fat/min!) becomes active in the presence of Ca^{2+} and a **co-lipase**, the latter arising from the action of trypsin on a *proco-lipase* from the pancreatic juice. The *triglycerides* (1. and 3. ester linkages; → p. 197, B) are hydrolyzed into **free fatty acids** and **2-monoglycerides**. Around the enzyme a so-called *viscous-isotropic phase* containing both aqueous and hydrophobic regions is formed (→ **B 2**).

In the presence of an excess of Ca^{2+} ions, or too low a concentration of monoglycerides, some of the fatty acids react to form Ca^{2+} soaps that escape absorption and leave the body in the feces.

With the aid of the **bile salts** (→ p. 214) so-called **micelles** spontaneously form from the *monoglycerides* and *long-chain fatty acids* (→ **B 3**). With a diameter of only 3–6 nm (about 300 times smaller than the emulsion drops mentioned above), the micelles make possible intimate contact between the lipophilic breakdown products of fats and the intestinal wall, and thus represent an essential step in normal lipid absorption (→ p. 220). The polar portions of the molecules constituting the micelles point towards the aqueous surroundings, and the apolar portions face the interior of the micelle. Totally apolar lipids (e. g. *fat-soluble vitamins* as well as lipophilic toxins) are always embeded in a lipophilic milieu during these events (so-called *hydrocarbon continuum*), and in this way reach the lipophilic brush border membrane of the intestinal epithelium for absorption. The relatively polar *short-chain fatty acids* do not require bile salts for their absorption.

Phospholipids (chiefly *phosphatidyl choline* = lecithin) from the food and bile are broken down in the presence of bile salts and Ca^{2+} by **phospholipase A_2** (from prophospholipase A_2 of the pancreatic juice, activated by trypsin). On the other hand, *cholesterol esters* (e.g. from egg yolk and milk), the 2. ester linkage of the triglycerides, esters of vitamins A, D, E, and many other lipid esters (xenobiotics as well) are attacked by the *cholesterol esterase* of the pancreatic juice, which is therefore more accurately referred to as **nonspecific lipase**.

This lipase also occurs in human **milk** (not in cow's milk), thus providing the infant at the same time with the enzyme necessary for breaking down the milk fat. Since it is not a heat-stable enzyme, pasteurization considerably reduces the digestibility of the milk fat of human milk for premature infants.

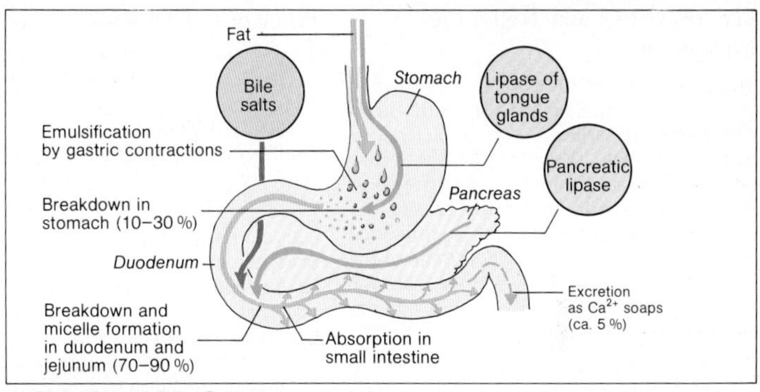

A. Digestion of fats: Survey

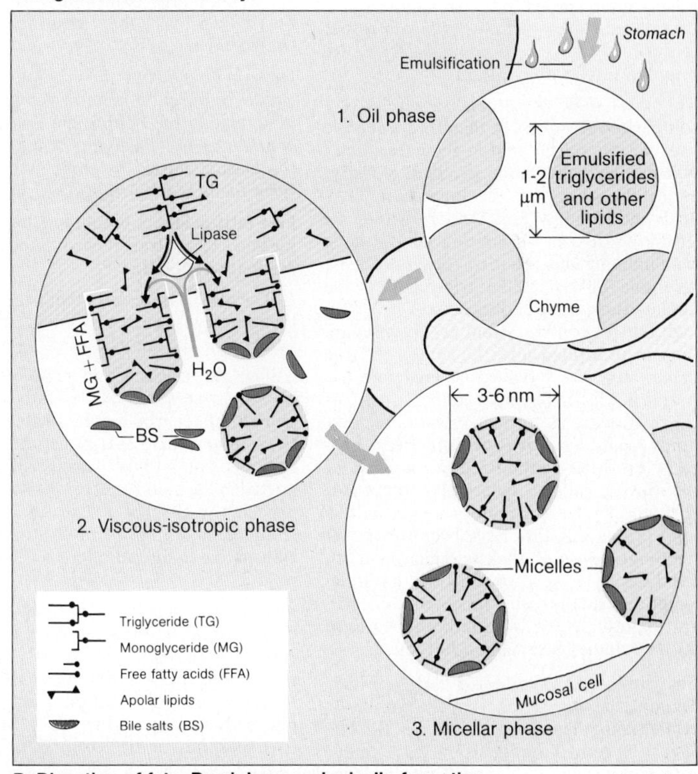

B. Digestion of fats: Breakdown and micelle formation (after Patton)

Fat Absorption and Triglyceride Metabolism

Triglycerides in the food are broken down in the gastrointestinal tract to **free fatty acids** and **2-monoglycerides** (→ p. 218). Embedded in **micelles**, they contact the brush border of the small intestine and are passively absorbed into the epithelial cells (→ **A**). The absorption of the fat is completed by the time the chyme reaches the end of the jejunum, but the **bile salts** set free from the micelles are absorbed in the ileum (secondary-active cotransport with Na^+).

The total bile salts in the body amount to about 6 g; this quantity must undergo **enterohepatic circulation** (bile – small intestine – portal vein – liver – bile; → p. 223, B) about four times daily since around 24 g/d are required for fat absorption.

The *short-chain fatty acids* are relatively soluble in water and can therefore reach the liver via the portal vein in their free form, whereas the hydrophobic products of triglyceride digestion (i.e. the long-chain fatty acids and monoglycerides) are *resynthesized to triglycerides* in the smooth endoplasmic reticulum of the *intestinal mucosa*. Being insoluble in water, the triglycerides, the apolar esters of the cholesterol (→ p. 222 f.), and the fat-soluble vitamins likewise are then incorporated into the center of large *lipoproteins*, the so-called **chylomicrons** (→ **A** and p. 222 f.). The hydrophilic "shell" of the chylomicron is formed by the more polar lipids (*cholesterol, phospholipids*) and proteins, the latter synthesized in the RER of the mucosa cells as **apoproteins** (types A I, A II, and B). After *secretory vesicles* of this lipid protein mixture have been formed at the *Golgi apparatus*, the chylomicrons reach the ECS by *exocytosis*. Via the **intestinal lymph** they finally reach the systemic *plasma*, and are responsible for its milky appearance 20–30 minutes following a fatty meal.

The **liver** also synthesizes triglycerides, draining the fatty acids from the plasma or synthesizing them from glucose (→ **B**). The liver triglycerides are incorporated together with apoproteins B, C, and E into another kind of lipoprotein, the **v**ery-**l**ow-**d**ensity **l**ipoproteins (**VLDL**; → p. 222 f.) and enter the plasma in this form.

Triglycerides and their breakdown products, the free fatty acids, are *high-energy substrates* for energy metabolism (→ p. 198 ff.). **Lipoprotein lipase** (**LPL**) of the capillary endothelium of many organs splits off fatty acids from the triglycerides of the chylomicrons and the VLDL (→ **B**).

This step is preceded by a lively exchange of protein components between the various lipoproteins. One of these proteins, apoprotein CII, reaches the chylomicrons where it plays an important role as a cofactor in triglyceride breakdown. The **insulin** released following a meal activates the lipoprotein lipase, and thus contributes to the rapid breakdown of the absorbed dietary triglycerides.

Heparin (e.g. from the basophilic granulocytes or the endothelium) also plays a role in activating the lipoprotein lipases, which contributes to "clarifying" the milkiness of the plasma caused by chylomicrons (*clarifying factor*).

Free fatty acids are bound to albumin in the plasma and are transported to the following **destinations** (→ **B**):

1. The **musculature** and other organs, where their energy is set free by oxidation to CO_2 and H_2O in the mitochondria (β-oxidation).

2. The **fat cells**, where triglycerides are resynthesized from the free fatty acids and stored until energy requirements rise or the food intake drops. Triglycerides are then broken down into fatty acids and are transported in the blood to the site where they are needed (→ **B**). The release of fatty acids from their storage form is stimulated by **epinephrine** and inhibited by **insulin** (→ p. 246 ff.).

3. The **liver**, where fatty acids either undergo oxidation or are turned into triglycerides once more. Since only a limited quantity of triglycerides can be exported in VLDL, an excessive supply of fatty acids (or indirectly via excess glucose, → **B**) can lead to deposition of triglycerides in the liver (*fatty liver*).

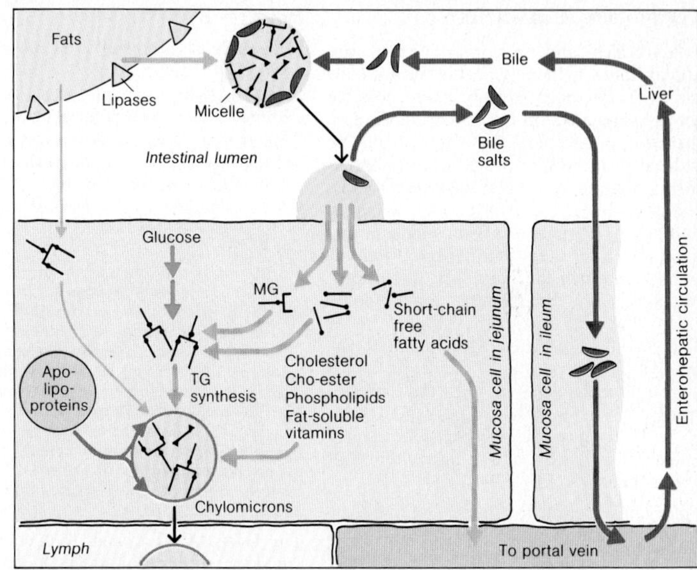

A. Absorption of fats

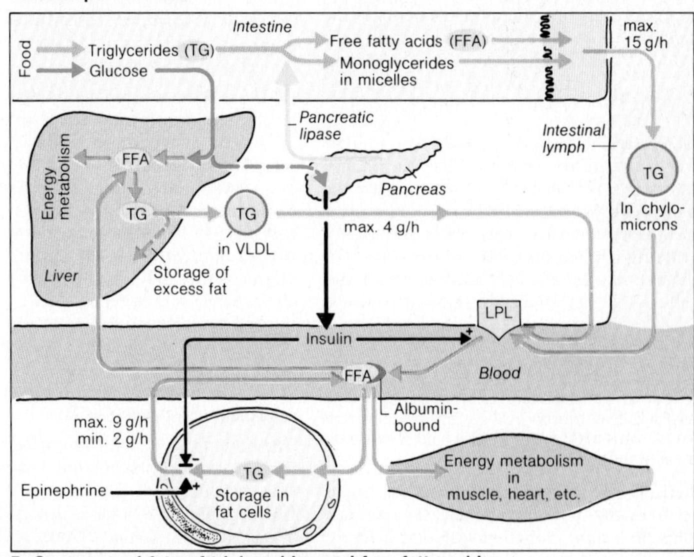

B. Sources and fate of triglycerides and free fatty acids

Lipoproteins, Cholesterol

Triglycerides and *cholesterol esters* are apolar lipids; their transport in the aqueous milieu of the body is only made possible through the mediation of other substances (proteins, polar lipids), and they can only be utilized in metabolic processes after transformation into more **polar lipids** (fatty acids, cholesterol). Thus, triglycerides act predominantly as *reservoirs* from which, at any time, *free fatty acids* can be mobilized (→ p. 220). In much the same way, cholesterol esters represent the storage-, and to some extent the transport-form of *cholesterol*. Triglycerides are transported in the intestinal lymph and plasma in the center of the large lipoproteins (→ p. 220); the **chylomicrons** consist to 86%, and the **VLDL** to 55% of triglycerides (→ **A**). Cholesterol esters (Cho-E) are found in the center of all lipoproteins (→ **A**).

Polar lipids include the *long-chain free fatty acids* and the "shell" lipids of the lipoproteins, that is, **phosphatidyl choline (lecithin)** and **cholesterol (cholesterin)**. Both are fundamental constructional elements of the *cell membranes*. In addition, important substances such as the *bile salts* (→ **B** and p. 214) and the *steroid hormones* (→ p. 258 ff.) are synthesized from cholesterol.

Cholesterol in the *food* is partly present in its free form, partly esterified. The cholesterol esters are hydrolysed to cholesterol by the **nonspecific pancreatic lipase** (→ p. 218) prior to absorption. Together with cholesterol entering the duodenum in the *bile*, the dietary cholesterol is absorbed from micelles (→ p. 218) in the upper small intestine.

The **mucosa cell** contains at least one enzyme that re-esterifies a part of the cholesterol after absorption (**ACTA** = **a**cyl-CoA-**c**holesterol **a**cyl **t**ransferase), so that cholesterol esters as well as cholesterol are then incorporated into the chylomicrons (→ **A**). (Part of this cholesterol is synthesized in the mucosa cell).

In the **liver**, *acid lipases* hydrolyse the cholesterol esters in the *chylomicron remnants*. This cholesterol, together with that from the "shell" of the chylomicrons and from other sources (neosynthesis, **HDL** ⟨**h**igh-**d**ensity **l**ipoproteins⟩), takes one of the following **routes** (→ **B**):

1. Excretion of cholesterol in the bile (see above and p. 214).

2. Conversion of cholesterol into bile salts, which are a major constituent of the bile (→ p. 214).

3. Incorporation of cholesterol into **VLDL**, from which, by the action of *lipoprotein lipase* (LPL, → p. 220), *VLDL remnants* and, finally, **LDL** (**l**ow-**d**ensity **l**ipoproteins) arise. The **LDL** deliver the cholesterol esters to cells that have LDL receptors (see below).

4. Incorporation of cholesterol into the disklike "*pre*"-*HDL* and its conversion in the plasma to cholesterol esters by the enzyme **LCAT** (**l**ecithin **c**holesterol **a**cyl **t**ransferase). The cholesterol esters fill the preHDL and convert it to (spherical) **HDL**. Cholesterol for this esterification is also drawn from *chylomicron remnants*, VLDL remnants, and from the membranes of dead cells. At the same time, lecithin is degraded to *lysolecithin*, which is bound to albumin and carried off in the plasma for use elsewhere in lecithin resynthesis. Most of the cholesterol esters in the HDL are, in the end (via VLDL remnants), passed on to LDL (→ **B**).

The **HDL-LCAT system** thus collects and converts cholesterol from a variety of sources and, apart from the brief intestinal phase of absorption, constitutes the most important source of cholesterol esters for the cells of the organism.

LDL are the main vehicle by which cholesterol esters are transported to extrahepatic cells. The latter are equipped with **LDL receptors**, the density of which on the cell surface varies according to the cell's need for cholesterol esters. LDL is taken up endocytotically and lysosomal enzymes accomplish the breakdown of (1) the apoproteins into amino acids and (2) cholesterol esters to cholesterol. The latter is then available to the cell for incorporation into membranes or for steroid synthesis (→ p. 258). If more cholesterol is present than is required, ACAT is activated and the cholesterol is esterified and stored (→ **B**).

The daily *loss of cholesterol* via the stool in the form of coprostanol and in sloughed-off skin amounts to approximately 0.6 g; a further 0.4 g is lost in the form of bile salts. These losses (minus dietary cholesterol) have to be made good by continuous neosynthesis in intestine and liver (→ **B**).

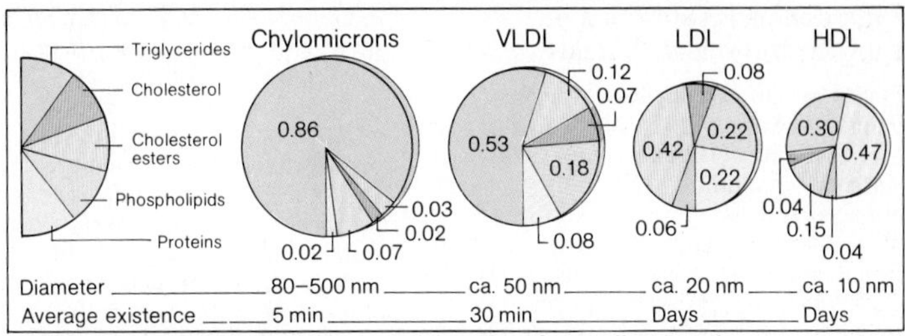

A. Proportion of lipid and protein (g/g) in the plasma lipoproteins

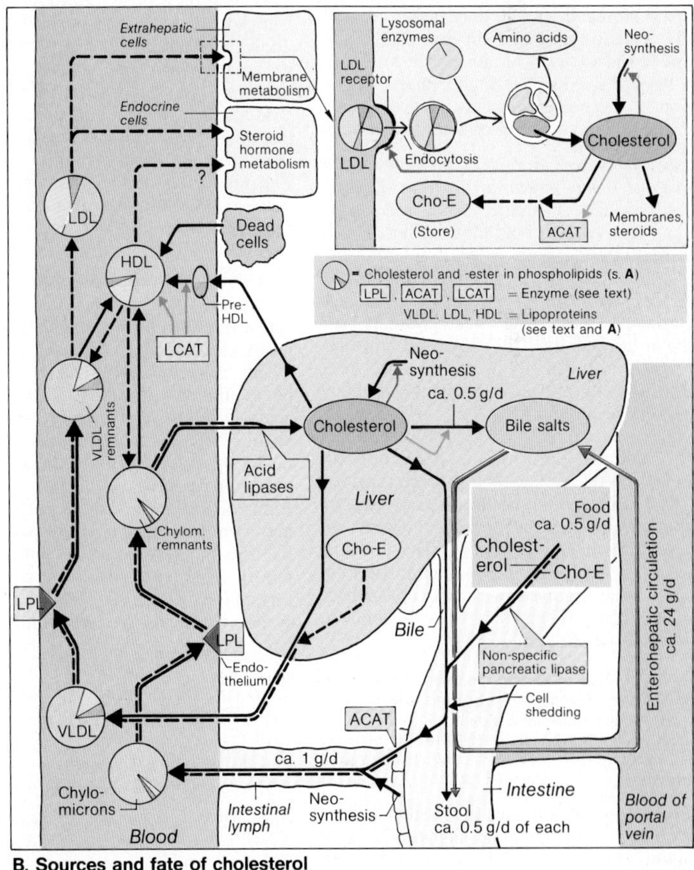

B. Sources and fate of cholesterol

Digestion and Absorption of Carbohydrates and Proteins

Digestion of **carbohydrates** begins in the **mouth** (→ **A** and p. 202), where an *α-amylase – ptyalin* – from the saliva breaks down the **starches** (including the polysaccharides amylose and amylopectin) into oligo- and di-saccharides (maltose, isomaltose, maltotriose, α-limit dextrin) at a pH of about 7. This process continues in the proximal **stomach** (→ p. 206) and is inhibited by mixing with the acid gastric juice in the distal stomach. In the **duodenum**, the acid pH is neutralized and digestion continues due to the addition of pancreatic α-amylase (→ p. 212) to the intestinal contents. Digestion of the polysaccharides now continues to the above-named intermediate products.

Carbohydrates can only be absorbed in the form of *monosaccharides*, so that the maltose, isomaltose, maltrotriose, and α-limit dextrin have to be degraded further. This is done with the help of *maltases* and *isomaltases* from the pancreatic juice and, in larger quantities, from the ileal mucosa. Intestinal 1,6-*glucosidase* is needed to split the branched dextrins. The end product, **glucose**, is taken up by the mucosal cells against a concentration gradient (*secondary active cotransport with* Na^+; → p. 229, D) and is passed down a concentration gradient by facilitated diffusion (→ p. 11) into the portal blood. Other dietary disaccharides, such as **lactose** and **saccharose** (sucrose), are split by the corresponding mucosal enzymes: *lactases, saccharases*. The resulting **galactose** is absorbed by a similar mechanism to that involved in glucose absorption, whereas **fructose** has only been shown to be passively absorbed (→ **A**).

In the absence of lactase, lactose cannot be degraded and is therefore not absorbed. Deficiency of the enzyme results in diarrhea, since the nonresorbed lactose (1) pulls water into the intestinal lumen osmotically; and (2) is converted into toxic substances by the intestinal bacteria.

Protein digestion begins in the **stomach** (→ **B**), where the several **pepsinogens** produced mainly by the chief cells of the gastric mucosa are activated to several isozymes of **pepsin** by gastric **hydrochloric acid**. At pH values of 2–5 (HCl!) pepsin splits proteins preferably at the sites where tyrosine or phenylalanine follow towards the carboxyterminal of the peptide chain (→ p. 197, B). The pepsin isozymes are partly inactivated in the duodenum when HCO_3^- from the bile and pancreas neutralizes the gastric acid to about pH 6.5. Protein digestion is then carried on by the action of the endopeptidases **trypsin** and **chymotrypsin**, which are both capable of splitting the polypeptides and proteins down to dipeptides. Trypsin is produced in the duodenum from pancreatic *trypsinogen* by the action of duodenal *enteropeptidase* and then activates pancreatic *chymotrypsinogen* to chymotrypsin (→ p. 213, A). Pancreatic *carboxypeptidase* and *aminopeptidases* produced by the intestinal mucosa split **L-amino acids** off the free ends of the peptide chains (→ p. 197, B). The final breakdown of the peptides into individual amino acids is accomplished by the *dipeptidases* localized on the brush border membrane of the intestinal mucosa.

Specific Na^+ cotransport systems (→ **B** and p. 223, D 3) are responsible for secondary active **transport of the amino acids** from the intestinal lumen into the mucosal cell, after which they enter the portal blood probably by facilitated diffusion. "Dibasic" amino acids (arginine, lysine, ornithine) and the "acidic" amino acids (glutamic acid, aspartic acid), which are largely broken down in the mucosa cells, each have their own transport systems. It is still not known how many transport mechanisms are available to the "neutral" amino acids. Whether or not the β- and γ-amino acids, proline, and certain others each have their own transport system is still under discussion (→ **B**).

In a number of congenital disorders the absorption of certain amino acid groups is defective, often accompanied by similar defects in the renal tubules (renal amino acidurias) as for example in *cystinuria*.

In addition, the small intestine is able to absorb certain **dipeptides** and **tripeptides** as intact molecules. As in the kidney (→ p. 128), this is achieved by carriers present in the luminal cell membrane. Where peptide transport is active, it is driven by a H^+ gradient (H^+ co-transport).

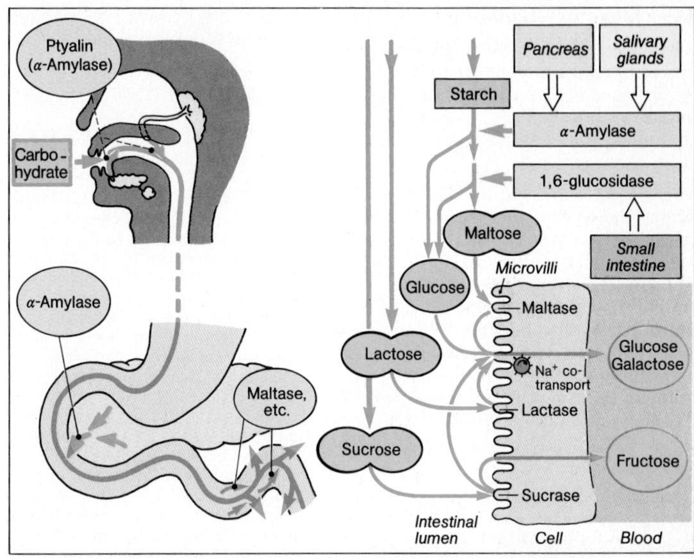

A. Digestion of carbohydrates and absorption of monosaccharides

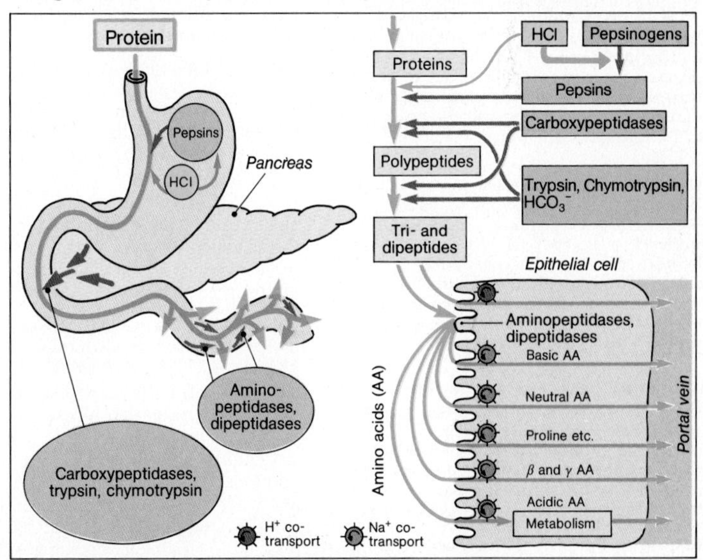

B. Digestion of proteins and absorption of amino acids and oligopeptides

Absorption of Vitamins

Cobalamines (B_{12} vitamins) are synthesized by microorganisms but have to be taken up by higher animals in their food. The best sources of cobalamines for human beings are animal products such as liver, kidney, meat, fish, eggs, milk.

On account of the relatively high molecular weight and low fat solubility of the cobalamines (CN-, OH-, methyl-, and adenosyl cobalamines) their intestinal absorption requires special *transport mechanisms* (→ **A**). During their passage through the intestine and in the plasma, cobalamines are bound to one of three types of **transport proteins**: (1) **intrinsic factor (IF)** of the gastric juice (from the parietal cells); (2) **transcobalamine II (TC II)** in the plasma; (3) **R-proteins** in the plasma (**TCI**), granulocytes (**TC III**), saliva, bile, milk, etc.

Cobalamines are released from the dietary proteins by the gastric acid and are for the most part then bound to the R-protein from the saliva or (at a high pH) to IF. The R-protein is digested by *trypsin* in the duodenum, the cobalamine is set free and taken up by the trypsin-resistent IF. The mucosa of the **ileum** possesses highly specific receptors for binding this cobalamine-IF complex, which is then taken up by the mucosal cells by endocytosis. This **transport process** requires the presence of Ca^{2+} and a pH > 5.6 (→ **A**). The density of the receptors, and consequently absorption, is increased during pregnancy.

Once in the blood the cobalamine is bound to TC I, II, and III; TC II is responsible for its distribution to all *actively dividing cells* in the body (*TC II receptors; endocytosis*). TC III from the granulocytes transports surplus cobalamine and undesirable cobalamine derivatives to the **liver** (*TC III receptors*), where they are either *stored* or *excreted*. *TC I* (t/2 about 10 days) serves for the temporary storage of cobalamine in the plasma (→ **A**).

An exclusively vegetarian diet, or disorders in cobalamine absorption (e.g. from lack of IF secretion) can lead to severe *deficiency symptoms* such as *pernicious anemia*, damage to the spinal cord (*funicular myelosis*) etc. However, since the body contains about 1000 times the daily requirement of 1 µg the above symptoms may only become apparent after several years (→ p. 62 f.).

Folic acid or **pteroyl glutamic acid (Pte-Glu$_1$)** is required in its metabolically active form (*tetrahydrofolic acid*) for DNA synthesis (*daily requirements 0.1–0.2 mg*). The folic acid reserves in the body (about 7 mg) are sufficient for several months (→ p. 62 f.).

Symptoms of deficiency are anemia, leucopenia, thrombopenia, diarrhea, and disorders of the skin and hair growth, etc.

Folic acid is chiefly present in the diet in forms consisting of up to 7 glutamyl residues (γ-linked peptide chains; **Pte-Glu$_7$**) instead of merely one pteroyl glutamic acid (Pte-Glu$_1$), but since only the latter can be absorbed from the intestinal lumen (proximal **jejunum**; → **B**) the polyglutamyl chain has to be shortened by specific enzymes (*pteroyl polyglutamate hydrolases*) before absorption can take place. These enzymes are probably located on the luminal membrane of the small intestine. Pte-Glu$_1$ is absorbed by means of a specific active **transport mechanism**, after which metabolites of Pte-Glu$_1$, including *5-methyltetrahydrofolic acid (5-Me-H_4-Pte-Glu)*, form in the cell. If these breakdown products are already present in the food, they are absorbed as such from the intestinal lumen. The same holds for the cytostatic drug, *methotrexate*, which uses the same transport mechanism as Pte-Glu$_1$. If the specific absorption mechanism in the jejunum is absent, the daily intake of folic acid has to be raised 100-fold in order to ensure adequate absorption (by passive diffusion). Cobalamine is necessary for the transformation of 5-Me-H_4-Pte-Glu$_1$ into the metabolically active tetrahydrofolic acid.

Other **water-soluble vitamins** (B_1 ⟨thiamine⟩, B_2 ⟨riboflavin⟩, C ⟨ascorbic acid⟩, H ⟨biotin, niacin⟩) are absorbed together with Na^+ by a secondary active process (*cotransport*), that is, like glucose and the amino acids (→ **C**). Except for vitamin C, which is absorbed in the **ileum**, absorption of these vitamins takes place in the **jejunum**. The B_6 vitamins (pyridoxal, pyridoxine, pyridoxamine) are thought to be absorbed passively by a diffusion process.

Absorption of **fat-soluble vitamins** (A ⟨retinol⟩, D_2 ⟨cholecalciferol⟩, E ⟨tocopherol⟩, K_1 ⟨phyllochinone⟩, K_2 ⟨farnochinone⟩), like the absorption of fats, involves the formation of *micelles* and a *hydrocarbon continuum* (→ p. 218). The mechanisms of absorption are not fully understood, although they are known to be partly saturable and energy-dependent. Fat-soluble vitamins are transported in the plasma in chylomicrons and VLDL (→ p. 220 ff.).

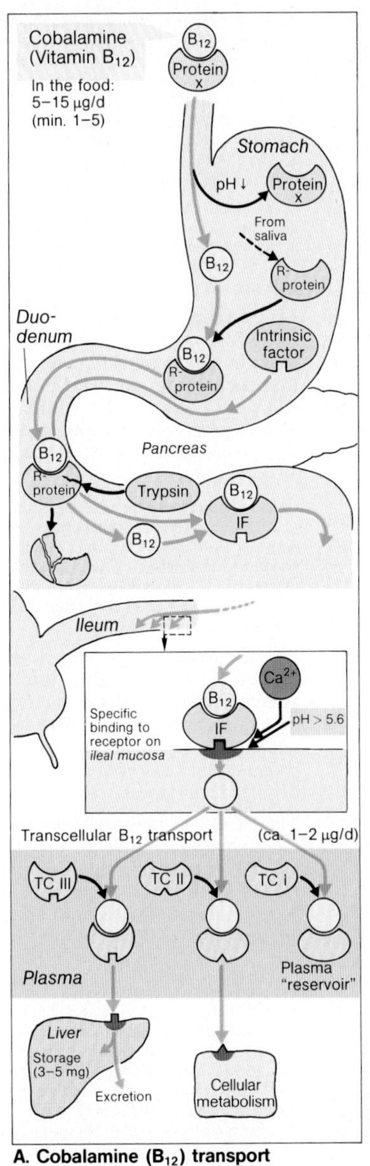

A. Cobalamine (B_{12}) transport

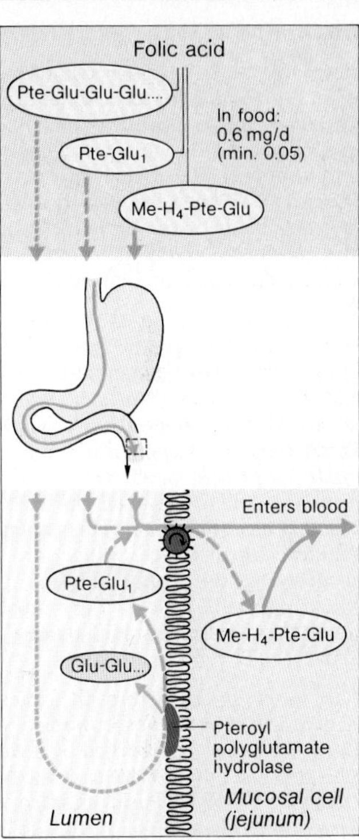

B. Folic acid absorption

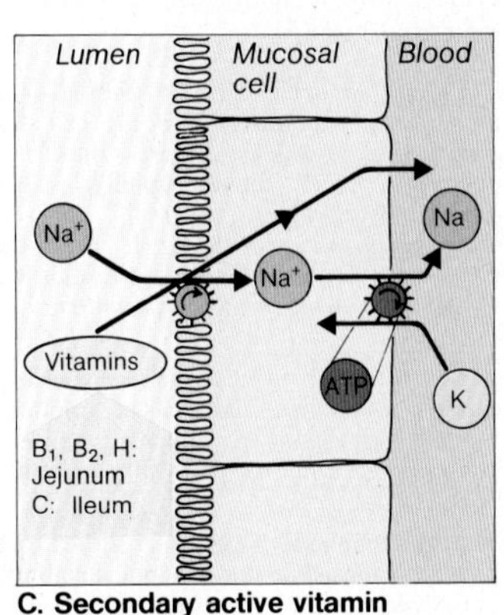

C. Secondary active vitamin absorption

Absorption of Water and Electrolytes

The daily intake of **water** averages **1.5 l**. In addition, about **6 l** of water enter the gastrointestinal tract as *saliva, gastric juice, bile,* and *pancreatic* and *intestinal juices.* Since only about 0.1 l is excreted daily in the feces, at least **7.4 l** must be **absorbed** from the intestine in the same period of time. This absorption takes place mainly in the *jejunum* and *ileum*, a small amount in the *colon* (→ **A**).

Water moves across the intestinal wall by *osmosis*. If osmotically active particles (e.g. Na^+ and Cl^-) are absorbed, water follows in their wake. Conversely, if substances are excreted into the lumen or if nonabsorbable matter is present in the food, water flows from the cell or between the cells (paracellular) into the lumen. This explains why sulfates, for instance, which are poorly absorbed, act as *laxatives*. The intestine as a whole normally absorbs very much more water than it secretes (difference 7.4 l; see above).

The water uptake from the intestine is largely driven by the absorption of $\mathbf{Na^+}$ and $\mathbf{Cl^-}$ (→ **B**). Na^+ is absorbed by a variety of mechanisms, but in every case the $\mathbf{Na^+}$**-**$\mathbf{K^+}$ "**pump**" (Na-K-ATPase) situated on the basolateral membrane of the cell is the primary active event, keeping the Na^+ concentration in the cell low and the cell potential high.

1. **Cotransport of** $\mathbf{Na^+}$ **with** $\mathbf{Cl^-}$: At the luminal cell membrane Na^+ moves into the cell down an electrochemical gradient, whereas the Cl^- on their common "carrier" moves "uphill" (→ **D 2**), and leaves the cell again "downhill". This kind of transport takes care of the larger part of the Na^+, Cl^-, and H_2O absorption from the intestine and is influenced by *hormones* and *transmitters* via cAMP (→ p. 242).

2. $\mathbf{Na^+}$ **cotransport with organic substances**: In this case the Na^+ influx into the cell is exploited for the "uphill" transport of glucose, amino acids, vitamins, bile salts, and many other substances (→ **D 3**).

3. A small amount of $\mathbf{Na^+}$ is absorbed on **its own** through **channels** in the luminal membrane (rectum, ileum; → **D 1**). Aldosterone has a regulatory effect on this type of transport (→ p. 150). The transfer of the positive ion into the cell leaves a net negative charge on the luminal side and gives rise to a transcellular potential (electrogenic or rheogenic transport: → p. 15), along which either Cl^- is absorbed (upper small intestine) or K^+ secreted (ileum, → **C**). Due to the relative ease with which water can penetrate the mucosa of the small intestine at the "tight" junctions, the flux of Cl^-, K^+, and water mainly takes place *between* the cells (*paracellular*) (→ **D**, left: Cl^-).

4. $\mathbf{Na^+}$ and other low-molecular substances are also dragged through the mucosal epithelium along with the absorptive flow of H_2O involved in 1.–3. (so-called "**solvent drag**", taking place paracellularly; → p. 10).

$\mathbf{Cl^-}$ **secretion** takes place at the epithelial cells of the *crypts of Lieberkühn* (→ p. 211, A 8). In this case Na^+ and Cl^- are cotransported from the **blood to the cell**. The efflux of Cl^- into the lumen (Cl^- channels) is accelerated by *cAMP* and is controlled by **VIP** (**v**asocative **i**ntestinal **p**eptide) and *prostaglandins.*

Cholera toxin blocks the GTPase of the G_s-protein and thereby keeps the adenyl cyclase continuously activated and, therefore, brings about a maximum rise in cAMP levels (→ p. 242). With the resulting increase in Cl^- secretion, other ions and large amounts of water enter the lumen and can lead to *diarrhea* of as much as 1 l per hour.

Possible *physiological roles* for this water "secretion" may be (a) to thin down too viscous chyme; (b) to flush out the products of the Paneth cells (→ p. 210); (c) to recirculate H_2O (crytps – lumen – villi- crypts) to promote the absorption of insufficiently dissolved substances.

The $\mathbf{HCO_3^-}$ of the pancreatic juice buffers the chyme (gastric acid), and any excess HCO_3^- is absorbed in the jejunum (→ **A**). HCO_3^- is also secreted in both small and large intestine (protection against acid; constant pH). HCO_3^- is therefore *lost* in *diarrhea* (*nonrespiratory acidosis*; → p. 114).

The stool contains almost no Na^+ or Cl^-, whereas at least 1/3 of the $\mathbf{Ca^{2+}}$ intake is lost via this route. $\mathbf{K^+}$ is secreted in the ileum and colon (→ **A** and **C**) and is present in the stool in high concentrations (about 90 mmol/l; K^+ losses in diarrhea!). The Ca^{2+} absorption by the intestine is reduced in rickets (*vitamin D deficiency*) and by substances that form water-insoluble compounds with Ca^{2+} (e.g. phytin, oxalate, fatty acids). $\mathbf{Mg^+}$ absorption from the intestine is similar to that of Ca^{2+}; **iron (Fe)** is absorbed by a special mechanism (→ p. 62).

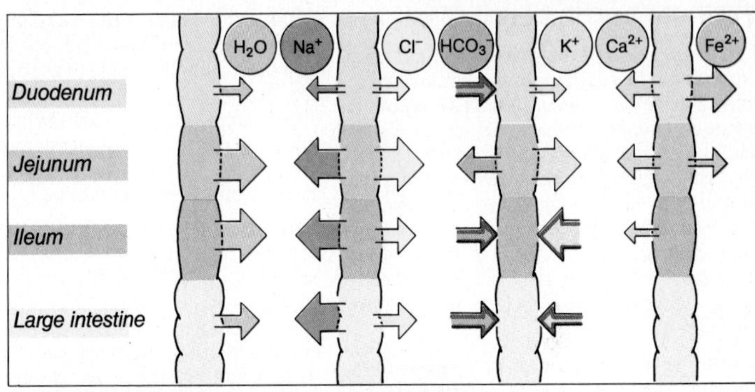

A. Intestinal absorption of water and electrolytes

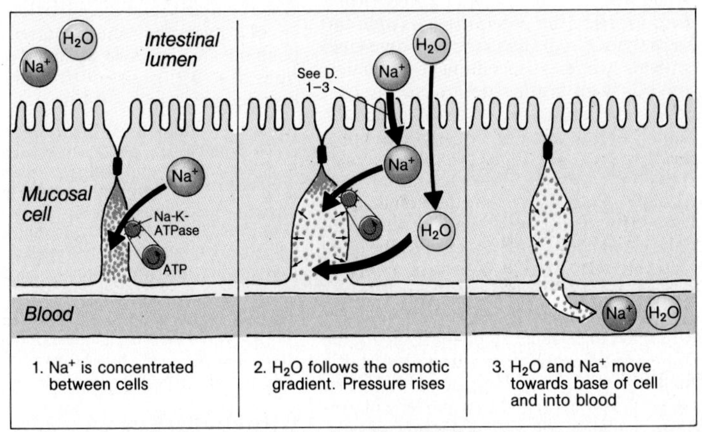

B. Absorption of salt and water from intestine (model)

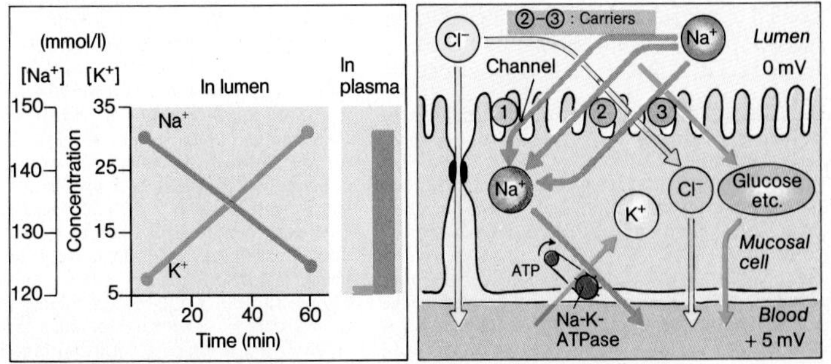

C. Na^+ - K^+ exchange in ileum
(after Code et al.)

D. Na^+ and Cl^- absorption

Colon, Rectum, Defecation, Feces

The last portion of the gastrointestinal tract consists of the **colon** (roughly 1.3 m long) and the **rectum** (→ p. 200). The colon is larger in diameter than the ileum. Its longitudinal muscles form three bands (*teniae coli*) that create pouches (*haustra*) along the length of the colon. The colonic **mucosa** has no villi; there are, however, deep *crypts* lined predominantly with *goblet cells* that produce mucus. Some of the superficial cells have a *brush border* (→ p. 210) for absorption.

The large intestine serves as a respository for the intestinal contents (1. respository: cecum and ascending colon; 2. repository: rectum). Here the *absorption of water* and electrolytes (→ p. 228) from the intestinal contents (chyme) is continued. Vitamins, some of them synthesized by colonic bacteria, seem to be partly absorbed here. Each day approximately 100–200 g feces are produced from about 500–1500 ml of chyme.

Motility. When food enters the stomach, the ileocecal valve relaxes, allowing the small intestine to empty its contents into the large intestine (*gastroileal reflex*). Several *mixing movements* of the colon can be distinguished, chief among which is *haustration* or formation of pouches along the length of the colon. *Mass movements*, strong peristaltic waves, occur repeatedly throughout the day (4- to 6-h intervals). Mass movements require integrity of the myenteric plexus. Normally, three to four such movements are needed to propel the colonic contents to the rectum; however, they serve only for transfer and are not directly related to defecation.

A typical sequence is shown in **A**, which depicts the movement of a barium mass as seen during an X-ray examination. The meal is taken at 7 : 00 AM. At 12 : 00 noon, the barium contrast medium has passed to the terminal ileum and cecum. The noon meal hastens emptying into the cecum. An annular constriction develops (→ **A 3**). The transverse colon is filled (→ **A 4**). Haustrations and annular contractions are seen throughout the length of the colon; their action mixes the colonic contents (→ **A 5**). Some minutes later, a gastrocolic reflex (and GI-hormones?) initiates a mass movement (→ **A 6**), which moves the contents into the sigmoid (→ **A 7**) and rectum (→ **A 8**).

Final water absorption takes place in the **rectum**. Absorbable substances introduced into the rectum by enema or as suppositories are not exposed to the gastric acid or digestive enzymes; they also bypass the liver since they enter the systemic venous drainage. This method of drug administration is preferred for infants.

Defecation. When the upper rectum (→ **B 6**) becomes distended by its contents, pressure receptors (→ **B 7**) stimulate a sensation of urgency for defecation. The act of defecation is initiated voluntarily (in most cases): the longitudinal muscles (→ **B 8**) of the rectum contract; the two anal sphincters, the inner (→ **B 3**; involuntary) and the outer (→ **B 4**; voluntary), and the puborectal muscles (→ **B 2**) relax; the rectum shortens; and the contents are expressed by annular contractions (→ **B 9**) assisted by increased abdominal pressure (→ **B 10**).

Dry weight of **feces** is 25 % of its total weight (→ **C**); one third of the dry weight is contributed by bacteria. *Frequency* of defecation varies widely, from three times daily to three times weekly, and depends upon the bulk conent of the diet ("fibers"; chiefly cellulose; → p. 196). Cellulose is metabolized by intestinal bacteria to methane and other gases, which accounts for the flatulence that follows, for example, a meal of beans.

Diarrhea (> 200 g feces/day), if extreme (e.g. in cholera; → p. 228) can lead to dangerous losses of water and K^+, and to acid-base disturbances (HCO_3^- loss; → p. 102).

Intestinal Bacteria

The intestinal tract is sterile at birth, but during the first weeks of life it becomes colonized orally by bacteria. The adult **colon** contains 10^{10}–10^{12} bacteria (almost exclusively anaerobic) per ml of intestinal contents. The bacteria enhance the activity of intestinal immune defense ("*physiological inflammation*"), and their metabolism is important for the "host." Bile salts and sexual hormones, e.g., are deconjugated (increases their enterohepatic circulation: → p. 214), and disaccharides that have not been absorbed higher up are converted into short-chain, absorbable fatty acids. In the **ileum** (mainly on account of the much more rapid chyme transport), the bacterial density is roughly four orders of magnitude smaller. The low pH value in the stomach is an important barrier for bacteria, so that here and in the upper small intestine their numbers are very low (0–10^4/ml).

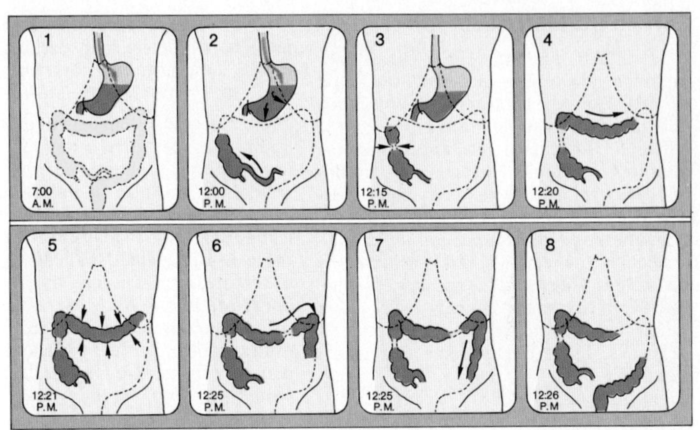

A. Motility of large intestine (after Hertz and Newton)

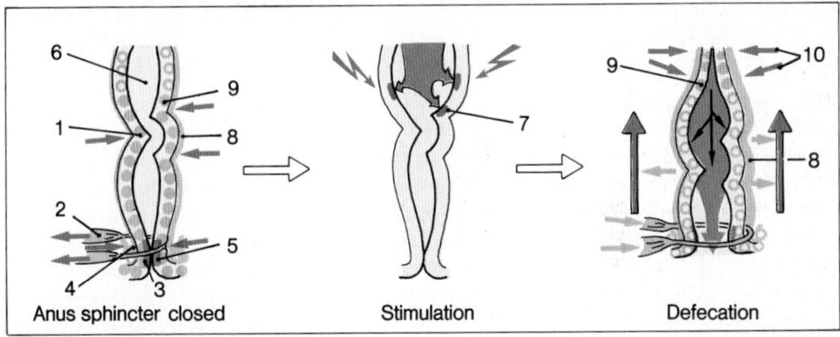

B. Defecation

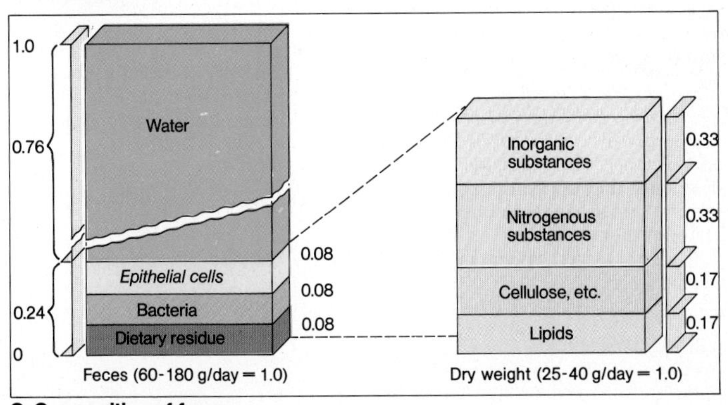

C. Composition of feces

Physiological Integration

In unicellular organisms, such as bacteria or protozoa, the entire cell responds to a stimulus from the environment. Within the cell, the signal can be transmitted by the diffusion of chemical substances since the distances are small. In contrast, multicellular organisms, with their many specialized *groups of cells* and *organs*, require some means of central *integration* and *coordination*. In mammals, this is achieved by the **nervous system** and the **endocrine system**, each of which transmits **signals** (→ **A**), either via electric-nervous or hormonal-humoral pathways. These signals serve to regulate *metabolism*, the "*internal milieu*" (circulation, pH value, water and electrolyte balance, temperature etc.), and to control *growth, maturation, reproduction*, etc. Stimuli from the environment, psychic-emotional factors and *feedback mechanisms* are all involved in this regulation.

The **nerves** are specialized for the *rapid transmission* of, as a rule, *finely graded signals*. Peripherally, a distinction can be made between (1) a **somatic nervous system** (→ p. 272 ff.), which primarily regulates the skeletal musculature and transmits the signals from the sensory cells towards the center; and (2) an **autonomic** or **vegetative nervous system** (→ p. 50 ff.), the chief role of which is to regulate the circulation, internal organs, sexual functions, to mention only a few.

The **endocrine system** is specialized for the *slow, chronic transmission of signals*, employing the *circulatory system* for covering the larger distances within the body. The **messengers** of the endocrine system are the **hormones**; they are synthesized in hormone-producing cells and act either on a subordinate hormone gland or on non-endocrine cells (the **target organ** or **target cells**, respectively).

Since all hormones circulate more or less together in the blood, there has to be some way of ensuring that the hormone and its specific target cell recognize one another. For this purpose the target cell is equipped with *specific binding sites* (**receptors**) for the hormone in question. The affinities (→ p. 11) of these receptors for the hormone must be extremely high because the hormone concentrations amount to only 10^{-8}–10^{-12} mol/l.

The endocrine system acts in close cooperation with vegetative centers in the brain and with the autonomic nervous system in regulating **nutrition, metabolism, growth**, psychical and physical **development** and **maturation**, the **reproductive mechanisms**, the necessary **adaptations in the performance** of the organism, and the "internal milieu" (**homeostasis**) (→ **A**). Most of these principally vegetative functions are subject to central control by the **hypothalamus**, which is itself influenced by *higher centers of the brain* (→ p. 290).

In the hypothalamus, electric (neural) stimuli are converted into chemical (hormonal) signals. Special nerve cells within the hypothalamus (**neuroendocrine cells**) synthesize hormones that are released in response to stimulation and are carried in the blood to their target.

The substances released at other nerve endings (acetylcholine, norepinephrine, etc) are known as transmitter substances or **neurotransmitters** since they only transmit the signal over a short distance (i.e. across the synaptic cleft), to the next cell, usually a muscle or nerve cell (→ p. 30 ff.).

The **adrenal medulla** (→ p. 58) occupies an intermediate position: its epinephrine and norepinephrine are released into the blood (although, chemically, they belong to the class of transmitter substances), and act as hormones in the organism.

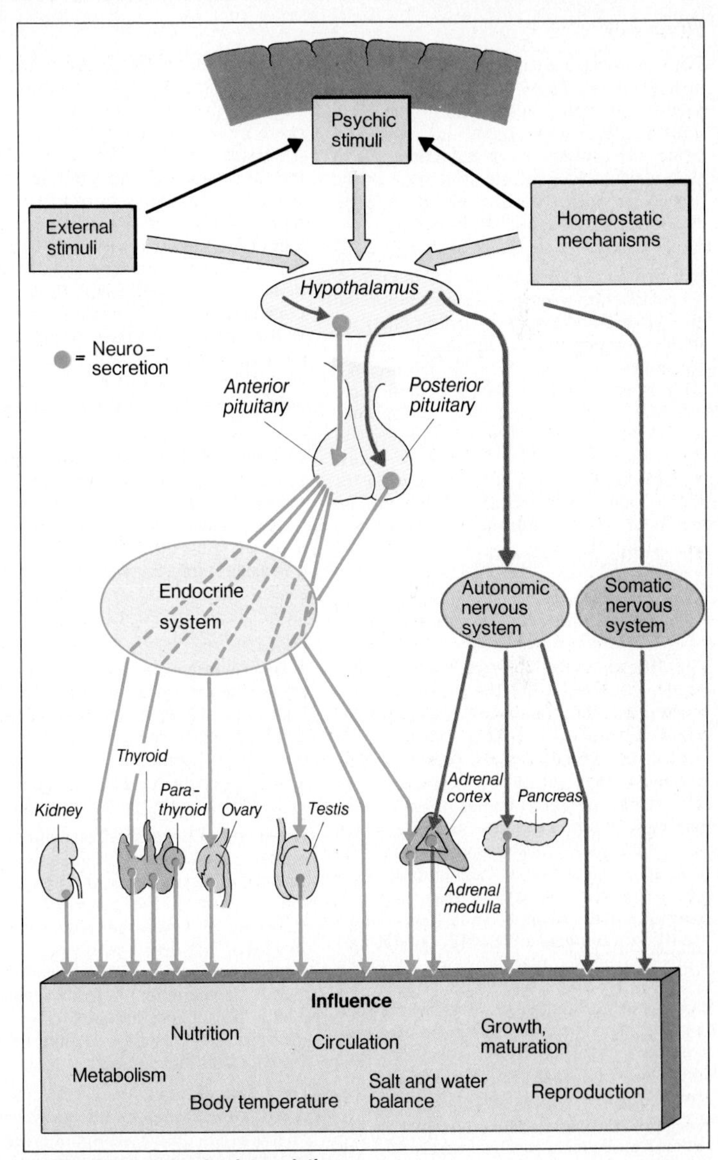

A. Neural and endocrine interrelations

Hormones

The hormones are the chemical messengers of the body. They **transmit information** necessary for the regulations of the functions of the various organs and cells. Hormones are synthesized in **endocrine cells** or even **glands** and (with the exception of the tissue hormones) reach the cells of the target organ (**target cells**) via the blood.

On the basis of *chemical structure*, three groups of hormones can be distinguished: (1) **peptide hormones** (→ **A**, dark blue) and **glycoprotein hormones** (→ **A**, light blue); (2) **steroid hormones** and chemically related hormones (D hormone) (→ **A**, yellow); and (3) hormones derived from the amino acid *tyrosine* (→ **A**, orange). The hydrophobic steroid hormones are bound to proteins in the blood, some of which are specific transport proteins, for example, *transcortin* for cortisol and progesterone, or the *globulin* that binds the sex hormones, testosterone and estrogens.

Most hormones are metabolized before their action is recognizable. Although 50% of the growth hormone (STH, GH) is broken down after only 20 minutes, its action continues for a week.

The **receptors** for the glycoprotein and peptide hormones and for catecholamines are situated on the *outer* surface of the target cell membrane (→ p. 242). As far as is known, these receptors are peptide chains (around 50000 Dalton) that penetrate the cell membrane many times in a zig-zag manner. When hormone binding takes place, an *intracellular* "**second messenger**" is released on the inside of the membrane and passes on the message within the cell. Examples of second (and sometimes "third") messengers are **cAMP, cGMP, inositoltrisphosphate, diacylglycerol**, or $\mathbf{Ca^{2+}}$ (→ p. 242ff.). *Steroid hormones*, on the other hand, enter the cells *themselves*, where they bind with specific **cytoplasmic receptor proteins** (→ p. 244). The thyroid hormones, too, enter the cells and probably bind to receptors on the cell nucleus.

A target cell may possess different receptors for one and the same hormone (e.g. for epinephrine, which binds to α_1, α_2, β_1 and β_2 receptors), or may possess receptors for different hormones (e.g. insulin and glucagon).

Hormonal hierarchy (→ **A**). In many cases, the initial stimulus for a hormonal response begins in the *CNS*. Input to the higher brain centers is connected synaptically with the **hypothalamus** (→ p. 240 and p. 290) and initiates a neurosecretory output. The hypothalamic message is transduced into hormone release from the **anterior** or **posterior lobe of the pituitary** (APit and PPit). Most of the APit hormones (so-called *tropic hormones*) control **peripheral** or **target endocrine glands** (→ **A**, green), from which the **end hormone** is then released (→ **A**). In each of these steps the original signal can not only be *amplified*, but can also undergo *modulation* (e.g. to the feed back regulation; → p. 238).

The **release of hormones from the APit** is governed by other controlling hormones released by the *hypothalamus* (→ **A** and p. 240). These are low-molecular weight peptides, some of which promote the release (**releasing hormone, RH**) and others which inhibit the release (**release-inhibitor hormone, IH**) of the APit hormones (→ **A** and table on p. 235).

The **hormones of the PPit** (ADH, ocytocin) are synthesized in the hypothalamus, transported to the PPit axoplasmatically, and then released by nervous signals (→ p. 240). Both PPit hormones act (like the APit hormones prolactin, LPH, and partly, STH; → p. 240) directly on the target cell.

Hormones of the adrenal medulla (→ **A** and p. 58) are released via the mediation of autonomic nerve fibers; this also applies to some extent to the **pancreatic hormones**. Primarily, however, the latter are under the control of humoral signals (→ p. 246) in the form of metabolites. **Parathyrin** (→ p. 254ff.), **calcitonin** (→ p. 256), **aldosterone** (→ p. 152), and **erythropoietin** (→ p. 60), are also released in response to humorally transmitted metabolic signals.

The **GI hormones** (→ p. 200ff.) and the so-called **tissue hormones** are synthesized elsewhere than in the classical endocrine system. The action of tissue hormones is often local or **paracrine**. Angiotensin II (→ p. 152), bradykinin (→ p. 176 and p. 202), histamine

(→ p. 72 and p. 208), serotonin (→ p. 74), and the prostaglandins belong to this group. Some hormones also act on the cell they were released from: **autocrine action** (e. g. interleukin 2).

In humans, **prostaglandins (PG)** and related compounds are derived from the fatty acid **arachidonic acid** (**AA**; index 2 is used for prostaglandins derived from AA), which is either ingested (meat) or derived from the dietary essential fatty acid, linoleic acid. In the body, AA is esterified as a component of cell membrane phospholipids from which it is released by *phospholipase* A_2. Three main pathways for **PG synthesis** from AA exist:

(a) *Lipoxygenase* pathway: the main end product is 12-OH-AA (HETE), which is chemotactic and involved in invasion by leukocytes during inflammation.

(b) Formation of *leukotrienes A, B,* and *C*. The latter is identical with the "slow reacting substance of anaphylaxis" (→ p. 72).

(c) *Fatty acid cyclooxygenase* pathway (inhibited by nonsteroid antiinflammatory drugs like aspirin): via intermediates (PGG_2, PGH_2), the biologically active compounds PGE_2, PGD_2, $PGF_{2\alpha}$, thromboxane (TXA_2, TXB_2), and prostacyclin (PGI_2) are formed.

Some typical **effects of prostaglandins** are: (1) *bronchial muscle*: $PGF_{2\alpha}$ contracts, PGE_2 relaxes; (2) *gastrointestinal secretion*: PGE_2 usually inhibits; (3) *uterus*: PGE_2 and $PGF_{2\alpha}$ contract; (4) *kidney*: PGE_2 and PGI_2 provoke natriuresis etc.; (5) *pain*: PGE_2 and PGI_2 sensitize nociceptive nerve endings (inflammation); (6) *platelets*: TXA_2 causes aggregation, PGI_2 inhibits.

Recommended name*	Other names	Abbreviations
Hypothalamus		
The ending **liberin** is used for releasing hormones (RH) or factors (RF)		
The ending **statin** is used for release-inhibiting hormones (IH) of factors (IF)		
Corticoliberin	Corticotropin RF	CRF, CRH, ACTH-RH
Gonadoliberin	Follicle-stimulating hormone RG**	FSH-RH, FSH-RF
	= Luteinizing hormone RH** = FSH/LH-RH	LH-RH, LH-RF
Melanostatin	Melanotropin release-IF	MIH, MIF
Melanoliberin	Melanotropin RH	MRH, MRF
Prolactostatin	Prolactin release-IF (= Dopamine)	PIH, PIF
Somatostatin***	Somatotropin (or growth hormone) release-IH	STS, SIH, GH-IH
Somatoliberin	Somatotropin (or growth hormone) RF	SRH, SRF, GH-RH
Thyroliberin	Thyrotropin RH	TRH, TRF
Anterior lobe of pituitary		
The ending **tropin** is used for anterior pituitary hormones		
Corticotropin	Adrenocorticotropic hormone	ACTH
Follitropin	Follicle-stimulating hormone	FSH
Lutropin	Luteinizing hormone or	LH
	interstitial cell-stimulating hormone	ICSH
Melanotropin	Melanocyte-stimulating hormone	MSH
Somatotropin	Growth hormone	STH, GH
Thyrotropin	Thyroid-stimulating hormone	TSH
Prolactin	Luteotropic or mammatropic hormone	PRL, LTH
Posterior lobe of pituitary		
Ocytocin or Oxytocin (→ p. 268)		
Vasopressin, antidiuretic hormone, adiuretin		ADH, AVP

* Most names are recommended by IUPAC-IUB Comm. on Biochemical Nomenclature (1974).
** The releasing hormones for LH and FSH are identical.
*** Also released by the pancreas and other GI organs.

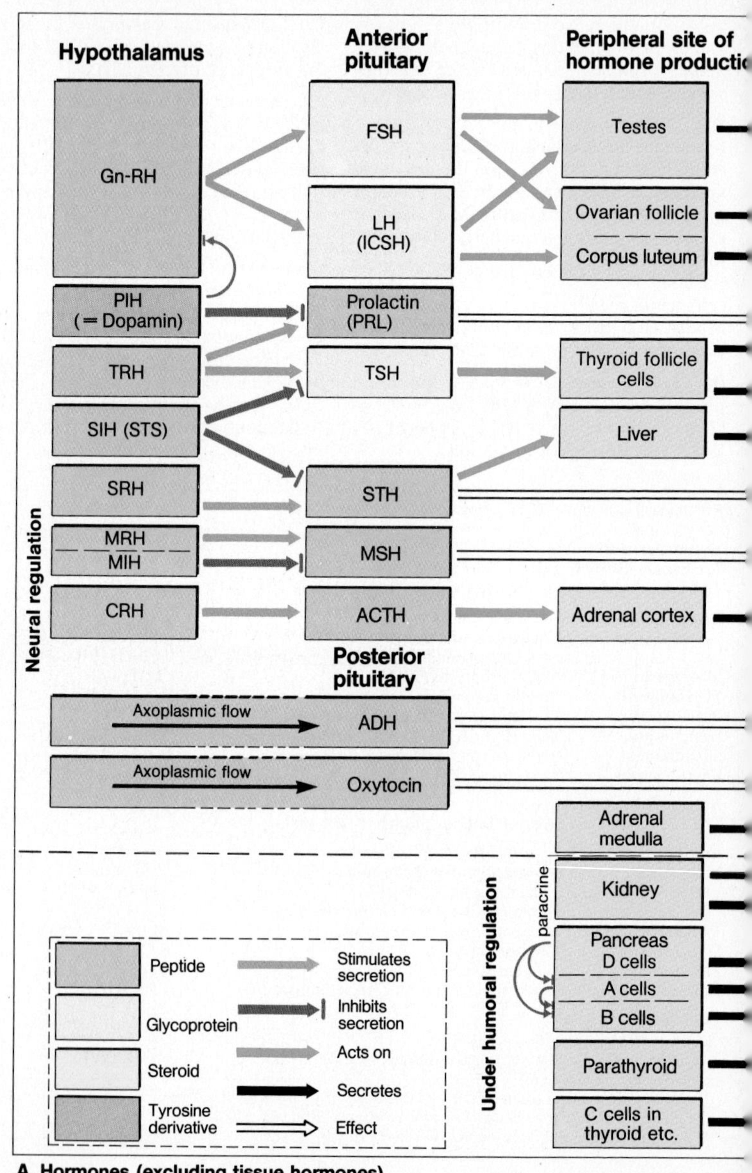

A. Hormones (excluding tissue hormones)

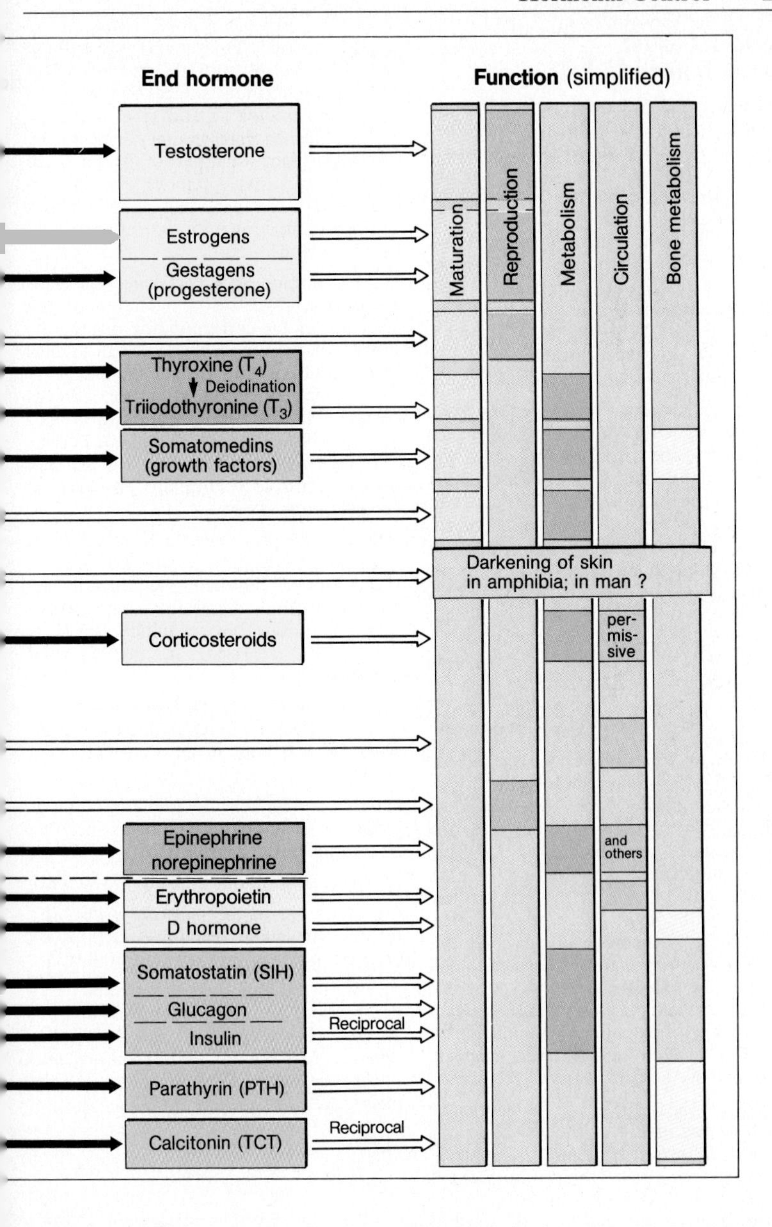

End hormone
Function (simplified)
Testosterone
Estrogens
Gestagens (progesterone)
Thyroxine (T_4)
Deiodination
Triiodothyronine (T_3)
Somatomedins (growth factors)
Corticosteroids
Epinephrine norepinephrine
Erythropoietin
D hormone
Somatostatin (SIH)
Glucagon
Insulin
Reciprocal
Parathyrin (PTH)
Calcitonin (TCT)
Reciprocal
Maturation
Reproduction
Metabolism
Circulation
Bone metabolism
Darkening of skin in amphibia; in man ?
per-mis-sive
and others

Feedback Control. General Hormone Effects

Feedback is the circumstance in which the response to a signal (e.g. of a cell to a hormonal stimulus) influences the source of the signal (in this case, the hormone-producing gland). In **positive feedback** (rare) the stimulus is *amplified* by the signal. This results in a larger subsequent response, which feeds back to produce an even greater stimulus (e.g. → p. 264). In **negative feedback**, the stimulus is *decreased* by the response. Like most other regulatory mechanisms in the organism, hormonal action is mostly subject to negative feedback.

The *releasing hormones (RH) of the hypothalamus* (e.g. CRH or corticoliberin) elicit the release of the corresponding *tropic hormones* from the anterior lobe of the pituitary gland (in our example, ACTH or corticotropin), which, in turn, influence the peripheral endocrine glands (in our example, adrenal cortex; → **A 1**). The **end hormone** (in our example, cortisol) not only acts on the target cell, it also *inhibits secretion of the RH* in the hypothalamus (→ **A 3** and **A 4**), which results in a decrease in the amount of end hormone set free (→ **A 5–A 7**). Inhibition of secretion of the RH is in this way reduced (→ **A 7**) and so on.

Feedback can, for example, also come about if the APit hormone inhibits the hypothalamus, or if the APit or the cells producing the end hormone are inhibited by the end hormone itself, for example, by TSH and ACTH (**autoinhibition**; → **A**, below, right). A further possibility is for the release of the hormone to be regulated (e.g. parathyrin; → p. 254ff.) by the *metabolic level* (in this case, Ca^{2+} concentration in plasma) that the hormone regulates. Feedback can also involve nervous signals (**neuroendocrine control circuit**), for example, in the endocrine control of behavior (glucose level → hunger; osmo- and H_2O-homeostasis → thirst, etc).

Not only do the tropic hormones govern the *formation* and *release* of the end hormone, they also influence the **growth** of the peripheral endocrine gland the end hormone is released from. If the concentration of the end hormone in the blood is still too low despite maximal synthesis and release, these cells then multiply until the feedback effect of their end hormone suffices to slow down the "commanding" gland (e.g. goiter → p. 252). Such a **compensatory hypertrophy** also takes place when part of an endocrine gland is operatively removed, as in thyroidectomy. The remaining gland increases in size and function until its original rate of secretion is restored.

When a hormone (e.g. cortisone) is given as *medication*, it causes a reduction in the secretion of the corresponding tropic hormone (e.g. ACTH), in the same way as does the natural hormone (in our example, cortisol, which is produced in the adrenal cortex). Chronic administration of an end hormone or of one of its synthetic derivatives thus results in inhibition and atrophy of the gland that normally produces it: **compensatory atrophy**.

Rebound is a phenomenon observed when hormonal medication is stopped. During this period of "rebound" the feedback system has not fully recovered and the tropic hormone (in the above example, ACTH) is temporarily released in hypernormal quantities.

The **principal effects of the hormones** on their target cells consist in regulating the metabolism in these cells in the three following ways:

(1) changes in enzyme configuration (**allosteric mechanisms**), which result in direct alterations in the *enzyme activity*;

(2) changes in *enzyme synthesis* (**induction**);

(3) alterations in the *substrate availability* for the enzymatic reactions, such as by changes in cell membrane **permeability**.

Insulin, for example, employs all three ways in governing the intracellular availability of *glucose*. For this purpose a cellular "program" is initiated by the intracellular phosphorylation of the β-subunit of the membranal insulin receptor when insulin binds to it (→ p. 248).

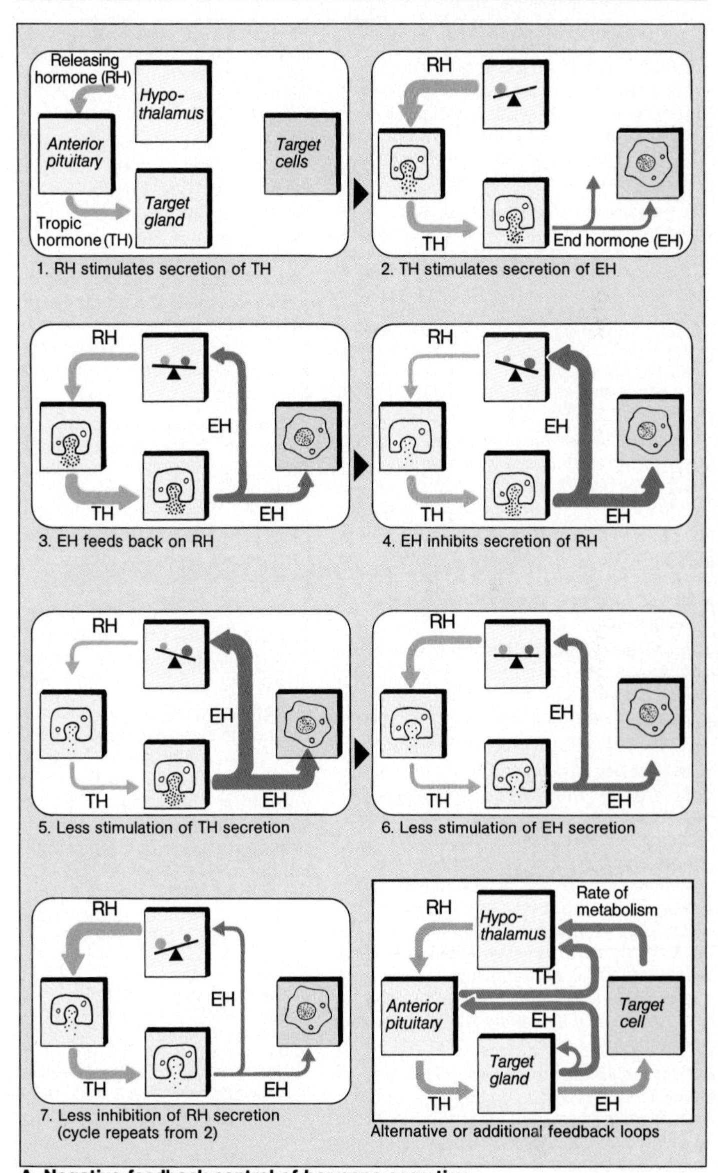

A. Negative feedback control of hormone secretion

Hypothalamic-Hypophyseal System

Certain neurons of the hypothalamus secrete hormones; this is termed **neurosecretion**. The hormones synthesized in the nerve cells are not released into the synaptic cleft like the transmitter substances (→ e.g. p. 54ff.), but *enter the blood directly*.

The neurosecretory neurons of the medial **hypothalamus** synthesize hormones in the endoplasmic reticulum of the soma and pass them on to the Golgi apparatus where the hormones are incorporated into *granules* (100 to 300 nm in diameter), each surrounded by a membrane. The granules are transported by **axoplasmic flow** (→ p. 22) to the nerve terminals. In this way, ocytocin and ADH are transported to the posterior lobe of the anterior pituitary, and the RH and IH to the *median eminence* of the hypothalamus (see below).

Release of the hormone granules by exocytosis occurs after an **action potential** (AP; → p. 26) reaches the nerve ending. This is accompanied by an influx of Ca^{2+} into the nerve ending, as with the release of neurotransmitters. The AP in neurosecretory nerves lasts up to 10 times longer than in other nerves, to allow sustained release of the hormone.

The **hormones of the posterior lobe of the pituitary**, that is, adiuretin (= vasopressin = ADH) and ocytocin (= oxytocin), are released from the neurosecretory nerves directly into the *systemic circulation.*

The **releasing hormones, RH**, for the *anterior lobe of the pituitary* (*APit, adenohypophysis*) are secreted from the neurosecretory neurons of the hypothalamus into a kind of **portal blood system**, or "short-cut" to the vascular network of the APit, where they bring about release (via second messengers; → p. 242) of the APit hormones into the general circulation (→ **A**). *Regulation of the release of RH* is accomplished by feedback (→ p. 238) via the plasma concentration of the tropic APit hormone or of the corresponding end hormone.

Release-inhibiting hormones (IH) exist for some of the APit hormones. These IH also pass from the hypothalamus to the APit via the hypophyseal portal system. A reduction in the amount of IH released leads to an increased liberation of the corresponding APit hormone (→ p. 236f.).

For the normal release of some of the APit hormones, the presence of additional hormones in the blood is essential. For example, the release of STH requires the participation of glucocorticoids and thyroid hormone in addition to its own RH.

The hypothalamus is intimately connected with the **limbic system**, the **formatio reticularis**, and (via the thalamus) with the **cerebral cortex** (→ p. 290). The hormone balance is thus not only concerned with purely vegetative regulation (energy and water balance; circulation; respiration), but it is also connected with the *sleeping-waking rhythm* and with *psychic-emotional factors.* Stress situations can, for example, be the reason for the absence of menstrual bleeding (→ p. 262ff.).

The **hormones released by the APit** are as follows: **STH** (see below), **ACTH** (acts on the adrenal cortex; → p. 246ff. and p. 260), **TSH** (acts on the thyroid gland; → p. 250ff.), **FSH** and **LH** (ICSH) (act on ovaries and testes; → p. 262ff.), and **prolactin** (acts mainly on the mammary glands; → p. 264).

ACTH, TSH, FSH, and LH act on subordinated endocrine glands, and are therefore termed **glandotropic**. Prolactin is nonglandotropic; STH acts both ways.

Somatotropin (STH, growth hormone) governs skeletal growth and other metabolic processes (→ p. 246ff.), usually with the mediation of **somatomedins** or **growth factors** (from the liver), as for example in sulfate incorporation in cartilage, and in protein synthesis. Somatomedin C (= insulinlike growth factor = IGF) also inhibits the release of STH in the APit (negative feedback). Without the intermediation of somatomedins, STH is lipolytic and glycogenolytic.

Pro-opiomelanocortin (**POMC**) is the precursor polypeptide of ACTH and a lipotropic hormone (**β-lipotropin**) in the APit. In the intermediate lobe of the pituitary, the ACTH formed in this way is further processed to **melanocyte-stimulating hormone (MSH)**, **γ-lipotropin**, and **β-endorphin**. Their physiological significance is unclear.

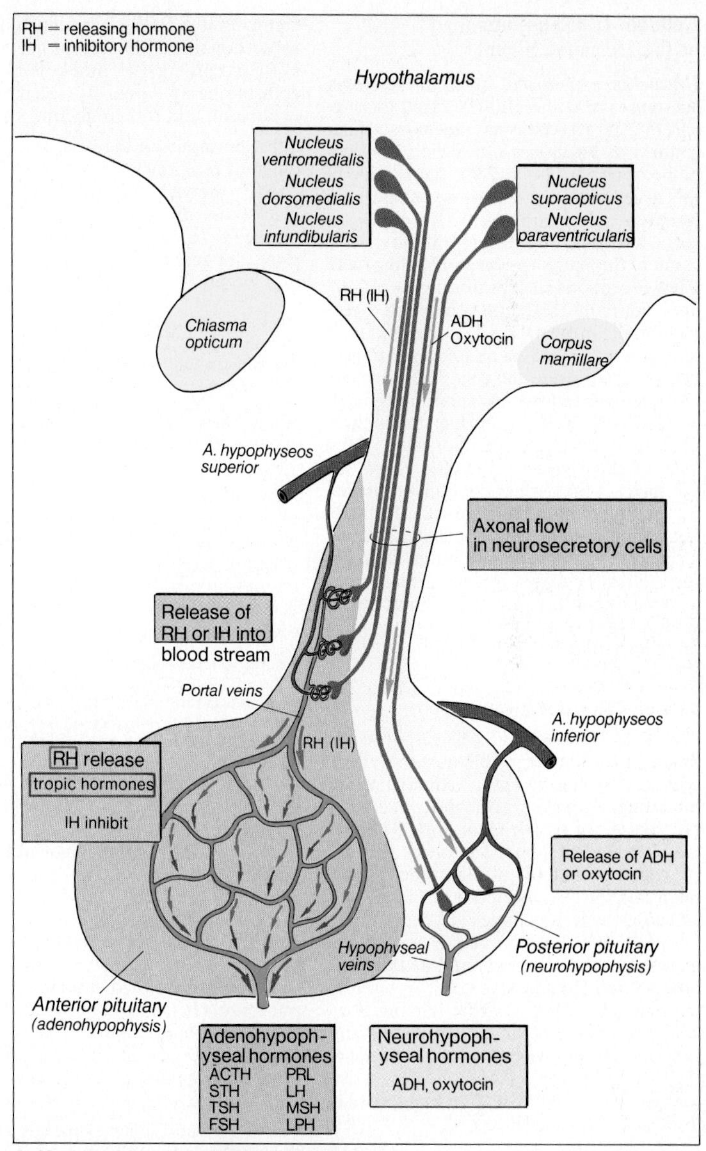

A. Hypothalamic and pituitary hormones

Cellular Transmission of the Hormone Signal

Hormones, as humoral signal or **messenger substances** (first messenger), reach their respective target cells via the extracellular route. For hormones other than the lipophilic hormones (→ p. 244), the outside of the target cell membrane possesses **hormone receptors**, which are **specific** for each hormone and bind it with high affinity. As a result of this **hormone-receptor binding** (with a few exceptions such as insulin; → p. 246), and certain reactions between cell-membrane proteins (and sometimes also phospholipids), **second messengers** are usually released inside the cell. These include **cAMP** (cylic adenosine monophosphate) and **cGMP** (cyclic guanosine monophosphate), inositol-1,4,5-trisphosphate (**IP_3**) and 1,2-diacylglycerol (**DAG**). Since the **specificity** of the hormone action is provided by the receptors on the target cell, many hormones can employ the same second messenger. In addition, its concentration in the cell can be raised by one hormone but lowered by another. The cell often possesses several types of receptors for one and the same hormone.

cAMP as Second Messenger

For a cAMP-mediated cell response the membrane of the target cell must contain, in addition to the receptor, stimulating or inhibiting guanyl nucleotide-regulatory proteins, **G_s** or **G_i** (→ **A**), or both. They are made up of the 3 subunits α_s (or α_i), β, and γ. At rest, α is bound to guanosine diphosphate (**GDP**). If the **h**ormone reacts with the **r**eceptor, the H-R complex attaches itself to G_s-GDP (or G_i-GDP). GDP is now replaced by cytosolic GTP, and at the same time β-γ and H-R are split off. This requires the presence of **Mg^{2+}**. What remains is α_s-GTP or α_i-GTP, of which the former *activates* the **adenyl(ate) cyclase** on the inside of the membrane (i.e. cAMP, rises), whereas α_i-GTP (with unknown cofactors, possibly β-γ) *inhibits* it (cAMP drops).

Hormones acting via **G_s** and a **rise in cAMP** are: glucagon, VIP, ocytocin, adenosine (A_2 receptors), serotonin (S_2 receptors), secretin, PGE_2, PGI_2, histamine (H_2 receptors), dopamine (D_1 receptors), adiuretin (VP_2 receptors), LH, FSH, TSH, ACTH, epinephrine (β_1 and β_2 receptors), corticoliberin, and somatoliberin.

Other hormones or the same hormones on a different receptor act via **G_i** and **lower the cAMP**: acetylcholine (M_2 receptors), somatostatin, opioids, angiotensin II, epinephrine (α_2 receptors), adenosine (A_1 receptors), dopamine (D_2 receptors), serotonin (S_{1a} receptors), and others.

Cholera toxin blocks the GTPase, thus cutting out its "turning off" effect on the adenyl cyclase, and the cellular cAMP concentration rises to extremely high values (→ p. 228) for consequences for the intestinal cells). **Pertussis toxin** inhibits the G_i protein, thus *deinhibiting* its effect on the adenyl cyclase, and thus also causes a rise in cAMP in the cell.

cAMP activates **protein kinase A** with whose help **proteins** (mostly enzymes or membrane proteins, including the receptor itself) are **phosphorylated** (→ **A**). The specific cell response depends on the nature of the phosphorylated protein, which in turn is determined by the particular protein kinase present in the target cell.

Another kind of specificity is provided since some enzymes are activated by the phosphorylation whereas others are inactivated. Thus cAMP exerts a glycolytic effect in two ways: the enzyme glycogen synthetase, which catalyses glycogen formation, is inactivated by phosphorylation, whereas the enzyme catalysing glycogen breakdown, phosphorylase, is activated by the cAMP-mediated phosphorylation.

In order to **switch off the chain of signals,** α-GTP is reconverted by the hormone-activated GTPase to α-GDP, which eventually joins up again with β-γ to form G-GDP. Further, cAMP is inactivated by a phosphodiesterase to 5'-AMP, and also the previously phosphorylated proteins can be dephosphorylated by phosphatases. Inhibition of the reaction cAMP → 5'-AMP (e.g. by theophyllin or caffeine), on the other hand, prolongs the life of cAMP and thus also the hormone action.

Via G_s, G_i, and other G-proteins (G_o; G_k) the **ion channels** and **ion pumps** (K^+, Ca^{2+}) can also be regulated *without* the intervention of adenyl cyclase.

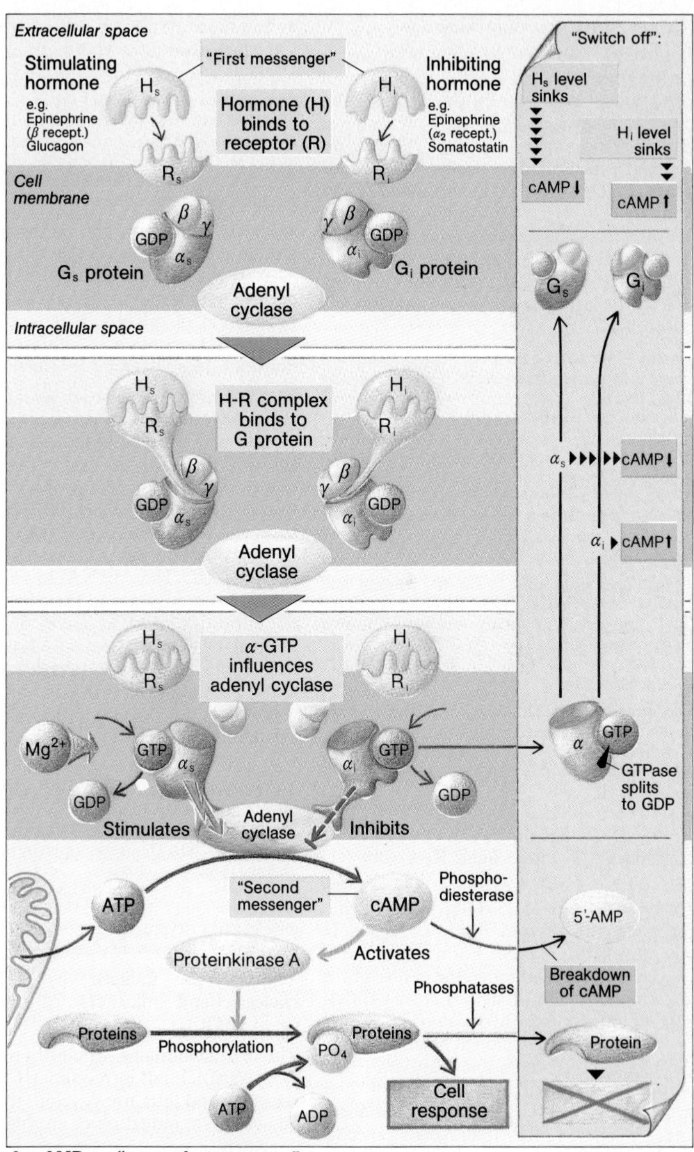

A. cAMP as "second messenger"

IP_3 and DAG as second messengers

Following the extracellular hormone-receptor binding, and again with the help of G-proteins (→ p. 242) (in this case G_p and others), the **phospholipase C** on the inside of the cell membrane is activated. This enzyme splits the **phosphatidyl inositol-4,5-biphosphate (PIP_2)** of the cell membrane into IP_3 and **DAG** (→ p. 242), which as parallel second messengers have different effects (→ **B**). The effect of DAG persists considerably longer than that of IP_3 since the Ca^{2+} set free by the latter (see below) is immediately pumped off.

The lipophilic **DAG** remains in the cell membrane where it activates **protein kinase C** which, among other functions, also phosphorylates, and thus activates, the carrier protein for the Na^+/H^+ exchange. One of the results is a rise in **cell pH**, which is another important signal for many cell processes. The *arachidonic acid* that can be released by (possibly also from) DAG exerts, via its metabolites, the prostaglandins and prostacyclines (→ p. 234f.), a variety of other effects on cellular metabolism.

Via the cytoplasm most of the IP_3 reaches, and empties, the Ca^{2+} stores of the cell (ER) so that the Ca^{2+} can now, as a *third messenger*, influence numerous cell functions. The Ca^{2+} can bind to **calmodulin** (→ p. 17f. and p. 44) as a possible intermediary step.

The **hormones** epinephrine (α_1 receptors), acetylcholine (M_1 receptors), serotonin (S_1 receptors), thyroliberin, CCK, gastrin, adiuretin (VP_1 receptors), histamine (H_1 receptors), and thromboxane act via IP_3 and DAG.

Hormones with Intracellular Receptors

The steroid hormones (→ p. 237, yellow areas), calcitriol (1,25-$(OH)_2$-cholecalciferol), and the thyroid hormones have in common with the other hormones the **specific cell response**, although the nature of the intracellular chain of signals is very different. In contrast to the hydrophilic hormones (→ p. 242) the above-named hormones pass through the cell membrane relatively easily on account of their *fat solubility*. In their respective target cells the **steroid hormones** and calcitriol find their own specific **cytoplasmic binding protein** ("receptor" protein, → **C**), to which they bind: **transformation**. The hormone-receptor binding is essential for hormonal activity; neither of the partners on its own has an effect.

For some hormones (e.g. estradiol) there is more than one receptor protein in the target cell; other cells may have receptors for several hormones (e.g. for estradiol and progesterone). The concentration of the receptor protein is variable; estradiol, for example, can increase the number of progesterone receptors in progesterone target cells.

After its formation, the **hormone-receptor protein complex** migrates into the **cell nucleus** (**translocation**), where, after binding to **nuclear receptors**, it stimulates increased production of mRNA, that is, the DNA-mRNA **transcription** is influenced by the hormone-receptor complex (**induction**).

The mRNA-producing or so-called *structural genes* of a chromosome are "switched on and off" by what is termed an *operator gene*. A **repressor**, formed by a so-called *regulator gene*, switches the operator gene to "off". Probably the **effect of the hormone** is to inactivate this repressor; this "switches on" the operator gene, that is, more mRNA is produced. The mRNA leaves the cell nucleus and migrates to the **ribosomes**, the site of protein synthesis. Here, the increased number of templates (mRNA) makes possible increased copying (**translation**) of proteins. The increased number of proteins (→ e.g. AIP, p. 151) resulting from this induction then lead to the actual cell response (→ **C**).

The **glucocorticoids**, for example, induce, among other substances, a number of enzymes that bring about a rise in the glucose concentration in the blood (→ p. 260). This is supported by induction of gluconeogenetic enzymes (e.g. glucose-6-phosphatase, pyruvate carboxylase) and of enzymes promoting the conversion of amino acids into glucose (tryptophane pyrrolase, tyrosine-α-ketoglutarate transaminase, etc.).

The protein induced by **1,25$(OH)_2$-cholecalciferol** influences Ca^{2+} transport (→ p. 254ff.).

The **thyroid hormone**, triiodothyronine (T_3; → p. 250ff.), binds intracellularly to *nuclear receptors* and thus also develops its metabolic action via induction of enzymes.

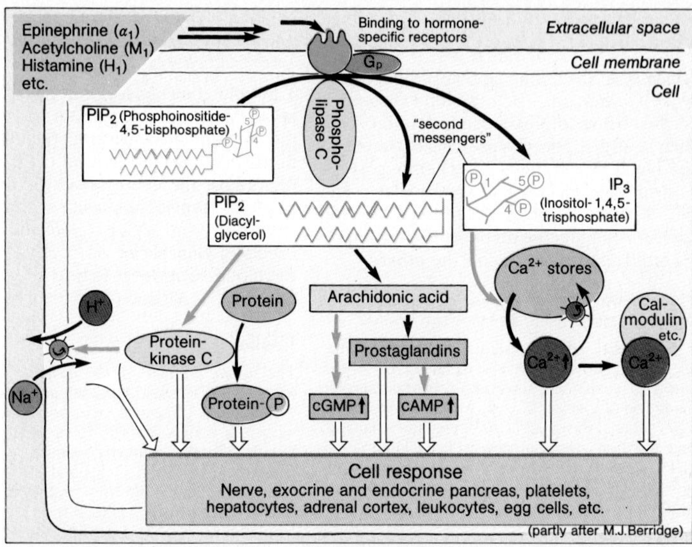

B. Diacylglycerol and inositol-1,4,5-trisphosphate as "second messengers"

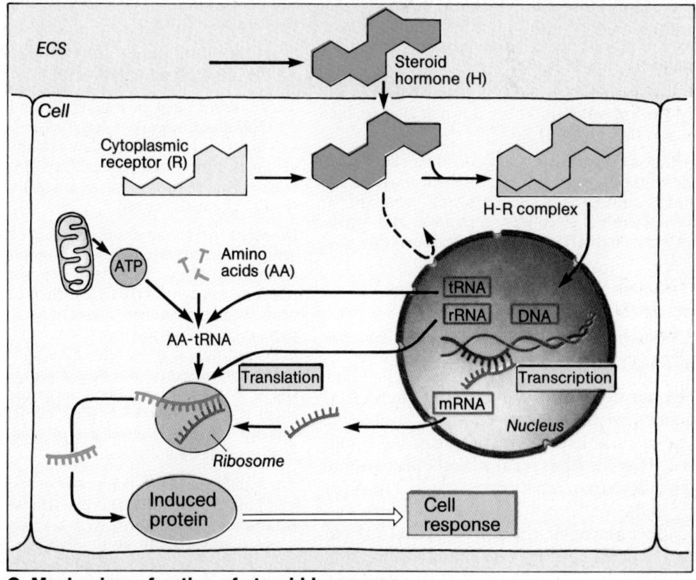

C. Mechanism of action of steroid hormones

Carbohydrate Metabolism. Pancreatic Hormones

Glucose is the *major source of metabolic energy* in humans, the brain and the erythrocytes being totally dependent upon it for their energy. The **glucose concentration in the blood** (*blood sugar level*) thus plays a central role in energy metabolism. The blood sugar level is determined by the *utilization* of glucose on the one hand and by its *intake* and *synthesis* on the other.

The following definitions are of importance in discussing carbohydrate metabolism (→ **A**):

1. **Glycolysis**, strictly speaking, refers to the anaerobic breakdown of glucose to lactate (→ p. 46). The term is extended, however, to cover the aerobic oxidation of glucose as well. Anaerobic glycolysis takes place in the erythrocytes, renal medulla, and, under certain circumstances, in skeletal muscle (→ p. 46). Aerobic breakdown of glucose occurs mainly in the CNS, skeletal muscle, and in most other organs.

2. **Glycogenesis** is the process of glycogen formation from glucose (in the liver and muscle). Glycogenesis is important for the storage of glucose and for the maintenance of the blood sugar level. Glycogen stored in muscle is for the exclusive use of the muscle itself.

3. **Glycogenolysis** is the term given to the process by which glycogen is broken down to glucose, that is, it is the reverse of 2.

4. **Gluconeogenesis** (in the liver and renal cortex) is the synthesis of glucose from non-sugars, i.e. from amino acids (from muscle protein), lactate (from the anaerobic glycolysis in muscle and erythrocytes), and glycerol (from the breakdown of fats).

5. **Lipolysis** is the breakdown of fats to glycerol and free fatty acids, whereas

6. **Lipogenesis** refers to formation of fats (for storage in fat depots).

The hormone-producing cells in the *islets of Langerhans* in the **pancreas** play a decisive role in carbohydrate metabolism. A, B and D cells of the islets form a kind of functional syncytium (gap junctions; → p. 7). The **A** (or α-) **cells** make up 25% of the cells and produce **glucagon**, the **B** (or β-) **cells** (60% of the islet cells) produce **insulin**, and the **D** (or δ-) **cells** (10%) produce **somatostatin** (SIH). In addition, **pancreatic polypeptide** is released from the islets of the head of the pancreas. These hormones also influence one another locally at the production stage (*paracrine effect*).

The chief **functions of the pancreatic hormones** are (1) to bring about the storage, in the form of glycogen and fat, of the nutrients taken up in the food (insulin); (2) to remobilize the energy reserves during the phase of hunger or during work, in stress situations, and so on (glucagon; see also effects of **epinephrine** on p. 58); (3) to keep the blood sugar level as near to constant as possible (→ **A**); and (4) to promote growth.

Insulin

The *insulin content of the pancreas* is roughly 6–10 mg, of which about 2 mg are released daily. The half-life of insulin is approximately 10–30 min; it is mainly broken down in the liver and kidneys. **Insulin synthesis**: insulin is a peptide consisting of 51 amino acids and is formed by the removal of the C chain from proinsulin (84 amino acids), which in turn is split off from preproinsulin. Insulin consists of two peptide chains, A and B, connected by two disulfide bonds (S-S bonds). Proinsulin and, subsequently, insulin-containing *granules* are formed.

The chief stimulus for the **release of insulin** is an **elevated blood sugar level** (→ **B**).

These are the steps of insulin release from the B-cells: plasma glucose ↑ → cell glucose ↑ → cell ATP ↑ → K^+ channels close → depolarization → Ca^{2+} channels open → cytoplasmic Ca^{2+} ↑ → (a) insulin release by *exocytosis*, and, as a negative feedback, (b) K^+ channels reopen.

Glucagon (paracrine action, see above) and certain *gastrointestinal hormones*, too, promote the release of insulin. There may also be a special *insulin releasing polypeptide* (IRP). Further, certain *amino acids* (lysine, arginine, leucine) and a number of hypophyseal and steroid hormones lead to an increased release of insulin. *Epinephrine* and *norepinephrine* (α-receptors) *slow down* insulin release (→ **A, B**). If, for example, a large drop in blood sugar level is registered in the CNS (chemoreceptors for glucose), the release of epinephrine (→ p. 58) is reflexly increased.

The **insulin receptor** is composed of two α-subunits that bind the hormone and two (transmembranal) β-subunits, which are *tyrosine-specific protein kinases* activated by insulin binding within one minute. As a further step, the hormone-receptor complex is *internalized*. The subsequent (or parallel?) steps mediating metabolic and growth effects of insulin are not well understood.

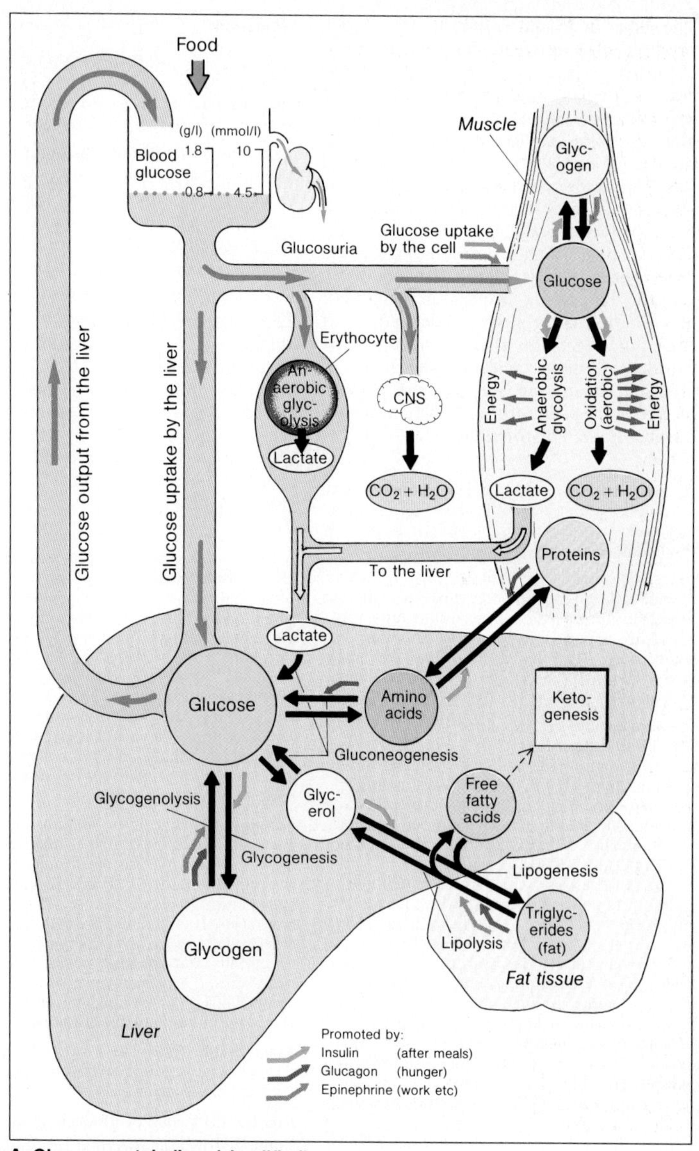

A. Glucose metabolism (simplified)

The **effect of insulin** (→ **A, B, C**): Insulin promotes the **storage of glucose**, principally in the liver, where it leads to an intracellular rise in glycolysis and glycogenesis. In this way, the elevated plasma concentration of glucose following the intake of food is rapidly *lowered* again. About two thirds of the glucose absorbed postprandially in the intestine is temporarily stored in this way, for stepwise remobilization during the interdigestive phase. This ensures that, above all, the CNS, which is highly dependent on glucose, has at its disposal a supply of glucose that is relatively independent of food intake. Insulin also provides for the storage of **amino acids** in the form of proteins (mainly in the skeletal muscle; *anabolism*), promotes **growth**, and influences the $\mathbf{K^+}$ **distribution** in the body (→ p. 148).

An **excess of insulin** results in **hypoglycemia**, which at values < about 2 mmol/l (< 0.35 g/l) results in metabolic disturbances in the brain (and possibly results in *coma*): *hypoglycemic shock*.

Excessive intake of carbohydrate (*fattening*) overtaxes the glycogen storage capacity so that the liver then also converts glucose into fatty acids. These are exported to fat tissue and stored as **triglycerides**. Their mobilization and the breakdown of fat to free fatty acids (*lipolysis*) is inhibited by insulin (→ p. 220ff.).

Diabetes mellitus can be caused by:
(1) insulin deficiency (**type I**);
(2) reduction of the number of intact insulin receptors (**type II**, e.g. in obesity and uremia);
(3) a reduced affinity of the insulin receptors (e.g. in acidosis or due to excessive glucocorticoids);
(4) dominance of hormones that raise the blood sugar (glucagon, STH; see below),
and is characterized by an elevated blood glucose concentration (**hyperglycemia**), which may lead to **glucosuria** (→ p. 128 and 142). In addition, in cases (1) – (3) there is no inhibition of lipolysis (see above), which means that large quantities of **fatty acids** are set free. Although some of these can be used for energy production via acetyl CoA, the latter gives rise to acetic acid and, from it, β-oxybutyric acid (*metabolic acidosis* → p. 114) and acetone (*ketosis*). Since fat synthesis in the liver is independent of insulin, and large amounts of free fatty acids are available, it also accumulates triglycerides (*fatty liver*).

Glucagon

Glucagon is a peptide hormone with 29 amino acids, produced in the A cells of the pancreas from proglucagon. Like insulin (see above) it is stored in *granules* and released by exocytosis.

The main **stimuli for glucagon release** are *hunger* (hypoglycemia, → **B**), or a surfeit of *amino acids*, although its release is also brought about by sympathetic stimulation (via β-receptors; → **A**) or a lowering of the plasma concentration of free fatty acids. *Hyperglycemia inhibits* glucagon release.

The main **effect of glucagon** (→ **A, B, C**) is antagonistic to that of insulin and consists of raising the blood sugar level, thus ensuring an adequate supply of glucose throughout the organism. The ways in which this is accomplished are (a) by increased glycogenolysis (liver, not muscle) and (b) increased gluconeogenesis from lactate, amino acids (= catabolism), and glycerol (from lipolysis).

A rise in the concentration of **amino acids** in the plasma leads to the release of more insulin, and if additional glucose were not available *hypoglycemia* would result. This is prevented, however, because the amino acids also stimulate the release of glucagon, which elevates the blood sugar. Additionally, however, glucagon increases gluconeogenesis from amino acids, that is, the latter are to some extent channeled into *energy metabolism*. Therefore, if a patient is given an infusion of amino acids with the object of building up protein it is essential to provide *glucose simultaneously* in order to prevent the oxidation of the amino acids.

Somatostatin (SIH) is an inhibitor of insulin and glucagon release (paracrine action) and decreases the rate of assimilation of all nutrients from the gastrointestinal tract. SIH release is stimulated by high plasma levels of glucose, amino acids, and fatty acids. It is inhibited by catecholamines and others. SIH also has an endocrine inhibitory effect on motility and secretion in the gastrointestinal tract. Thus, it is probably part of a feedback loop preventing a rapid overload of nutrients. SIH may also act as an antiobesity hormone.

Somatotropin (STH) has a short-term insulin-like action (somatomedin), but its long-term effect is to raise the blood sugar level (growth-promoting).

For the influence of **glucocorticosteroids** on carbohydrate metabolism (→ **C**) see p. 260.

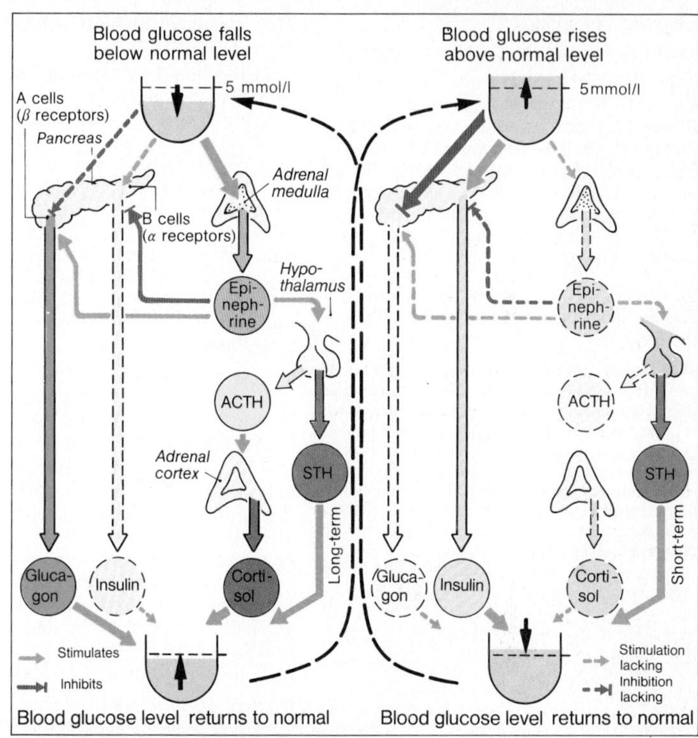

B. Hormonal control of blood glucose level

Hormone	Insulin	Glucagon	Epinephrine	Cortisol
Function	Satiated ← Buffer	→ Hungry	Emergency, exercise	Restorative
Glucose				
Uptake into cells	+ M, F		+ M	− M, F
Glycolysis	+	−	+	−
Gluconeogenesis (L)	−	+	+	+
Glycogen				
Synthesis ⇄ lysis	L, M ←	L →	L, M →	L ←
Fat				
Synthesis ⇄ lysis	L, F ←	F →	F →	F →

L = Liver M = Muscle F = Fat tissue

C. Hormonal influences on carbohydrate and fat metabolism

Thyroid Hormones

The thyroid gland contains spherical **follicles** (50–500 μm diameter), the cells of which synthesize the thyroid hormones **thyroxine** (**T_4**; prohormone) and **triiodothyronine** (**T_3**; active hormone). The thyroid gland also contains clear cells (parafolicular or C cells) that synthesize **calcitonin** (→ p. 256). T_3 influences growth, differentiation, and metabolism, the latter in a wide variety of ways.

T_3 and T_4 are *stored* (→ **B**), bound to a glycoprotein, **thyroglobulin**, in the **colloid** of the follicles.

Biochemistry: thyroglobulin (660,000 Dalton) is synthesized on the *ribosomes* of the thyroid gland cells, from amino acids. Carbohydrates are added at the *Golgi apparatus*. The thyroglobulin is "packed" into vesicles and is transferred by **exocytosis** (→ **A** and p. 12) to the follicular space where it is incorporated ino the colloid. The **iodination** of the **tyrosinyl** residues of the thyroglobulin takes place probably on the outer surface of the cell membrane. This requires the presence of **iodine**, which is actively taken up from the blood as the iodide ion, I^-, into the thyroid gland cells. I^- uptake (accumulation about 25-fold as compared to plasma) is ATP-dependent and linked to Na^+ transport (→ **B**). **Thyrotropin (TSH)** from the adenohypophysis, the controlling hormone for the thyroid gland (cAMP as second messenger), promotes this uptake of I^- by raising the transport capacity (I^- accumulation up to 250 ×), whereas other anions (in order of their efficiency: ClO_4^-, SCN^-, NO_2^-) competitively *inhibit* I^--uptake. I^- is continuously withdrawn from the intracellular I^- pool; with the help of a **peroxidase** it is *oxidized* to elementary I^0 or I_3^- which, bound to an **iodine transferase**, is expelled by exocytosis into the follicular space where it immediately reacts with about 10% of the 110 tyrosyl residues of the thyroglobulin.

In this process, the phenol ring of the tyrosyl residue is iodinated in the 3- or 5-position, or both, so that the protein chain now contains **diiodotyrosyl** (**DIT**) and **monoiodotyrosyl** (**MIT**) **residues**. These steps in synthesis are promoted by TSH, but inhibited by thiouracil, thiocyanate, glutathione, and other reducing substances.

The tertiary structure of the thyroglobulin is such that the iodinated tyrosyl residues (still in the colloid) can react with one another: the phenol ring of one DIT (or MIT) becomes **coupled** to another DIT by an ether bond, so that the thyroglobulin chain now contains **tetraiodothyronyl residues** and (to a smaller extent) **triiodothyronyl residues** (→ **D**). These are the **storage forms** of the thyroid hormones T_4 and T_3.

TSH also stimulates the **release of T_3 and T_4** from the thyroglobulin of the colloid reentering the cell by **endocytosis** (→ **C** and p. 12). These vesicles fuse with primary lysosomes to form *phagolysosomes*, in which the thyroglobulin is hydrolyzed by proteases. In the process, T_3 and T_4 are released (about 0.2 and 1–3 mol per mol thyroglobulin, respectively) and enter the blood, whereas the **I**- of the MIT and DIT that are also released is split off by a deiodinase and thus once more becomes available for resynthesis.

T_3 and T_4 metabolism: T_3 is two to four times as potent as T_4 and also acts more promptly, producing its effects within hours, whereas T_4 requires days for its maximal response. Of the circulating T_3 only about 20% are derived from the thyroid gland, 80% from the deiodination of T_4. For these reasons, T_3 is considered to be the actual, effective hormone, T_4 fulfilling the role of a *prohormone* (storage). **Conversion of T_4 to T_3** (primarily in liver and kidney) is catalyzed by a microsomal **5′-deiodinase**, which removes the 5′-iodine from the outer ring of T_4 (→ **D**).

If instead, one iodine is split off from the inner ring (by a 5-iodinase), T_4 is converted to the inactive metabolite *reverse T_3* (**rT_3**). Normally, production of T_3 and rT_3 in the periphery is of similar rate (25 μg/d). Formation of T_3 decreases and that of rT_3 increases during fasting, as 5′-deiodinase is inhibited. The 5′-deiodinase of the hypophysis (see below), however, is an exception and is not inhibited, so that a release of TSH (in this case, undesired) due to negative feedback does not occur.

The ratio of T_3 and T_4 in plasma is 1 : 100, both of them bound to three different proteins (more T_4): (1) thyroxine-binding globulin (**TBG**), which transports two thirds of the T_4; (2) thyroxine-binding prealbumin (**TBPA**), which, together with (3) serum albumin, transports the rest of the T_4. Only traces of unbound T_3 and T_4 are present in the circulating blood.

Regulation of secretion of the thyroid hormones. The thyroid hormones are unusual in that the T_3 and T_4 plasma concentrations remain remarkably constant. The controlling hormone is **TSH** (adenohypophysis).

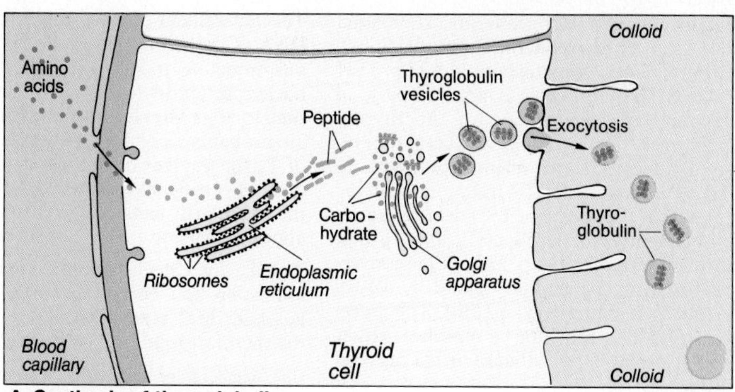

A. Synthesis of thyroglobulin

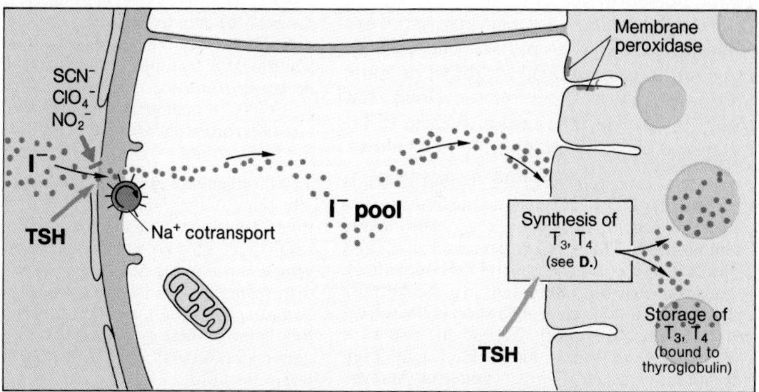

B. Iodine uptake, synthesis and storage of thyroid hormones

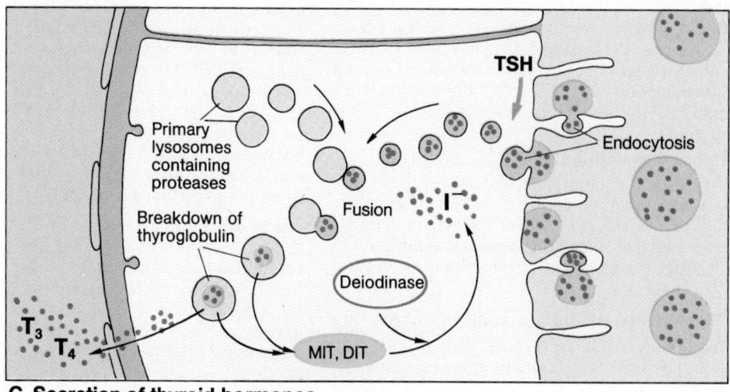

C. Secretion of thyroid hormones

Its synthesis and release are stimulated by **TRH** or **thyroliberin** (hypothalamus). Hypothalamic **somatostatin** inhibits TSH release. The TRH effect is modified by T_3: a rise in the concentration of the thyroid hormones, for example, reduces the responsiveness of the adenohypophysis to TRH (reduction of TRH receptors) with the result that the release of TSH decreases and, consequently, the levels of T_3 and T_4 sink (*negative feedback*, → p. 238). (T_4 is converted to T_3 in the pituitary gland by a very active 5′-deiodinase; → p. 250). The secretion of TRH, too, can be modified, either negatively by T_3 (feedback) or via nervous influences.

In the newborn, cold appears to stimulate the release of TRH via nervous pathways (thermoregulation, → p. 194). The T_3 does not seem to be employed for direct or acute thermoregulation (reaction too slow), but rather to "adjust" the responsiveness of fat tissue and heart to epinephrine (see below).

Goiter is an enlargement of the thyroid gland. It may be either *diffuse* or *nodular*. One of the causes of diffuse goiter is a *deficiency of iodine* in the drinking water. The resulting deficiency of T_3 and T_4 leads to an increased release of TSH (see above). A chronic increase in TSH results in goiter because the TSH also stimulates proliferation of the follicle cells (*hyperplastic goiter*), T_3 and T_4 rise as a consequence, and their blood level may even normalize (*euthyroid goiter*). A goiter of this type often persists even after its cause has been removed (e.g. lack of iodine).

Hypothyroidism, or too little T_3 and T_4, occurs when even the enlarged thyroid cannot deliver sufficient hormones: *hypothyroid goiter*. Among the many possible causes are inborn errors in T_3 and T_4 synthesis or inflammatory diseases of the thyroid.

In **hyperthyroidism**, too much T_3 and T_4 are synthesized, usually independently of TSH, for example, by a thyroid tumor, or by a diffuse goiter (*Basedow's syndrome, Graves' disease*). In the latter case, a **thyroid-stimulating immunoglobulin** binds to the TSH receptors and thereby stimulates T_3 and T_4 production.

The **effects of the thyroid hormones** are manifold, although it is difficult to say whether they have specific target organs. T_3 and T_4 are taken up by the target cells in the same way as steroid hormones, although they do *not* require specific *receptor proteins*. The intracellular point of attack for T_3 is the **DNA** of the **cell nucleus** (influence on transcription; → p. 3). (The affinity of the nuclear receptors for T_4 is 10 times smaller than for T_3.) The effects on the **mitochondria** are probably secondary; under the influence of T_3 the number of mitochondria and the cristae (→ p. 5) increase, which is the basis of the **increased metabolic activity** brought about by T_3 and its precursor T_4.

T_3, like the catecholamines, **stimulates O_2 utilization** and **energy turnover**, and thus **increases heat production**. T_3 is important for thermoregulation (→ p. 194).

Core temperature and basal metabolic rate are reduced in thyroid hormone deficiency, and elevated in severe hyperthyroidism. Thus, T_3 influences the activity of other hormones. In hypothyroidism, insulin, glucagon, STH, and epinephrine, for example, lose their stimulating effect on energy turnover, whereas in hyperthyroidism sensitivity to epinephrine is increased (e.g. heart rate rises) probably because T_3 induces synthesis of β-adrenoceptors.

T_3 also **stimulates growth** and **development**, especially that of the *brain* and *bones*. Deficiency of thyroid hormones in the newborn can therefore result in retarded growth and development (*stunted growth, retarded sexual development*, etc) and to disturbances in the nervous system (*mental retardation, epilepsy*, etc), a condition known as **cretinism**. Some of these symptoms can be prevented by administration of thyroid hormones within the first 6 months of life.

Iodine metabolism and requirements (→ **E**): circulating iodine exists in three forms: (1) anorganic I^- (2–10 μg/l); (2) organic nonhormonal iodine (traces) in the form of iodinated thyroglobulin, MIT, and DIT; (3) iodine contained in T_3 and T_4, which in turn are bound to plasma proteins (protein-bound iodine (**PBI**; 35–80 μg iodine/l). Of the latter, 90% are T_4, and are also known as butanol extractable iodine (**BEI**). About 150 μg T_3 and T_4, in the ratio of 5 : 2, are utilized daily (in fever and hyperthyroidism 250–500 μg).

Iodine lost by excretion (→ **E**) is replaced from dietary sources such as sea food, and plants (grain, bread) that have grown on iodine-containing soils. Insufficient iodine in food is often supplemented by the addition of iodine to common salt. The milk of nursing mothers contains iodine so that their requirements rise to about 200 μg daily.

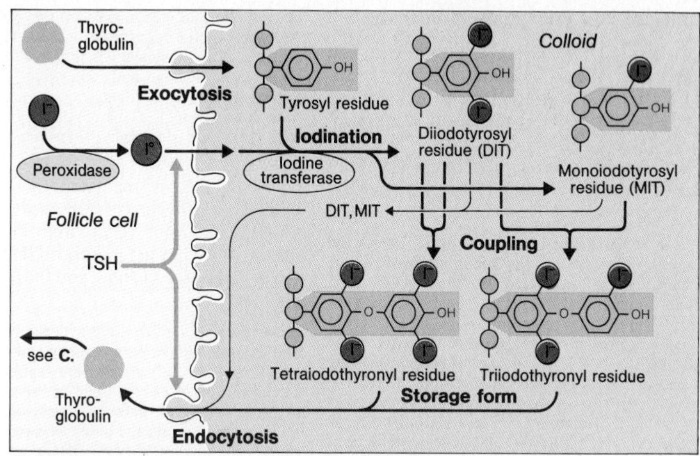

D. Synthesis, storage and mobilization of the thyroid hormones

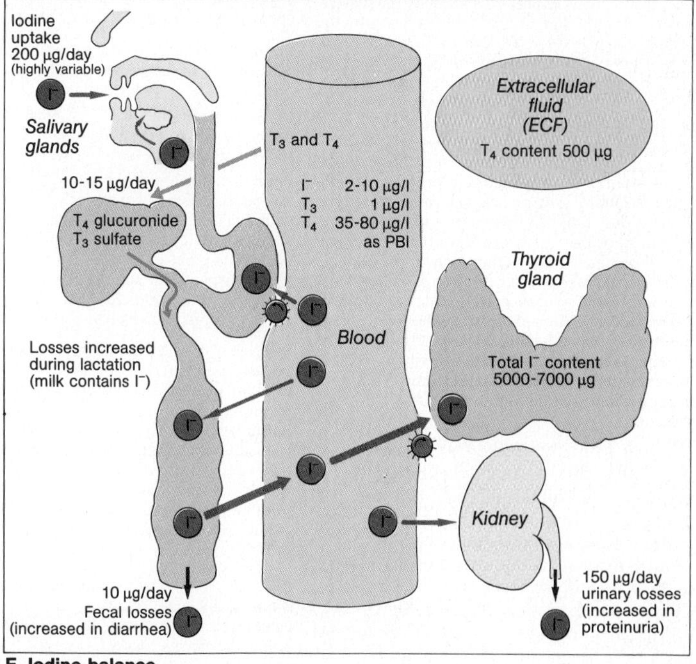

E. Iodine balance

Calcium and Phosphate Homeostasis

Calcium (Ca^{2+}) plays a critical role in the regulation of cell functions (→ pp. 15, 36, 44, 74ff., 242f.). It accounts for 2% of the body weight; 99% of this are in the bones, 1% dissolved in the body fluids. The **total Ca^{2+} concentration in the serum** is normally **2,3–2,7 mmol/l** (= 4.6–5.4 meq/l = 9.2–10.8 mg/dl).

About 60% of this is freely filtrable (across the capillary wall, as in the renal glomerulus), of which, again, 4/5 is in the form of **ionized Ca^{2+}**, 1/5 bound in complexes (calcium phosphate, calcium citrate, etc.). The remaining 40% of the serum Ca^{2+} are protein-bound and are thus not freely filtrable (→ p. 10). The degree of protein-binding depends upon the **pH value** of the blood (→ p. 100ff.): it rises in alkalosis and sinks in acidosis (by about 0.21 mmol/l Ca^{2+} per pH unit). This is the reason why alkalosis (e.g. by hyperventilation) may lead to *tetany*.

The **phosphate balance** of the body is closely **connected with that of Ca^2**, but is not strictly regulated as the latter. The daily intake of phosphate is about 1.4 g, of which 0.9 g are absorbed and, on an average, also excreted again via the kidneys. Normally, the *serum phosphate concentration* amounts to 0.8–1.4 mmol/l (= 2.5–4.3 mg/dl). Calcium phosphate has a low solubility. If the product of the Ca^{2+} concentration and the phosphate concentration exceeds a certain value (**solubility product**), calcium phosphate *precipitates* out of the solution; in the living organism, calcium phosphate is mainly deposited in the *bones* but, in extreme cases, in other parts of the body as well. If a patient is given an infusion of a phosphate solution, the Ca^{2+} concentration in the serum is lowered since calcium phosphate is deposited in the bones (or even in other organs) due to the solubility product having been exceeded. Conversely, a drop in the serum phosphate concentration leads to hypercalcemia due to release of Ca^{2+} from the bones.

In order to maintain the **Ca^{2+} balance** (→ **A**), Ca^{2+} intake and loss have to be matched. The **intake of Ca^{2+}** amounts to about 12–35 mmol/d. Milk, cheese, eggs, "hard" water are particularly rich in Ca^{2+}. Under normal conditions approximately 9/10 of this intake leave the body again in the **feces** and the rest in the **urine** (→ **A**), although if the Ca^{2+} intake is low up to 90% of it can be absorbed (→ **A**).

During **pregnancy** and **lactation** a higher Ca^{2+} intake is necessary, owing to its uptake via the placenta (about 625 mmol) or the milk (up to 2000 mmol), for incorporation into the infant's skeleton. **Ca^{2+} deficiency** is thus a very common occurrence during and following pregnancy. Pathologically, Ca^{2+} deficiency is seen in *rickets* (one cause of which is lack of vitamin D), parathyroid hormone deficiency (*hypoparathyroidism*), and a number of other conditions.

Ca^{2+} balance is regulated by **three hormones**; parathyrin (parathyroid hormone, PTH), calcitonin (CT), and D-hormone (calcitriol). They act mainly on **three organs**; the *intestine*, the *kidneys*, and the *bones* (→ **B** and **D**).

PTH is a peptide hormone with 84 amino acids, and is formed in the **parathyroid glands**,. Synthesis and release of the hormone are regulated by the concentration of *ionized Ca^{2+}* in the plasma. If it drops below the normal value (*hypocalcemia*) more PTH is released into the blood, whereas a rise has exactly the reverse effect (→ **D**).

The **PTH effects** are all aimed at **raising the** (previously lowered) **Ca^{2+} level** (→ **D**):

a) In the **bones**, the osteoclasts (see below) are activated and *bone resorption* takes place with the release of Ca^{2+} (and phosphate).

b) In the **intestine**, Ca^{2+} uptake is indirectly increased due to the stimulating effect of PTH on the formation of D hormone in the kidneys.

c) In the **kidneys** the reabsorption of Ca^{2+} is increased, which is particularly important on account of the rising Ca^{2+} concentrations in the serum due to (a) and (b). In addition, PTH inhibits phosphate reabsorption (→ p. 151). The resulting hypophosphatemia stimulates Ca^{2+} release from the bone tissue and prevents the deposition of Ca^{2+} phosphate (solubility product, see above).

Deficiency or inactivity of PTH (*hypoparathyroidism and pseudohypoparathyroidism*) lead to **hypocalcemia** (destabilized resting potential →cramps: tetanus) and to secondary deficiency of D-hormone. However, excess PTH (hyperparathyroidism), or malign osteolysis, which overtaxes Ca^{2+} regulation, results in **hypercalcemia**; over longer periods of time this can lead to calcification (kidneys etc.) and, if $[Ca^{2+}] > 3.5$ mmol/l, to coma and disturbances in cardiac rhythm (→ p. 168).

CT, like PTH, is a peptide hormone (32 amino acids) and is mainly synthesized in the parafollicular or C cells of the thyroid gland. In hypercalcemia, the plasma concentration of CT is increased many times above its normal value, whereas at Ca^{2+} concentrations below 2 mmol/l CT is no longer detectable. *CT lowers* the (elevated) Ca^{2+}

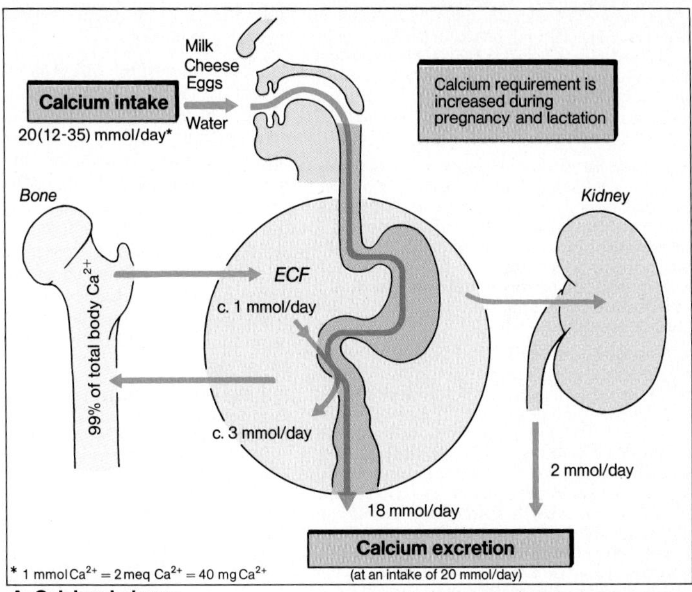

A. Calcium balance

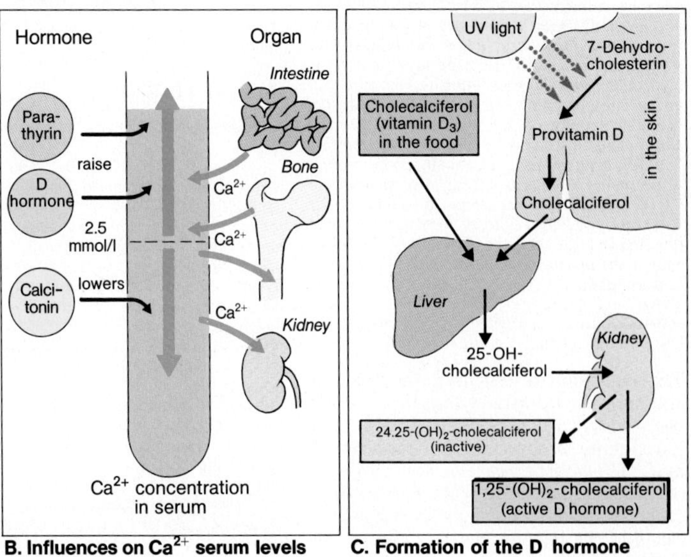

B. Influences on Ca^{2+} serum levels

C. Formation of the D hormone

concentration of the serum mainly by its action on **bone** tissue where CT inhibits the osteoclast activity which is stimulated by PTH, that is, it leads (at least temporarily) to an *increased incorporation of* Ca^{2+} into bone tissue (→ **D**). CT-sensitive adenylate cyclase is present in the **kidneys**, but renal effects of CT are not well understood.

Some **gastrointestinal hormones** increase the release of CT, which promotes the incorporation of postprandially absorbed Ca^{2+} in the bones. This effect, together with a possible retarding influence of CT on digestion, prevents a postprandial hypercalcemia, which would result in an (in this case, undesirable), inhibition of PTH release and, as a consequence, increased renal excretion of the recently absorbed Ca^{2+}.

D-hormone (= **calcitriol** = 1,25-$(OH)_2$-cholecalciferol). Several organs are involved in the synthesis of this lipophilic hormone, which is closely related to the steroid hormones (→ **C**). From *7-dehydrocholesterol*, **UV irradiation** (sun, solar lamps) brings about the formation of **cholecalciferol** (= **vitamin** D_3 = **calciol**) in the skin, via an intermediate step (*previtamin D*). Both products are bound in the blood to the **vitamin-D-transport protein** (α-globulin), whereby calciol has the higher affinity and is therefore better transported. Previtamin D, therefore, remains in the skin for some time after UV irradiation (short-term storage). Calcidiol (see below) and calcitriol are also bound to the transport protein, which is produced in greater quantities during pregnancy, under the influence of estrogen. Particularly if exposure to UV is inadequate, insufficient vitamin D is synthesized and it must be administered **orally** as a *vitamin*. Children require about 400 units = 10 μg/d; adults, half this amount. Instead of vitamin D_3 obtained from animals, the equally effective *ergocalciferol* (= *Vitamin* D_2) from plant sources can be used. The following steps are identical for Vitamin D_2 and D_3: Cholecalciferol is converted in the **liver** to 25-OH-cholecalciferol (= **calcidiol**). This is the main *storage form*, with plasma concentrations of 25 μg/l and a half-life of 15 d. But the active D-hormone itself, 1,25-$(OH)_2$-cholecalciferol (= **calcitriol**; →**C**), is formed in the **kidney** (and sometimes in the placenta).

The **regulation of calcitriol formation** is effected at the 1-*α-hydroxylase*, that is, at the last stage of synthesis. **PTH** (which is released in larger amounts in hypocalcemia) as well as *phosphate deficiency*, and *prolactin* (lactation) promote this synthesis. Calcitriol synthesis is slowed down again because the calcitriol (a) restores the plasma concentrations of Ca^{2+} and phosphate by boosting intestinal absorption (see below), and (b) also directly inhibits the release of PTH (negative feedback).

The most important **target organ** of calcitriol is the **intestine**, although it also has an effect on **bones**, kidneys, placenta, and mammary glands (binding to intracellular receptor protein, alterations in gene expression; → p. 244). In physiological concentrations it increases *Ca^{2+} absorption* in the intestine and the *mineralization* of the skeleton. In excessive doses, however, it *decalcifies* the bones, an effect that is potentiated by PTH. In kidney (→ p. 151), placenta, and mammary glands, calcitriol appears to increase the transport of Ca^{2+} and phosphate.

In transient **hypocalcemia** the bones can serve as a short-term buffer for Ca^{2+} (→ **D**), but in the end the Ca^{2+} deficit is met by increased uptake from the intestine mediated by calcitriol. If, on the other hand, insufficient vitamin D is available, as for example in **vitamin D deficiency** caused by inadequate intake or absorption (impaired fat digestion), by lack of UV light, or by a reduction in the synthesis of calcitriol (renal insufficiency), demineralization of the skeleton occurs (*osteomalacia; rickets* in children). The main reason for this is the continued excessive release of PTH due to the chronic hypocalcemia (compensatory *hyperparathyroidism*).

Bone Metabolism

Bone consists of an organic **matrix** containing mineral deposits: $\mathbf{Ca^2}$; **phosphate** (as $Ca_{10}(PO_4)_6(OH)_2$); $\mathbf{Mg^2}$; $\mathbf{Na^+}$; and others. The major portion of the matrix is **collagen**, a protein containing large quantities of the amino acid *OH-proline* (in connection with breakdown of the matrix more OH-proline appears in plasma and urine). Normally, mature bone tissue is in a state of dynamic equilibrium of continuous resorption and deposition, although this may be temporarily interrupted by periods of imbalance. *Undifferentiated cells* at the bone surface can be activated (e.g. by PTH) to **osteoclasts** (bring about bone resorption). If their activity is suppressed (e.g. by CT, estrogens) they are transformed to **osteoblasts**) that lay down new bone surface. The osteoblast effect is based upon their content of *alkaline phosphatase*, an enzyme that can create a high phosphate concentration locally, resulting in deposition of Ca^{2+} (when the solubility product is exceeded). Calcitriol increases the activity of this enzyme.

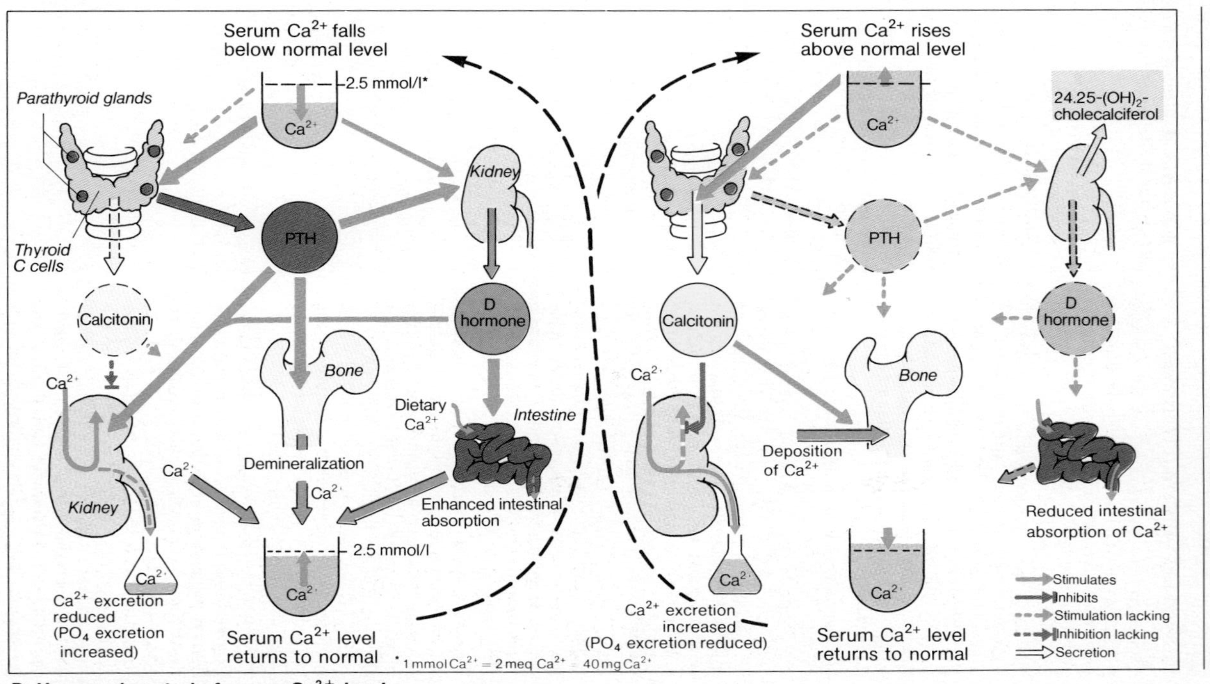

D. Hormonal control of serum Ca^{2+} level

Biosynthesis of Steroid Hormones

Cholesterol is the source of all steroid hormones (→ **A**). It is formed in the *liver* and *endocrine glands* via a number of intermediate steps (*squalene, lanosterol*, etc.) from *acetyl CoA*. Although the placenta can synthesize steroid hormones (→ p. 268), it cannot form cholesterol, and therefore must take it up from the blood (→ p. 222). Steroid hormones are stored only in small quantities at their site of production (adrenal cortex, testes, ovaries), which means that they have to be synthesized from cellular reserves of cholesterol when needed.

Cholsterol contains 27 C atoms (for numbering → **A**, top left). Via a number of intermediate steps it is converted into **pregnenolone** (21 C atoms; → **A, a**) from which all steroid hormones derive. Pregnenolone gives rise to **progesterone** (→ **A, b**) which is a potent female sex hormone (→ p. 262 ff.), and can also be converted into every other steroid hormone: (1) the hormones of the adrenal cortex, with 21 C atoms (→ **A**, orange and yellow); (2) the male sex hormones (androgens) with 19 C atoms, in the testes (→ p. 270), ovaries, and adrenal cortex (→ **A**, green and blue); and (3) the female sex hormones (estrogens; → p. 262 ff.) with 18 C atoms (→ **A**, red).

The starting substances for the synthesis of steroid hormones are present in all steroid hormone glands; which hormone is in fact produced depends on (1) which *receptors* for the controlling tropic hormone (ACTH, FSH, LH, etc) are present and (2) which *enzymes* for changing the structure of the steroid molecule predominate in the endocrine cell in question.

The adrenal cortex contains **17-, 21-**, and **11-hydroxylases** (enzymes which introduce a hydroxyl group into the steroid molecule at the carbon atom indicated by the number). Hydroxylation at C atom 21 (→ **A, c**) renders the steroid immune to attack by the 17-hydroxylase, so that only (e.g. in the *zona glomerulosa* of the adrenal cortex) **mineralocorticoids** (→ p. 150), that is, **corticosterone** and **aldosterone** (→ **A, d** and **e**) can be formed. If the C 17 is hydroxylated first (→ **A, f** or **g**), synthesis can continue either to **glucocorticoids** (*zona fasciculata*; → **A, h-j-k**) or the **17-ketosteroids** (keto group at C 17; *zona reticularis*; → **A, l** and **m**). Both groups of hormones can also be synthesized from **17α-OH-pregnenolone**, bypassing progesterone (glucocorticoids: → **A, g–n–h** etc; 17-ketosteroids; → **A, g–m** or **g–n–l**).

A direct pathway leads from the 17-ketosteroids to the two estrogens (→ p. 266) **estrone** and **estradiol** (→ **A, o–p**) or an indirect route via the androgenic hormone **testosterone** (→ **A, q–r–p**). In some target cells for androgens (e.g. the prostate) **dihydrotestosterone** or **estradiol** are the effective hormones. Both are formed from testosterone (→ **A, s** or **r**, respectively).

17-ketosteroids are formed in the gonads (testes, ovaries) and in the adrenal cortex. They also appear in the urine, where their presence forms the basis of the *metopirone* (*methopyrapone-*) *test* for ACTH reserves. Normally, ACTH release is controlled (feedback) by the glucocorticoids (→ p. 260). Metopirone inhibits the 11-hydroxylase (→ **A, d** and **j**), which means that in healthy subjects it leads to the deinhibition of ACTH release and more 17-ketosteroids are produced. If this does not happen, and the adrenal cortex is healthy, the conclusion can be drawn that there is a disturbance in ACTH release.

Breakdown of the steroid hormones takes place mainly in the **liver**. Usually, they are coupled to sulfate or to glucuronic acid by their hydroxyl group (→ p. 130 and p. 214) and subsequently excreted in the urine or bile. Estrogens are chiefly excreted as *estriol*; gestagens (progesterone, 17-α-OH progesterone), as *pregnandiol*. The estimation of the latter substance in the *urine* serves as a *pregnancy test* (→ p. 268). An elevated estrogen level in the *male* (normal values, see table on p. 266), for example, due to decreased estrogen breakdown (liver damage), leads to the development of breasts (*gynecomasty*).

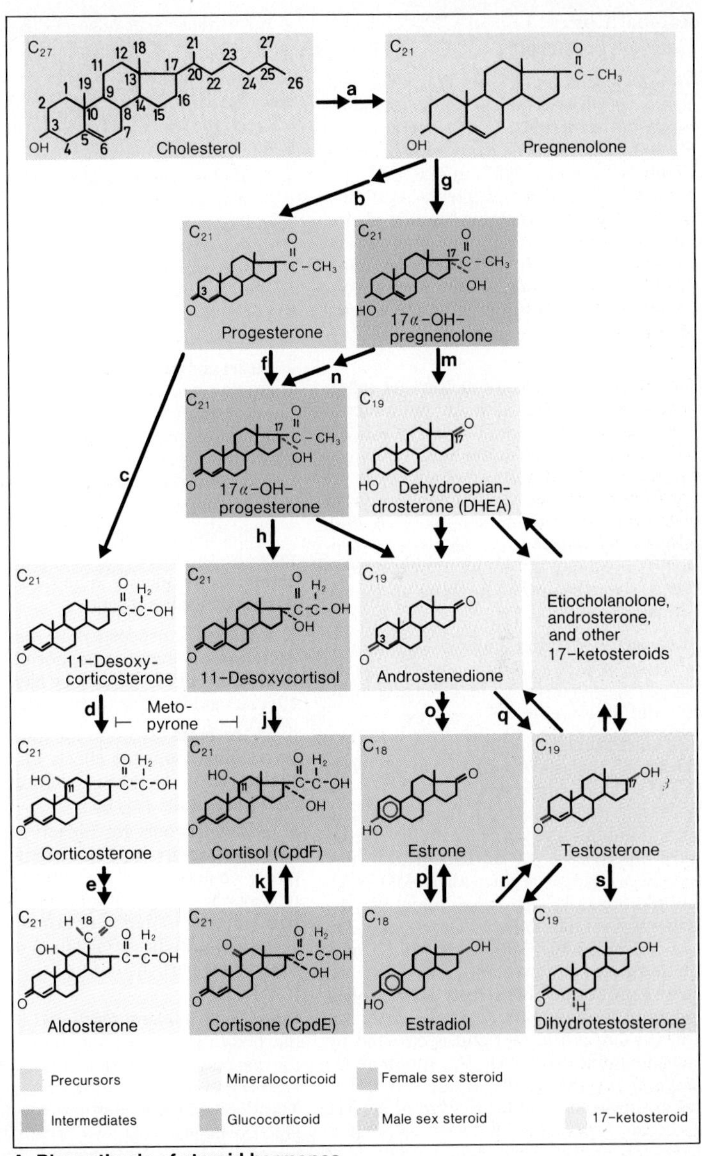

A. Biosynthesis of steroid hormones

Adrenal Cortex: Glucocorticosteroids

The *zona glomerulosa* (→ **A**) of the adrenal cortex produces the **mineralocortico(stero)ids** (aldosterone, corticosterone; → p. 150 and p. 259), the primary function of which is to retain Na^+ in the body (→ p. 140). The *zona fasciculata* (→ **A**) synthesizes chiefly the **glucocortico(stero)id cortisol** (hydrocortisone), and, to a small extent, cortisone (→ p. 259); the *zona reticularis* is the main site of production of the anabolic (tissue building) **androgens of the adrenal cortex** (dehydroepiandrosterone, etc; → p. 268 ff.).

The physiological role of the **adrenal androgens** is still unclear. Inborn pathological deficiency of 11- or 21-hydroxylase (→ p. 258) in the adrenal cortex is accompanied by an increased production of androgens by the adrenal cortex, one of the effects of which is a *virilization* in females (→ p. 270). Athletes are sometimes administered synthetic, anabolically active steroids (*anabolica*) with the aim of stepping up muscle formation.

Transport of glucocorticoids in the blood is achieved by binding to *transcortin* (a specific transport protein with a high binding affinity) and to albumin.

The hormones responsible for regulating the **release of glucocorticoids** are **corticoliberin (CRH)** and **corticotropin (ACTH**; → p. 234 ff.). ACTH (split off from POMC in the anterior pituitary lobe; → p. 240), stimulates the release of adrenocortical hormones, primarily the glucocorticoids (→ **A**). In addition, ACTH helps maintain the structure of the adrenal cortex and the availability of the substances from which the hormones are synthesized (cholesterol, etc; → p. 258). **ACTH release** is controlled (*negative feedback*) by cortisol (partly via **CRH**; → **A** and p. 238) and is elevated by catecholamines (→ **A**). A spontaneous *night-day* rhythm in the release of CRH is responsible for a similar rhythm in ACTH and cortisol release (→ **B**, "average value"). Measurements of the hormone level made at short intervals show that the release of ACTH and thus of cortisol takes place in 2- to 3-hourly episodes (→ **B**, "broken" curve).

Receptor proteins (→ p. 244) for **glucocorticoids** have been found in nearly every organ. Clearly, the (essential) glucocorticoids have a wide variety of **effects**:

1. **Carbohydrate and amino acid metabolism** (→ see also p. 247, A and p. 249, C): cortisol raises the glucose concentration in the blood ("*steroid diabetes*"), for which *amino acids* are required. These are provided, in turn, by protein metabolism (i.e. *catabolic effect* of the glucocorticoids). As a result, the *excretion of urea* is also increased (→ p. 146).

2. **Heart and circulation:** the glucocorticoids strengthen the action of the heart and lead to peripheral vasoconstriction, in both cases via *reinforcement of the catecholamine effects* (→ p. 176). Glucocorticoids also lead to increased production of epinephrine and angiotensinogen (→ p. 152).

3. In the **stomach**, the production of gastric juice is enhanced by the glucocorticoid action. Consequently, high doses bring the danger of gastric ulcers.

4. **Kidneys:** glucocorticoids delay water excretion and maintain the normal GFR (→ p. 124). At higher doses they have the same effect as aldosterone (→ p. 150).

5. In the **brain**, a rise in the glucocorticoid level, apart from the effects on the hypothalamus (→ **A**), causes alterations in the EEG and in the psyche.

6. Glucocorticoids (in higher doses) have **antiinflammatory** and **antiallergic** effects, partly on account of the inhibition of protein synthesis and lymphocyte formation, partly due to inhibition of histamine release (→ p. 72), and the stabilization of the lysosomes that participate in phagocytosis (→ p. 66).

Stress leads to **alarm reactions** (→ p. 290) in the body. The catecholamines released in the process (→ p. 58) increase release of ACTH and hence of glucocorticoids (→ **A**). Many of the above-listed effects of the glucocorticoids thus may form part of the alarm reaction (mobilization of energy metabolism, increased cardiac efficiency, etc).

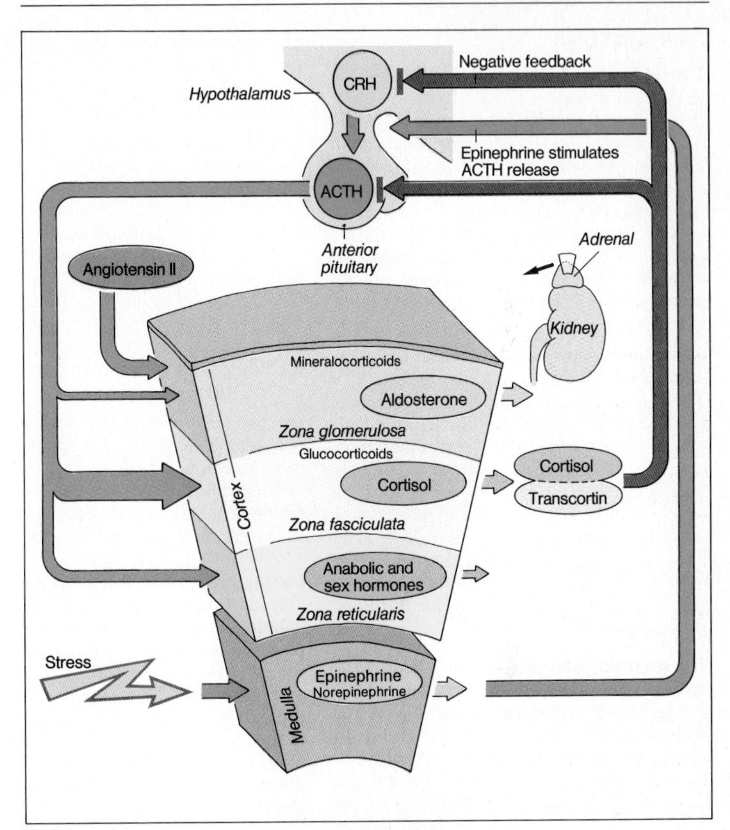

A. Adrenal

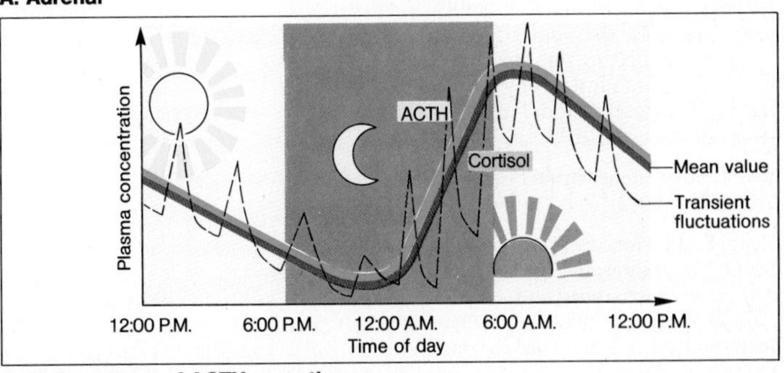

B. Daily pattern of ACTH secretion

Menstrual Cycle

The release of *follitropin* (**FSH**), *lutropin* (**LH**), and *prolactin* (**PRL**) from the anterior lobe of the pituitary gland is relatively *constant* (and low) in the *male*, whereas in *females*, following puberty, the release of these hormones follows a cyclic pattern (frequency about 1/month). Female sexuality is thus subject to periodic hormonal control. Characteristic of the **menstrual cycle** in women is the recurrent vaginal **menstrual bleeding** at approximately monthly intervals.

Several **hormones** are involved in the menstrual cycle (see also p. 234ff. and p. 264): The gonadotropic **releasing hormone FSH/LH-RH** or **gonadoliberin (Gn-RH)** and *prolactostatin* (**PIH = dopamine**) regulate the release of the anterior pituitary hormones **FSH, LH**, and **prolactin**. The latter mainly influences the hypothalamus and the *mammary glands* (→ p. 264); FSH and LH govern the **ovarian hormones**, that is, primarily the *estrogenic hormone* **estradiol** (**E_2**; → p. 266) and the *gestagenic* hormone **progesterone** (**P**; → p. 267).

The **length of the cycle** may vary from **21** to **35 days**. The second, **secretory phase** of the cycle, or **corpus luteal phase** (→ **A**) is relatively uniform in duration (about **14 days**) whereas the first, **proliferative phase** or **follicular phase** (→ **A**) is variable and can last from **7–21 days**. The duration of the follicular phase depends on the time required for maturation of the follicle. Unlike many animals, the "clock" governing the cycle is situated in the ovary. During the cycle (apart from other physical and psychical adjustments) periodic changes affect the *ovary, uterus*, and *cervix* as follows (→ **A**):

Day 1: onset of **menstrual bleeding** (duration about 2–6 days).

Days 5–14 (variable, see above): the **follicular** or **proliferative phase** begins at the end of menstruation and continues until ovulation takes place. Uterine mucosa (endometrium) is built up for reception of the fertilized ovum, that is, it is prepared for pregnancy. During this phase increasing secretion of FSH stimulates development and maturation of a single ovarian **follicle** (→ **A**), which produces increasing quantities of *estradiol* (E_2; → p. 266). The **cervical os** is small and closed by a highly viscous mucus.

Day 14 (variable, see above): **ovulation**. The E_2 production of the follicle rises steeply from about the 13th day (→ **A** and p. 264) and the resulting highly increased release of LH leads to ovulation. Shortly after ovulation the *basal body temperature* (measured in the morning, before rising) rises by about 0.5 °C (→ **A**). During ovulation the cervical mucus is less viscous (It can be drawn out into long threads: "Spinnbarkeit"). The os is enlarged to receive sperm (see also p. 266 and p. 270).

Days 14 to 28: the **luteal** or **secretory phase** is characterized by the development of the ovarian **corpus luteum** and of the secretory endometrium of the uterus (→ **A**). The endometrial glands become tortuous and the arterioles spiral, and the glands produce a secretion. The strongest reaction of the uterine mucosa to progesterone is seen on the 22nd day, which is the day on which **nidation** would occur. If this does not take place, E_2 and P bring about an inhibition of the *Gn-RH* release (→ p. 264), which leads to degeneration of the corpus luteum. The rapid drop in the E_2 and progesterone levels in the blood cause constriction of the endometrial blood vessels and ischemia; the uterine mucosa is shed, that is, **menstrual bleeding** begins.

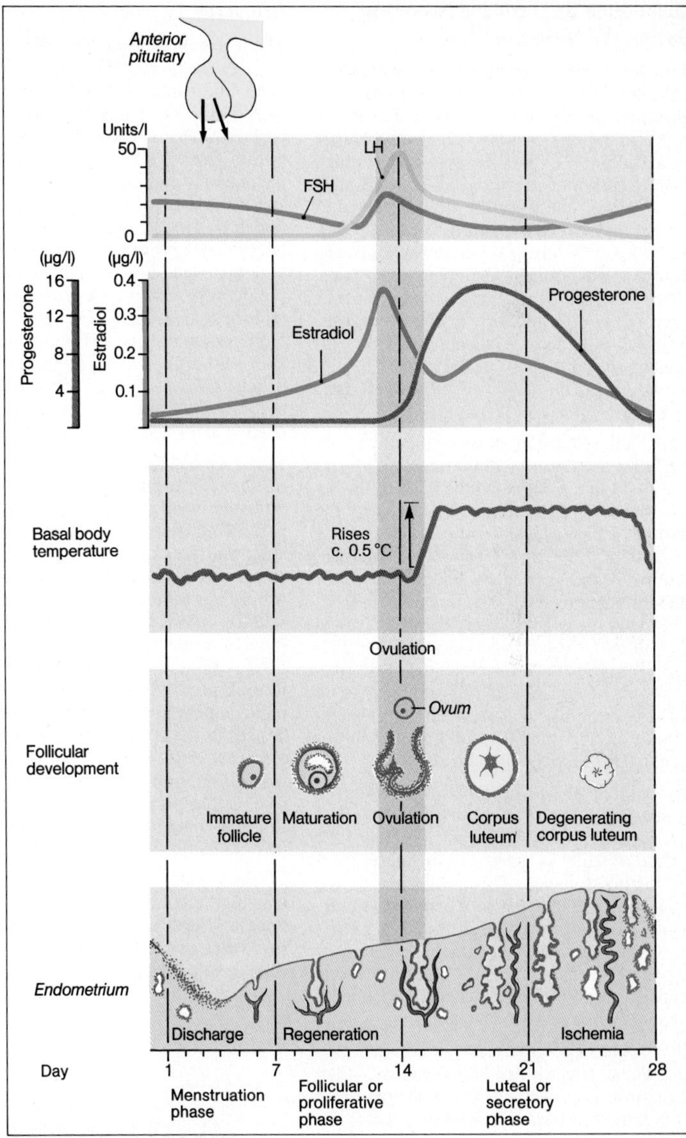

A. Human menstrual cycle

Regulation of Hormone Secretion during the Menstrual Cycle

In the female, **gonadoliberin** or **Gn-RH** enhances the release of **FSH** and **LH** from the anterior lobe of the pituitary. Gn-RH is released in bursts at intervals of 1.5 h before ovulation and of 3–4 h thereafter. A much faster ryhthm or continual release considerably reduces the secretion of FSH and LH (infertility). Since the amounts of FSH and LH released, *relative to one another*, change continuously during the menstrual cycle, there must be other factors that influence their release. In addition to **central nervous influences** (including psychic effects), **estradiol** (**E_2**) plays a special role. This E_2 action in turn is modified by **progesterone** (**P**).

During the **follicular phase** of the menstrual cycle the secretion of **LH** remains relatively low (→ **A** and p. 262). On the **12th–13th day** (→ **A**) the E_2 production, enhanced by the action of FSH, leads to stimulation of the release of FSH and LH, which, in turn, stimulate the release of E_2 (and later of progesterone as well). Via this **positive feedback loop** (→ p. 238), a very **high LH** level in the blood is rapidly established (→ p. 262) and initiates **ovulation** on the **14th day** (variable, → p. 262). If this sudden rise im LH does not take place, or if it is too small, ovulation and therefore pregnancy cannot occur (*anovulatory infertility*).

In the **luteal phase** of the menstrual cycle (→ **A, day 20**), **E_2** and **P** have an *inhibitory* effect on the secretion of FSH and LH, one of the results of which is to prevent the maturation of further follicles. This inhibition of the release of gonadotropic hormones by E_2 and P in the second half of the cycle results in a *negative feedback*, that is, secretion of E_2 and P drops towards the end of the cycle and on the 26th day it sinks drastically, probably initiating menstrual bleeding.

If estrogens and gestagens are administered *together* during the first half of the cycle then ovulation is prevented, the principle upon which most **ovulation inhibitors** ("**pill**") are based.

Prolactin

The secretion of **prolactin** (**PRL**) is inhibited by *prolactostatin* (**PIH**), identical with **dopamine**, and stimulated by *thyroliberin* (**TRH**; → p. 235 f.). Opinions differ as to whether a **prolactin-release hormone** exists. E_2 and P inhibit the release of PIH (→ **A**) so that, particularly during the second half of the cycle and during pregnancy, more PRL is released. In the female, PRL (and other hormones) promotes **breast development** and **lactogenesis** (milk formation) during pregnancy. The *sucking stimulus* via the maternal mamilla is the signal for a very high production of PRL during lactation. (For *milk let-down*, ocytocin is required; → p. 240 f.). An additional effect of PRL is that in both male and female it increases the release of PIH (negative feedback).

Stress and *drugs*, for example morphine, reserpine, phenothiazine, and some tranquilizers, inhibit the release of PIH, thus permitting release of PRL. Abnormally large amounts of PRL in the blood (**hyperprolactinemia**) may also be due to a PRL-producing tumor or to **hypothyroidism** (→ p. 252), where the increased TRH in the blood stimulates the release of PRL. In the *female*, hyperprolactinemia leads not only to milk production (independently of pregnancy; *galactorrhea*), but also to disturbances in the cycle: *amenorrhea* and *anovulatory infertility*. This condition acts as a natural method of birth control for primitive peoples, since they usually breast feed for years on end (see above) and are therefore (usually) infertile during this time.

In the *male*, whose PRL plasma level is usually the same as that of the nonpregnant female, hyperprolactinemia leads to disturbances in the function of the testes and thus to *impotence*.

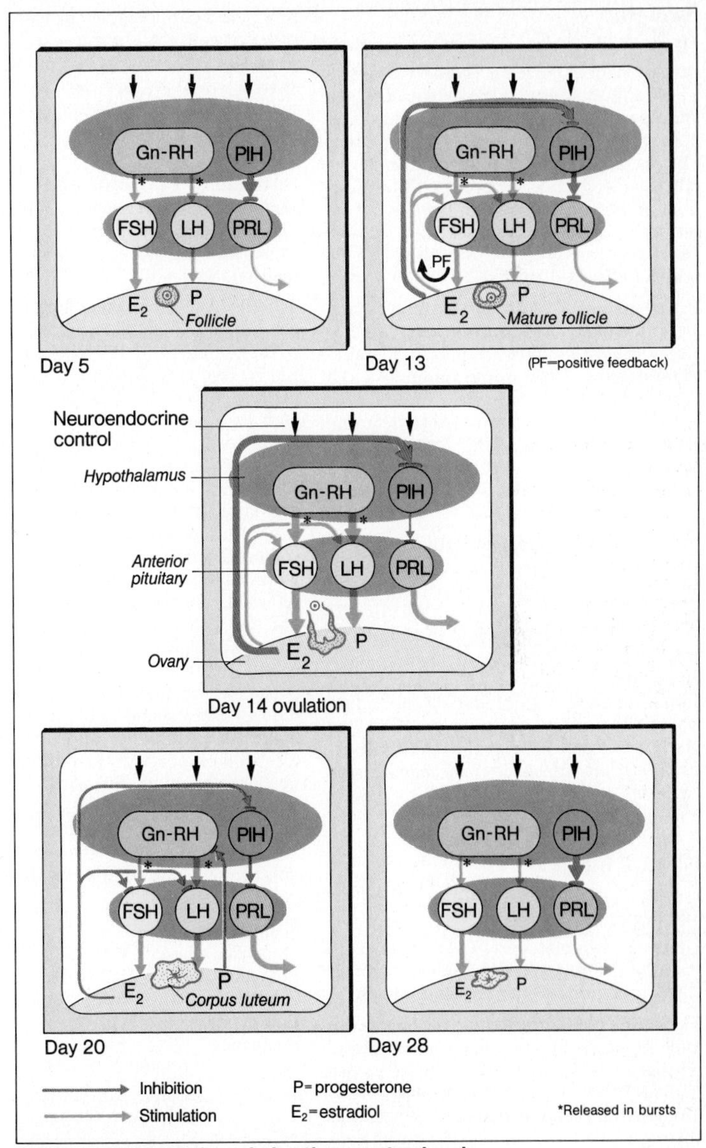

A. Hormonal interactions during the menstrual cycle

Estrogens

Estrogenic hormones are responsible for the development of the female sexual characteristics (although not to the extent to which androgens are responsible for sexual development in the male; → p. 270). Estrogens also promote the development of the uterine mucosa (→ p. 262), the process of fertilization, and so on. The preparatory action of estrogens is in many cases (e.g. uterus) necessary for progesterone to exert its optimal effect (→ p. 267).

Estrogens are C 18 steroid hormones and are principally derived from the 17-ketosteroid *androstenedione* (→ p. 259). The **sites of formation** are the *ovaries* (granulosa and theka cells), the *placenta* (→ p. 268), the *adrenal cortex*, and the Leydig *interstitial cells in the testes* (→ p. 270). Testosterone is converted into estradiol in some of the testosterone target cells, where it then exerts its effect.

In addition to the most important estrogen, **estradiol (E_2)**, *estrion* (E_1), and *estriol* (E_3) also possess estrogenic activity, even if weaker (relative activity $E_2 : E_1 : E_3 = 10 : 5 : 1$). For **transport** in the blood, E_2 is bound to a specific protein. The main **breakdown** product of E_2 is E_3. Orally administered E_2 has practically no action whatever since it is largely removed from the blood by its first passage through the liver. In order to be effective, *orally administered estrogens* must therefore have a slightly different structure.

Effects of Estrogens:

Ovary: E_2 promotes maturation of *follicle* and *egg* (→ p. 262 ff.).

Uterus: E_2 stimulates *proliferation* (development) of the *uterine mucosa* and strengthens the contractions of the uterine muscles.

Vagina: E_2 causes thickening of the mucosa and increases the sloughing of epithelial glycogen-containing cells. The glycogen permits more lactic acid production by Döderlein bacteria, which lowers the pH of the vagina to 3.5–5.5, thus decreasing danger of infection.

Cervix: the cervix and its *plug of mucus* represent a considerable barrier to the entry of sperm into the uterus (→ p. 267). Estrogen changes the consistency of the mucus in such a way that, especially at the time of ovulation, the *migration of sperm* is facilitated and their *survival time* is increased (→ p. 262).

Fertilization: E_2 controls the *speed of migration of the ovum* along the Fallopian tube and prepares the sperm (in the female organism) for penetration of the ovum (*capacitation*).

The effects of E_2 on **other hormone cells:** E_2 acts on the "higher" hormone organs (→ p. 265), modulating the release of FSH and LH by feedback. **Blood:** estrogens raise the *clotting power of the blood*, which brings with it a somewhat increased danger of thrombosis connected with the "pill". **Salt and water balance:** E_2 leads to retention of water and salt in the kidneys and locally. The latter leads to local edema (→ p. 158), which has been exploited as the basis of some cosmetic creams intended to eradicate wrinkles. **Bone:** the rate of linear growth is retarded, epiphyseal closure is hastened, and the osteoblast activity (→ p. 256) is enhanced. **Lipid metabolism:** atherosclerosis is less frequent in females before menopause, which may be partly because E_2 reduces the plasma cholesterol. **Skin:** E_2 makes the skin softer and thinner, causes recession of sebaceous glands and increases the amount of fat deposited beneath the skin. **CNS:** estrogens affect sexual and social behavior, and psychic reactions.

Average secretion rates of estradiol (mg/d)

Females	
Menstrual phase	0.1
Follicular phase	0.2–0.3
Ovulation	0.7
Luteal phase	0.3
Pregnancy	8–15
Males	*0.1*

Gestagens

By far the most potent gestagenic hormone is **progesterone (P)**. It is mainly produced during the **secretory** or **luteal phase** of the menstrual cycle. The **chief function** of P is to prepare the genital tract of the female for the reception and maturation of the fertilized ovum and to maintain pregnancy.

Progesterone is a steroid hormone with 21 C atoms (→ p. 259). The **sites of formation** are the *corpus luteum*, the *follicle* (→ p. 264), the *placenta* (→ p. 268) and, in males as well, the *adrenal cortex*.

Biosynthesis of P proceeds from cholesterol via pregnenolone (→ p. 259). As with estradiol (→ p. 266), P is to a large extent broken down during its first passage through the liver, so that orally administered P is practically ineffective.

For its **transport** in plasma a specific *progesterone-binding protein* is employed. The main *breakdown product* of P is pregnandiol.

Actions of progesterone: for almost all of the effects of P a *preceding* or *simultaneous action of estradiol* (E_2) is *essential*. In the follicular phase of the menstrual cycle, E_2 increases the intracellular receptor proteins (→ p. 244) for P. Their numbers drop again during the luteal phase.

The **uterus** is the most important target organ for P. Following a preparatory effect of E_2, P stimulates the growth of the *uterine musculature* (*myometrium*), brings about the secretory alterations (→ p. 262) in the *uterine mucosa* (*endometrium*) that has been developed by the action of E_2, and also changes its glycogen content and vascular supply (i. e. changeover from proliferative to secretory endometrium). These alterations reach their peak on about the 22nd day of the cycle. At this point, P also plays a considerable role if *nidation* of a fertilized ovum occurs. *Prolonged action of progesterone* results in retrogression of the endometrium, which is then unsuitable for nidation (→ p. 262). P also reduces the myometrial activity, especially during pregnancy.

At the **cervix**, P alters the *consistency of the mucus plug* in such a way that it becomes almost impenetrable for sperm. This effect is the basis of the partially contraceptive action of P in the first phase of the cycle. In the **breast**, P promotes (together with prolactin, STH, and other hormones) the development of the milk ducts.

Influence of progesterone on **other hormone cells:** in the luteal phase P *inhibits the release of LH* (→ p. 264). If progesterone-like gestagens are also administered during the follicular phase, the inhibitory effect interferes to some extent with ovulation. Together with the effect on the cervix and an inhibitory action on the capacitation of sperm (→ p. 266), this has a *contraceptive* effect ("**mini pill**").

CNS: large doses of P produce anesthesia (via the breakdown product pregnenolone). P raises the threshold for epileptic seizures, has a *thermogenic effect* that leads to a rise in basal temperature at midcycle (→ p. 263), and is probably the reason for behavioral disturbances and depressions premenstrually and toward the end of pregnancy.

In the **kidneys** P inhibits aldosterone action (→ p. 140 and p. 150), which results in increased NaCl excretion.

Source	Progesterone	
	Secretion rate (mg/day)	Concentration (μg/l plasma)
Females, follicular phase	4	0.3
Luteal phase	30	15
Early pregnancy	90	40
Late pregnancy	320	130
First day postpartum	–	20
Males	0.7	0.3

Endocrinology of Pregnancy and Parturition

The **placenta** has several functions: to nourish the fetus and provide it with O_2 (→ p. 190f.), to carry away its waste products, and to meet the great hormonal needs of pregnancy. Especially at the beginning of pregnancy, the hormones of the maternal ovaries are also important (→ **A**).

The placenta produces the following **hormones**: *estradiol* (**E_2**), *estriol* (**E_3**), *progesterone* (**P**), *human chorionic gonadotropin* (**hCG**), a hormone known as **h**uman **c**horionic **s**omatotropin (**hCS**) or **h**uman **p**lacental **l**actogen (**hPL**), POMC (→ p. 240) etc.

As an endocrine organ, the human placenta has several *unique characteristics*; its hormone production is (probably) independent of the usual feedback control (→ p. 238); it synthesizes both steroid *and* protein hormones in the same tissue; the **protein hormone phase** (→ **A**) predominates during the 1st trimester, the **steroid hormone phase** (→ **A**) during the 2nd and 3rd trimesters.

The placenta secretes its hormones into two compartments, the fetal and the maternal. On account of the intimate connections between the hormone production in mother, fetus, and placenta, the term **fetoplacental unit** is also employed.

In contrast to other endocrine glands, the placenta cannot synthesize the steroid hormones P and E_2 unless the steroid precursors (→ p. 259) are provided by the maternal and fetal *adrenal cortex* (→ **A**). (At certain phases of fetal development this organ is larger than the kidneys.) Placental **progesterone** thus produced from *cholesterol* gives rise to *dehydroepiandrosterone* (*DHEA*) and other substances in the fetal adrenal cortex; from *DHEA* the placenta forms E_2, which is largely converted to *estriol* (E_3) in the fetal liver. *Progesterone* is converted to *testosterone* in the testes of the male fetus (→ p. 270).

The release of large amounts of the proteohormone **hCG** begins *immediately on implantation of the fertilized ovum* (→ **A** and **B**); its major roles are (a) to stimulate the production of DHEA and other steroids in the *fetal zone* of the fetal adrenal cortex and (b) to suppress the formation of follicles in the maternal ovaries and to maintain the function of the corpus luteum, that is, the production of P and E_2 (→ **B**). From the 6th week of pregnancy onwards this is no longer necessary because the placenta by this time can produce enough P and E_2 itself.

Most **pregnancy tests** are based on the biological or immunological detection of hCG in the urine (starting from the 6th–8th day after fertilization). The fact that the rate of secretion of E_2 and P rises drastically during pregnancy (→ tables on p. 266f.) and therefore increased amounts of E_2 and P and of their breakdown products, estriol and pregnandiol, are excreted in the urine of pregnant women, can also be used to diagnose pregnancy.

The concentration of **hCS** (identical with **hPL**) rises continuously during pregnancy. One suggestion is that hCS influences *mammogenesis* (development of the breasts) and that it regulates the synthesis of steroid hormones in the adrenal glands and placenta.

The **hormonal regulation of birth** is still not fully understood. It is assumed that at the end of pregnancy an increased release of ACTH in the fetus stimulates its adrenal cortex to secrete *cortisol* (→ p. 260), which in turn inhibits the placental production of P and promotes that of estrogens. The result is a depolarization of the uterine musculature, increased formation of *gap junctions* (→ p. 7), and an increase in the numbers of receptors for ocytocin and catecholamines (α receptors), i. e. reactions that enhance the excitability of the uterus. Stretch receptors in the uterus respond to increase in size and movements of the fetus, and the resulting nervous signals to the hypothalamus elicit a greater release of **ocytocin**, which in turn leads to enhancement of uterine contraction (positive feedback). The name "oxytocin" has been replaced, therefore, by the more accurate name ocytocin (from Greek ὠκυτόκος = quickly bearing). In addition, ocytocin raises the production of **prostaglandins** (→ p. 235) in the endometrium, which, on arrival in the myometrium also have an activating effect on it. The *gap junctions* ensure that the spontaneous excitation of the individual pacemaker cells in the fundus can spread in a "concerted" manner over the entire myometrium (about 2 cm/s) (→ p. 44).

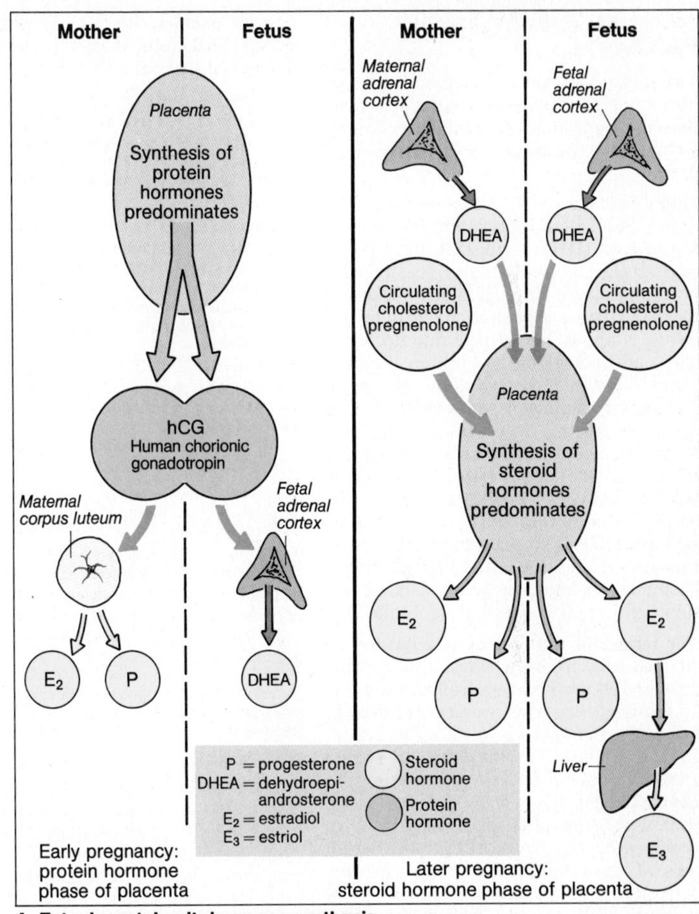

A. Fetoplacental unit: hormone synthesis

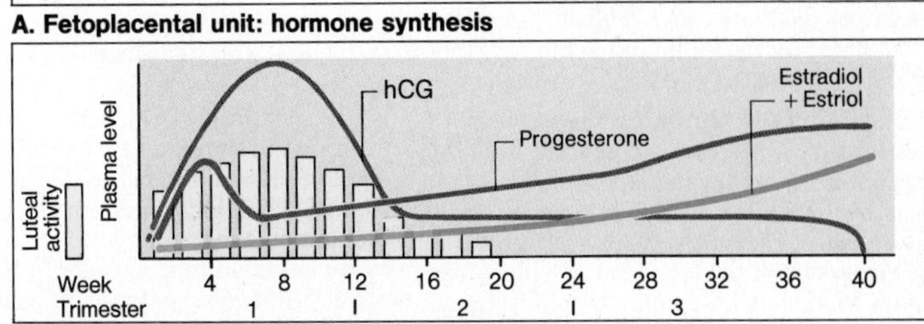

B. Plasma levels of hormones during pregnancy

Androgens, Testicular Function, Ejaculate

The most important androgen (male sex hormone) is **testosterone**. Its principal **functions** are to promote sexual differentiation, formation of sperm, and sexual drive in the male.

Androgens are *steroid hormones* with 19 C atoms. Besides *testosterone* and 5-α-*dihydrotestosterone* (**DHT**; → p. 259) the 17-ketosteroids (DHEA, etc) are also, strictly speaking, androgens, although much weaker. In the male (**testes**), the secretion rate of testosterone is about 7 mg/d (the rate decreases with advancing age), and in females (*ovaries, adrenal cortex*) about 0.3 mg/d. The corresponding plasma concentrations are about 7 and 0.5 μg/l, respectively. Testosterone, like other steroids, is bound to proteins in the blood: *testosterone-binding globulin*, **TeBG** (→ **A**). Testosterone can be converted into *estradiol* (E_2) (→ p. 259) as well as to DHT, so that, besides DHT, E_2 can also act as the intracellularly active hormone in some cases (e.g. in the Sertoli cells before puberty, and in the brain).

The **release of testosterone** is **regulated** by LH and FSH from the anterior lobe of the pituitary gland. Their controlling hormone — Gn-Rh — is, as in women, released in bursts (→ p. 264) in a 2- to 4-hourly rhythm. **LH** (also known in the male as **ICSH**) *stimulates release of testosterone* from the **Leydig (interstitial) cells** of the testes (→ **A**), while **FSH** stimulates the formation of an *androgen-binding protein* (BP) in the **Sertoli cells** of the seminiferous tubules (→ **A**). Testosterone inhibits the secretion of LH by negative feedback: an "**inhibin**" has been postulated for the inhibition of the release of FSH in the male (→ **A**).

In addition to the important effects of testosterone on sexual differentiation of the male, on spermatogenesis and spermiogenesis (see below) and on growth of prostate and seminal vesicles, it also governs the development of the *secondary sex characteristics* in the male: genital growth, size of larynx (voice breaks), facial hair, pubic and axillary hair, skin thickness, and sebaceous vesicles (acne). Sufficient testosterone secretion is essential for the development of normal **libido** (sexual activity), **fertility**, and **potency** of the male.

Testosterone *stimulates blood formation* (→ p. 60 ff.) and has an **anabolic** effect (see also p. 260), one of the results of which is the more strongly developed musculature of the male. In the **CNS**, testosterone also has an effect on *behavior* (aggressiveness, etc).

Sex differentiation. The **genetic (chromosomal) sex** (→ **B**) determination of the embryo is followed by the development of the specific *gonads* (reproductive glands), which are later invaded by the *primary germ cells*. The subsequent **somatic sex development** and **differentiation** is female in the absence of testosterone (→ **C**). For male sex development the presence of testosterone at both stages is essential (→ **C**), and some steps (e.g. descent of the testes into the scrotum) require a further, unidentified factor. Overproduction of androgens or the administration of testosterone may produce *virilization* of the female organism (→ **C**).

Functions of the testes: in addition to producing testosterone, the gonads produce the male germ cells (**sperms**) via a series of developmental stages (early steps: *spermatogenesis*; late steps: *spermiogenesis*). **Spermiogenesis** takes place in the *seminiferous tubules*, which are strictly protected from the surroundings by the *blood-testes barrier* (i.e. the Sertoli cells). The testosterone required for the *development of sperm in the testes and epididymus* can only cross this barrier and can only be taken up into responsive cells in the reproductive tract if *bound* to androgen-binding protein (BP; → **A**).

The **ejaculate** (semen fluid; 2–6 ml) contains 35–200 million sperms/ml as well as the sperm plasma, which is produced by the prostate gland and the seminal gland; it also contains fructose as nutrient for the sperms and prostaglandins to promote uterine contraction. At the climax of sexual excitation (**orgasm**) the sperm is conveyed reflexly to the posterior urethra (*emission*), whose distension leads to reflex contraction of, chiefly, the bulbocavernosus muscle and thus to expulsion of the ejaculate (**ejaculation**). The alkaline seminal plasma raises the pH in the vagina, which is essential for the motility of the sperms; the sperms have to travel as far as the fallopian tube in order to fertilize the egg cell (only one sperm cell is necessary for fertilization).

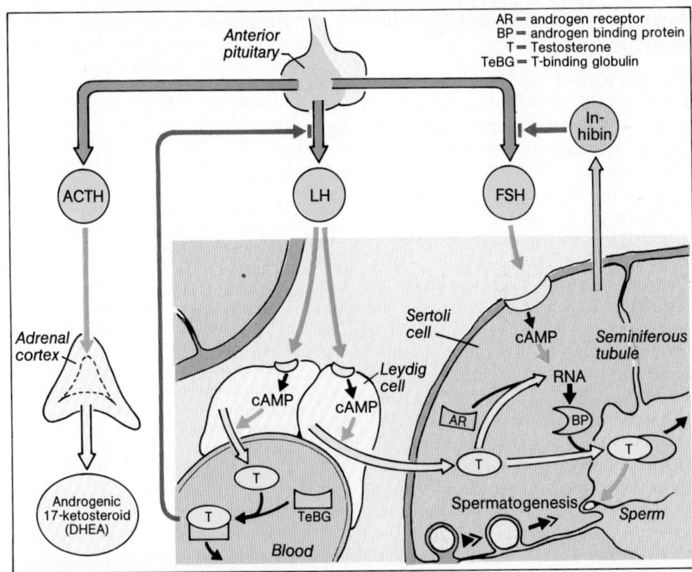

A. Control and transport of androgenic hormones. Effects of testosterone.

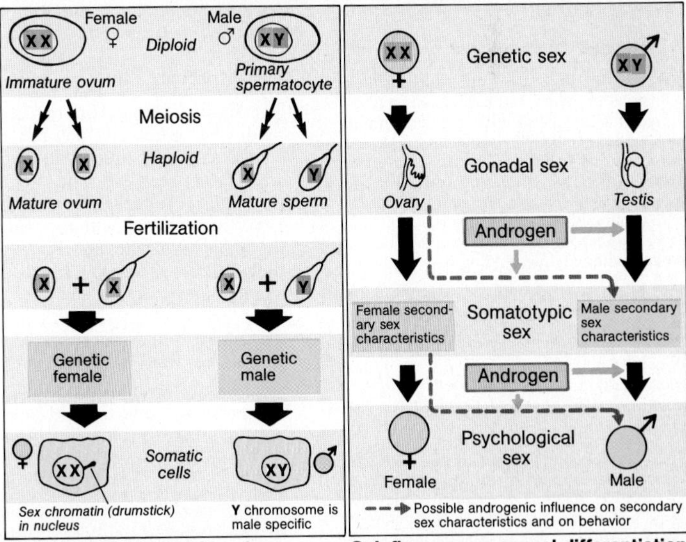

B. Genetic determination of sex

C. Influences on sexual differentiation

Anatomy of the Central Nervous System (CNS)

The peripheral nervous system (i.e. the somatic and the autonomic nervous systems) conveys the information input (afferent) to the CNS and transmits information away from it to the periphery (efferent). The role of the CNS is to scan, evaluate (e.g. comparing with stored information), and process the information received and to respond with efferent impulses. The CNS is thus an organ with integrating and coordinating functions. It consists of the **brain** and **spinal cord** (→ **A**). The latter can be divided into *segments* corresponding to the vertebrae, although it is shorter than the vertebral column (→ **A**); nevertheless, the **spinal nerves** (→ **B**) leave the vertebral canal at the level of their corresponding vertebra. A spinal nerve contains *afferent fibers* that enter the spinal cord at the **dorsal root**, and *efferent fibers* that leave the spinal cord for the periphery at the **ventral root** (→ **B**). A **nerve** is therefore a bundle of nerve fibers running in opposing directions and fulfilling different functions (→ p. 22).

A **cross section of the spinal cord** (→ **B**) contains two fields. First, a central, darker, butterfly-shaped area, the *gray matter*, composed chiefly of the nerve cell bodies of the efferent pathways (mainly to the musculature: *motoneurons*) in the *ventral* (anterior) *horn*, and the cells of the *interneurons* in the *dorsal* (posterior) *horn*. The cell bodies of most of the afferent fibers lie *outside* the spinal cord in the corresponding *spinal ganglion*. The rest of the cross-section of the spinal cord is occupied by the *white matter*, composed predominantly of axons of the ascending and descending pathways.

The **brain** is a specialized extension of the spinal cord. In ascending sequence it is composed of the *medulla oblongata* (→ **E 1**), the *pons* (→ **E 2**), *mesencephalon* (→ **E 3**), *cerebellum* (→ **C, E**), *diencephalon* and *telencephalon* (cerebral hemispheres, → **C, D, E**).

The first three components make up the **brain stem** (similar construction to the spinal cord), which contains the cell bodies (nuclei) of the *cerebral nerves* and, among other things, the regions ("*centers*") controlling *respiration* (→ p. 104) and *circulation* (→ p. 176ff.). The brain stem as well as the **cerebellum** play important roles in motor control (→ p. 284ff.).

The **diencephalon** includes the *thalamus* (→ **D 4**), an important switchboard for all afferent sensory input (skin, eyes, ears, etc., and from other parts of the brain), and the *hypothalamus* (→ **D 5**), a center for autonomic functions (→ p. 290). The latter plays an important role in the integration of the nervous system with the endocrine system (→ p. 240) via the hypophysis (→ **E 6**), to which it is attached.

The **telencephalon** consists of **nuclei** and **cortical regions**. The former include the **basal ganglia**, which are important for motor control: *nucleus caudatus* (→ **D 7**), *putamen* (→ **D 8**), and *globus pallidus* (→ **D 9**) and also, in part, the *corpus amygdaloideum* (→ **D 10**). The latter, together with other brain components (e.g. cingulate gyrus; → **E 11**), belongs to the **limbic system** (→ p. 290). The external surface of the telencephalon, the **cerebral cortex**, is divided by *sulci* or grooves (e.g. the sulcus centralis; → **C 12, D 12, E 12**, and the sulcus lateralis; → **C 13**) into four major *lobes* (→ **C, D, E**): frontal, parietal, occipital, and temporal lobes. The two hemispheres are intimately connected with one another by the *corpus callosum* (→ **D 14** and **E 14**).

The cortex is the site of origin of all conscious and many subconscious actions; it is the collecting and processing station for sensory impressions, sensations, and perception, and is the seat of memory, etc.

Cerebrospinal Fluid

The brain is bathed in the cerebrospinal fluid, **CSF**, a nutritive and protective fluid, which also fills the internal cavities or *ventricles* of the CNS. The lateral ventricles (→ **D 15, F**) are connected with the IIIrd and IVth ventricles and with the *central canal* of the spinal cord. The *choroid plexus*, a collection of blood vessels and tissue (→ **D 16, F**), secretes approximately 650 ml CSF daily. The CSF exits from the IVth ventricle to fill the subarachnoid space, and is reabsorbed via the *arachnoid villi*.

The exchange of substances between blood and CSF or blood and brain, apart from H_2O, CO_2, and O_2, is strictly controlled (*blood-CSF-* and *blood-brain barrier*): certain substances such as glucose or amino acids are transported by special mechanisms, whereas others (e.g. proteins) are unable to cross the barrier; this should be considered when administering drugs. Obstruction of the outflow of the CSF leads to its accumulation and to brain compression (*hydrocephalus* in infants).

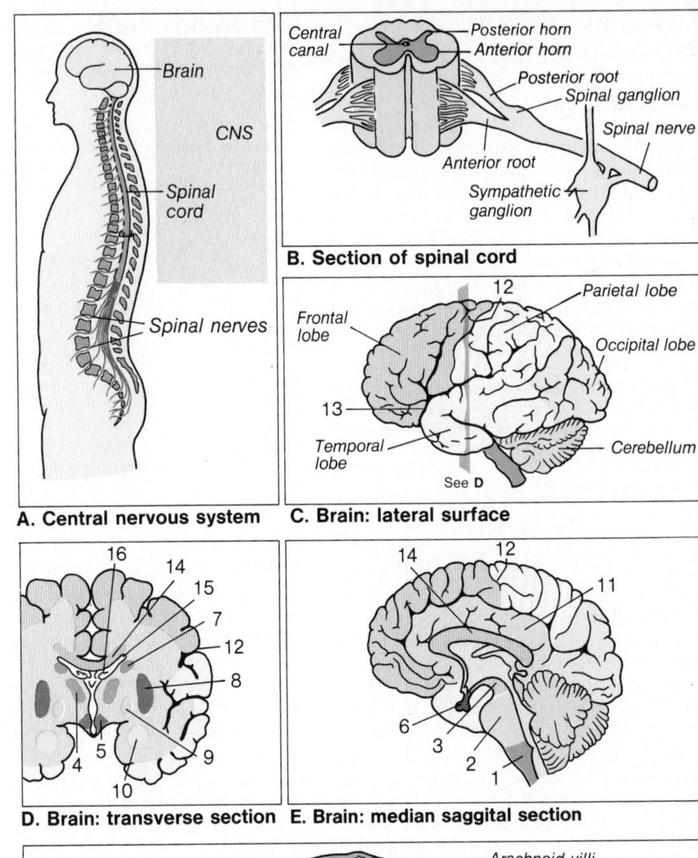

A. Central nervous system

B. Section of spinal cord

C. Brain: lateral surface

D. Brain: transverse section

E. Brain: median saggital section

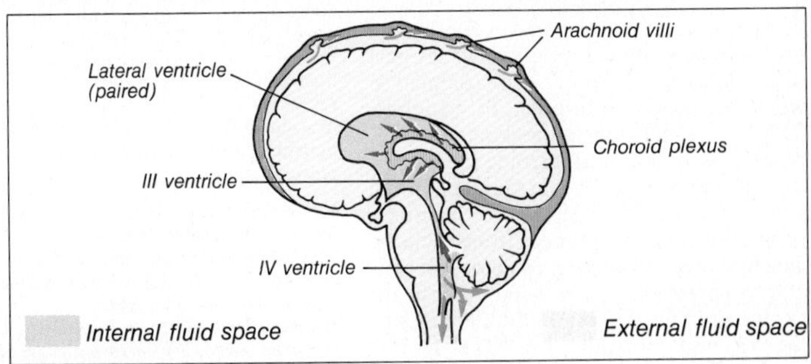

F. Brain: fluid spaces

Processing of Informational Input

We receive large quantities of **information** (10^9 bit/s) from the surroundings via our **sense organs**, although only a very small proportion of this information (10^1–10^2 bits/s) is consciously registered; the remainder is either subconsciously processed or not used at all. In other words, information that is important ("interesting") to the cerebral cortex (consciousness) is *selected* (especially noticeable in the cases of intent listening or watching). Conversely, we *deliver* information to the extent of about 10^7 bits/s to our surroundings by means of *speech* and *motor activity* (e.g. *mimicry*; → **A**).

The unit of information content is the binary digit or **bit** (→ **A**). Flow of information is a time function (bits/s). One printed letter contains 4.5 bits; a book page 1000 bits. If a page is read in 20 seconds, intake is 1000/20 = 50 bits/s. A television picture conveys more than 10^6 bits/s.

Stimuli reach the body in *various energy forms* (electromagnetic energy in the case of visual stimuli, mechanical energy in touch, and so on). Each type of stimulus has its own **specific receptors**, which may be collected together into **sense organs** or scattered over the surface (skin receptors) or interior of the body (pressure receptors, etc.). Each type of sensory cell is spezialized to register its own **adequate stimulus** leading to specific sensory impressions (*modality*). In many cases different *qualities* can be distinguished within one modality (e.g. sound level and sound frequency in hearing; → p. 316ff.).

The **reception of the stimulus** (→ **B**) involves the "*selection*" by the receptor of that part of the information from the environment to which it is suited to respond, for example, the pressure receptors in the skin (→ p. 276) register information concerning the magnitude of the pressure. Within the receptor the stimulus causes changes in the memrane properties of the receptor cell (**transduction**), which in turn provoke a **receptor** (or generator) **potential** (*local response*): the stronger the stimulus the greater the *amplitude* of the receptor potential (→ **C 1**). If this reaches a certain **threshold value (→ B 1)**, an **action potential**, **AP**, is set off (→ p. 26ff.) and transmitted along the nerve fiber: **transformation** of the stimulus. The *stronger* the stimulus and the *higher* the receptor potential, the *greater the frequency* with which APs are fired and passed on (→ **C 2**).

Thus, the original information is **encoded in the form of the frequency** (impulses/s) of the APs. The information is *decoded* at the next *synapse* (→ p. 30); the greater the frequency of the APs arriving at the synapse, the larger the quantity of transmitter substance released (→ e.g. p. 54) and the higher is the *excitatory postsynaptic potential* (→ p. 30). When this potential reaches *threshold* (→ **B 2**), APs are again fired and transmitted, that is, the information is *recoded*. Encoding as the frequency instead of as the magnitude of the potential has the advantage that the information, which may have to travel as much as 1 m along the axon, loses less of its content in this form over the relatively large distances covered. On the other hand, for the reinforcement or weakening of the information at the synapse (by other neurons), the magnitude of the potential is more suitable, and therefore the information is *decoded* presynaptically. **Other methods of coding** are time of occurrence, temporal pattern of impulses or duration of burst, instant of firing, increment and decrement from background, and many others.

Information is modulated at the synapse, for example, due to its ability to elicit either **excitatory** or **inhibitory** impulses that influence the **contrast** of information (→ **D** and p. 312) on its way to and in the CNS. Stimuli being conducted in adjoining fibers are weakened in the process (*lateral inhibition*).

Objectively, the processing of sensory stimuli (e.g. by measuring cell potentials) can be traced as far as their integration in the CNS. The jump into consciousness, however, can only be observed **subjectively**. As a first step we are able to describe *sensory impressions* and *sensations* (e.g. "brown lines with small green dashes"). Experience and reasoning interpret the event, and *recognition* (of a fir twig) follows. Puzzle pictures show that one individual may even recognize (interpret) the same sensory impression in different ways.

Further important definitions in sensory physiology are: a) **absolute threshold** (→ e.g. pp. 296, 306 and 316), b) **difference thresholds** (→ e.g. pp. 296, 306 and 322), c) **spatial and temporal summation** (→ e.g. p. 306), d) **adaptation** (→ e.g. p. 306), e) **receptive field** (→ e.g. p. 312), f) **P-**, **D-** and **PD-receptors** (→ p. 276). These terms are applicable to the function of receptors in general, although on the pages cited they are only explained on the basis of specific examples.

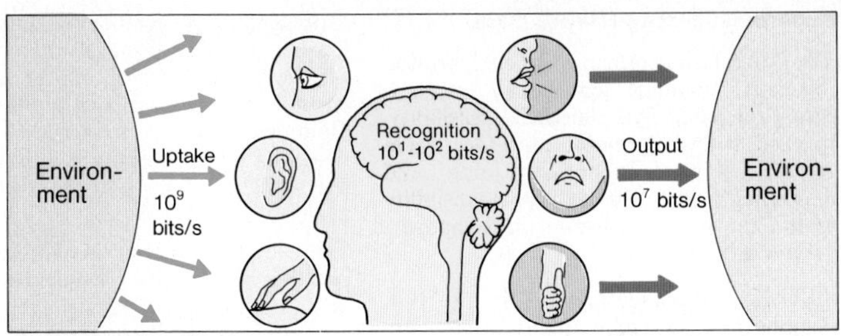

A. Information flow

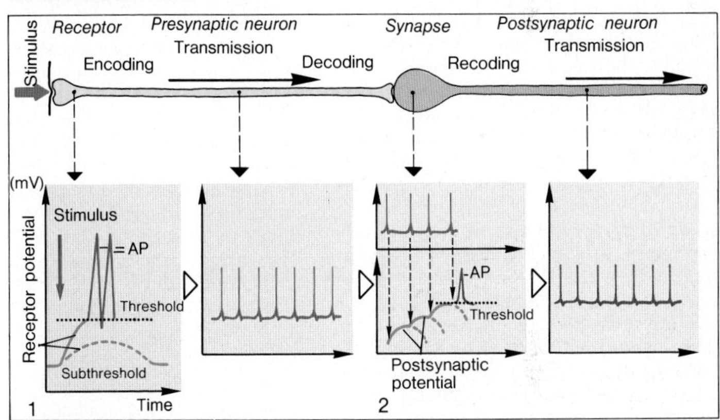

B. Processing of information; encoding

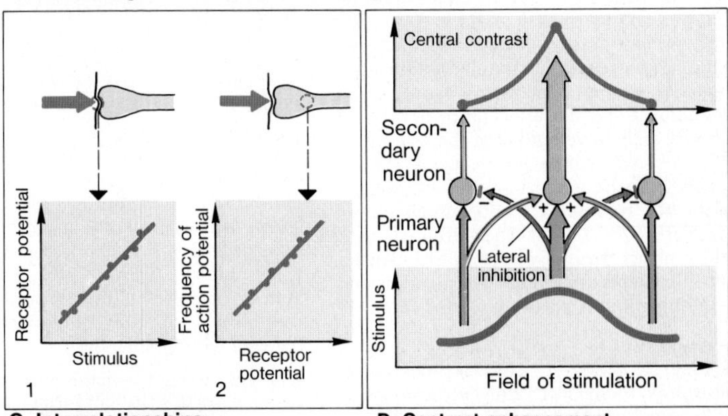

C. Interrelationships: stimulus, potentials

D. Contrast enhancement

Superficial Senses (Skin), Pain

The skin registers pressure, touch, vibration (collectively: tactile sensations), temperature, and pain. This **superficial sensibility**, together with the **proprioceptive** (deep) **sensations** (receptors in muscles, joints, and ligaments; → p. 278) and the **pain** sensibility within the body constitute the **somatovisceral senses**.

Specialized **mechanoreceptors** in the skin respond to three tactile qualities: *pressure, touch*, and *vibration*. If the *Merkel's cells* (→ **A 2**) or *disks* (→ **A 5**) in the skin are stimulated by the application of different weights, the frequency (impulses/s) of the APs in the nerve fibers leading away from them is *proportional* to the pressure (→ **B 1**). In other words, the *intensity* of the **pressure** is measured (slowly adapting **intensity detectors**). In the hairy skin, Merkel's cells form *dome corpuscles*. Deeper in the skin, other intensity receptors are found: *Ruffini corpuscles*. In contrast, *Meissner's corpuscles* (→ **A 1**) and the *hair follicle receptors* (→ **A 4**) respond to **touch**. In this case it is not the intensity (e.g. the degree of hair deformation) but the velocity at which the stimulus alters that is registered (moderately adapting **velocity detectors**), and the number of impulses is proportional to this rate (→ **B 2**). Because of their rapid adaptation, the *Pacinian corpuscles* (→ **A 3**) are specialized to register **vibrations**: they react to a *single* change in strength of stimulus with a *single* impulse, whatever the rate of change of the stimulus. If this rate alters (as in vibrations), that is, if the **acceleration** of the skin movements becomes greater or smaller, the frequency of the nerve impulses transmitted from the receptor is proportional to this acceleration (→ **B 3**). Apart from those in the skin, **acceleration detectors** are located in the ligaments, muscles, and joints, where they function as *proprioceptive detectors.*

Receptors of the intensity detector type are frequently referred to as **proportional**- or **P-receptors**, those of the rate detector type as **differential-** or **D-receptors**. A mixed form, the **PD**-receptors, measure, for example, the position of joints (part of the proprioceptive sensations): the rate of change of position is expressed in the temporary, high impulse frequency (→ **D**; peaks of the curves), the final position of the joint in the subsequent constant frequency of the impulses (→ **D**). Opinions differ as to whether this PD reception takes place in the joint itself or in the spindles of the attached muscles (→ p. 278).

There are two types of **thermoreceptors**: the *cold receptors* register temperatures below 36 °C, and the *warm receptors* register those above this value. The lower the temperature (in the range 36°–20 °C) the higher is the frequency of impulses in the nerve fibers leading from the **cold receptors**. The situation is reversed for the **warm receptors** (within the range 30°–43 °C; → **C**). Between 20 °C and 40 °C *adaptation of thermoreception* soon sets in (= PD reception: water at 25 °C feels cold only at first). More extreme temperatures, on the other hand, are continuously experienced as hot or cold, which serves to protect the skin from damage. There are probably special *heat receptors* for temperatures above 45 °C, in all likelihood thermospecific nociceptors.

Pain is an unpleasant sensory impression accompanied by awareness of a disagreeable experience. It is the reaction to the message that *damage* to the organism threatens or has already occurred (**nociception**). Identification of the cause is less important than recognition of the effect. Receptors report *visceral* pain (→ p. 282), "deep" pain (e.g. headache), and superficial *skin pain* (see also *referred pain*, p. 282). Superficial pain is sensed in two stages: there is an initial *sharp* sensation of pain, which stimulates in most cases a *flight or an escape reflex* (→ p. 280), and a *second* continuous ache (0.5–1 s later), which provokes protection of the damaged part. The pain receptors are free nerve endings. They *do not adapt* to pain stimuli, as is exemplified by an unremitting toothache (→ p. 262), otherwise continued damage would be ignored.

Damage along the pain pathways is experienced as if it had originated at the periphery; *projected pain* (e.g. backache when a nerve is crushed by a slipped disk).

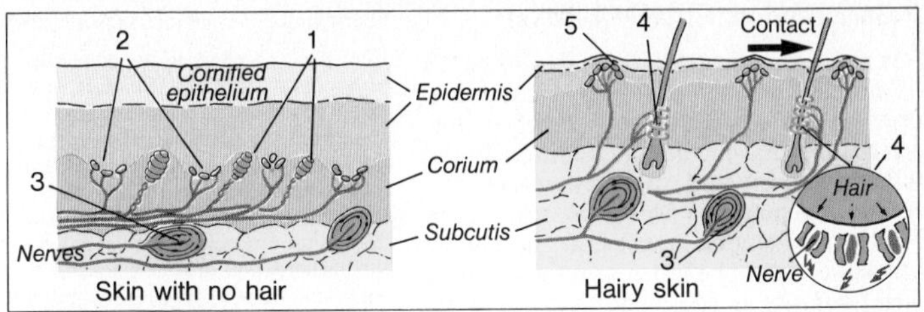

A. Skin receptors

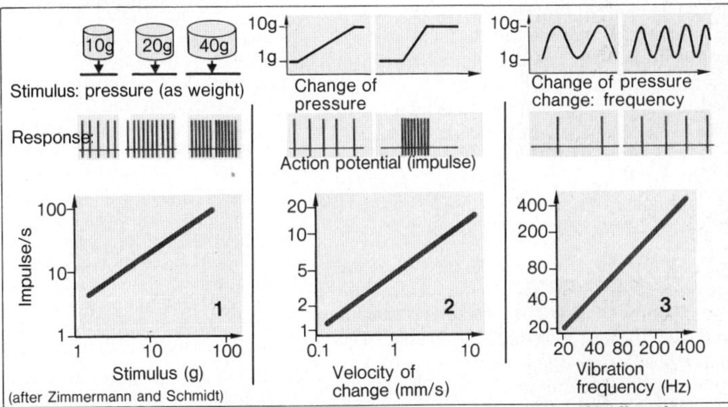

B. Responses of skin receptors to (1) pressure, (2) touch, (3) vibration

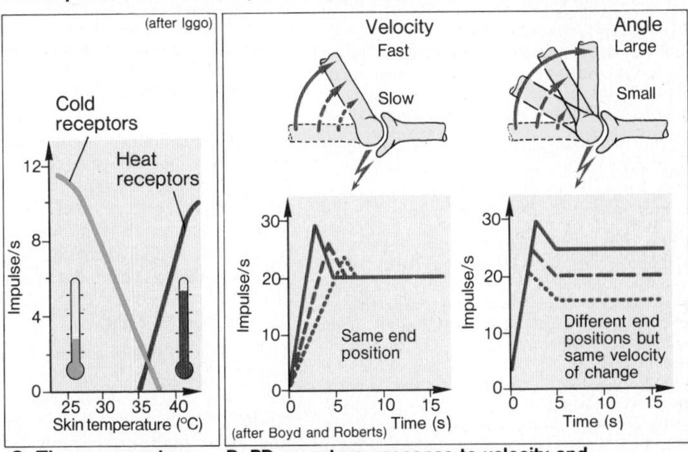

C. Thermoreceptors

D. PD receptors: response to velocity and to joint position

Proprioception, Stretch Reflexes

For measuring the position of joints, length of muscles, and so on ("deep" sensations), the body is equipped with so-called **proprioceptors** that include, in addition to the joint receptors and other receptors, the **tendon receptors** (= Golgi organs; where muscle and tendon meet) and the **muscle spindles** (→ **A**). The latter contain so-called *intrafusal muscle fibers* and lie *parallel* between the (*extrafusal*) contractile musculature. The efferent innervation of the ends of the intrafusal fibers comes from the **γ-motoneurons**. The middle of the intrafusal fibers (2 types: nuclear bag fibers and nuclear chain fibers) is surrounded by a spiral of nerve endings (so-called *anulospiral ending*), which report the *state of stretch* of the intrafusal fibers to the spinal cord (**Ia fibers** of the nuclear bag fibers; → **A–C**; group II fibers of the nuclear chain fibers).

Although certain of the messages from the proprioceptors also reach the cerebellum and cerebral cortex, subconscious reactions (**reflexes**) to these messages can be mediated at the *level of the spinal cord.*

For example, if a skeletal muscle is suddenly stretched by a blow on its tendon, the muscle spindles are also stretched, which leads to stimulation of the *Ia* fibers (→ **B** and **C**) and II fibers. Both run via the dorsal root to the anterior horn of the spinal cord, where they directly stimulate the motoneurons of the *same* muscle, causing its *contraction.* This is a **monosynaptic stretch reflex**, involving a *single synapse* between the *afferent* and *efferent* neuron. The *reflex time* of such a monosynaptic reflex is extremely short (about 20 ms).

Stretch reflexes can also be tested (e.g. with skin electrodes) by electrical stimulation of the (mixed) nerves of the muscles and simultaneous recording of muscle excitation; **H**(offmann)-**Reflex**. Weak stimulation (about 25 V) excites only the Ia fibers of the nerves, which results in muscle activation after a latency of about 30 ms (*H-wave*). If the stimulus is increased (about 60 V), the α-motoneurons are also directly activated (*M-wave* after 5–10 ms). With even stronger stimuli (95 V) the M-wave increases, whereas the H-wave disappears. The main reason for this is the antidromic excitation of the α-fibers, which collides at the α-somata with incoming Ia-fiber impulses, rendering them inactive. Further reasons are the Renshaw inhibition (→ p. 281, C) elicited by the antidromic α-fiber excitation, and the stimulation of the Ib fibers (autogenic inhibition, see below).

The stretch reflex is supplemented by several polysynaptic circuits. Efficient stretching of an extensor muscle can only take place if the corresponding antagonistic flexor muscle is *inhibited* (reciprocal inhibition). This is effected via an *inhibitory interneuron* (→ **B 1**). For termination of the reflex response, inhibition of the contraction of the extensor muscle is also necessary. This involves the participation of four mechanisms: (a) *Stretching of the muscle spindle ceases* and the *Ia fibers* are hence no longer stimulated. (b) *Strong contraction* causes inhibition of the α-motoneuron (*autogenic inhibition*) via the *Ib fibers* of the *tendon receptors* and an interneuron (→ **B 2**). (c) The *Ib fibers* also stimulate the α-motoneurons of the antagonistic muscle (*reciprocal innervation*; → **B**, flexor). (d) The α-motoneurons undergo "feedback" inhibition via collaterals and the Renshaw cells, which act as interneurons (*recurrent inhibition*; → **B** and p. 281, C).

The reflex excitability of the motoneurons, on which hundreds of other neurons synapse, is also under the control of the *supraspinal centers* in the brain. Disturbances in the latter may lead to abnormal activity of the monosynaptic reflexes (→ p. 284). The absence of monosynaptic reflexes indicates abnormalities in the corresponding region of the spinal cord or in the peripheral nerve.

The anulospiral ending, in addition to responding to extension of the whole muscle (→ **C, left**), can also be stimulated if its **intrafusal muscle fiber** contracts due to excitation of the **γ-motoneurons**, which, via Ia fibers, leads indirectly to α-motoneuron activation (→ **C, right**). This so-called *γ loop* and the direct activation of α-fibers may contribute to the precision of muscular movement (*α-γ linkages*).

The **muscle spindles** serve mainly to **control the length** of the muscle. Undesirable changes in muscle length can be corrected via monosynaptic reflexes. Desired alterations in the length of muscle are achieved by centrally regulated changes in the activity of the γ-fibers, which determine the *prestretching* of the muscle spindles (principle of γ loop; → **C, right**).

The **tendon receptors** are arranged *in series* with the contractile muscle fibers. They serve primarily for the **regulation of muscle tension** (**Ib** afferents) whereby the stimulation of a single motor unit (→ p. 32) is sufficient to stimulate them. They also **protect** the muscle from excessively high tension.

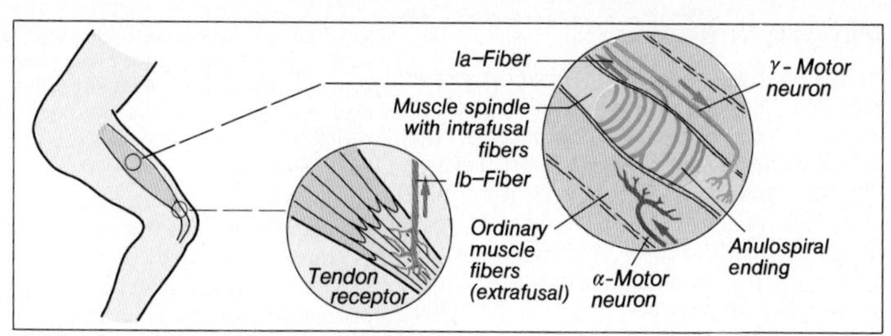

A. Muscle spindle and tendon receptors

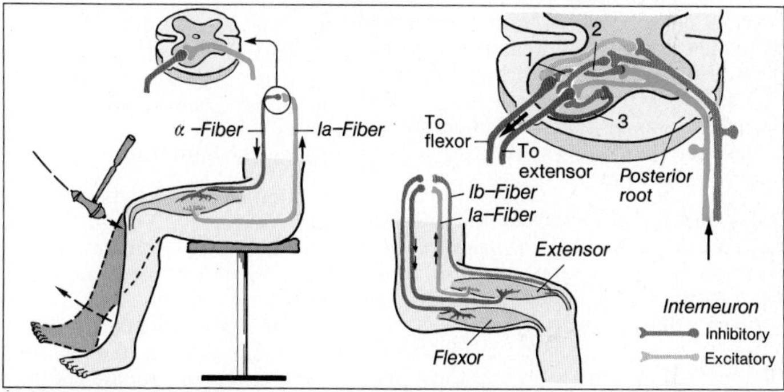

B. Monosynaptic reflex

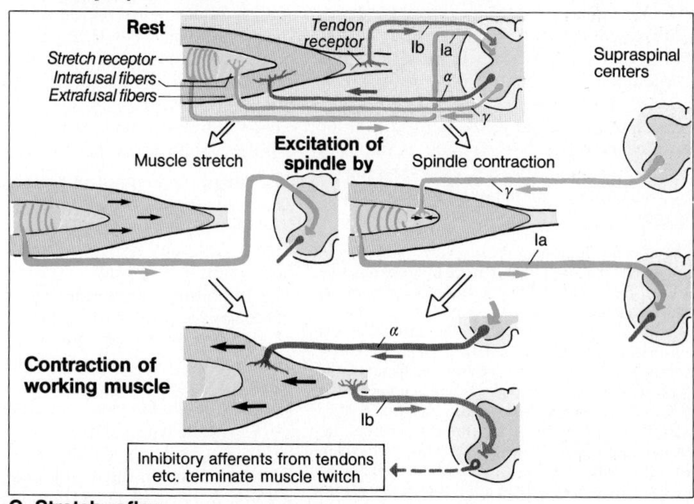

C. Stretch reflex

Polysynaptic Reflexes

In contrast to the simple, monosynaptic reflex (→ p. 278) that originates and terminates in one and the same muscle, the **polysynaptic reflex**, which crosses several synapses, may originate in one organ and terminate in another. The reflex arc may contain somatic (motor, sensory) or autonomic neurons, or a mixture of both. Due to the number of synapses involved, the **reflex time** is longer than in the monosynaptic reflex, as well as being dependent upon the *intensity of the stimulus* (variable **temporal summation** in the CNS). Example: itching in the nose → sneezing. A further characteristic of certain polysynaptic reflexes is that the reflex response can spread in proportion to the intensity of the stimulus (e. g. a slight cough → choking).

Polysynaptic reflexes include the **protective reflexes**, for example, the *withdrawal reflex* (see below), *corneal reflex, tear flow, coughing, sneezing*, reflexes serving nutrition (**nutritional reflexes**) such as *swallowing* and *sucking*, reflexes involved in locomotion (**locomotion reflexes**), as well as the numerous **vegetative reflexes** (circulation, respiration, stomach, intestines, sexual functions, bladder, etc.). Polysynaptic reflexes can be used diagnostically to localize lesions in the nervous system (e. g. *Babinski, cremasteric*, and *abdominal skin reflexes*). A typical polysynaptic reflex is the **flexor** or **withdrawal reflex** (→ A). For example, a *painful stimulus* in the foot, as in stepping on a tack, activates *flexion* in all muscle groups in that leg.

The afferent impulses follow several pathways in the spinal cord: (a) via excitatory interneurons (→ **A 1**) to the ipsilateral (same side) flexor muscles, which contract; (b) via inhibitory interneurons (→ **A 2**) to the motoneurons of the ipsilateral extensors, which relax (→ **A 3**); (c) via excitatory interneurons (→ **A 4**) to the motoneurons of the contralateral extensor muscles, which contract (→ **A 5**) (this **crossed extensor reflex** magnifies withdrawal from the damaging stimulus, thereby increasing the distance between the cause of the pain and the pain receptors, as well as helping to support the body); (d) via inhibitory interneurons to the motoneurons of the contralateral flexors, which relax (→ **A 6**); (e) via ascending and descending fibers to other spinal segments (→ **A 7** and **A 8**) since not all flexors and extensors are supplied from one and the same segment, and in this way activity can spread to all four extremities. Simultaneously, pain sensations are conducted to the cortex where they are *consciously* recognized (→ p. 276 and p. 282).

In contrast to the motosynaptic reflex, in which only the α motoneurons are activated (γ neurons inhibited), excitation, in the case of the polysynaptic reflex, runs parallel in both types of motoneurons. This ensures that the fibers of the muscle spindle (→ p. 278) shorten at the same time as those of the working muscle, so that, in spite of their shortening, the tension and the ability of the muscle spindle receptors to respond remains more or less constant.

Synaptic Inhibition

Inhibition of transmission across the synapse (see also p. 30) can be **presynaptic** (→ **B**). An impulse from an excited neuron (→ **B, c**) depolarizes the terminal portion of the presynaptic axon (→ **B, a**). This depolarization reduces the *amplitude* of the action potential (AP) in the neuron **a** (→ **B**) (steepness and amplitude of the AP depend on the preceding resting potential; → p. 26) so that *less neurotransmitter* is liberated into the synaptic cleft (→ **B, d**). As a result, the postsynaptic neuron (→ **B, b**) is less depolarized and the excitatory postsynaptic potential (EPSP; → p. 30) may be sufficiently reduced to prevent the generation of a postsynaptic AP in neuron **b** (→ **B 2**).

The situation is completely different in the case of **postsynaptic inhibition** (→ **C**). An impulse from an *inhibitory interneuron* causes transient hyperpolarization of the postsynaptic membrane (IPSP; → p. 30). The inhibiting interneuron can be (a) activated by recurrent axon collaterals (→ **C 1**) of the neuron to be inhibited (*recurrent inhibition via Renshaw cells*; negative feedback → **C 2**), or (b) *directly* activated (*"feed-forward" inhibition*) via a parallel pathway (→ **C**, right). An example of forward inhibition is seen in the inhibition of the ipsilateral extensor in the withdrawal reflex (→ **A 2** and **A 3**). Since the antagonistic muscle is inhibited at the same time, this is also termed **antagonistic inhibition**.

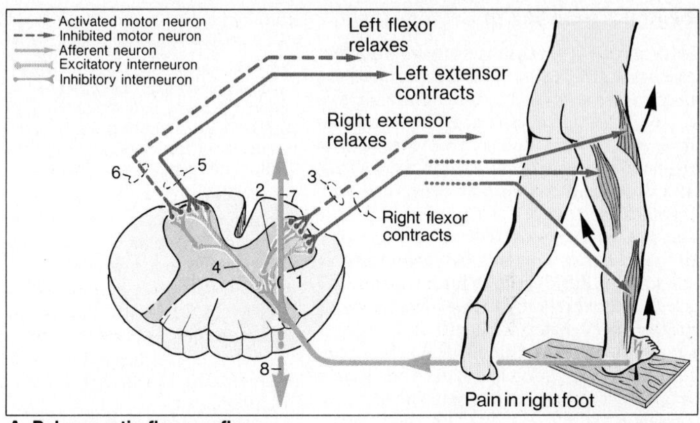

A. Polysynaptic flexor reflex

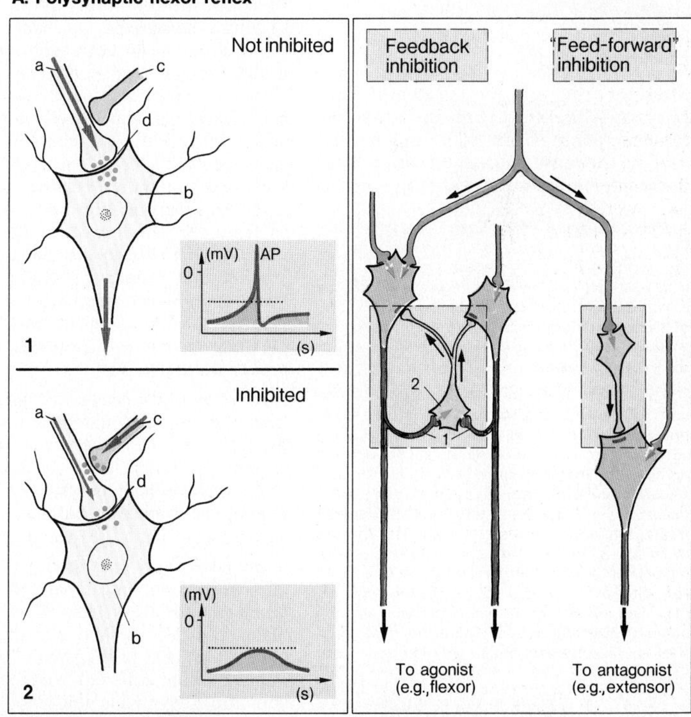

B. Presynaptic inhibition

C. Postsynaptic inhibition

Central Processing of Sensory Input

Information from the external environment reaches the CNS via the receptors of the major *sense organs* and via the *skin receptors* (→ p. 276). Messages concerning *body posture, muscle tone*, and so on are passed on from receptors in the *muscles*, in the *tendons*, and in *joints* of the organs concerned with locomotion (→ p. 278 ff.), and from the *vestibular organ* (→ p. 298). The major part of these messages reaches the **somatosensory** (or sensomotoric) **centers** in the *postcentral gyrus* of the cerebral cortex, where (as in the motosensory cortex) each part of the body is represented in a specific cortical field (*somatotopic representation*; → **A** and **B**), in proportion to the richness of its innervation. Peripheral stimulation is reflected in the so-called *evoked potentials*, which can be detected in the corresponding cortical area and indicate its excitation.

Messages from the skin (*superficial sensibility*) and from the organs of locomotion (*proprioception*) are passed on to the spinal cord via the **dorsal root** and proceed centrally along the following pathways:

1. **Dorsal columns** (→ **C 1**): These fibers synapse for the first time in the *nuclei of the dorsal columns* in the medulla oblongata, whence the second order neurons either pass to the *cerebellum* (→ p. 286) or cross the midline tor reach the *thalamus*. The fibers of the dorsal column carry information concerning *pressure, touch*, and *proprioception*, and are thus the pathways for the *sense of posture*.

2. The sensory fibers from the receptors for *pain* and *temperature*, and some of the pathways for pressure and touch, cross to the opposite side in the corresponding segment of the spinal cord and run in the **ventrolateral column** of the spinal cord as the *spinothalamic tract*, through the brain stem to the *thalamus* (→ **C 2**). Since this tract also carries messages concerning visceral pain, the latter is often felt in the skin areas served by the same segment of the spinal cord: **referred pain** (e. g. in the left shoulder and arm in connection with O_2 deficiency of the heart: *angina pectoris*). Consequently, these skin areas are frequently hypersensitive to touch and pain (*hyperesthesia* and *hyperalgesia*).

3. The sensory fibers from the *head region* (**trigeminal V**) also terminate in the thalamus.

4. Two other pathways (carrying mainly proprioceptive information) run to the **cerebellum**: the *posterior spinocerebellar tract* (→ **C 3**) and the *anterior spinocerebellar tract* (→ **C 4**).

Hemisection of the spinal cord leads to the *Brown-Séquard syndrome* and causes, caudally to the injured segment, initial limb paralysis followed by spastic motor paralysis and disturbances in the tactile senses ipsilaterally, and contralateral loss of pain and temperature sensation (*dissociated paralysis*).

In the ventrobasal nuclei of the **thalamus**, the afferent sensory pathways are relayed to third order neurons that run to various cortical sensory regions, chiefly the postcentral gyrus. These are specific thalamocortical **projection pathways** (→ **D**), which, like the optical and acoustical projections, are also allocated to strictly defined portions of the thalamus.

In contrast, from other portions of the thalamus **nonspecific** (*reticular*) **pathways** run to almost all cortical regions (but chiefly to the frontal cortex; → **E**). The impulses in these pathways come mainly from the **reticular formation**, which, besides playing an important role in motor control (→ p. 284), also receives afferent messages from all sense organs and the ascending spinal tracts (eye, ear, superficial sensibility, etc), as well as from the basal ganglia. The reticular pathways seem to play a fundamental role in *alertness* and *awareness* (ARAS; → p. 292), mediate *emotional aspects* (for example, of pain) to the limbic system, and fulfil complex *vegetative functions* (circulation, respiration, hormonal, etc).

In addition to the primary projection areas, there are a number of association areas in the cortex (→ p. 290) that are also connected in both directions with *nonspecific nuclei* (e. g. Nucl. medialis dorsalis) of the thalamus by the **association pathways**.

The sensory input to the cortex can be inhibited at every relay station (spinal ganglion, medulla oblongata, and thalamus) by **descending pathways** (from the cortex). This makes possible alterations in the receptive field, adjustment of thresholds, and (where impulses from different sources run in a common afferent) the ability to "pick out" sensory input from a specific region.

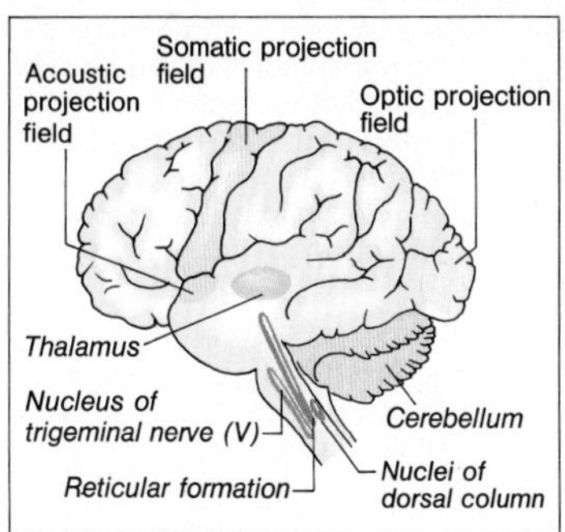

A. Sensory centers

Genitals
Rectum
Bladder
Thalamus
Tongue
Throat
(after Penfield and Rasmussen)

B. Somatotopic representation in somatosensory cortex

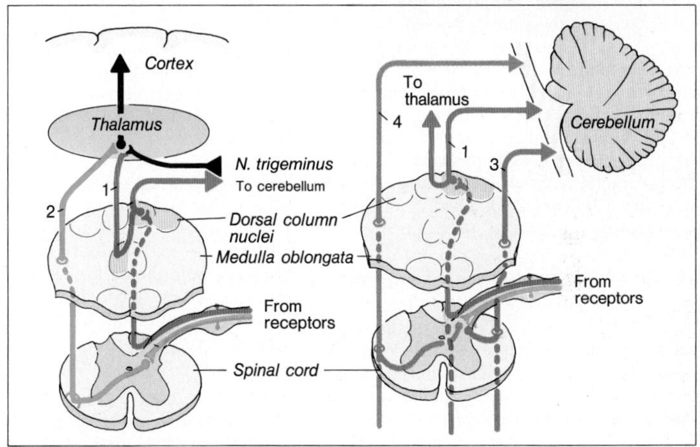

C. Spinal afferent pathways

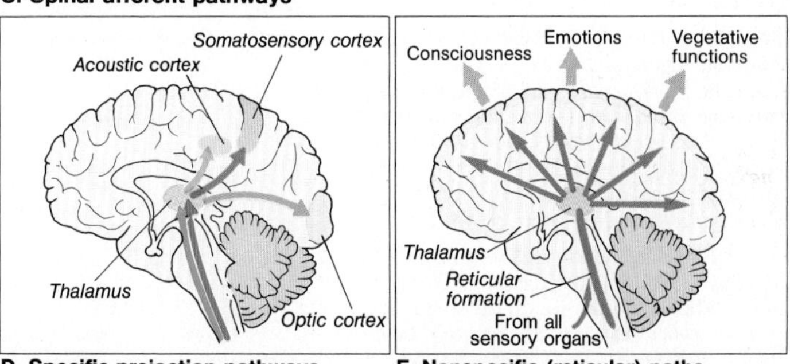

D. Specific projection pathways to cortex

E. Nonspecific (reticular) paths to cortex

Motor Activity – Motor Hold System

With few exceptions, all activities of the CNS, receiving, processing, and integrating information, ultimately find expression in contraction of muscles (→ p. 275, A). Muscle activity may occur in terms of **purposeful motion** (*motor move system*; → p. 288) or of **postural control** (*motor hold system*), that is, control of **equilibrium** and **spatial position**

The anatomical distinction drawn between *pyramidal* and *extrapyramidal motor systems* is somewhat misleading from the point of view of function because, functionally, these "systems" are intimately interconnected.

The motoneurons supplying the skeletal muscle arise in the **anterior horn of the spinal cord** (→ p. 273, B). At the **spinal level**, in addition to the relatively simple *monosynaptic reflexes* (→ p. 278), complicated motor events such as the *withdrawal reflex* (→ p. 280) and a number of *standing* and *running reflexes* can also be mediated. Transection of the spinal cord first produces a loss of peripheral reflexes (*spinal shock*), followed by recovery in spite of persistence of the spinal lesion.

In the intact organism, the spinal reflex program is subordinated to the *higher (supraspinal) centers*. **Control of the motor hold system** is primarily effected by the **motor centers of the brain stem**: the *nucleus ruber* (→ **A 1**), *vestibular nuclei* (→ **A 4**) (particularly the *lateral nucleus of Deiter*; → p. 286), and parts of the *reticular formation* (→ **A 2** and **A 3**).

The **main afferents** to these centers come from the *organs of balance* (→ **A** and p. 298), from the *proprioceptors* of the *neck* (→ **A**), from the *cerebellum* (→ p. 286), and from the *motor cortex* (directly and via the *basal ganglia*; → **A**). The fibers from the cortex are provided by collaterals of the pyramidal (→ **C** and p. 288) and by other pathways.

The **descending tracts** leading from the nucleus ruber and from the medullary parts of the reticular formation to the spinal cord (*tractus rubrospinalis* and *tractus reticulospinalis lateralis*; → **A**) exert a predominantly inhibitory influence on the α- and γ-motoneurons (→ p. 278) of the extensor muscles and a stimulating effect on the flexor muscles. Conversely, the pathways originating in Deiter's nucleus and in the pontine reticular formation (*tractus vestibulospinalis* and *tractus reticulospinalis medialis*) inhibit the flexor muscles and stimulate the α- and γ-fibers of the extensors (→ **A**).

Transection of the brain stem below the nucleus ruber leads to a state of *decerebrate rigidity*, since the tonic extensor influence of Deiter's nucleus now predominates.

The motor centers of the brain stem coordinate the *postural* and *positioning* or *righting reflexes*, whose (involuntary) function is to maintain **body posture** and **balance**. Postural reflexes serve to maintain **muscular tone** (→ p. 40) and *coordinate eye movements* (→ p. 299, C). The sensory input comes from the *organ of balance* (tonic labyrinth reflexes) and from the *neck proprioceptors* (tonic neck reflexes). The same afferents are involved in the righting reflexes (*labyrinth-* and *neck-righting reflexes*), which serve to return the body to its normal position. Firstly, the *head* (in response to labyrinth afferents; → p. 298), then (in response to afferents from the neck proprioceptors) the *rump* are returned to their normal positions. In addition, afferents from the eyes, ears, olfactory organs, and the skin receptors modify these righting reflexes.

Statokinetic reflexes play an important role in posture and positioning of the moving body, for example, in preparation for jumping, and in rotatory nystagmus (→ p. 314).

(Text to **B** and **C** → p. 288).

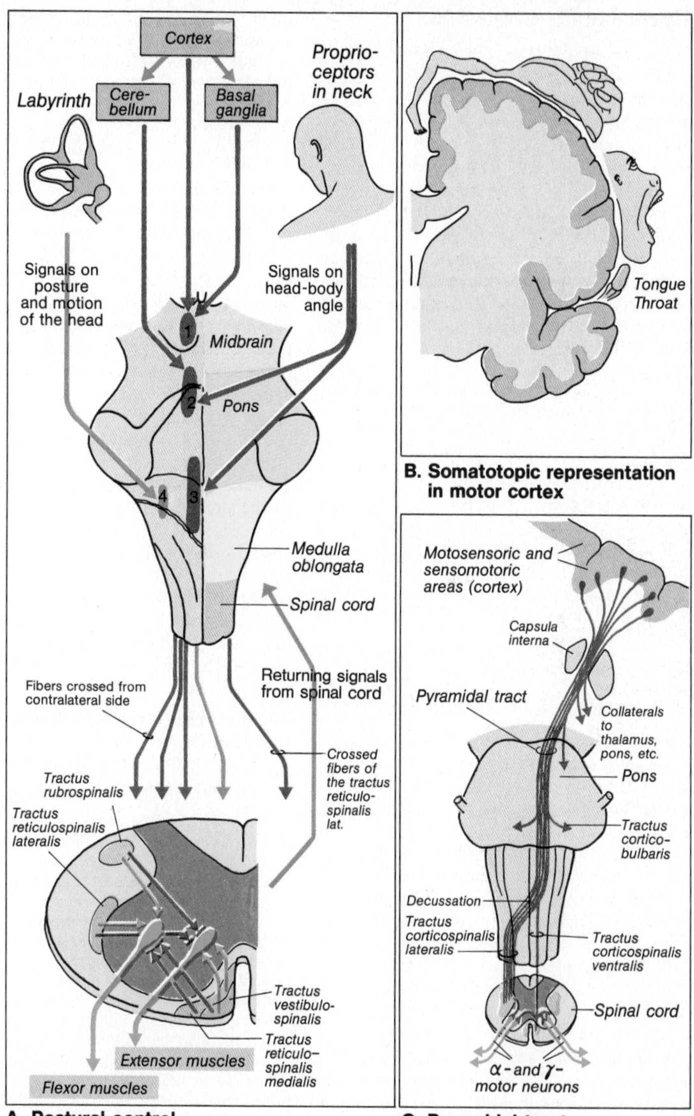

A. Postural control

B. Somatotopic representation in motor cortex

C. Pyramidal tract

Function of the Cerebellum

The cerebellum is an important control center for motor functions. It coordinates movement and posture and is involved in programing movements. The phylogenetically older parts of the cerebellum, the *archicerebellum* and *paleocerebellum*, are situated **medially**. They consist, respectively, of *nodulus* and *flocculus*, and of *pyramis, uvula, paraflocculus*, and parts of the *anterior lobe*. The *pars intermedia* can also be considered as part of the median cerebellum. The more recent, and in the human brain, highly developed *neocerebellum* is situated **laterally**.

Median cerebellum (→ **A 1,2**). This is primarily concerned with the control of the motor hold system (→ p. 284) and with visual motor control. It receives **afferent copies** of spinal, vestibular, and visual origin as well as **efferent copies** of the motor signals for skeletal movements. The **outgoing fibers** from the median cerebellum run via the *globose, emboliform*, and *fastigial nuclei* to the motor centers of the spinal medulla and brain stem and to the vestibular nuclei (*Deiters' nucleus*).

Lesions of the median cerebellum result in disturbances in balance and eye movements (pendular nystagmus) and in trunk and gait ataxia.

Lateral cerebellum (hemispheres). This is chiefly concerned with programing movements (→ **A 3** and p. 288). Its functional flexibility makes possible motor adaptation and the learning of motor processes. It is bidirectionally connected with the cerebral cortex. **Afferently**, it is connected with the cortical regions (parietal, prefrontal, and premotor *association cortex*, sensomotor and visual cortex) that are chiefly concerned with the "*planning phase*" of movements (→ p. 288) via the *pontine nuclei* and the mossy fibers (see below), whereas from the *interferior olive* and climbing fibers (see below) it receives afferents from cortical and subcortical motor centers. The **efferents** from the lateral cerebellum mainly run via the motor thalamus to the motocortex.

Lesions of the cerebellar hemispheres result in disturbances in beginning, coordinating, and terminating purposeful motor movements and in rapid "reprograming" to the opposite movement (diadochokinesis). The results are the development of tremor as the intended target is approached (*intention tremor*); deterioration in the ability to gauge distances (*dysmetria*) and to terminate movements (*rebound phenomena*), and *adiadochokinesis*. In addition, speech is slow, monotonous, and slurred (*dysarthria*).

The efferent pathways of the cerebellar cortex consist of neurites of the more than 10^7 **Purkinje cells**. They have an *inhibiting* effect on the cerebellar nuclei with which they are connected. Afferents from the spinal cord relaying in the olive end as **climbing fibers** on the **modulating functional units** of the cerebellar cortex (longitudinal microstrips). Via their (multiple) excitatory synapses, they *enhance the inhibitory effect of the Purkinje cells*. All other afferents to the cerebellum end as **mossy fibers**. They can either, via excitation of the innumerable **granular cells** and their **parallel fibers**, enhance the inhibiting effect of the Purkinje cells, or, via inhibitory intermediate cells (**Golgi cells**), bring about *deinhibition*. Direct deinhibition can take place via the **stellate** and **basket cells**: *convergence* (roughly 10^5 parallel fibers → 1 Purkinje cell) and *divergence* (collaterals of one climbing fiber → 10–15 Purkinje cells) occur simultaneously in the signal chain.

The role of the cerebellum in integration and coordination with other motor centers (→ pp. 278 ff., 284, and 288) and sense organs can be seen in the behavior of a tennis player (→ **B**).

When one partner serves, the body of the other player moves to meet the ball (*motor move*), whereby adequate support (right leg) and balance (left arm) have to be maintained (*motor hold*). *Eye movements* keep the ball within view; the optical cortex analyses the path and speed of the ball. The "associative" cerebral cortex plans the return movements, taking into considerations the ball, net, opponent's position, and, among other things, that the backlash resulting from impact of racket and ball must be compensated by motor hold movements. Movement programs from the cerebellar cortex and basal ganglia assist the motosensory cortex in initiating the motion of hitting. The ball is not only hit and returned to the opponent's court, but is as a rule also rotated by a tangential stroke ("cut"): *learned, rapid motor move*.

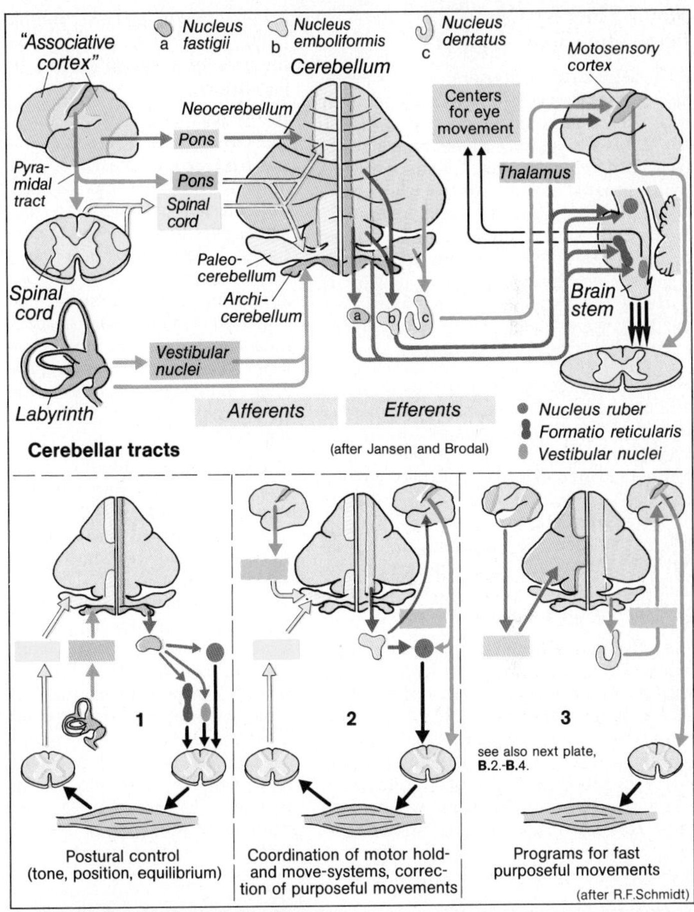

A. Cerebellar tracts and functions

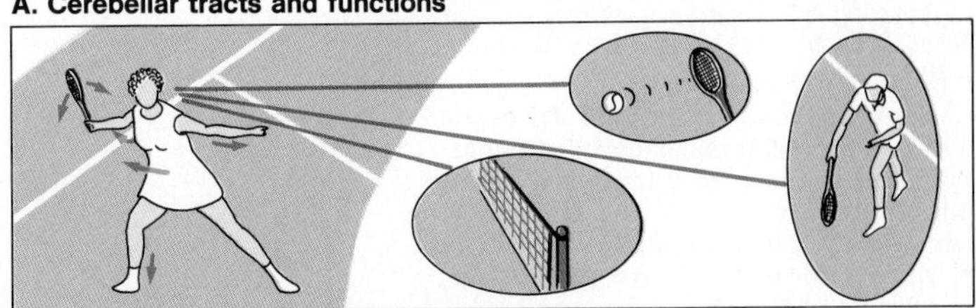

B. Motor hold and motor move integration (see text)

Motor Move System – Purposeful Motor Control

Purposeful movements are controlled by the **moto(sensory) cortex** (in close cooperation with the motor hold system; → p. 284 ff.). It comprises the **primary motor region** in the *Gyrus praecentralis* (area 4) and, in a wider sense, situated rostrally to this, the **secondary motor region** (area 6). Both are, like the motor thalamus and the striatum, organized according to different regions of the body (*somatotopic representation*; → p. 285, B). Parts of the body performing fine movements (fingers, face) are relatively better represented.

Whereas the neurons of area 4 can only set off contractions of smaller groups of muscles (e.g. bending the fingers), a (longer-lasting) excitation in area 6 (probably superordinate to area 4) leads to the performance of more complex movements (e.g. rotation of the trunk). The longer *latency* of up to 100 ms between the beginning of the impulse in area 4 and the (voluntary) movement is apparently necessary for the temporal summation at the motoneuron.

The **efferents** of the primary motor cortex reach the motoneuron via the **pyramidal tract** (→ p. 285, C) and via the motor centers of the brain stem (→ p. 285, A). Of the purposeful motor impulse signals in the pyramidal tract, only those for *fine movements* (fingers) are transmitted monosynaptically to the α-motoneurons, whereas the large majority reach them via spinal *interneurons*; segmentally organized patterns of movements (*reflex arcs*) are also mobilized.

The pyramidal tract fibers from the somatosensory cortex (*Gyrus postcentralis*; → p. 283, B) probably modulate the transmission of sensory signals (→ p. 282).

Copies of the motor commands are transmitted to the motor thalamus, the striatum, the nuclei of the inferior olive, and to the pons (from here, on to the lateral cerebellum; → p. 287, A 3) and the brain stem (→ p. 285, A); these tracts are part of the **supraspinal feedback loops**.

The **basal ganglia** are incorporatd into several corticocortical (skeletomotor and oculomotor as well as "complex") signal loops (→ **A**), which mainly serve to convert the *plan* for a movement into its *execution* (see below).

In the *skeletomotor loop*, for example, the signals (above all from areas 4 and 6) arise in the **striatum** (glutamate as stimulating transmitter), from which two parallel tracts (via the internal **pallidum** and the **substantia nigra**) pass to the motor thalamus (→ **A**). Each of these two tracts consists of two serially arranged *inhibitory* neurons (GABA as transmitter), which in each case leads to *deinhibition* in the secondary neuron. Thalamocortical tracts to area 6 complete the loop. Reverse pathways (e.g. the dopaminergic pathway from substantia nigra to striatum) primarily serve the modulation of the signals.

The steps from "intention" to performance of a voluntary movement are probably, in a simplified form, as follows: In predominantly subcortical parts of the brain, in some unknown way, an **impetus for motor activity** arises (→ **B 1**). This signal reaches the "associative" cortex (skeletomotor, chiefly area 6), where the **plan for movement** is formed (→ **B 2**). At this site, a *stand-by potential* can be led off from the exterior (about 1 second before the – even if only thought about – movement). Via the loops involving *cerebellum* (→ p. 286) or *basal ganglia*, for the most part, **programs for movement** can be recalled (→ **B 3**); via the thalamus they reach areas 4 and 6, which control the **execution of the movement** (→ **B 4**).

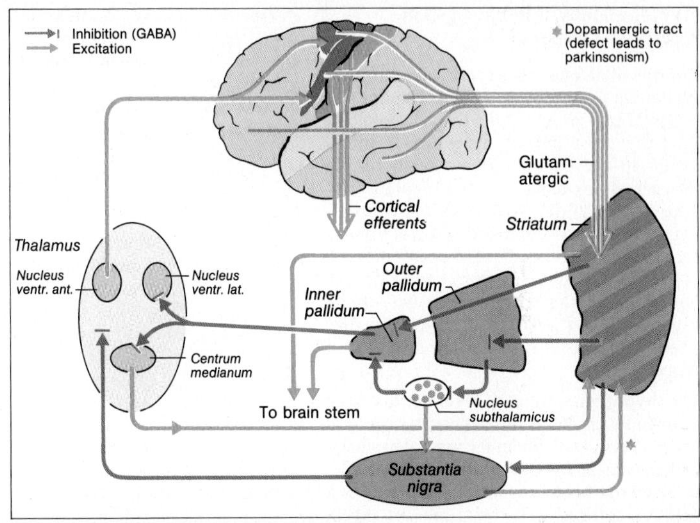

A. Basal ganglia: afferents and efferents (partly after Delong)

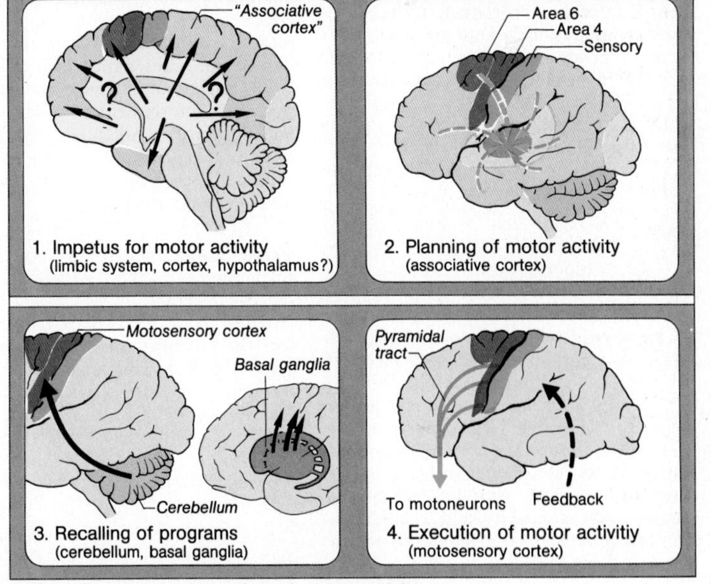

B. From impetus to execution of motor activity

Hypothalamus, Limbic System, Associative Cortex

The **hypothalamus** is the center governing all vegetative (→ p. 50ff.) and most endocrine (→ p. 232ff.) processes in the body, and is thus the most important organ of integration in the homeostatic control of the **internal environment**. For the regulation of body temperature (→ p. 194), the (medial) hypothalamus is equipped with *thermoreceptors*; for the regulation of osmolality (→ p. 335), with *osmoreceptors*; and for the regulation of the hormone balance of the organism (→ p. 240), with receptors that (feedback) register the hormone levels in the blood.

The hypothalamus can regulate the hormonal as well as the autonomic and somatic nervous processes under its control in such a way that the organism exhibits, for example, (a) **defensive behavior** (*alarm reaction*), (b) a reaction conductive to nutrition and digestion (**nutritive behavior**), (c) **thermoregulatory reaction** (→ p. 194), or (d) behavior serving the ends of reproduction (**reproductive reaction**). Which kind of reaction is elicited depends upon the "*program*" selected in the hypothalamus: according to the requirements of this program, hormones and the autonomic and somatic nervous systems are brought into play as the "*tools*" that activate or inhibit the appropriate peripheral organs and structures (→ **A**).

Defensive behavior (fight-or-flight reaction) involves increased blood flow to the muscles, rise of blood pressure and respiratory rate, vasoconstriction in the gastrointestinal tract, etc. A similar program is activated during **physical work**. A rise in blood pressure also occurs in connection with *nutrition*, although in this case the blood supply and motility of the gastrointestinal tract are enhanced, while the blood supply to the skeletal muscles is cut back. The *sexual response* and *reproduction* involve central control of courtship, neuronal mechanisms mediating sexual activity, hormonal regulation of pregnancy (→ p. 268), and so on.

The **limbic system** chiefly regulates inborn and acquired **behavior** („choice of program") and is the site of origin of instinctive behavior, motivation, and emotion ("*inner world*"). *The limbic system has* **cortical** (*hippocampus, parahippocampal gyrus, cingulate gyrus*, parts of the olfactory cortex) and **subcortical divisions** (*amygdaloid body, septal area, anterior thalamic nucleus*). Reciprocal **connections** exist with the (lateral) hypothalamus (mainly *recall of programs*, see above) and with the temporal and frontal cortex. The latter serve, above all, for the integration (important for determining behavior) of perception and evaluation of signals from the "*outer world*" and of *memory content*. The limbic system governs the **expression of emotions** (fear, anger, rage, boredom, happiness, etc.), which is important for its *signal effect* on the social environment. Conversely, *smells* act as signals from the environment and are intimately connected with behavior. This is reflected in expressions such as "familiar atmosphere" (alarm reaction unnecessary).

Especially important for the overall regulation of behavior are the **monoaminergic pathway systems** (noradrenergic, dopaminergic, and serotonergic neurons), which radiate from the brain stem to almost all parts of the brain. Experimental autostimulation, mainly of adrenergic areas, leads to positive reinforcement (arousal of interest and reward), whereas serotoninergic neurons may be part of a "disinterest" system. The monoaminergic systems are also points of attack for *psychopharmaceutical agents*.

The **nonspecific** or **associative** cortex consists of (1) prefrontal and (2) limbic portions of the **frontal brain**, as well as (3) temporal, parietal, and occipital areas. It is responsible for certain **integrative functions** of the cerebral cortex. Thus area 3 is responsible for higher sensory activities, whereas the function of area 2 is to effect the subordination of inborn behavior to *acquired controls* (i.e. to *intentions* and *plans*) and to reconcile internal and external motivations with one another.

Lesions in the frontal cortex may result in perseveration (persistent continuation of an activity), a decreased attention span, an altered sense of time, listlessness, pronounced irritability, a state of euphory, etc.

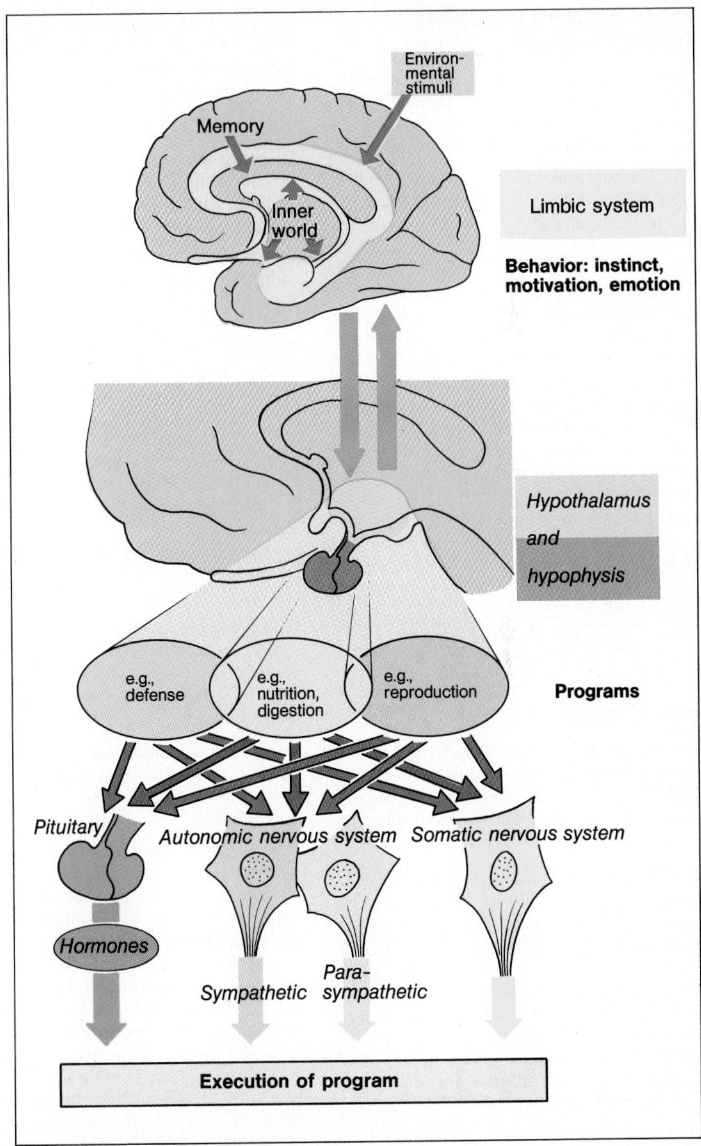

A. Limbic system and hypothalamus

Electroencephalogram (EEG), Waking-Sleeping

The cerebral cortex generates electric potentials that can be led off from the skin covering the cranium and recorded as the electroencephalogram (**EEG**; → **A**).

The fluctuations of the potentials (EEG "curves") normally depend upon the *degree of alertness* and vary in amplitude (a) and frequency (f) (→ **B**); *α-waves* occur in the relaxed adult (awake, with *closed eyes*; $f \approx 10$ Hz; $a \approx 50$ μV). The recording obtained is termed a *synchronized EEG*. If the eyes are opened, other sense organs stimulated, or if, for example, a difficult calculation is performed, the waves disappear (*α-blockade*) and are replaced by *β-waves* ($f \approx 20$ Hz, smaller a than in α-waves) (*desynchronized EEG*). EEG phases of this kind express an increase in alertness and (e.g. via epinephrine) increased activity (*arousal activity*) of the so-called *ascending reticular activating system* (ARAS; → see also p. 282).

The EEG is an important diagnostic tool in the clinic, as for example in epilepsy (localized or generalized *convulsion waves*; → **B**), in judging the degree of maturity of the brain, in monitoring anesthesia, and in the diagnosis of brain death (*zero line EEG*).

Falling asleep (stages A/B/C [→ **C**]) is accompanied by the occurrence of low-frequency *ϑ-waves* (→ **A**), which gradually give way to even slower waves (*δ waves*), in *deep sleep* (stages D/E).

Normally, sleep is a cyclic sequence (→ **C**), progressing from light to deeper stages and returning to light sleep. The cycle is repeated four to five times each night, whereby stage "REM" plays a special role. In this phase, although most of the skeletal musculature is atonic, sudden twitches of fingers or face, excitation of the penis, and, most important of all, **r**apid **e**ye **m**ovements occur (**REM sleep**). All other stages are collectively termed **NREM** sleep (non-REM sleep). If sleeping test subjects are awakened (which is equally difficult from REM sleep and from deep sleep) from the two types of sleep, those awakened from REM sleep report (chiefly "sensory") **dreams** more frequently than those awakened from deep NREM sleep, whose dreams are more "intellectual." The proportion of the entire sleeping period occupied by REM sleep decreases with aging. Deprivation of REM sleep (due to being awakened) is followed by an increased duration of this phase during the ensuing nights. Apparently REM sleep is made up in this way.

The normal **rhythm of sleeping and waking** is subject to control by an "**internal clock**" (*circadian clock*), the basis of which is not understood. The endogenous circadian wake-sleep cycle of usually 25 hours only becomes apparent during total isolation from external environmental signals (windowless cellar, caves, etc.) (→ **D**). Under normal circumstances, *environmental zeitgebers* (including light [day] and dark [night]) **synchronize** this rhythm to the accustomed 24 hour periodicity. Temporary disturbances in synchronization, lasting several days, can be observed in connection with long journeys in an east-west direction (jet lag).

There appears to be more than one internal clock, since the periodicity of the body temperature (→ p. 331, A), for example, can uncouple from the sleep-wake rhythm if synchronization is faulty.

As the EEG shows, **sleep** is not simply a state in which the brain is at rest, but rather a *form of brain function that differs from wakefulness*.

The fundamental causes of the states of waking and sleeping are largely unknown. Of the many theories concerning sleep, those postulating the existence of **endogenous sleep factors** have again become attractive. According to such theories, either a substance that accumulates during the waking hours causes tiredness and sleep when it reaches a certain concentration, or at the onset of sleep a sleep-promoting substance is released. There is experimental evidence for both theories; the accumulation or release of specific oligopeptides appears to elicit the various forms and stages of sleep.

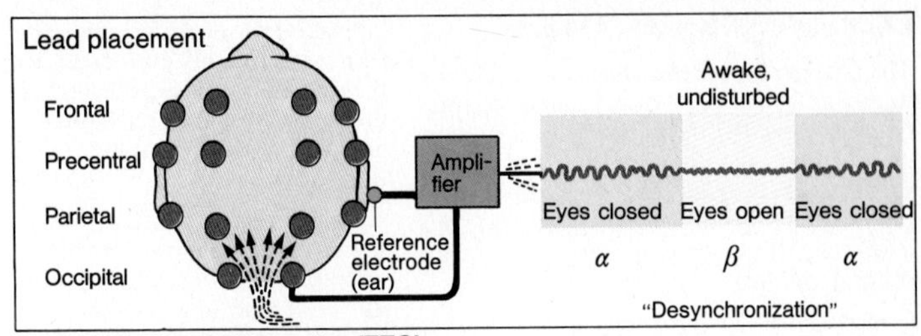

A. Electroencephalography (EEG)

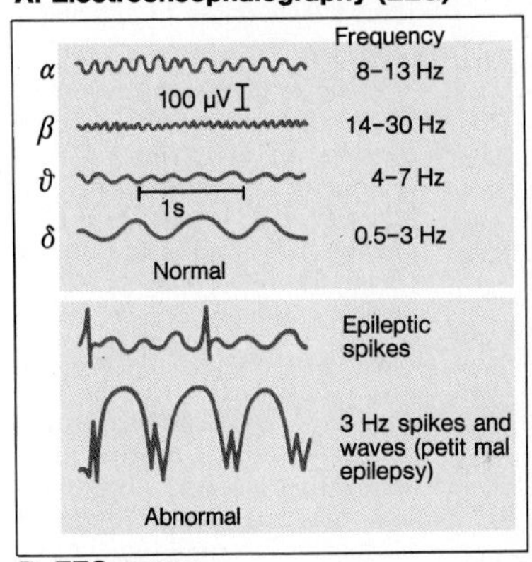

B. EEG waves

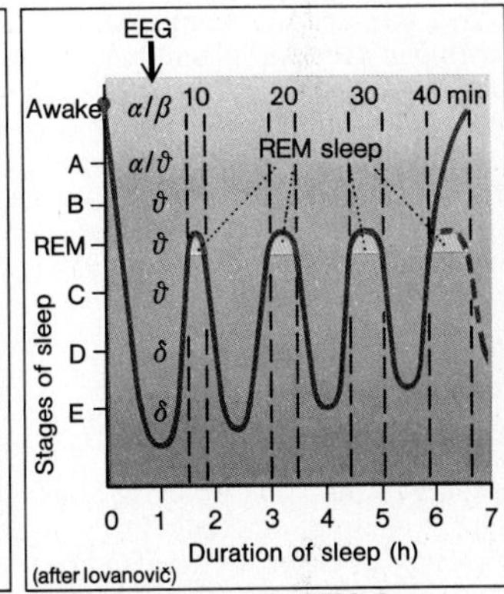

C. Stages of sleep, REM sleep

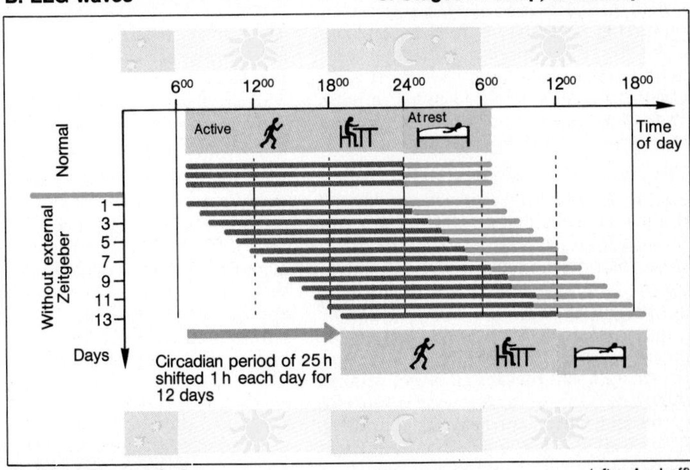

D. Circadian periods

(after Aschoff)

Consciousness, Speech, Memory

The **human state of consciousness** involves a number of aspects: (a) directed and organized *awareness*; (b) the capacity to deal with *abstractions*; (c) the ability to *verbalize* events; (d) to *plan* and to extract new relationships from observations; (e) *self awareness*; (f) a *sense of values*, and many other attributes.

Consciousness requires the presence of a highly developed nervous system, and bestows the ability of coping with situations arising in the environment (*adaptation*) that could not be dealt with by reflexes alone. We have only a fragmentary knowledge of the neural activities essential for consciousness. It is thought, for example, that cooperation between the cerebral cortex and the reticular formation (→ p. 282 and p. 292) is necessary.

Speech is an important product of the brain and is an essential component of human consciousness. On the one hand, it provides a means of communication with other human beings; *intake of information* via eyes, ears, and, especially in the blind, via the sense of touch; *output of information* via speech and writing. On the other hand, speech is also necessary for the conscious processing of sensory impressions, that is, in order to create *concepts*, to fashion *ideas*, and then to express them *verbally*. It is this ability to form thoughts and to verbalize them that makes possible an *economical storage* in the memory (see below).

The development of ideas and of speech, and their processing, are *unequally* distributed between the *two halves of the brain*. Judging from the behavior of patients after the connections between the two hemispheres have been severed (*split brain*), it can be concluded that it is almost always the *left hemisphere* that contains the seat of the powers of speech ("*dominant*" *hemisphere*). If such a patient touches an object with the right hand, for example (message reported to the left hemisphere), he is able to name it. If he uses the left hand, however (right hemisphere), this is impossible. Nevertheless, the right hemisphere also possesses highly developed qualities (e.g. memory). Appreciation of music and recognition of form are probably even better developed in the right hemisphere, although their conscious appreciation is only possible with the help of the left hemisphere (if the connections are intact).

Aphasia or speech disturbances can be due to defects in the motor functions connected with speech (*motor aphasia*) or to a defect in its understanding (*sensory aphasia*; → p. 324). A third form of disability is characterized by inability to find words (*amnestic aphasia*).

A component of consciousness is **memory** (→ **A**): the *very brief* (< 1 second) *sensory memory* serves to retain sensory signals (automatically). A small part of this information (→ **A**) is passed on to the *primary memory*, which is able to store roughly 7 bits (→ p. 274) for several seconds. This information is usually already *verbalized*.

Long-term storage in the *secondary memory* is the result of frequent repetition: *consolidation* (→ **A**). Recall from the secondary memory is usually relatively slow. The *tertiary memory* (→ **A**) stores facts that have been extremely well ingrained (writing, reading, own name); they can be rapidly recalled throughout a lifetime.

The primary (short-term) memory is probably correlated with **circuits of impulses** in the local groups of neurons concerned, whereas the long-term memory is principally dependent upon **biochemical mechanisms** (protein synthesis).

Loss of memory (amnesias): *retrograde amnesia* is recognized clinically as loss of the primary memory associated with (temporary) difficulty in recall from the secondary memory (causes: concussion, electrical shock, etc). *Anterograde amnesia* is the term applied to the inability to transfer new information from the primary to the secondary memory (*Korsakoff syndrome*).

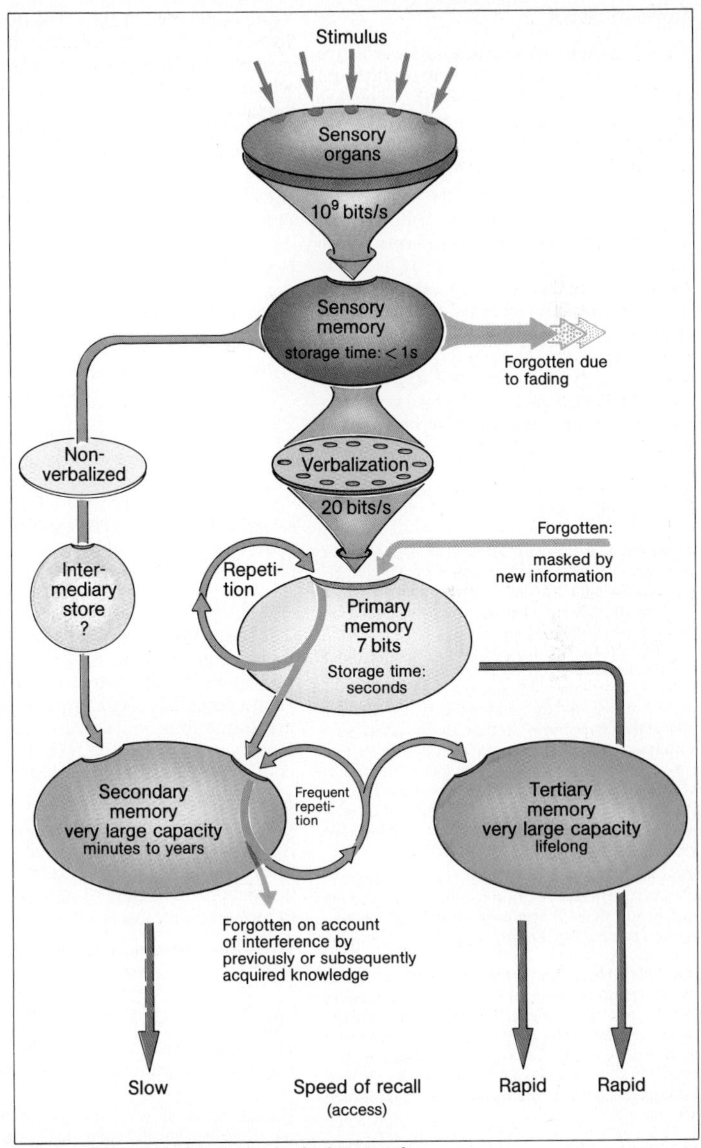

A. Information storage in the brain (memory)

Sense of Smell

The 10^7 **primary sense cells** that are sensitive to smell are situated in the neuroepithelium of the *regio olfactoria* (→ **A**). They are bipolar; the end of their dendrite bears 5–20 cilia, covered by a mucus layer; the axons are directed centrally (→ **B**).

Odorous substances reach the regio olfactoria via the air (helped by sniffing) and must be dissolved in its **mucus layer** before they can reach the receptor proteins of the **ciliary membrane**. Corresponding to the numerous **qualities of odors**, the number of types of receptors is estimated at several dozens to several thousands. One sensory cell is sensitive to many (not all) odorous substances, and each one has its own *spectrum*, which only partially overlaps that of others. A particular "odor" thus stimulates a highly specific population of receptors, whose combined stimulation "portrays" the smell to the CNS.

In just the same way as hormones (→ p. 242), the odorous substance binds to its own specific **receptor protein** on the ciliary membrane, after which (with many, but not all such substances), via a G_s-protein, the adenyl cyclase is activated. The **cAMP** thus formed appears to open cation channels in the membrane (directly or by phosphorylation?); this finally leads, via an influx of Na^+ (and Ca^{2+}?) to **depolarization** of the sensory cell (→ **C**). An alternative hypothesis is that the odorous substance itself directly opens the cation channel, and the cAMP closes it again by phosphorylation. For other odorous substances, cGMP and IP_3 are under discussion as transducers. Some stimuli (e.g. acids) also elicit a reaction in **free nerve endings** (N. trigeminus) in the olfactory epithelium.

Thresholds. A mere $4 \cdot 10^{-15}$ g methyl mercaptan (in garlic) per liter of air are sufficient to give the impression that "it smells of something" (= perception or *absolute threshold*). At $2 \cdot 10^{-13}$ g/l, the odorous substance is recognized (*recognition threshold*). These thresholds depend on air humidity, and temperature, and for other substances are up to 10^{10} times higher. The relative *intensity difference threshold* $\Delta I/I$ of 0.25 (→ p. 306) is comparatively high. The rapid **adaptation** (→ **D**) is probably due to a chemical receptor desensitization; after about 1 minute, neuronal adaptation also occurs.

Olfactory pathway. The axons of the sensory cells project upwards in bundles as the *Fila olfactoria* to the **olfactory bulb** where, in the "glomeruli" region, 100–1000 of them **converge** on each *mitral cell* (transmitter: carnosine?). *Periglomerular* and *granular cells*, which are subject to **efferent inhibition**, connect the mitral cells with one another and inhibit them. In the opposite direction, the mitral cells have an excitatory effect on the same (reciprocal) synapses. These "circuits" make possible **auto** and **lateral inhibition**, as well as **deinhibition** by higher centers. The axons of the mitral cells project centrally, reaching, among other sites, the *cortex, hypothalamus, limbic system*, and *reticular formation*.

The **role** of the sense of smell includes: (1) eliciting *salivary* and *gastric secretions* in response to pleasant smells and warning against spoiled food by registering unpleasant smells (→ p. 202 ff.); (2) that of a *hygiene monitor* (sweat, excrement); (3) providing *social information* via smells connected with "family" or "outsiders"; (4) influencing *sexual behavior*; (5) affecting the *emotional state* (feelings of enthusiasm, listlessness, etc).

Sense of Taste

The (**secondary**) **taste receptors** are collected together in the *taste buds* (→ **E**), situated principally on the **tongue** and palate. The action potentials elicited by a taste stimulus are transmitted via nerve endings on the sensory cells along the VII, IX, and Xth cranial nerves to the *nucleus tractus solitarii*.

Four basic **taste qualities** are recognized: *salt, sour, sweet*, and *bitter*. The receptors for these sensations are unevenly distributed over the tongue (→ **G**). A large part of the distinctions in "taste" (e.g. apple/pear) are in fact made on the basis of *smell*.

Thresholds. The taste *recognition threshold* for quinine (bitter) is 4 mg/l H_2O and for NaCl 1 g/l, i.e. much higher than for the sense of smell. The relative *intensity difference threshold* is at most about 0.2 (→ p. 306). The concentration of the odorous substance determines whether the taste of a particular substance is pleasant or unpleasant (→ **F**).

The **role** of the sense of taste is, for example, to guard against spoiled food (bad taste: retching reflex; bitter taste [low threshold]: warning, usually poisonous) and to elicit the secretion of salivary and gastric juices (→ p. 202 and p. 208).

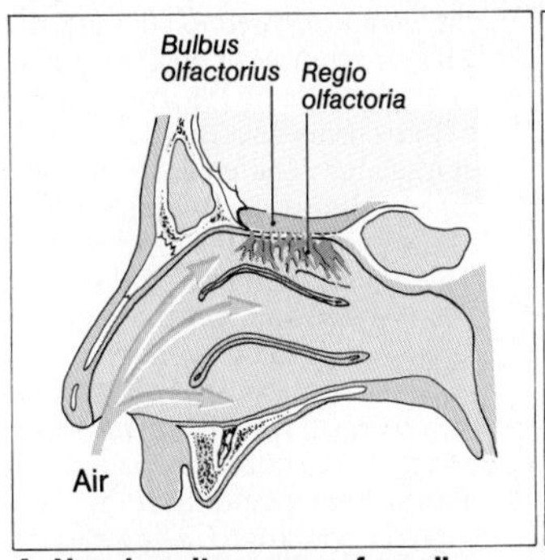

A. Nasal cavity: organ of smell

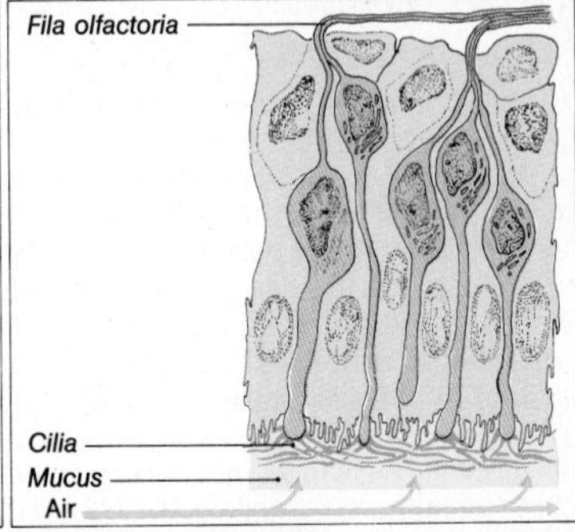

B. Olfactory epthelium (after Andres)

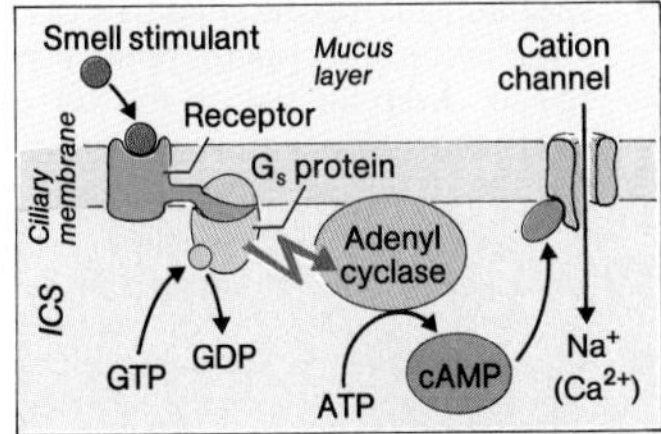

C. Transduction of olfactory stimulus

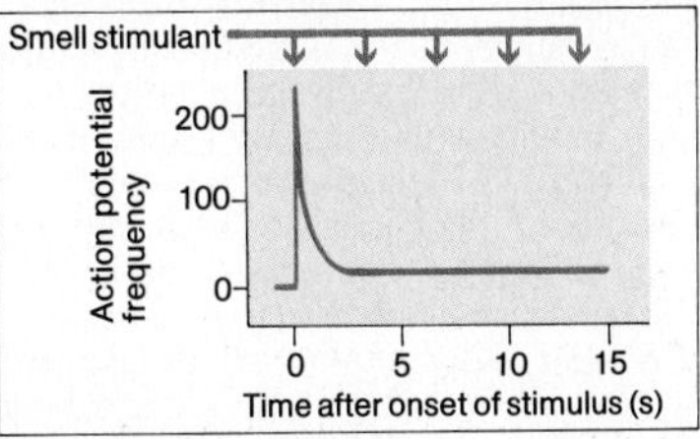

D. Adaptation to smell

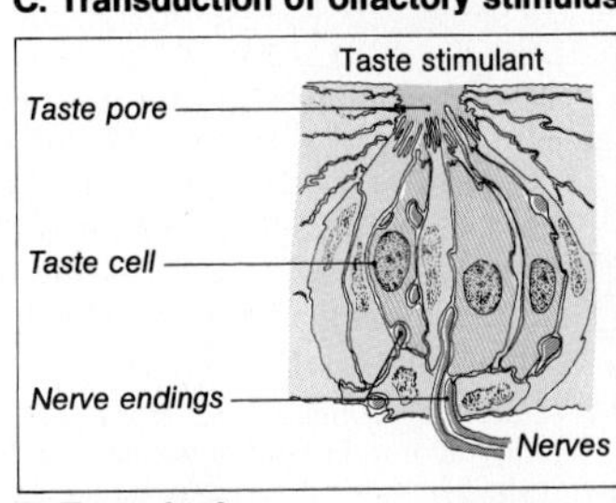

E. Taste buds (after Andres)

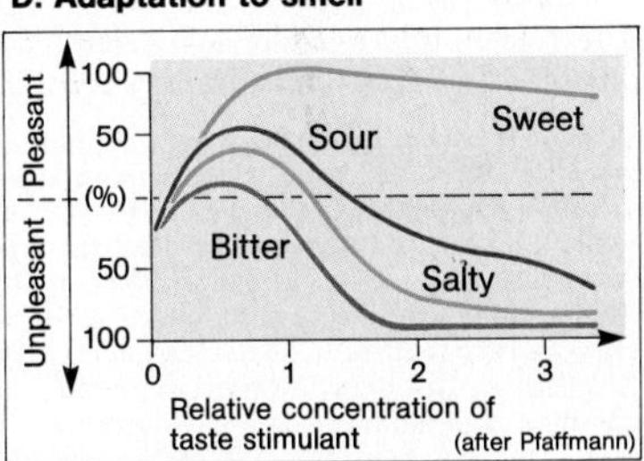

F. Evaluation of taste stimuli

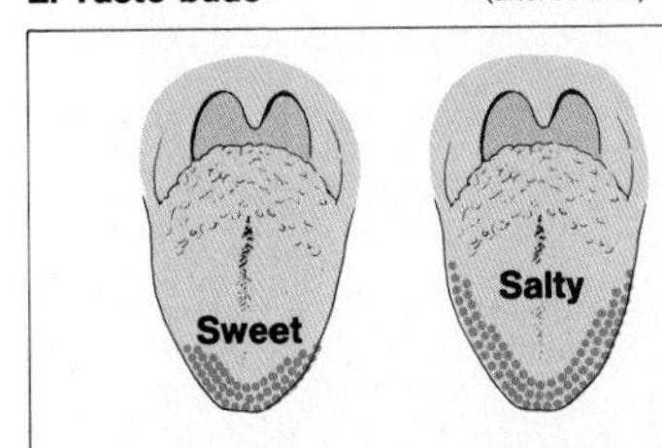

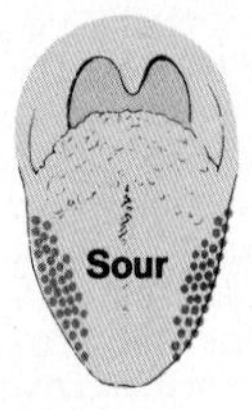

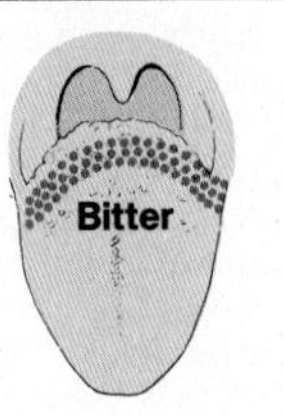

G. Localization of taste qualities on the tongue

Sense of Balance – Vestibular Function

The *organ of balance* or *vestibular apparatus* lies near the cochlea in the inner ear (→ p. 319). It consists of the three **semicircular canals** (→ **A 1**), which each widen to form an *ampulla* containing a *crista* (→ **A 2**) equipped with the ciliated receptor cells. The **cilia** (→ **A 3**) of these receptor cells (one long kinocilium at the cell margin and about 80 short stereocilia) are embedded in the gelatinous, swaying, *cupula* (→ **A 4**). When the head is turned the semicircular canal automatically moves with it. The canal contains **endolymph** (has the same density as the cupula) that, on account of its inertia, cannot immediately follow this movement, that is, a short-lived *current* arises in the semicircular canal and bends the cupula, and with it all of the cilia in, for example, the direction of the kinocilia. This in turn leads to excitation of the sensory nerve fibers. *Three* semicircular canals are necessary in order to register rotation about all possible spatial axes (nodding, turning, and sideways bending of the head).

The **role of the semicircular organs** is to measure *angular* (*rotational*) *acceleration.*

If the body rotates for a longer stretch of time at a constant speed (i.e. when a constant rotational velocity is attained), fluid and canal move at the same rate and no signal is generated. When rotation slows or stops, the endolymph continues to move, but in the reverse direction. If the swinging of the cupula at the start has increased the frequency of the action potentials generated, slowing down now inhibits them, and vice versa.

The vestibular organ possesses two other sensory epithelia, the **saccule** (→ **A 5**) and the **utricle** (→ **A 6**), each of which contains an otolithic organ, the **macula**. The latter contains cells bearing kinocilia and stereocilia surmounted by a gelatinous membrane (→ **A 7**) in which relatively heavy calcium carbonate crystals (density ≈ 3.0), the *statoliths* (→ **A 8**) are embedded. When the head is moved, these "weights", on account of their *inertia*, displace the membrane with the cilia and stimulate the hair cells (→ **A 3**). At different resting positions of the head relative to gravity, the statoliths move the membrane and also its cilia in response to the change in direction of the pull of gravity. The recognition of difference in direction is made possible by the irregular arrangement of the sensory cells (kinocilia) in the macular epithelium. The **role of the maculae** is to register linear (**translational**) **accelerations** and *deviations* of the head *from the vertical.* The bipolar neurons of the *vestibular ganglia* (→ **A 9**) conduct the impulses from the sensory cells to the *vestibular nuclei*, whence important tracts lead to the nuclei of the *ocular muscles*, to the *cerebellum* (→ p. 286), to the motoneurons of the *skeletal muscle*, and to the postcentral gyrus (conscious spatial orientation). The **reflexes** elicited by the vestibular organ principally serve two mechanisms: (a) maintenance of the equilibrium of the body (**motor hold**; → p. 284), and (b) to "keep an eye on the surroundings", independently of movements of the head and body (**visual motor control**).

If, for example, the body axis of a test subject is tilted (→ **B**), the resulting stimulation of the vestibular organ leads to extension of the arm and thigh muscles on the "downhill" side (supporting) and flexion of the arm on the "uphill" side (→ **B 2**). In vestibular disturbance, the patient cannot react in this manner (righting reflex is lost) and, therefore, tips over (→ **B 3**).

The intimate connections between the vestibular organ and the nuclei of the ocular muscles (→ **C**) are reflected in the fact that any alteration in the position of the head is at once corrected for by *counter-directional eye movements*; this is a great help in **spatial orientation**.

Since the vestibular organ alone cannot recognize whether only the head or the entire body is moving (which is of course very important for motor hold), the muscle spindles and the receptors in the cervical vertebral joint capsules are very closely connected with the vestibular nuclei and the cerebellum.

The functioning of the vestibular apparatus can be clinically tested by means of its effects on visual motor control. When a body that has previously been rotating about its own axis (swivel chair) for about 30 s (eyes closed to avoid optokinetic nystagmus; → p. 314), comes to an abrupt stop, stimulation of the horizontal semicircular canals causes a **postrotatory nystagmus**: the eyes move horizontally and *slowly in the direction of the rotation*, followed by *rapid return*, whereby rotation to the right produces nystagmus to the left, and vice versa (→ p. 314).

(**Kinetoses** → p. 204).

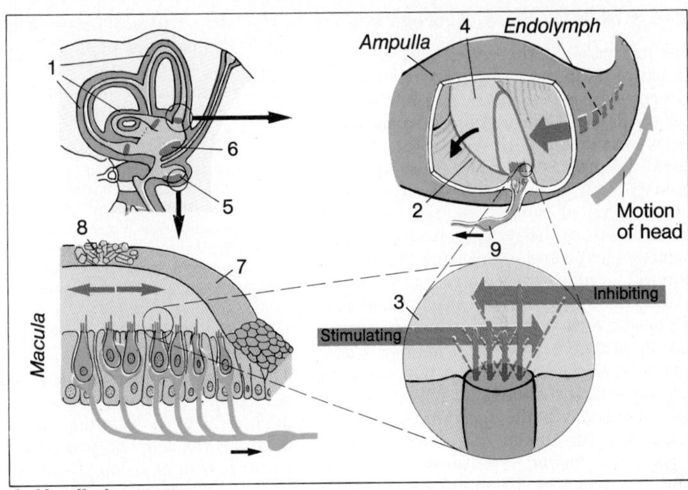

A. Vestibular organ

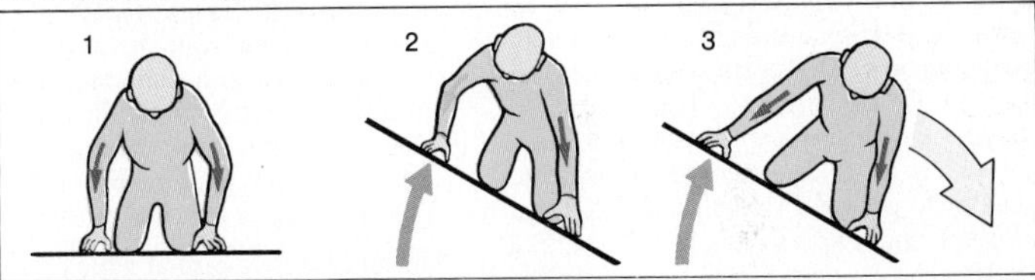

B. Vestibular organ: effect on equilibration

(after Kornhuber)

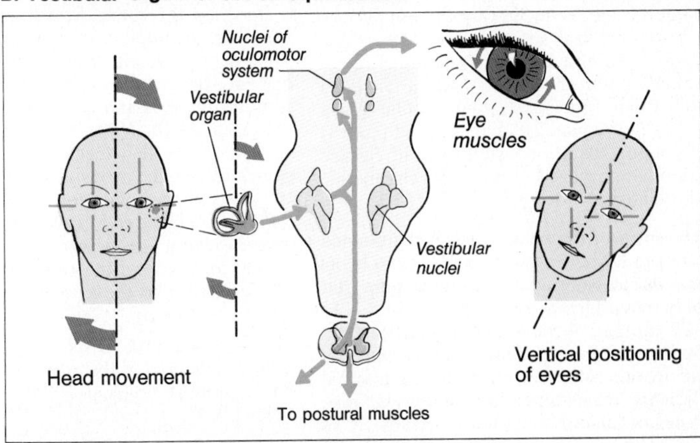

C. Vestibular organ: effect on eye movement

Eye Structure, Tears, Aqueous Humor

Before reaching the *retina* and its light-sensitive receptors, light must pass through the *cornea, aqueous humor, lens*, and *vitreous gel*. With the help of this **optical apparatus** a *reduced* and *inverted image* of the environment is projected onto the retina. The transparency, constancy of form, and smoothness of surface of the individual components of the system have to be maintained at an optimum to ensure the production of an undistorted image. In the case of the **cornea** this is to a large extent taken care of by the **lachrymal fluid**, which is continuously secreted by the *lachrymal glands* (situated in the upper, outer corner of the eye socket), distributed by reflex *blinking*, and drained via the *lachrymal canaliculi* (openings on upper and lower eyelids; *puncta lachrymalia*; → **B**), the *lachrymal sac*, and the duct into the nasal cavity. Tears serve to lubricate the *lids*, to wash away irritating particles and chemicals, to correct unevenness of the corneal surface, and to prevent its drying out. They contain lysozyme (→ p. 66) and immunoglobulin A (→ p. 64) as a *defense* against infections. Tears also provide a much-exploited means of emotional expression.

Entry of light into the eye is regulated to some extent (→ p. 306) by the **iris** (→ **A**), which functions in the same way as the diaphragm of a camera. It contains radial smooth muscle fibers (*M. dilatator pupillae*) that dilate (*mydriasis*), and circular fibers (*N. sphincter pupillae*) that constrict (*miosis*) the pupil. Dilatation results from adrenergic stimulation; constriction, from cholinergic stimulation.

The form of the ocular bulb is maintained by its tough **sclera** (→ **A** and **C 1**) and by its **internal pressure**, which is higher than that of the environment (normally 2–3 kPa = 15–22 mmHg). A balanced production and outflow of the aqueous humor is essential for maintaining constancy in this internal pressure, which tends to increase with age. **Aqueous humor** is a fluid produced by active transport of electrolytes (carbonic anhydrase is involved) in the *ciliary process* (→ **C 2**) of the *posterior chamber* (→ **C 3**) and is completely replaced every 60 minutes. It flows through the pupil into the *anterior chamber* (→ **C 4**) and exits at its angle through the *canal of Schlemm* (→ **C 5**), where it is drained away by the venous blood flow.

The angle is made smaller when the iris dilates (mydriasis), which may partially impede efflux via the canals. If production exceeds outflow the ocular pressure rises, causing *glaucoma* (*pain* and *visual disturbances*). Ultimately, the high pressure may cause blindness by damaging the retina. Treatment of glaucoma may involve (1) inhibition of the formation of aqueous humor (e.g. by carbonic anhydrase inhibitors); and (2) stimulation of miosis to increase the size of the drainage angle.

The **lens** is suspended on the *zonular fibers* or suspensory ligaments (→ **C 6**), which influence the tension on the lens and thereby its shape, its dioptric power, and its focal point. In distant vision (**distant accommodation**) the **ciliary muscle** (→ **C 7**) is relaxed, the zonular fibers are stretched, and the lens, particularly its front surface, is flattened. In near vision (**near accommodation**) the ciliary muscle contracts, thus allowing the zonula fibers to relax and the lens, due to its elasticity, to resume its original, more curved form (→ **D** and p. 302).

Except for the point at which the optic nerve (→ **A**) leaves the eye (*optic disk*), the eyeball is lined by the **retina**, which ends at the margin of the pupil. Near the optic disk is a slight depression, the *fovea centralis* (→ **A**), opposite the pupillar aperture. The retina contains the **rods** and **cones**, both of which are receptors for light. They connect with *bipolar cells*, which then synapse with the *ganglion cells*, the axons of which form the 10^6 fibers of the **optic nerve**. The *horizontal cells*, the *amacrine cells*, and a third cell type (*interplexiform cells*) provide additional connections among the receptor cells, the bipolar cells, and the ganglion cells of the retina (→ **E** and p. 312).

The light-sensitive visual pigments (→ p. 304) are stored in the membrane disks of the outermost segments of the rods and cones (→ **F**).

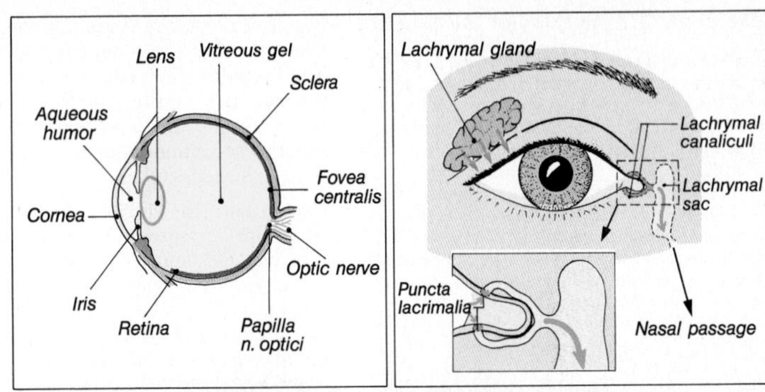

A. Right eye (horizontal section)

B. Right eye: tear flow

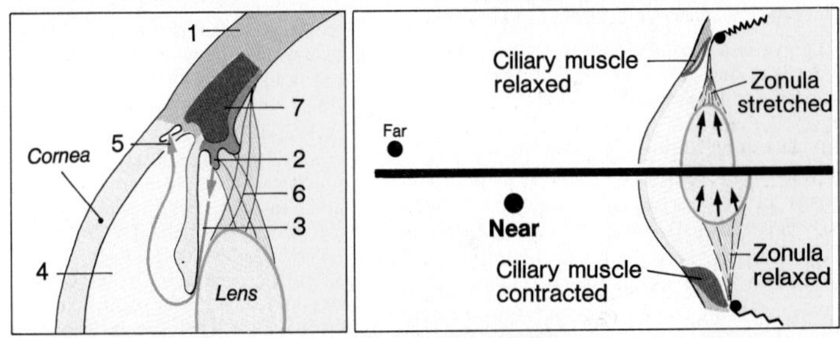

C. Aqueous humor

D. Accommodation

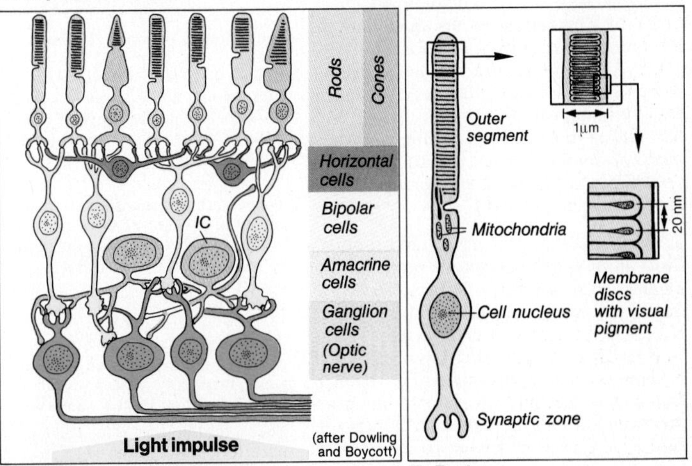

E. Retina (schematic)

F. Rods

The Optical Apparatus of the Eye

Light rays that cross an interface, such as from air to another medium, are refracted (bent). If they pass through a spherical interface, an *image* is produced (i.e. all rays originating in a point source meet again at a point on the other side of the interface). A **simple optical system** of this kind (→ **A**) has an *anterior focal point* (in air) ($\mathbf{F_a}$), a *posterior focal point* ($\mathbf{F_p}$), a *principal point* (**P**), and a *nodal point* (**N**) (see textbooks of physics). Light rays from a *very distant point* (∞) can be considered to be *parallel*. If they also enter the system parallel to its *optical axis*, they meet at F_p (→ **A 1**, red dot). If they enter at an angle to this axis, however, they form an image to one side of F_p, but in the same (*focal*) *plane*. (→ **A 1**, mauve dot). Light rays from a *near point* are *not* parallel and, therefore, do not form an image in the focal plane, but rather *behind* it (→ **A 2**, green and brown dots).

The eye is a complex optical system with many interfaces and media (→ p. 300), but it can be treated as a single optical system ("*reduced eye*").

In **accommodation for distant vision**, the parallel rays originating from a far distant point produce a point ("sharp") image at F_p (→ **B 1**, red dot). In the eye accommodated for distant vision, F_p lies exactly *on the retina* so that the receptors receive a sharp image. To the eye accommodated to distant vision, a near object appears blurred since it is focused *beyond the retina* (→ **B 1**, green dots). In **accommodation for near vision** the lens must increase its curvature (→ p. 301, D), that is, its *refractive power* (see below), to bend the rays onto the retinal plane so that the image of a near point is "sharp" (→ **B 2**, green dots). In this state, a distant object is focussed in front of the retina and is seen as blurred because F_p no longer lies in the retinal plane (→ **B 2**: F'_p).

Refraction of the eye is measured in diopters (dpt) and is the *reciprocal value of the anterior focal length* ($1/f_a$) in meters (f_a = distance between F_a and P = 0.017 m when the eye is accommodated for distant vision; → **B 1**). Thus, the refractive power of the eye accommodated to maximum far vision is 1 : 0.017 = 58.8 dpt. In maximum near-accommodation, the lens can increase its refractive power by about 10 dpt. This is known as the **range of accommodation**, and can be calculated from 1/near point – 1/far point. The **near point** is the nearest point to which the lens can accommodate (normal value in young adults approximately 0.1 m). The **far point** is the farthest point to which the lens can accommodate to produce a sharp image, and is normally at ∞. The range of accommodation of the normal eye, since $1/\infty = 0$, is therefore 1/near point.

The lens stiffens with age and its range of curvature decreases. The near point therefore becomes more distant as one grows older and the range of accommodation decreases. This condition is known as **presbyopia** (→ **C 1–C 3**). Distant vision remains unaffected (→ **C 1**). Correction for near vision (reading) in presbyopia is achieved with a (convex) *convergent lens* (→ **C 3**).

A **cataract** is a decrease in transparency of the lens. The lens may be removed surgically and replaced by a convex lens of at least + 15 dpt (glasses, contact lenses, or an artificial lens in the eye itself).

In **myopia**, or short sightedness, the light rays focus *in front* of the retina (usually because the eye ball is too long; → **C 4**). Normally the far point is at ∞, but in myopia it is much closer (→ **C 5**). The myopia can be corrected by a (concave) *divergent lens* with minus diopters (→ **C 6** and **C 7**); it must scatter the parallel rays so that they appear to come from this (too near) far point. The refractive power of the lens needed can be calculated from the reciprocal of the far point in meters. Example: far point 0.5 m, lens of – 2 dpt required.

In **hyperopia**, or far sightedness, the eye ball is relatively too short and the image is focussed behind the retina. Accommodation changes for near vision occur at longer than normal distances (→ **C 8**), so that the range of accommodation is insufficient for near distances (→ **C 9**). A *convergent lens* (+ diopters) is needed for correction (→ **C 10, C 11**).

The surface of the cornea is often more strongly curved in one direction (usually the vertical) than in the other. The result is a difference in refraction between the two planes so that a point appears as a line (one plane out of focus): (regular) **astigmatism**. It can be corrected by *cylindrical lenses*. An *irregular astigmatism* with irregularly deformed images arises, for example, from scarring of the cornea; it can be corrected with spherical contact lenses, beneath which lachrymal fluid compensates for the deviation from a spherical form.

The optical apparatus of the eye has a greater refractory power at the edges than in the optical axis. This **spherical aberration** is reponsible for the fact that the wider the pupil the less sharp the image.

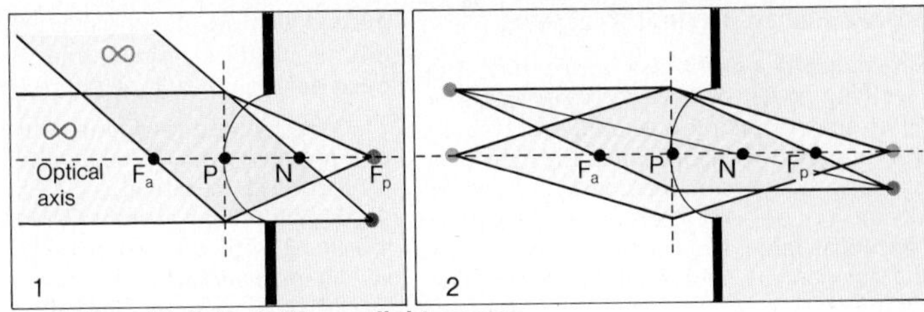

A. Optics: (1) distant, (2) near light source

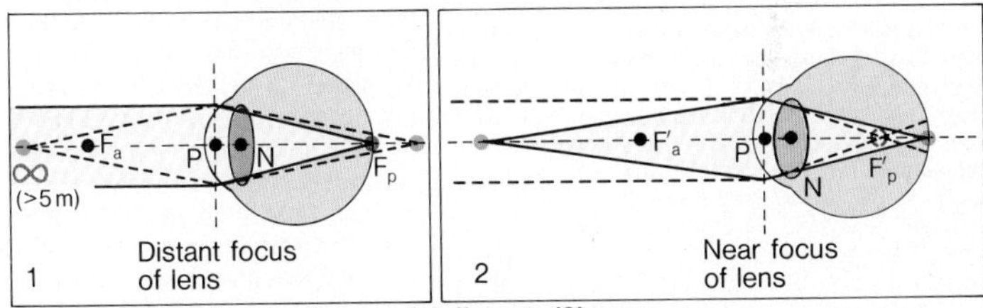

B. Accommodation in the eye (1) distant, (2) near

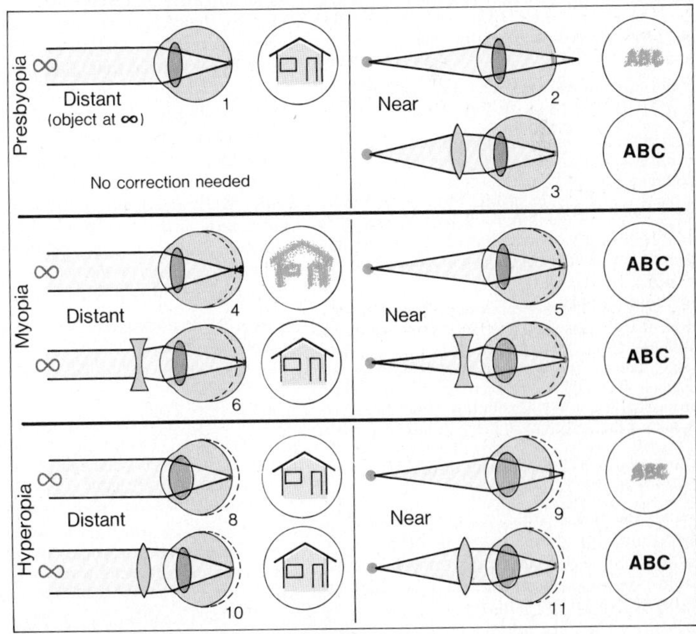

C. Visual defects

Visual Acuity, Retinal Receptors

Visual acuity (visus) is a measure of the resolving power of the eye. With *adequate light*, the normal eye should be able to discriminate two points when the light rays reaching the eye from them form an angle of 1 min (α; → **A**; 1 min = 1/60°). The reciprocal value, 1/α, (in min^{-1}), represents the visual acuity, and in the normal eye it is around 1.

Clinical **testing of visual actuity** makes use of tables with large letters, the details of which, at the given distance (e.g. 5 m; → **A**), are seen as subtending the angle of 1 min. Instead of letters, for example, split rings in which the split subtends the angle of 1 min can be employed (→ **A**). Visual acuity can be calculated from the actual distance/normal distance at which the letters (or gaps) can be recognized. Example: Normally, it should be possible to recognize the gap in the ring to the right in (**A**) at a distance of 3.3 m. In this case the acuity is 3.3/3.3 = (normal). If, however, only the gap in the ring to the left can be recognized at this distance then the acuity is only 3.3/8.5 = 0.39, since the normal eye would recognize the split in the latter ring at 8.5 m.

Rods and **cones** are the light sensitive receptors of the retina (→ p. 301, E).

They are not uniformly distributed: the *fovea centralis* contains *exclusively cones*, their density declining rapidly as the distance from the fovea increases, whereas the density of the *rods* is at its highest 20° from the fovea (→ **B, left**). No receptors are present on the *optic disk*, which corresponds therefore to a *blind spot* in the visual field (→ p. 310).

In order to study an object closely, the gaze adjusts to the foveal axis so that its image is thrown directly onto the fovea centralis, which has the greatest visual acuity. The latter declines rapidly as the distance from the fovea increases (→ **B, right**), corresponding to the distribution of the cones (→ **B, left**). If the sensitivity of the dark-adapted retina is tested, however (→ p. 306), the distribution of visual acuity corresponds to the distribution of the rods (→ **B, right**).

Thus, the **cones** are specialized for highly discriminating color vision in bright light (**phototopic vision**), the **rods** for (black and white) vision in poor light (**scotopic vision** or **twilight vision**). In the latter, visual acuity is sacrificed for dark adaptation.

The rods and cones contain the **visual pigments** in their disk membranes, which mediate the transduction of a light stimulus into a chemical signal and finally into electrical stimulation of the receptor.

The **rods** contain **rhodopsin**, which consists of a protein (**opsin**, 38.000 Daltons) and an aldehyde, **11-*cis*-retinal** (→ **C**). Light produces a steric change at the C-11 atom of the aldehyde, with the production, via *bathorhodopsin, lumirhodopsin* (opsin + 11-*trans*-retinal), and *metarhodopsin I*, of *metarhodopsin II* (whole reaction time is only 1 ms!). The latter substance, like the hormone-receptor complex (→ p. 243), reacts with a **G_s-protein** ("**transducin**"), whereby (after replacement of GDP by GTP) the subunit α_s-GTP splits off (→ **C**). This now activates (not as on p. 243) a *phosphodiesterase*, which in turn *lowers* the **cGMP** concentration in the cell. Activation of one rhodopsin molecule can result in the hydrolysis of up to 10^6 cGMP/second (amplifying effect of the enzyme cascade). Subsequently, cGMP dissociates from (previously open) cation channels in the cell membrane, thus closing them: **hyperpolarization** takes place (*secondary receptor potential*; → p. 312). During these events the Ca^{2+} concentration of the cell also drops (cation channels close), which may be a part of the switching-off or adaptation of the transduction process, or both. Metharodopsin II ultimately breaks down into opsin and its aldehyde component. Rhodopsin is bleached in the process and can be *regenerated* with the expenditure of energy (see also p. 306).

In order for rhodopsin to be bleached, light has to be **absorbed**. Since rhodopsin can absorb light over the entire range of visible light (→ p. 309, D), the rods are unable to distinguish between light of different wavelengths (colors). The three different types of cones, however, with their different pigments (11-*cis* retinal with different opsin components), each absorb only light from a certain, narrow range of wavelength (→ p. 309, E), which is a prerequisite for color vision (→ p. 308).

Retinal is the aldehyde of the alcohol *retinol*, also called **vitamin A_1**. A chronic deficiency of this vitamin or its precursors (carotinoids) leads to *night blindness* due to insufficient formation of rhodopsin (→ p. 306).

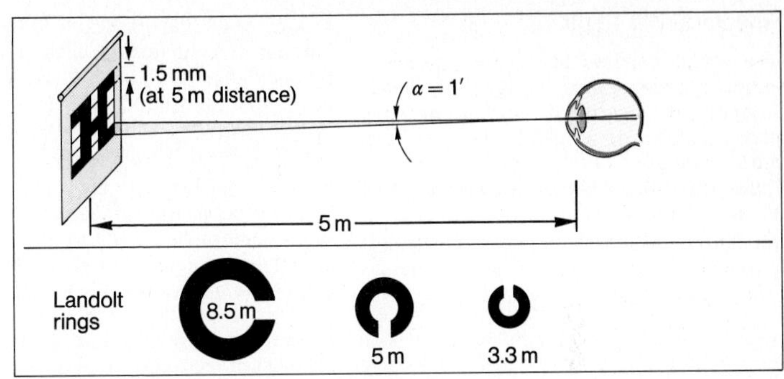

A. Visual acuity

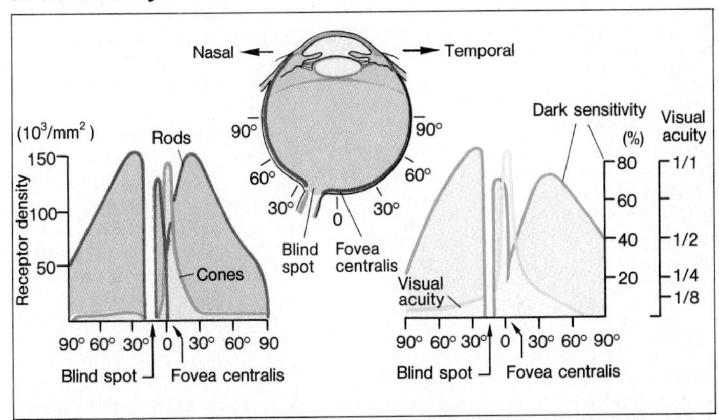

B. Retina: distribution of rods, cones, dark sensitivity and visual acuity

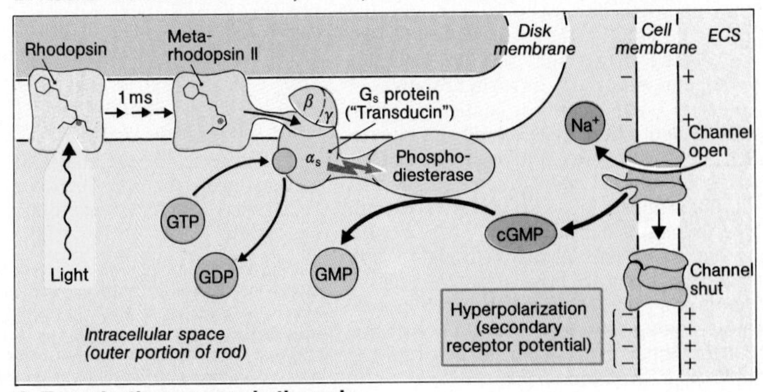

C. Transduction process in the rods

Adaptation to Light

The eye is capable of responding to an extremely wide range of light intensities, from the low intensity of a distant star to the glare of a sunlit snow field. Response to such a broad range ($1 : 10^{12}$) is made possible by the unusual flexibility in the **adaptation** of the eye to the prevailing light intensity.

For example, in going from daylight into a darkened theater, the eye is unable to perceive the low-intensity light signals because they are temporarily below the retinal threshold. After several minutes it is possible to distinguish the outlines of the seats etc. because the threshold has now become lower. In order to observe the stars, a longer period of adaptation is necessary.

Maximum adaptation is achieved after about 30 min (→ **A**), i.e. the minimum light intensity that can just be detected after maximum dark adaptation is the **absolute visual threshold** (in **A** and **B** is it taken to be 1). The time course of adaptation of the retina exhibits a break at about 2000 × the absolute threshold (→ **A**, mauve curve). This is the point at which the *threshold* of the *cones* is reached **(threshold for daylight vision)**, and the rest of the curve is governed by the somewhat slower *adaptation of the rods* (→ **A**, brown curve). The latter can be measured on its own in totally color-blind subjects *(rod monochromism)*; isolated adaptation of the cones can be observerd in subjects suffering from *night blindness* (→ p. 304).

For good vision not only the absolute threshold but also the capacity to discriminate between two light stimuli of similar strength is important. If two light sources of intensity I and I′ can just be distinguished, the **absolute difference threshold**, Δ I, is given by I−I′. The **relative difference threshold** is ΔI/I; in optimal light (about 10^9 × the absolute threshold; → **B**) this is very small (i.e. 0.01), which means that the capacity for resolution is very high, allowing recognition of a change in intensity of 1%. The relative resolution threshold becomes much greater in dark adaptation and in very bright light. In the latter case, sunglasses lower the difference threshold.

Several **mechanisms** are available to the eye **for adapting** to variations in light intensity:

1. The **pupil** can reflexly influence the amount of light falling on the retina by a factor of 16 (→ **C 1**). In darkness, the pupil is larger than in bright light. The main function of the pupil is the rapid adaptation of the eye to a *sudden* change in light intensity (pupillary reflex; → p. 310).

2. The **concentration of the visual pigment** in the receptors "adapts" to the degree of sensitivity required (→ **C 2**). *Bright light* causes the breakdown (bleaching) of more pigment molecules (→ p. 304), thus *lowering* its concentration (until equilibrium is re-established between breakdown and synthesis) and reducing the probability of its coming into contact with further light (photons). In the *dark*, the concentration of pigment is *high* and the probability of contact with a photon is maximal (increased sensitivity).

3. The sensitivity of the eye undergoes extensive adaptation because the retinal area (number of receptors) from which one fiber of the optic nerve receives its stimulus can be varied, possibly by a feedback mechanism (→ **C 3**). This is known as **spatial summation**, a phenomenon that decreases in bright light and increases in darkness. In effect, the organization of the receptive field is changed (→ p. 312). The "gain" is higher in dim light than in bright light.

4. By prolonging a subthreshold stimulus (looking at an object for a longer time), light intensity can be summated to exceed threshold and thus elicit an action potential (AP). This is termed **temporal summation** (→ **C 4**). The product of ligh intensity times duration of stimulus required to produce an AP is a relatively constant value.

5. The drop in intracellular Ca^{2+}-concentration during the transduction process (→ p. 304) may also have something to do with adaptation.

A local kind of adaptation is seen in the phenomenon of **successive contrast**. After staring at a black and white pattern for 20 seconds and then shifting the gaze to a plain field (→ **D**) the pattern reappears, but with a reversal of black and white because the areas of the retina that registered the black image in the first pattern have become more sensitive.

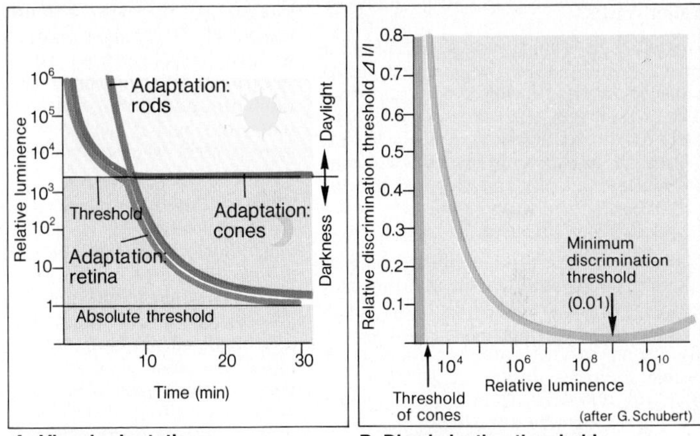

A. Visual adaptation

B. Discrimination threshold

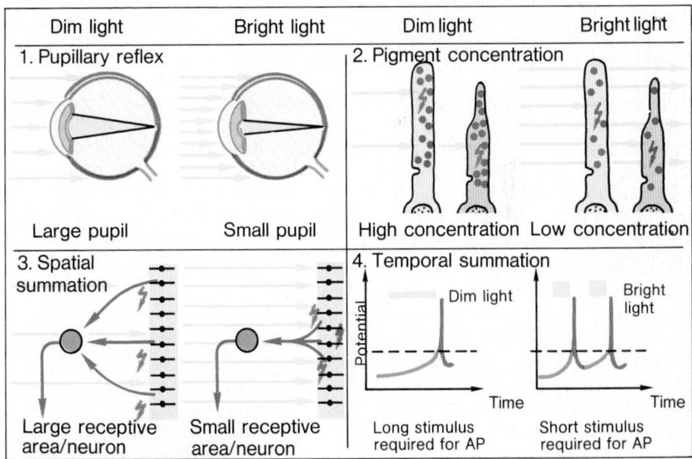

C. Mechanisms of adaptation

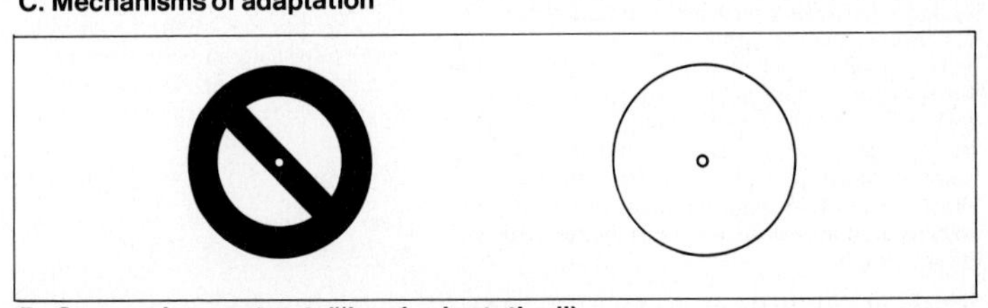

D. Successive contrast ("local adaptation") (see text)

Color Vision

If white light is split by passage through a prism, a color *spectrum* ranging from red to violet is obtained (colors of the rainbow). The wavelength (λ) of *red* is about 650–700 nm; that of *violet*, roughly 400 nm (→ **A**). The human retina is sensitive to light within this range; light rays of shorter (*ultraviolet*) or longer (*infrared*) wave lengths are not visible to the human eye.

White light can be produced if two **complementary colors**, such as blue (490 nm) and orange (612 nm) are *mixed additively*. This is demonstrated in the classic *color triangle* (→ **B**): the upper two limbs incorporate the visible spectrum. Inside the triangle is a point to represent white light. Any line passing through the white light point intersects the upper limbs of the triangle at two complementary wavelengths (e.g. 612 and 490 nm; → **B**). The additive mixture of roughly *equal quantities* of red and green yields yellow (→ **C**). If the proportion of red is larger the impression received is orange; less red yields yellow-green, that is, the color between red and green on the arms of the triangle. The same applies to mixtures of green and violet (→ **B** and **C**). (The combination of red and violet results in purple colors, which are not contained in the spectrum; → **B**).

This means that by varying the proportions of the *three "basic" colors – red, green and violet* – all other colors can be produced. White can also be produced since it is possible to mix every possible pair of complementary colors from the three basic colors.

Subtractive mixture of colors employs the opposite principle to additive mixture (→ **C**) and is used, for instance, in photography with color filters and in mixing paints. A yellow paint or filter absorbs the blue portion of white light, leaving the complementary color yellow. If red is added to the yellow paint, green is also absorbed, and the resulting color is orange. Using a yellow filter, the blue sky in black and white photographic prints appears darker, and therefore contrasts better with the bright clouds.

The **light sensitivity of the retinal receptors** depends upon the ability of the visual pigments within them to **absorb light**. The rhodopsin in the rods (→ p. 304), which is responsible for twilight vision (black and white) absorbs light from the *entire* visible spectrum. (The absorption maximum of rhodopsin is at 500 nm, which explains why blue-green light is the color mixture best seen at night whereas red appears as the darkest color; → **D**).

There are **three types** of **color sensitive cones** (→ **E**): those that strongly absorb blue-violet light, a second type that strongly absorbs green light, and a third type that, besides its strong absorption of yellow, also absorbs red light. The three types of cone are stimulated by "their" color, giving the retina its color sensitivity (**trichromatic theory of color vision** of Young and Helmholtz; → see also p. 312). Over a wide range of the visible spectrum, differences in wavelength of only 1–2 nm can be distinguished (absolute **color difference threshold**; → **F**, "normal").

However, **color perception** is even more complex because, for example, a "white" paper appears white no matter if white light (sun), yellow light (bulb), or even red light is used. Similarly, we do not see different colors when we look at the sunny or the shady side of a house. This *color constancy* (as well as size constancy, etc.) is the result of retinal and central processing of receptor signals.

In **color blindness** certain colors can only be distinguished with difficulty, or not at all (high color difference threshold; → **F**). Roughly 9% of the males and 0.5% of the females in a population are afflicted in this way (usually inherited). Red and green color blindness are known, respectively, as *protanopia* and *deuteranopia*, blue violet blindness as *tritanopia*. If there is only a color "weakness", the suffix *anopia* is replaced by *anomaly*. In order to test the ability to distinguish between colors (especially important in traffic, and in jobs involving paints or fashion) color tables or an anomaloscope are used. With the latter, the subject is required to mix a certain yellow from red and green. In protanomaly, for example, the subject uses too high an intensity of red. A person suffering from protanopia sees all colors of wavelengths above about 520 nm as yellow.

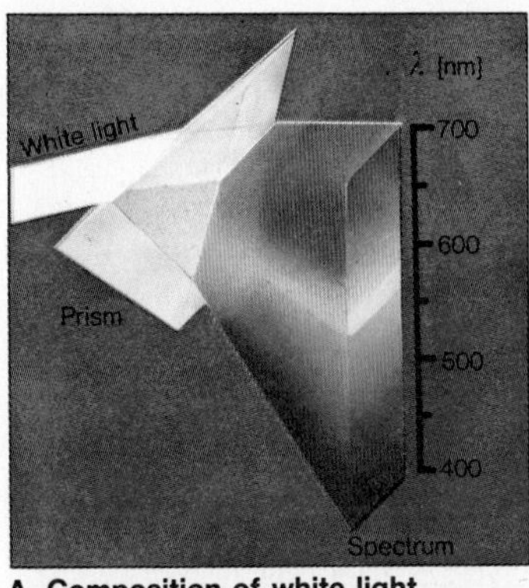

A. Composition of white light

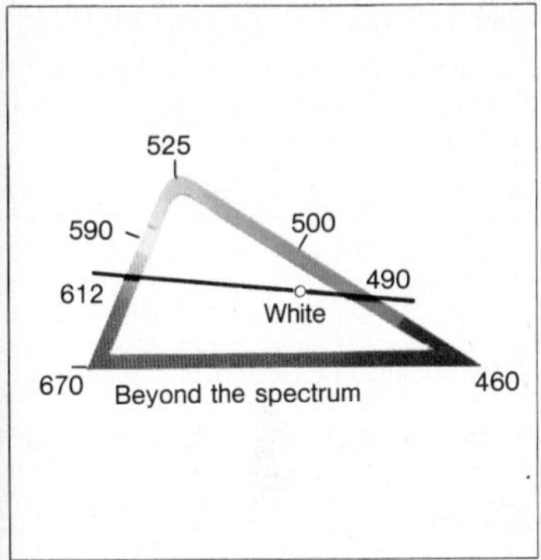

B. Color triangle (after Kries)

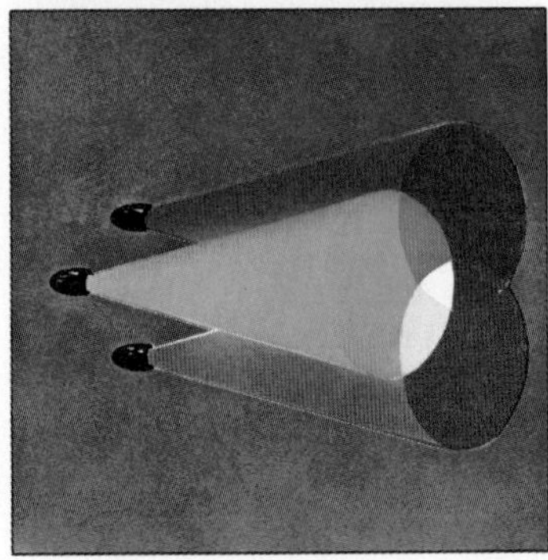

C. Additive color mixing

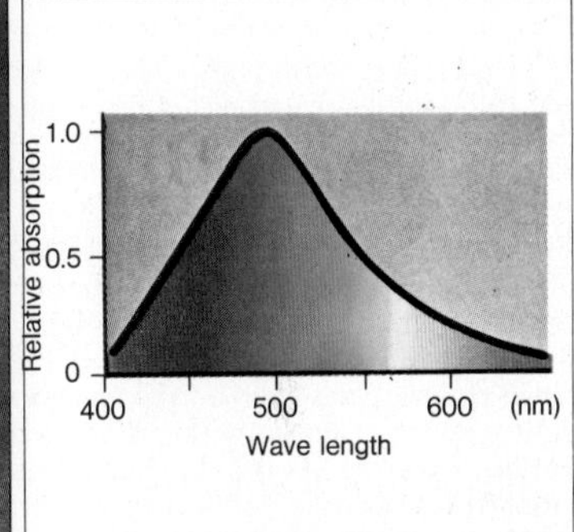

D. Spectral absorption curve: rhodopsin

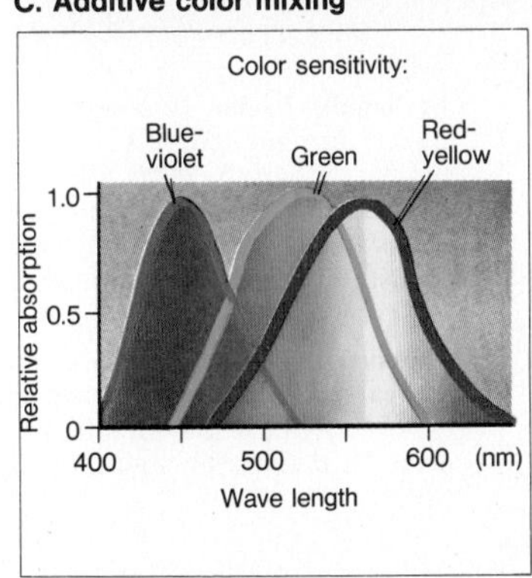

E. Spectral curves of cone pigments

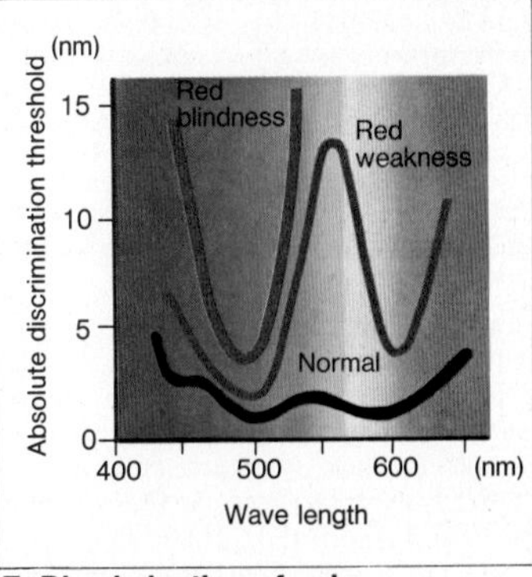

F. Discrimination of color

Visual Fields and Visual Pathways

The **visual field** of each eye is the segment of the external world seen by the fixed eye in a fixed head (→ **A 1**).

The visual fields are measured by a **perimeter**, which, in principle, is a hollow hemisphere. The eye is positioned at its center, and a light source (white or monochromatic) is moved into the field from the periphery. The points at which the light source are first recognized or disappear again are noted and mapped on the background. Partial interruptions in the visual field are termed **scotomas**, and may be due to a defect in the optical apparatus (e.g. cataracts; → p. 302), in the retina (e.g. inflammation) or at some point along the visual pathway (see below). The normal field of vision already contains a scotoma, the **blind spot** (→ **A 1**), which corresponds to the optic disk (→ p. 300). In the binocular (both eyes) visual field (→ p. 315, A) each blind spot is compensated by the other eye.

The **visual field for colored light** stimuli is smaller than that for black-white stimuli. If, for example, a red object is slowly moved from the periphery into the visual field the movement itself is perceived much sooner than the color is recognized.

The nasal part of the visual field of each eye registers on the temporal half of the corresponding retina and the temporal part on the nasal half (→ **A 2**, blue and green). In the **visual pathway**, the fibers of the *optic nerve* that come from the temporal retina remain on the same side (→ **A 2**, blue and green); those from the nasal half of the retina *cross to the other side* in the *optic chiasma* (→ **A 2**, orange and red).

Because the image is inverted on the retina, damage to the left optic nerve, for example, (→ **A 2, a** and **A 3, a**) results in loss (scotoma) of the entire visual field of the left eye, whereas a lesion in the left *optic tract* (→ **A 2, b** and **A 3, b**) results in loss of the right halves of the visual fields of both eyes. A lesion of the chiasma (→ **A 2, c** and **A 3, c**) results in loss of both temporal fields since the nasal halves of both retinas become nonreceptive.

Whereas the retina contains about 130 million receptors, the optic nerve contains only about 1 million fibers. Thus there is a very high degree of **convergence** of many receptors onto few neurons. This is much higher for the receptors at the periphery of the retina (more than 1000 : 1) than at the fovea where only a few cones converge onto one neuron and are thus well represented in the visual cortex. Low convergence, as in the fovea, produces a higher degree of visual acuity at a lower light sensitivity; high convergence of signals from the periphery has the reverse effect (→ see also *spatial summation*, p. 306 ff.).

Collaterals of the fibers of the **optic tract** proceed to the following stations:

1. **Lateral geniculate body (LGB)**. Most of its neurons pass as *visual radiation* to the primary visual cortex (V 1) and, following relay, to the secondary (V 2) and tertiary visual cortex (V 3, V 4), etc. (Function → p. 312 ff.).

2. **Visual motor centers** in the brain stem (after relaying in the "accessory" optic tract). It primarily controls vergence and vertical eye movements.

3. **Superior colliculi**. This connection and succeeding relays regulate saccades (→ p. 314).

4. **Hypothalamus** (Suprachiasmatic nucleus). The day-night alternation is registered here for synchronization of circadian rhythmicity (→ p. 292).

5. **Pretectum**, where, among other things, the width of the pupils is controlled.

6. **Nucleus of the optic tract**. Via these fibers visual signals reach the **cerebellum** (→ p. 286 and p. 298), which integrates vertical and horizontal motions of visual targets and their surroundings, with eye and head movements indexed to three-dimensional space.

The **pupillary reflex** is initiated by a sudden change in light intensity (→ p. 306). The efferent limb of the reflex follows parasympathetic fibers of nerve III and narrows the pupils (*miosis*). *Both* pupils react simultaneously even when the stimulus reaches only *one* eye (*consensual reflex*).

The **corneal reflex** is protective: touching the cornea (afferent: nerve V) or approaching the eye (a flying insect, for example; afferent: nerve II) initiates the reflex, and the eyelid closes.

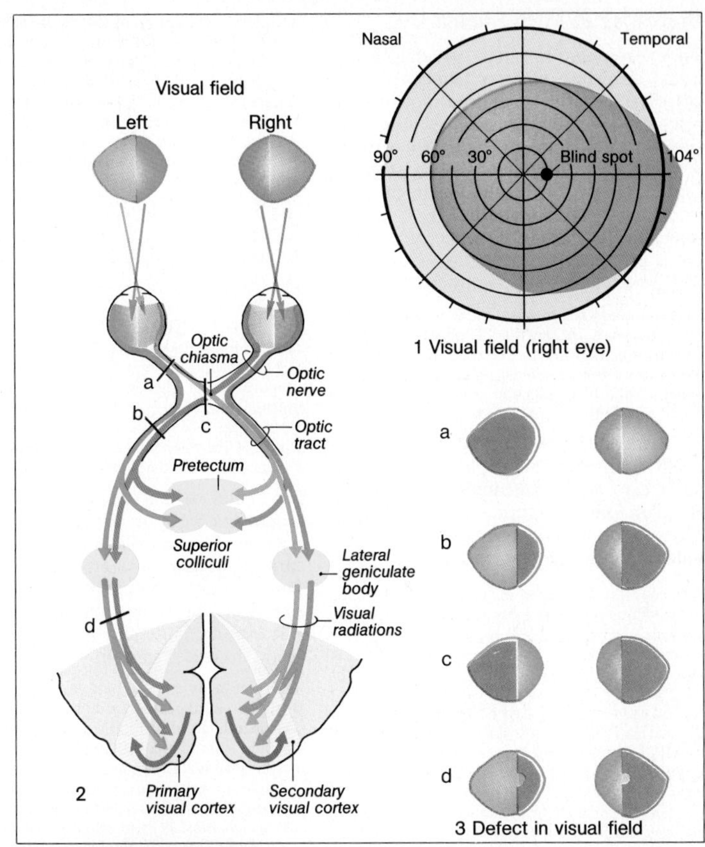

A. Visual pathway and field

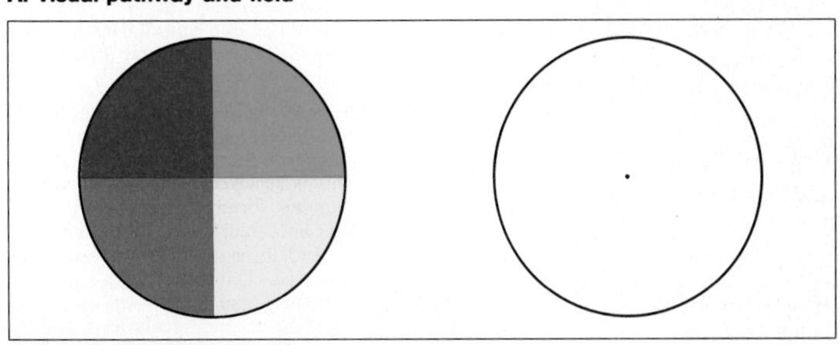

B. Successive contrast, color (see text, next page)

Processing of Light Stimuli

The eye responds to light by generating a so-called **secondary receptor potential** (→ **A, left**). The (negative) membrane potential (at rest − 30 to − 40 mV) becomes more negative (up to − 70 mV) in proportion to the strength of the stimulus. Unlike other receptors (→ p. 274), a stimulus in this case leads to *hyper*polarization. Over a wide range, the magnitude of the receptor potential is proportional to the *logarithm* of the relative strength of the stimulus.

This secondary receptor potential, which is caused by a decrease in the Na^+-conductivity of the receptor membrane, is preceded (roughly 1 ms after exposure) by a *primary fluctuation in receptor potential* due to changes in conformation of the visual pigments (→ p. 304).

Involving the whole circuit of the retina (→ p. 301, E), a sufficiently high receptor potential elicits **action potentials** (APs) in the ganglion cells (optic nerve; → **A, right**). Their *frequency* increases with the magnitude of the receptor potential (→ p. 274).

APs can only be elicited in the ganglion cells and the amacrine cells. The other cells communicate via graded, electrotonically propagated changes in potential (→ p. 28), which apparently suffice for the short distance involved within the retina. The advantage of this kind of propagation is that information involving either hyperpolarization or depolarization can be transmitted (similar to EPSP and IPSP; → p. 30). Because receptors react to light with *hyper*polarization, after which the ganglion cells are *de*polarized, the presence of inhibitory neurons at some point in the complicated retinal synaptic network is implied. A light stimulus causes deinhibition of these inhibitory neurons.

If APs are led off from the ganglion cells or subsequent neurons with microelectrodes, it is possible, by using the appropriate light stimuli, to delineate the area of the retina from which the stimulatory and inhibitory influences on the frequency of these APs originates. This area is the **receptive field** for that particular neuron.

The receptive fields of the **ganglion cells** of the retina are usually concentric, and, when light-adapted (→ p. 306), they exhibit two zones; a **central** zone and a circular **periphery** (→ **B**). Stimulation of the *center increases* the spike frequency (→ **B 1**), whereas stimulation of the *periphery inhibits* the propagation of APs. Turning off the peripheral light stimulus results in excitation (→ **B 2**). This type of receptive field is called **center on**, since light stimulates the center. In other receptive fields, the reverse situation occurs: light stimulates the periphery and suppresses the center; this type is called **center off** (→ **B 3** and **B 4**). The *functional organization of these receptive fields* is largely due to the neuronal connections within the retina (→ p. 301, E).

The function of the contrasting reaction of the center and periphery of the receptive field is to increase the **contrast** of a stimulus.

Where light and dark adjoin, the dark is perceived as darker and the light as lighter. A gray circle appears darker against a white field than against a dark field (**simultaneous contrast**; → **C, left**). In **C** (**right**) the white mesh appears darker at the points of intersection, whereas those of the dark mesh appear lighter. This phenomenon is explained by the balance sheet for stimuli within an receptive field (→ **C, middle**), which demonstrates that there is less contrast at the intersections.

The center of the receptive field increases progressively during dark adaptation at the expense of the peripheral zone until, in full dark adaptation, it completely disappears. Thus, spatial summation (→ p. 306) increases at the same time as contrast decreases (and consequently the sharpness of vision decreases as well; → p. 304).

Receptive fields can also be determined in neurons of higher visual centers, although here they often take on a different form; the direction and form of the light stimuli play an important role. Other receptive fields allow antagonistic responses to red-green or to yellow-violet sources. Functionally, this conforms with the **contrasting color theory** of *Hering*, whereby contrast also takes place (centrally) in connection with color vision. If a colored field is studied (→ p. 311, B) for about 30 s and the gaze then shifted to a neutral background, the complementary color will appear (**colored successive contrast**, → see also p. 306 ff.).

From the *LGB*, with its magnocellular and parvocellular regions, the **information on color, form, and movement** is conveyed by the visual radiations. This is achieved partly via separate information channels, forming, in cooperation with the LGB, V 1 (with "blobs" and "interblobs"), V 2, and V 4 (→ p. 310), a *tripartite processing system* for (a) color via "blob-channel", (b) highly resolved stationary perceptions of form via "parvo-interblob-channel" (colorless), and (c) movement and stereoscopic depth via "magno-channel" (colorless). Unified visual perception is only possible after integration of these individual aspects.

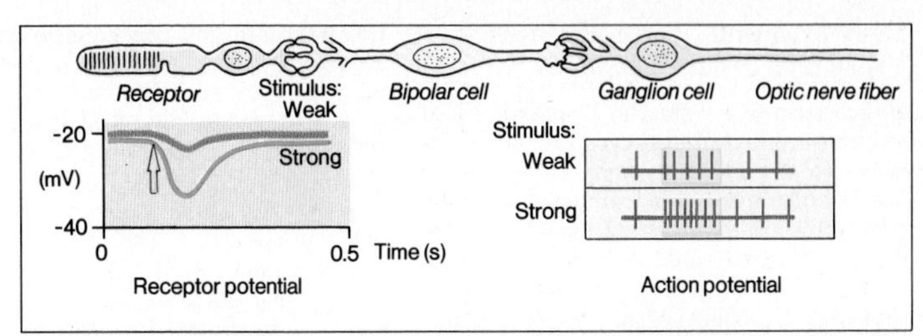

A. Receptor potential (1st and 2nd neuron) and action potentials of the retina

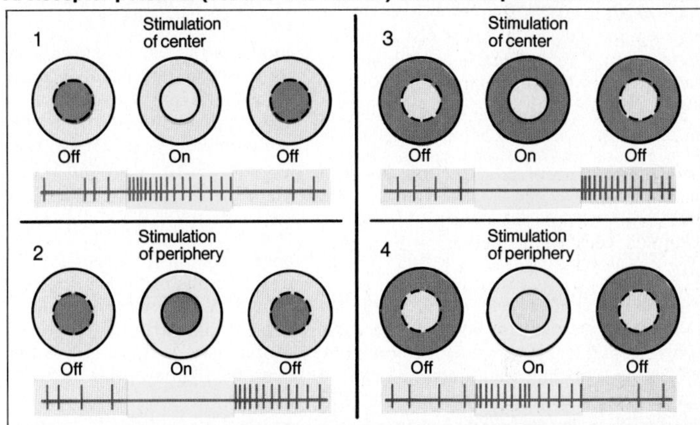

B. Receptive fields of the retina: center ON (1,2), center OFF (3,4)

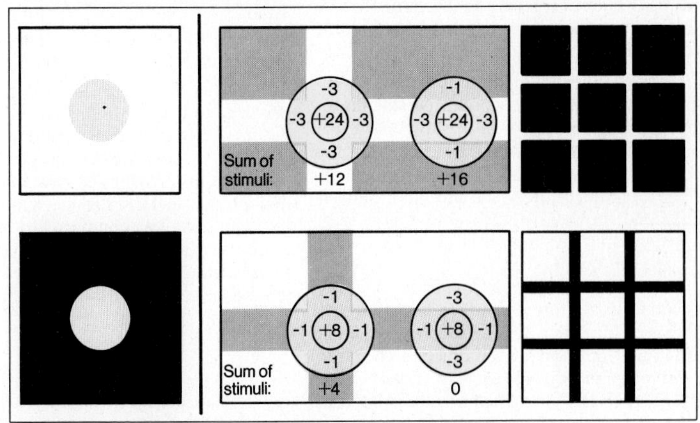

C. Contrast enhancement by receptive fields (center ON)

Eye Movements, Three-dimensional Vision and Distant Vision

If the **extrinsic eye muscles** (see textbook of anatomy) move both eyes in the *same* direction (from left to right, for example) this is termed *conjugate movement*; when the eyes move in contrasting directions, the movements are termed *vergence movements*. A changeover from diverging to converging movements occurs when the eyes move between far and near objects. For the **adjustment to near vision**, *convergence* of the two optical axes is accompanied by reflex *narrowing of the pupils* (→ p. 310), and by *accomodation* (→ p. 302).

The degrees of accomodation and of convergence in this complex reflex do not always match. In hyperopia, convergence becomes excessive because of the greater degree of accommodation demanded by the visual defect, and the patient **squints**. If the optical axes of the two eyes deviate too much from one another (**strabismus**), the image from one eye is suppressed centrally and blindness of this eye may develop.

In searching a given field, as in reading, the eye performs small, rapid, jerky **saccadic movements**. The eyes are fixed while a block of letters is being read, then they move quickly to the next fixation point. The shift of the image is *centrally suppressed* at the moment when the eyes move. (If one looks at both eyes alternately in a mirror the eye movements are only visible to an outside observer).

In order to keep a moving object "in sight" the eye makes slow, **smooth pursuit movements**. The combination of slow eye movements followed by rapid movements in the opposite direction is known as **nystagmus**. The direction of the nystagmus (left, right) is denoted as that of the rapid phase (e.g. postrotatory nystagmus; p. 298). An *optokinetic nystagmus* occurs, for example, if telephone poles are observed from a moving train (pursuit movement); when the eyes have made their rapid return movement they can focus on a fresh object, and so on. *Pathological nystagmus* can result from damage to the *cerebellum* (→ p. 286) and to the *vestibular organ* (→ p. 298).

Distant vision and **three-dimensional vision** are primarily achived by a combined effort of both eyes and are thus more or less limited to the **binocular field of vision** (→ **A**). When the gaze is fixed on a point (→ **B**, A), its image is thrown onto the fovea of both eyes (A_L, A_R) at the so-called *corresponding points of the retina*. The same holds for the points B and C (→ **B**) since they, too, lie on the **horopter circle** (in fact a spherical surface, the horopter) that can be described through the point A and the two nodal points N (→ p. 303, B).

In a hypothetical "**cyclopean**" **eye** in which the two retinas are brought functionally into coincidence (in the visual cortex), the corresponding points on the retinas are now represented by a single point (→ **C**; $A_L + A_R \triangleq A_M$). If a point D (→ **C, left**) lies *outside* the horopter, the cyclopean eye "sees" a double image (D′, D″), D′ coming from the left eye. If D and A are not too far apart, central processing of the double image results in the impression that D lies *behind* A (i.e. a perception of *depth* is gained). Similarly, point E (→ **C, right**) is nearer than A. Since in this case E′ comes from the right eye, E is recognized as being "nearer".

At great distances and in monocular vision, spatial interpretation requires the use of other *cues*: overlapping contours (→ **D 1**), haze (→ **D 2**), shadows (→ **D 3**, perspective (→ **D 4**), etc.

Movements of the head or the entire body provide additional information regarding depth of distance. Seen from a train, for instance, a near object appears to move relatively faster over the visual field than a distant one (→ **D**; station sign as compared to wall, or wall as compared to mountain). A similar example of *depth perception due to relative movement* is that the mountains are left behind whereas the moon appears to move with the train.

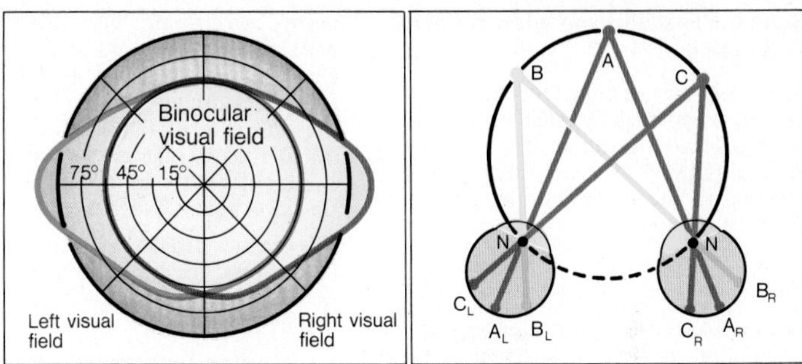

A. Binocular visual field

B. Horopter

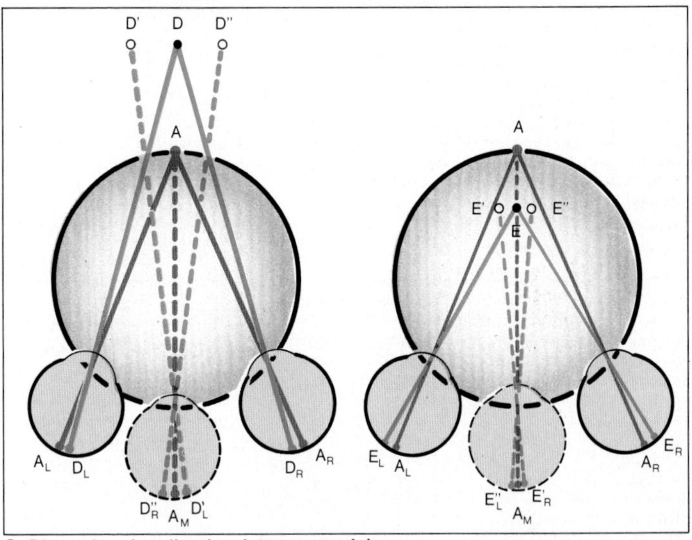

C. Binocular visualization (stereoscopic)

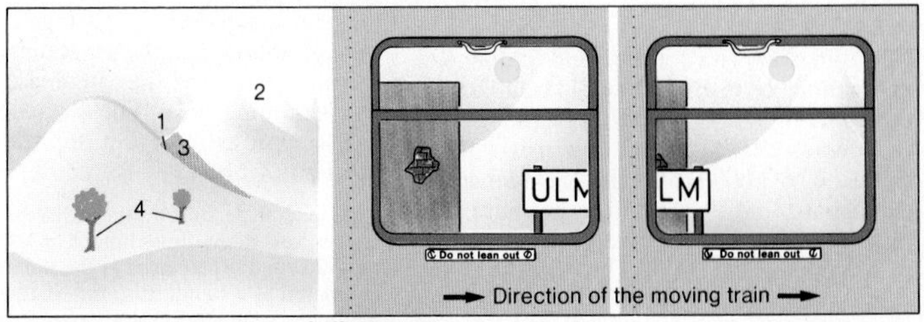

D. Cues for evaluation of distance

Sound: Physical Principles, Stimulus, Perception

The adequate stimulus for the organ of hearing is provided by **sound waves** emanating from a sound source (e.g. gong; → **A**) and propagated in gases, fluids, and solid substances. The most important conveyer of sound is the *air*.

At the sound source, the air is alternately compressed (rise in pressure) and rarefied (drop in pressure). These fluctuations in pressure (*sound waves*) spread with the **speed of sound** (**c**), which in air at 0 °C amounts to 332 m/s. If the fluctuations in sound pressure are graphically recorded (→ **A**) a series of waves is obtained. The distance between two corresponding points of a sound wave is the **wavelength** (λ), and the maximal deviation from the resting pressure is the **amplitude** (*a*) (→ **A**). If λ *increases* (*decreases*) the tone heard is *lower* (*higher*). A *reduction* (*increase*) in *a*, however, results in a *quieter* (*louder*) tone (→ **A**). The **pitch of a tone** is conventionally defined by its **frequency** (*f*), which indicates how many full cycles of pressure fluctuations per unit time occur in the sound field. Frequency, wavelength, and speed of sound are connected as follows ($Hz = s^{-1}$):

$f(Hz) \cdot \lambda(m) = c(m \cdot s^{-1})$.

Strictly speaking, only a pure, sinus-shaped oscillation is regarded as a **tone**. The "tone" emanating from the majority of sound sources (musical instruments, song), however, is much more complex, and is a superimposition of many frequencies and amplitudes. Although the *fundamental frequency* (the lowest tone contained in the complex) is identical, each such source (instrument) superimposes different higher frequencies (*overtones*) that constitute the "*timbre*" and allow the listener to distinguish the sources. An a^1 (440 Hz) sung by a tenor or played on a harp has a different sound from that produced on an organ or a piano. A combination of two similar tones of similar amplitude produces a third tone of lower frequency (*beat tone*; → **A**).

The human ear can detect sound frequencies in the range of **16 Hz to 20,000 Hz**, although the upper limit declines with age and may be as low as 5000 Hz (*presbyacusis*). At a frequency of 1000 Hz the **threshold for sound perception**, that is, the sound pressure that just suffices to elicit the sensation of hearing, is roughly $3 \cdot 10^{-5}$ Pa. The threshold for sound *depends on frequency* (→ **B**, red curve). The human ear is most sensitive in the frequency range of 2000–5000 Hz. The threshold for a tone rises considerably if it occurs concurrently with other tones. This **masking** is what makes it so difficult, for example, to carry on a conversation against loud background noises. Up to a sound pressure of 60 Pa (2 millionfold the threshold at 1000 Hz) the ear can cope with such stimuli without pain or damage, but above this level *pain* is experienced.

For practical reasons a logarithmic measure is employed for sound pressure, the **sound pressure level** (SPL), the unit being the **decibel** (**dB** SPL).

Starting from an arbitrary sound pressure $p_o = 2 \cdot 10^{-5}$ Pa, the sound pressure level (dB) at a higher sound pressure (p_x) is calculated from

$$SPL\ (dB) = 20 \cdot \log \frac{p_x}{p_o}$$

that is, a tenfold increase in sound pressure means that the SPL increases by 20 dB.

Sound intensity I ($J \cdot s^{-1} \cdot m^{-2}$) is the sound energy penetrating a unit area per time. I is proportional to the *square* of p_x.

dB values can not be calculated on a simple linear basis: two loudspeakers, each producing 70 dB ($p_x = 6.3 \cdot 10^{-2}$ Pa), do *not* produce a total of 140 dB. Since p_x (see formula above) only increases by a factor of $\sqrt{2}$ when I is doubled, the two loudspeakers together produce only about 73 dB (insert $\sqrt{2} \cdot 6.3 \cdot 10^{-2}$ for p_x, in the formula above).

Subjectively two sounds of different frequencies but the same sound pressure are not experienced as being equally *loud*. Example: a tone of 63 Hz seems to be as loud as one of 1000 Hz and 20 dB only when the sound pressure of the 63 Hz tone is increased about 30 fold (+29 dB). With the help of such *subjective* data, lines of equal **sound level** can be plotted on the dB-Hz diagram (**isophone**; → **B**, blue curve). The unit of sound level is the **phon**: at 1000 Hz the phon scale and the dB scale are identical by definition (→ **B**). The acoustic threshold is also represented by an isophone at a level of about 4 phon (→ **B**, red curve).

The **sone** scale, also obtained subjectively, describes how many times louder or less loud (**loudness**) a sound is perceived at the *same* frequency if compared to a standard sound of 1000 Hz and 40 phon (= 1 sone). For example, 2 sones means doubled loudness, 0.5 sone halved loudness.

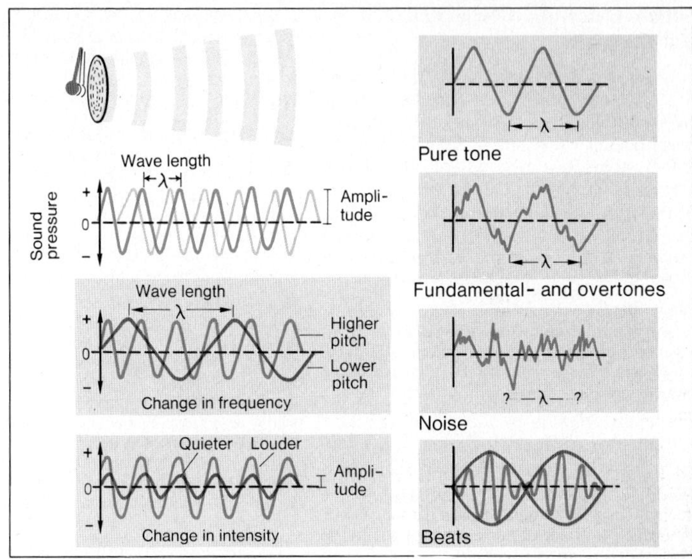

A. Characteristics of sound

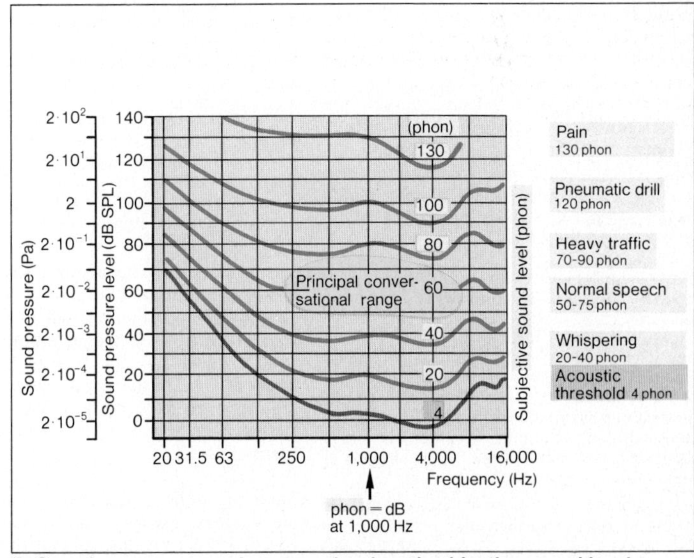

B. Sound pressure, sound pressure level, and subjective sound level

Sound: Detection, Conduction, and Receptors

Sound waves are transmitted to the acoustic organ via the *external ear* and the *auditory canal*, which terminates on the **tympanic membrane** or ear drum. The fluctuations in sound pressure cause vibrations of the tympanic membrane. These vibrations are transmitted via the **bones of the middle ear** (*malleus, incus*, and *stapes*; → **A**) to the **oval window** (→ **A**), which marks the beginning of the **inner ear** or **labyrinth**.

Outer ear: the external form of the ear and the funnel-shaped opening into the auditory canal assist in localizing the sound source (→ see also p. 322) and serve to reinforce the resonance of the tympanic membrane (range 2–7 kHz).

The main function of the **middle ear** is to transmit sound from a medium with a low impedance (air) for sound waves to one with a high impedance (fluid), with as little loss of energy as possible. Without the interpolation of this **impedance matching** by the ossicles, a large part of the sound energy would be reflected at the oval window if, for example, the small bones of the middle ear had undergone damage or destruction. This would result in a hearing loss of about 20 dB (conduction deafness). The matching of impedance is largely due to the fact that the sound pressure exerted on a larger area (tympanum, 50 mm^2) is transmitted to a smaller area (oval window, 3 mm^2), and that the small bones exert a leverage which increases the force about 1.3-fold (optimal effect at 1–2 kHz).

The two **muscles of the middle ear**, the *stapedius* and the *tensor tympani* muscles, can reduce the intensity of transmission of low-frequency sound to the cochlea to some extent. Some currently considered functions of these two muscles include the reflex maintenance of a constant intensity of this reduced sound, protection against very loud sounds, reduction of the noises produced by the hearer's own speech, damping of certain resonance vibrations in the middle ear, and reduction of the masking of high-frequency by low-frequency sounds.

Sound also causes vibrations in the entire skull. By means of **bone conduction** these vibrations are directly conveyed to the cochlea. Although this type of conduction is physiologically of little importance, it is exploited for the purposes of diagnosis. In **Weber's test** the handle of a vibrating tunig fork (customarily 256 Hz) is placed in contact with the vertex of the skull in the midline. A healthy subject localizes the source of the tone in the middle, due to the symmetry of the acoustic perception. A patient with *conduction deafness* on one side (e.g. middle ear disease) localizes the tuning fork on the deaf side (*lateralization*), since the absence of the masking effects of the environmental noise makes the tone seem louder. If deafness is due to a defect to the inner ear or to a neural (retrocochlear) defect in the acoustic pathway, the sound is reported as lateralized to the normal side since the diseased ear is deaf for both airborne and bone-conducted sounds.

Hearing acuity is tested quantitatively with the **audiometer**. The patient is presented, through earphones, with pure sound of varying frequency and intensity. The sound pressure, which is initially below threshold, is gradually raised at a certain constant speed until it is heard by the patient. If the sound required is louder than the normal threshold, there is a loss of hearing, which is measured in dB. (For diagnostic purposes the normal threshold for hearing [→ p. 317, B, red line] is usually taken as 0 dB for all frequencies, in contrast to the diagram on p. 317.) Hearing loss may result from inflammatory disease of the middle or inner ear, from pharmaceutical agents, from neurologic diseases etc., in addition to presbyacusis (→ p. 316).

The **inner ear** consists of the organ of equilibrium (→ p. 298) and the **cochlea**, a helical system of tubes in the temporal bone. The central tube, the **scala media** (ductus cochlearis), is filled with *endolymph*; it is accompanied on the one side by the **scala** vestibuli, which starts at the oval window, and on the other by the **scala tympani**, both containing *perilymph* and running with the scala media to the summit of the helix, the *helicotrema*, where they connect with each other. The scala tympani ends at the membrane of the **round window** in the wall of the middle ear (→ **A**).

The vibrations of the membrane of the oval window initiate pressure waves in the (incompressible) perilymph, which in turn lead to a distortion of the membrane of the round window (→ **A**). If Reissner's membrane and the basilar membrane (→ **A** and **D**) were completely rigid, pressure waves in the perilymph would run along the scala vestibuli to the helicotrema and hence along the scala tympani to the round window. Since the walls of the endolymphatic duct are flexible, however, the **travelling waves** (→ **B** and **C**) are transmitted to the scala tympani and in this way short-circuited to the round window without having to pass

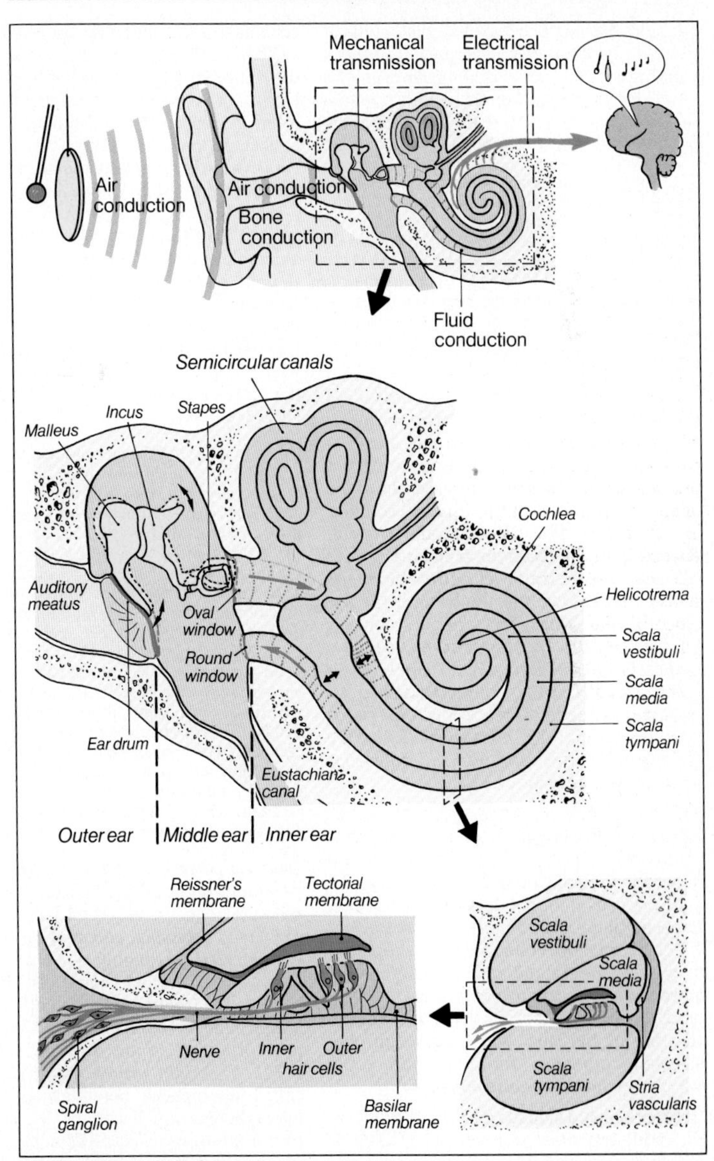

A. Sound transmission

the helicotrema. The wavelike distortion of the endolymphatic duct causes Reissner's membrane and the basilar membrane to swing from one side to the other, towards the scala tympani and scala vestibuli alternately (→ **C** and **D**).

The **speed of the travelling waves** (much less than the speed of sound!) and their **wavelength** gradually decrease with increasing distance from the oval window (→ **B**). One reason for this is that towards the helicotrema the basilar membrane gradually becomes broader and less rigid. The situation is comparable with that in the large blood vessels, that is, the less rigid they are, the greater is their "windkessel" effect (→ p. 156 and p. 163) and the slower the speed of the pulse waves.

While the wavelength of the travelling waves becomes shorter as they progress along the cochlea, their *amplitude* rises to a *maximum* (→ **B**, *bounding curve*) and then quickly falls off (attenuation because of the damping properties of the liquid-filled inner ear channels). The **site** at which maximal distortion of the endolymph canal occurs is **characteristic for the wavelength** of the initial sound: the shorter the wavelength (or the higher the initial frequency of the sound; → **C**), the nearer to the stapes or oval window is the *site at which maximum amplitude is reached.* Thus every sound frequency is allocated a definite place on the endolymphatic duct; high frequencies in the vicinity of the oval window, low frequencies near the helicotrema.

The oscillations in the endolympahtic canal cause a displacement of the tectorial membrane with respect to the basilar membrane in which the (secondary) receptors of the auditory organ, the **hair cells**, are embedded. Each of them has roughly 100 stereocilia, which are in close contact with the tectorial membrane (→ **D** and p. 319). The **relative movement** of the two membranes with respect to one another thus leads to a minute **shearing** of the cilia, which represents the *adequate stimulus* for the hair cells (*electromechanical transduction*).

On their ciliated side the hair cells border on the endolymphatic space, which has a standing *potential* (see below) of about + 80 mV with respect to the general ECS (→ p. 323, C). Since, at rest, the inner and the outer hair cells have a cell potential of − 70 mV and − 40 mV, respectively, there is a potential difference of 150 mV and 120 mV, respectively, across the ciliated cell membrane (cell interior negative). Additionally, the K^+ concentration of about 140 mmol/l in the endolymph is roughly the same as in the hair cells, so that the K^+ equilibrium potential (→ p. 14) amounts to 0 mV here. Thus the entire 150 mV and 120 mV, respectively, are available as driving forces for a K^+ influx. If the ciliar shearing opens up K^+ channels then there is an influx of K^+, and the cell is depolarized: **receptor potential**. This causes the release of a transmitter (glutamate?), which finally sets off the action potentials in the corresponding afferent fibers of the auditory nerve.

Of the roughly 25,000 hair cells, the smaller part is arranged in a single row along the spiral of the cochlea (**inner hair cells**), whereas the majority (the **outer hair cells**; → **A, bottom left**) are arranged in three to five rows. Nevertheless, roughly 95% of the 30,000 fibers of the auditory nerve originate in the inner hair cells. The two types of hair cells exert a reciprocal influence on one another, the mechanisms of which are not known; the inner hair cells are D receptors and the outer ones are P receptors (→ p. 276). The latter can contract synchronously upon stimulation, which may be part of an **amplifying process** that precedes the hair cells. This may be the explanation for the exceptionally low threshold within the narrow frequency range characteristic of each site. The approximately 1800 **efferent nerve fibers** (cholinergic) end on the hair cells and can, for example, inhibit the reception of certain frequencies. This mechanism may be employed in "filtering out" distracting noises from the surroundings.

Inner ear potentials. In addition to the action potentials in the afferent nerves fibers, other potentials can be recorded from the inner ear: (1) the **standing potential** or **endocochlear potential** of roughly + 80 mV, which is connected with the unequal distribution of Na^+ and K^+ between endolymph and perilymph, and is maintained by active transport processes in the *stria vascularis* (→ p. 323, C, bright yellow area); (2) the so-called **microphone potentials** or **cochlear microphonics** which can be led off at the round window reflect, like a microphone, the temporal course of the sound stimulus as fluctuations in voltage. It is not known how they arise.

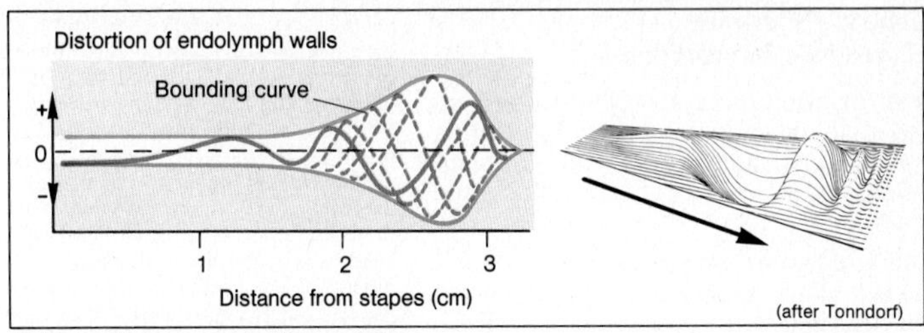

B. Traveling waves in the cochlea

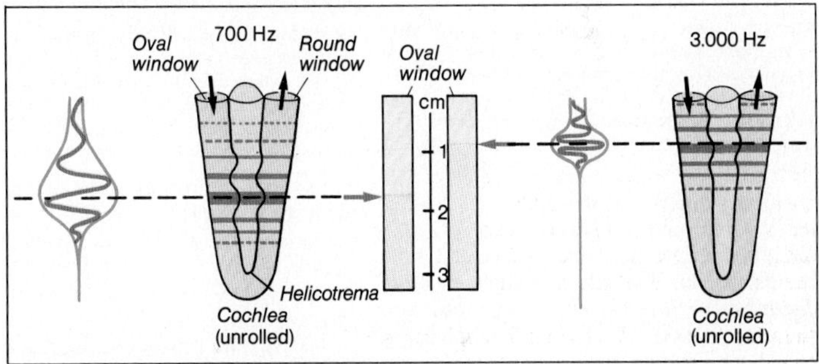

C. Frequency identification in the cochlea

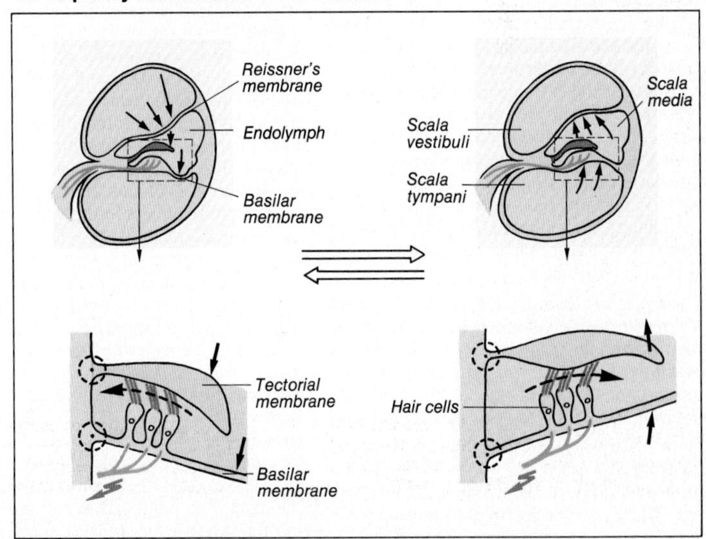

D. Excitation of hair cells by a shearing action on their hairs

Central Processing of Acoustic Information

The following properties of sound are encoded in the auditory nerve for further transmission: (1) frequency(ies); (2) intensity; (3) direction; (4) distance of the sound source.

Different frequencies are *individually "represented"* along the cochlea (→ p. 321, C) and are transmitted in *separate fibers* of the acoustic pathway.

There is another way of encoding sound frequencies in the auditory nerve. Higher (lower) sound frequencies move the receptor hairs back and forth more (less) often causing a certain *periodicity of the APs*. The one-electrode hearing prosthesis implanted in total cochlear deafness uses this type of encoding.

Frequency-difference threshold: A normal ear can just distinguish between 1000 Hz and 1003 Hz, which means that the difference of 3 Hz compared to 1000 Hz is the *relative difference threshold* (→ p. 306), and amounts to 0.003. Such fine discrimination is made possible by the individual representation of different frequencies in the cochlea, and by *contrast enhancement* (→ p. 275, D) along the acoustic pathway. However, there must be some additional mechanism(s) to account for the extreme precision of **tuning**. An active mobility of the outer hair cells seems to be involved. Just how fine this tuning is, is shown by the fact that the threshold of any single fiber in the auditory nerve is very low only for its "own" specific frequency. Recruitment of the adjacent fibers only takes place at higher sound pressures (see below).

The relative **intensity-difference threshold** (→ p. 306) of 0.1 is much higher than the relative frequency threshold, that is, sound is not experienced as being louder until its intensity has changed by a factor of 1.1 (i.e. the pressure of the sound has changed by a factor of $\sqrt{1.1} = 1.05$). Increased **sound intensity** (a) leads to an increase in the number of APs in the acoustic nerve fibers and (b) stimulates the involvement (*recruitment*) of neighboring nerve fibers (→ **A**).

The **direction** from which a sound comes can be recognized by two means: (a) sound waves reaching the ear *at an angle* to the midline reach one ear slightly *later* than the other. The difference in angle of arrival of 4° that can still just be detected (**direction threshold**) means a *sound delay* of about 10^{-5} s at the more distant ear (→ **B, left**). (b) The sound is *quieter* to the more distant ear. A lower sound pressure, however, means a slight delay in the firing of APs (increased latent period), so that the excitation originating in the more distant ear is later in arriving at the accessory nucleus (→ **B, right**). The effect of (a) and (b) are thus additive (→ **B**). The outer ear (→ p. 318) is of use in deciding whether a sound is in front of the head or comes from behind (or from above or below).

Recognition of the **distance of a sound source** is in part because high frequencies are damped to a larger extent during transmission than lower frequencies. The longer a sound is underway, the smaller is the contribution of higher frequencies on its arrival (e.g. near or distant thunder).

The most important relay stations of the **acoustic pathway** (→ **D**) and their probable functions are as follows: branches of the auditory nerve fibers run from the organ of Corti (→ **D 1**) to the anteroventral (→ **D 2**), posteroventral, and dorsal *cochlear nuclei* (→ **D 3**). The afferents are arranged in the three nuclei according to frequency (*tonotopic*) and in varying complexity. Here, lateral inhibition (→ p. 275, D) enhances *contrast* (i.e. *suppression of noise*). In the *superior olive* (→ **D 4**) and in the *accessory* nucleus (→ **D 5**), which also receive contralateral impulses, intensity and travelling time of sound are compared (direction, see above). The next stations are in the *lateral lemniscus* (→ **D 6**) and, after a predominant crossing of the fibers to the opposite side, in the *inferior colliculus* (→ **D 7**), which receives many afferents (→ **D**). In addition to relaying reflexes (e.g. middle ear muscles; →p. 318), the sensory analysis of the cochlear nuclei is compared here with the spatial analysis of the superior olive. Via the thalamus (*medial geniculate nucleus*, → **D 8**) the afferents finally reach the **primary acoustic cortex** (→ **D 9** and p. 283, A), which is surrounded by *secondary areas for hearing*. These centers are responsible for analysing complex sounds; they house the short-term memory for comparison of tones; they inhibit unwanted motor responses; and they are responsible for intent listening, etc.

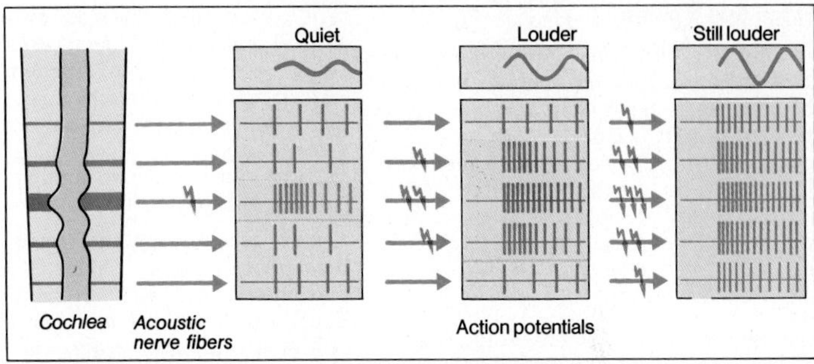

A. Encoding of intensity at constant sound frequency

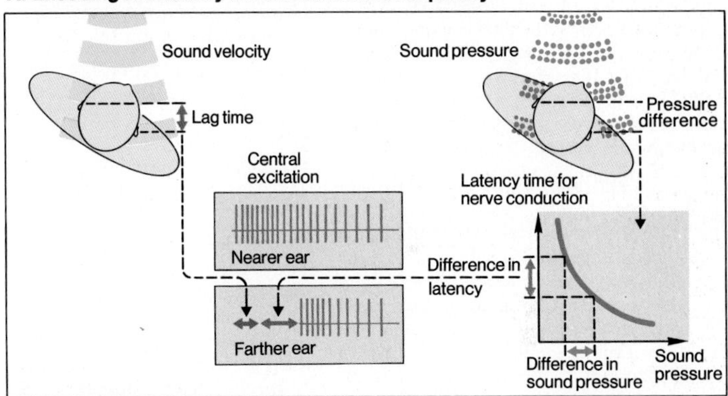

B. Directional hearing

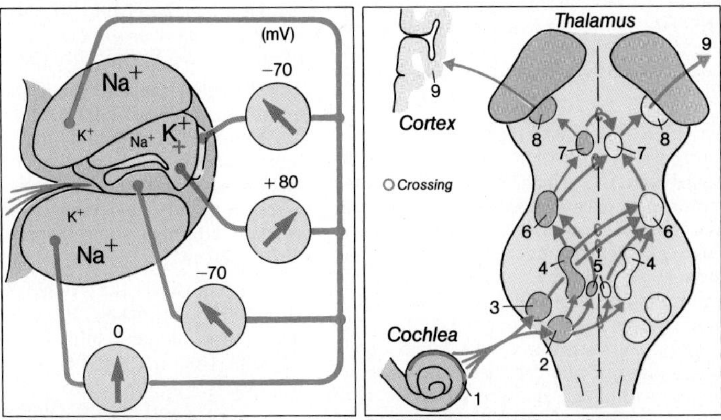

C. Cochlear potentials and electrolyte distributions

D. Afferent acoustic pathways

Voice and Speech

The human voice is primarily an **organ of communication**. Its performance matches the capacity of the human acoustic organ, the ear (→ p. 317, B). The voice is a wind instrument (reed). Air from *wind space* (lungs, bronchi, trachea) is driven past a narrow *slit* in a double reed (**vocal cords**), causing vibrations, the character of which is influenced by a *resonance chamber* (chest and oronasal cavity).

The wide range of variability of the voice is made possible by the participation of a large number of muscles by means of which velocity of air flow (**loudness** of the voice), tension of the vocal cords (→ **A 1**), shape and width of the vocal slit (→ **2**) (**fundamental tone** of the voice), and size and shape of the resonance chamber (**timbre, formants**) can be greatly varied.

The joints and muscles of the **larynx** (see textbooks of anatomy) adjust the position and tension of the vocal cords. When the vocal cords are set into vibration by the streaming air, the cords are not simply brought together (adduction), but move vertically in a "rolling" movement in the same direction as the air stream (→ **B**). In the production of low-frequency sounds, the cords are closed longer than they are open (5 : 1 at 100 Hz); for higher frequences (400 Hz) this ratio decreases to 1.4:1. In singing with a *head voice* (falsetto; → **C**, green) or *whispering*, they remain open all the time.

The efferent signals to the larynx arise in the motosensory cortex and travel to the nuclei of the vagus (X), which is responsible for the motor and sensory control of the larynx. The latter is important in the protective reflexes (*coughing*) and for voice production. Sensory fibers in the mucosa of the larynx and in its muscle spindles (→ p. 278) continuously send signals centrally, describing position and tension of the vocal cords. These reflexes, and especially the close associations via the acoustic pathways with bulbar and cortical speech centers, are essential for **fine control of the voice**.

The primary speech centers in the motosensory cortex are influenced by secondary cortical centers (e. g. *Brocas' center*). Failure of the secondary speech centers leads to **motor aphasia**, even though all of the motor pathways are intact. Failure of secondary *acoustic* centers (e.g. *Wernicke's center*) leads to a loss of understanding of speech (**sensory aphasia**).

Vowels, even though they have a similar basic frequency (100–130 Hz; → **D**), can be differentiated by their overtones (**formants**). The formants are determined by the distortion of the oronasal cavity. The three primary vowels ɑ:, i:, u: make up the vowel triangle; œ: ɔ: ø: y:, æ: and e: are the intermediates within this framework.

The phonetic notation used here is that of the *International Phonetic Society*. The symbols mentioned here are as follows; ɑ: as in glass; i: as in beat; u: as in food; œ: as in French peur; ɔ: as in bought; ø: as in French peu or in German hören; y: as in French menu or in German trüb; æ: as in bad; e: as in head.

The **consonants** are classified according to their site of origin, that is, *labial* (lips, teeth), for example, P, B, W, F, M; *dental* (teeth, tongue), for example, D, T, S, M; *lingual* (tongue, front of soft palate), for example, L; *guttural* (back of tongue and soft palate), for example, G, K. Consonants can also be distinguished according to their *mode of formation*; that is, *plosive* (P, B, T, D, G, K); *fricative* (F, W, S); *vibrative* (R), etc.

The **frequency range** of the human voice is approximately 40 to 2000 Hz. Sibilants (s, z) are high-frequency sounds that are poorly reproduced in radio and telephone. The **tonal range** (fundamental tones; → **C**) of *speech* is about one octave; of *singing*, two octaves, although some singers can span three octaves.

The usual musical scale is based on doubling of frequency, the **octave**. In equal-tempered tuning it is divided into 12 equal steps differing from each other by the factor 1,0595 ($\sqrt[12]{2}$).

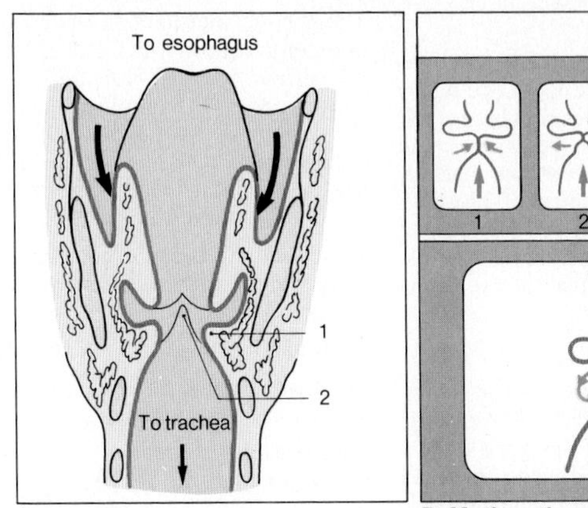

A. Larynx

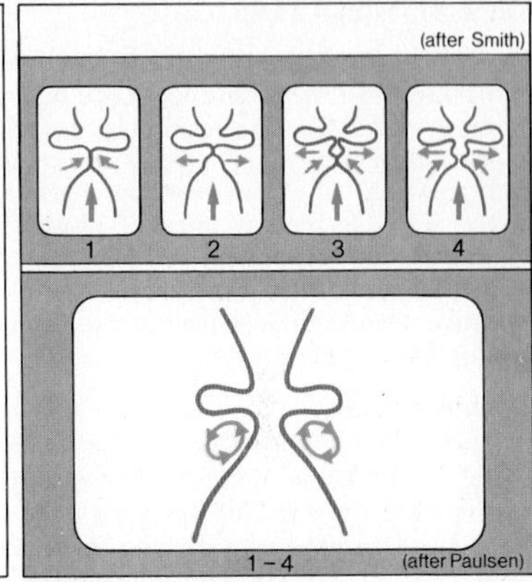

B. Motion of vocal cords

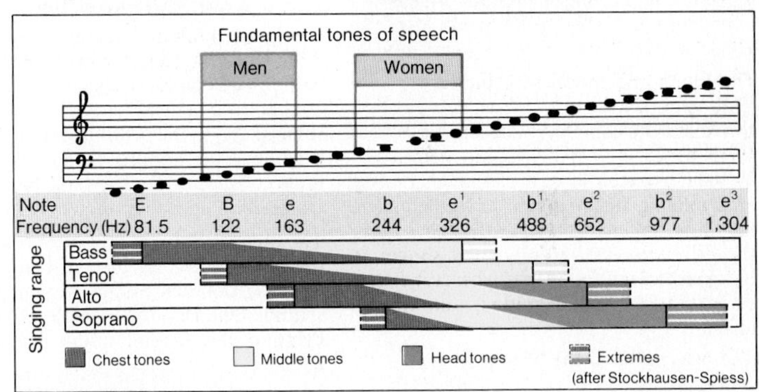

C. Vocal ranges

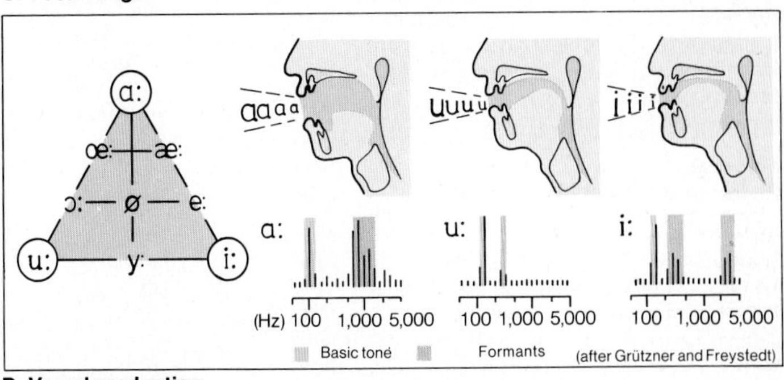

D. Vowel production

Dimensions and Units

Physiology is the science of life processes and body functions. Since these are partly based on physical and chemical laws, their investigation, learning, assessment, and manipulation of them are inseparably linked with the measurement of physical, chemical, and other dimensions, be it for the measurement of blood pressure, urine osmolality, or auditory acuity, or for determining the cardiac output.

Systems of Units

In medicine, and therefore also in physiology, a confusing number of units are encountered for one and the same dimension. For example, the dimension "concentration" can be expressed in the units g/l, g/100 ml, g/ml, mg%, pp, (w/v); or for expressing pressure, the units mm H_2O, cm H_2O, mmHg, Torr, at, bar, kg/cm^2, and so on can be used. To remedy this situation most countries now require by law the use of the international **SI-Units** (SI = Systeme International d'Unites).

With a few exceptions, the SI units are used in this book. In the period of transition, to facilitate the changeover to the new units, the old units are usually also shown in brackets. In the following, wherever necessary, conversions of individual units are also given.

The **basic units of the SI-system** are

for length:	m (meter)
for mass:	kg (kilogram)
for time:	s (second)
for quantity of a substance:	mol (mole)
for current strength:	A (Ampere)
for temperature:	K (Kelvin)
for luminous intensity:	c (candela)

These basic units are independent of one another and exactly defined; all other units are *derived from the basic units*, and in the majority of cases by multiplying or dividing the basic units with or by one another, for example:

for area (length · length): $(m \cdot m) = m^2$;
for speed (length/time): $m/s = m \cdot s^{-1}$.

If the new unit becomes too complicated, it is given a new name with its own symbol, for example:

for force: $kg \cdot m \cdot s^{-2}$ = N (Newton);

for energy, work, and quantity of heat: $kg \cdot m^2 \cdot s^{-2}$ = J (Joule);

for pressure: $N \cdot m^{-2} = kg \cdot m^{-1} \cdot s^{-2}$ = Pa (Pascal).

Fractions and Multiples of Units

Since it is both tedious and confusing to write, for example, 10000 g or 0.00001 g, *prefixes* are employed in front of the unit to denote *decimal multiples* and *fractions* (in steps of 1000); the previous examples would become 10 kg (kilogram) and 10 µg (microgram), respectively. The table below shows the prefixes with their factors and symbols.

The prefixes are not only placed in front of the basic units, but also in front of units derived from them and possessing their own symbol. One kPa, for example, is 1000 (10^3) Pascal; 1 µl is one-millionth (10^{-6}) liter, etc.

In addition, for some units, other prefixes for still smaller decimal steps are used (da, h, d, c; see table):

For *time* the familiar nondecimal multiples have been retained, that is, *second* (*s*), *minute* (*min*), *hour* (*h*), and *day* (*d*).

Multiples Prefix	Symbol	Value	**Submultiples** Prefix	Symbol	Value
deca-	da	10^1	deci-	d	10^{-1}
hecto-	h	10^2	centi-	c	10^{-2}
kilo-	k	10^3	milli-	m	10^{-3}
mega-	M	10^6	micro-	µ	10^{-6}
giga-	G	10^9	nano-	n	10^{-9}
tera-	T	10^{12}	pico-	p	10^{-12}
peta-	P	10^{15}	femto-	f	10^{-15}
exa-	E	10^{18}	atto-	a	10^{-18}

Length, Area, Volume

The basic unit of **length** is the **meter (m)**.

Conversions of other units are:
Ångström = 10^{-10} m = 0.1 nm
1 Micron = 10^{-6} m = 1 μm
1 Inch = 25.4 mm
1 Foot = 0.3048 m
1 Yard = 0.9144 m
1 (Statute) Mile = 1.609 km
1 Nautical Mile = 1.853 km

Area (m × m = **m²**) and **volume** (m × m × m = **m³**) are derived units.
Attention! 1 m = 10^3 mm, but
1 m^2 = 10^6 mm^2 and
1 m^3 = 10^9 mm^3!

Conversion of other units of area are:
1 square inch = 645.16 mm^2
1 square foot = 0.0929 m^2
1 square yard = 0.8361 m^2
1 acre = 4047 m^2
1 square mile = 2.59 km^2

For fluids and gases, the special unit, **liter (l)**, is used for volumes:
1 l = 10^{-3} m^3 = 1 dm^3
1 ml = 10^{-6} m^3 = 1 cm^3
1 μl = 10^{-9} m^3 = 1 mm^3

Conversions of other units are:
1 fluid ounce (USA) = 29.57 ml
1 fluid ounce (GB) = 28.41 ml
1 U.S. liquid gallon = 3.785 l
1 British or Imperial gallon = 4.546 l

Frequency, Velocity, Acceleration

Frequency refers to how often a periodic event occurs. Its unit is s^{-1} or Hertz (Hz). Expressed in minutes, min^{-1} = 1/60 Hz ≈ 0.0167 Hz.

Velocity of a mass (such as an automobile) is distance per time ($m \cdot s^{-1}$). Movement of fluids is described both by **linear velocity** as $m \cdot s^{-1}$ and by "volume velocity" or **rate of flow** as $l \cdot s^{-1}$ or $m^3 \cdot s^{-1}$.

Acceleration refers to rate of change of velocity or $m \cdot s^{-1} \cdot s^{-1} = m \cdot s^{-2}$. A negative value refers to deceleration.

Force and Pressure

Force is mass times acceleration (special case: "**weight**" = **weight force** = mass times acceleration due to gravity). Since the unit of mass is the kilogram (kg) and that of acceleration is $m \cdot s^{-2}$ (see above), the

unit of force: $kg \cdot m \cdot s^{-2}$ = *N (Newton)*.

Conversions of other units are:
1 dyne = 10^{-5} N = 10 μN
1 pond = 9.8 mN

Pressure is *force* exerted on an *area* or $N \cdot m^{-2}$; this equals the unit of pressure, that is, the **pascal (Pa)**. Conversions of other units are:

1 mmHg = 1 Torr = 133.3 Pa = 0.133 kPa
1 cmH_2O ≈ 98 Pa
1 technical atmosphere (at) = 98,067 kPa
1 physical atmosphere (atm) ≈ 101.324 kPa
1 dyne/cm^2 = 0.1 Pa
1 bar = 100 kPa

Work, Energy, Heat, Power

Work is *force* times *distance*, ($N \cdot m$), or *pressure* times *volume* ($N \cdot m^{-2} \cdot m^3 = N \cdot m$); this unit is the **joule (J)**.

Energy and **heat** have the same unit, the joule. Conversion of other units are:

1 erg = 10^{-7} J = 0.1 μJ
1 calorie ≈ 4.185 J
1 kcalorie ≈ 4.185 kJ
1 Watt second = 1 J
1 kilowatt hour = 3600 kJ = 3.6 MJ

Power is work per time, of $J \cdot s^{-1}$ = **watt (W)**. Flow of heat or of energy is also measured in W.

1 erg/s = 10^{-7} W = 0.1 μW
1 calorie/h = 1.163 mW

Mass, Quantity, Concentration

The unit of **mass** is the *kilogram* (kg). In this case an exception has been made and a prefix used in the basic unit. Furthermore, the metric *ton* (1 t = 10^3 kg) is used instead of Mg.

The mass of an object is usually determined by measuring the *gravitational force* (mass times gravitational acceleration = "weight"), although the scale of the balance is calibrated in units of mass (g, kg).

Other units for mass can be converted as follows:

Avoirdupois weight:
1 ounce (oz.) = 28.35 g
1 pound (lb.) = 453.6 g

Apothecaries and troy weight:
1 ounce = 31.1 g
1 pound = 373.2 g

The mass of a molecule or an atom (molecular or atomic "weight") is often expressed in **Dalton** (not an SI unit), where 1 Dalton (d) is 1/12 of the mass of one ^{12}C atom = 1 g/Avogadro's constant ≡ 1 g/Loschmidt's number = 1 g/($6.02252 \cdot 10^{23}$); thus,

1 Dalton = $1.66 \cdot 10^{-24}$ g

A dimension related to mass is the **amount of substance**, which is expressed in mole (symbol: *mol*). 1 mol is the mass of substance (in g) that indicates the relative mole-, ion-, or atomic mass of this substance, that is, by how much greater the mass of the atom, molecule, or ion is than 1/12 of the mass of the ^{12}C atom.

Examples:
relative mole mass of H_2O: 18
1 mol H_2O = 18 g H_2O

relative atomic mass of Na: 23
1 mol Na^+-ions = 23 g Na^+-ions.

relative mole mass of $CaCl_2$:
$(40 + 2 \cdot 35.5) = 111$
1 mol $CaCl_2$ = 111 g $CaCl_2$.

(1 mol $CaCl_2$ contains 2 mol Cl^--ions and 1 mol Ca^{2+}-ions).

If mol is divided by the *valency* of the ion in question, the result is the relative *equivalent mass* with the unit eq (= val):

For monovalent ions, mol and eq are identical:
1 eq Na^+ = 1/1 mol Na^+.

For bivalent substances (e.g. Ca^{2+}, see above):
1 eq Ca^{2+} = 1/2 mol Ca^{2+}
or
1 mol Ca^{2+} = 2 eq Ca^{2+}.

Another unit derived from mol is the Osmol (*osm*) (→ p. 335)

Concentration refers to various relationships:

Mass per volume = $g \cdot l^{-1}$
Amount per volume = $mol \cdot l^{-1}$ (molarity)
Amount per mass solvent = $mol \cdot kg^{-1}$ (molality)
Mass per mass = $g \cdot g^{-1} = 1$
Volume per volume = $l \cdot l^{-1} = 1$

The last two relationships are relative or *fractional concentrations.*

Conversions of other mass per volume units:

1 g/100 ml = 10 g/l
1 g% = 10 g/l
1 % (w/v) = 10 g/l
1 g‰ = 1 g/l
1 mg% = 0.01 g/l = 10 mg/l
1 mg/100 ml = 0.01 g/l = 10 mg/l
1 μg% = 10^{-5} g/l = 10 μg/l
1 γ % = 10^{-5} g/l–10 μg/l

Conversions of other quantity per volume units:

1 M (molar) = 1 mol/l
1 N (normal) = (1/valency) · mol/l
1 mM (mmolar) = 1 mmol/l
1 val/l = (1/valency) · mol/l
1 eq/l = (1/valency) · mol/l

Fractional concentrations (*mass ratio and volume ratio*) are based on the "unit" of 1 (or 10^{-3}, 10^{-6}, and so on). Conversions:

1 % = 0.01
1 ‰ = $1 \cdot 10^{-3}$
1 Vol% = 0.01
1 l/l = 1
1 ppm = $1 \cdot 10^{-6}$
1 ppb = $1 \cdot 10^{-9}$

The concentration of a solute is often related to the *volume of the solution* (molarity; mol/l). In biological fluids, however, the *volume of the solutes* frequently constitutes a significant fraction of the total volume. 1 l blood plasma, for instance, contains 0.07 l solutes (mostly proteins) and only 0.93 l water. Looking at chemical reactions and biological events (e.g. diffusion), it usually makes more sense to know the concentration given per *volume of the solvent* H_2O. Additional conversion of the temperature dependent term volume into mass yields the *molality* (mol/kg H_2O). The same applies to osmolarity and osmolality (→ p. 335).

The physico-chemically effective concentration, e.g. of ions as measured by means of ion-sensitive electrodes, is termed "**activity**". Activity and molal concentration are equal as long as the total **ionic strength** (μ) is very low (ideal solution).

$\mu \equiv 0.5\ (z_1^2 \cdot c_1 + z_2^2 \cdot c_2 + \ldots + z_i^2 \cdot c_i)$,

Here z_i is the valency of ion i, c_i its molal concentration and 1, 2 ... i the types of ions present in the solution.

At the high ionic strength of biological fluids, the activity (a) is always significantly lower than the molal concentration (c) and can be calculated from: $a = f \cdot c$, where f = **activity coefficient**. For instance, at an ionic strength of 0.1 (e.g. in a solution with 0.1 mol NaCl/kg H_2O or with 0.033 mol $Na_2\ HPO_4$/kg H_2O), f is 0.76 for Na^+.

Electrical Units

The movement of electrically charged particles, for example negatively charged electrons (→ textbooks of physics), along a wire, is called electrical current. **Current strength** is determined by the number of charges moving per unit time. The unit is the **ampere (A)**. The movement of ions (Na^+, K^+, etc), for example, through a cell membrane (*ionic current*) can also be expressed in A. Current can only flow from one point to another if a **potential difference** (abbreviated as **potential**) exists between the two points. A battery or a generator can produce such a potential. In the living organism, electrical potentials are usually produced by the transport of ions. If, for example, the K^+ concentration on the two sides of the cell membrane is different, the K^+ ions diffuse towards the side with the lower concentration (→ p. 8). As a result, the charges are unequally distributed and a potential arises at the membrane (*diffusion potential*, → p. 14). The *unit* of electrical potential is the **volt (V)**.

$V = W \cdot A^{-1} = kg \cdot m^2 \cdot s^{-3} \cdot A^{-1}$

The quantity of current flowing at a given potential depends upon the electrical **resistance** (*Ohm's Law*):

Potential = current · resistance

The *unit of resistance is the* **ohm (Ω)**.

$\Omega = V \cdot A^{-1} = kg \cdot m^2 \cdot s^{-3} \cdot A^{-2}$.

Its reciprocal (1/resistance) is **conductance**, and is measured in *Siemens* (*S*) or in '*mho*' ($= S = 1/\Omega$). The permeability of a membrane to ions can be converted to conductance (**ionic conductance**, → p. 9).

Electrical work or **energy** is expressed, like every form of work, in *Joules* (*J*; → p. 327). Electrical work/time or **power** is expressed in *watts* (*W* → p. 327).

Direct current (DC) flows only in one direction whereas in *alternating current* (AC) the direction of flow is continually changing. The number of times it changes per unit of time is termed the *frequency* (*Hertz*, → p. 327). Our lighting network usually has a frequency of 50 or 60 Hz.

Temperature

The SI unit for **temperature** is the *Kelvin* (*K*). 0 K, *absolute zero*, is the lowest possible temperature. The Celsius (centigrade) scale with the *degree celsius* (°C) as its unit is derived from the Kelvin scale;

Temperature in °C = temperature in K − 273.15.

In the U.S. the temperature is usually given in *degrees Fahrenheit* (°F). The conversion to degrees Celsius is as follows:

Temperature in °F = (9/5 × temperature in °C) + 32,

	°C	°F
Freezing point of water	0	+ 32
Room temperature	+ 20 to + 25	+ 68 to + 77
Body temperature	+ 37	+ 98.6
Fever	up to + 42	up to + 107.6
Boiling point of water (at sea level)	+ 100	+ 212

and in the reverse direction

temperature in °C =
5/9 × (temperature in °F − 32).

Some important temperatures are listed in the table on p. 329 in both Celsius and Fahrenheit.

Powers and Logarithms

Since numbers much larger or much smaller than 1 are inconvenient and complicated to write, so-called **powers of ten** can be employed. They are formed as follows:

$$100 = 10 \cdot 10 = 10^2$$
$$1000 = 10 \cdot 10 \cdot 10 = 10^3$$
$$10000 = 10 \cdot 10 \cdot 10 \cdot 10 = 10^4$$

How often the 10 occurs in these multiplications is thus more simply expressed by an **exponent**.

If the number is not exactly an even order of magnitude (e.g. 34500), it is divided by the next lower even order of magnitude (10000) and the result (3.45) is placed as a multiplication factor in front of this order of magnitude: $3.45 \cdot 10^4$.

10 can, as described above, be written as 10^1. Still smaller numbers can be denoted as follows:

$$1 = 10 : 10 = 10^0$$
$$0.1 = 10 : 10 : 10 = 10^{-1}$$
$$0.01 = 10 : 10 : 10 : 10 = 10^{-2}$$

and so on.

0.04, for example, can be split up as already described into $4 \cdot 0.01$ or $4 \cdot 10^{-2}$.

Note: For numbers smaller than 1, the (negative) exponent is calculated from the position occupied by the 1 *after the decimal point*; for 0.001, for example, the third place: $0.001 = 10^{-3}$.

In numbers from 10 upwards, 1 is subtracted from the number of places in front of the decimal point: the rest denote the (positive) power. Example: 100 has three places and is therefore 10^2; 1124.5 has four places in front of the decimal point and is thus $1.1245 \cdot 10^3$.

Units can also be raised to a power, for example m^3. In this example, this means that the base (i.e. m) is multiplied 3 times by itself ($m \cdot m \cdot m$ → p. 327). Negative powers are used in the same way: just as $1/10 = 10^{-1}$, $1/s = s^{-1}$ or $mol \cdot l^{-1}$ can be written instead of mol/l.

Calculating with powers has rules of its own. Addition and subtraction are only possible between the *same exponents*, for example:

$(2.5 \cdot 10^2) + (1.5 \cdot 10^2) = 4 \cdot 10^2$; but $(2 \cdot 10^3) + (3 \cdot 10^2)$ has to be converted into $(2 \cdot 10^3) + (0.3 \cdot 10^3) = 2.3 \cdot 10^3$.

Multiplication with powers involves addition of the exponents; division involves their substraction, for example:

$$10^2 \cdot 10^3 = 10^{2+3} = 10^5$$
$$10^4 : 10^2 = 10^{4-2} = 10^2$$
$$10^2 : 10^4 = 10^{2-4} = 10^{-2}$$

Numbers in front of the powers are treated in the usual way, for example:

$$(3 \cdot 10^2) \cdot (2 \cdot 10^3) = 2 \cdot 3 \cdot 10^{2+3} = 6 \cdot 10^5.$$

It is also possible to calculate with *exponents* alone: this is called **logarithmic calculation.** If a number (e.g. 100) is written as a power to the base 10 (10^2), then the exponent (2) is termed the (**decadic) logarithm** of 100 (abbreviated as **log** 100). Logarithms of this kind are used in physiology, for example, in defining the pH value (→ p. 334 and p. 110) or in plotting the pressure of sound using the decibel scale (→ p. 316).

For the **natural logarithm (ln)** the exponent to the base *e* is used, where $e = 2.71828\ldots$

Since $\log x = \ln x / \ln 10$ and $\ln 10 = 2.302585\ldots$, the conversion of ln to log and vice versa is as follows:

$$\log x = \ln x / 2.3$$
$$\ln x = 2.3 \cdot \log x$$

In **calculating with logarithms** the type of calculation is lowered by one step, that is, a multiplication becomes an addition, and a potentiation becomes a multiplication, and so on, thus:

$$\log a \cdot b = \log a + \log b$$
$$\log (a/b) = \log a - \log b$$
$$\log a^n = n \cdot \log a$$
$$\log \sqrt[n]{a} = (\log a)/n$$

Special cases are:

$\log 10 = \ln e = 1$
$\log 1 = \ln 1 = 0$
$\log 0 = \ln 0 = \pm \infty$

Graphic Representation of Data

In order to follow, for example, the body temperature of a patient over a longer period of time, the temperature and the corresponding time can be depicted graphically (→ **A**).

The two axes along which, in this case, temperature and time are plotted, are termed the *coordinates*, whereby the vertical axis is termed the *ordinate* (here, temperature) and the horizontal axis the *abscissa* (here, time).

It is customary to plot the first-chosen, variable dimension x (in this case the time of day) along the abscissa, and the other variable dimension y (in this case body temperature), which is dependent on the first, along the ordinate. This is why the abscissa (horizontal) is termed the x-axis and the ordinate (vertical) the y-axis.

Using this graphic method, every possible dimension can be plotted against other related dimensions, for example, body size against age, or lung volume against intrapulmonary pressure (→ p. 84).

It can then be seen whether the two dimensions change in the same direction, i.e. whether they are positively *correlated*. If, for example, body size is plotted along the ordinate and age along the abscissa, the curve rises during the phase of body growth; however, from roughly the age of 17 onwards, it is horizontal. This indicates that the body size in the first phase is dependent on age, but in the second (horizontal) phase it is largely independent of it. However, a correlation (see below) alone is not evidence of a *causal* connection. In Alsace, for example, for a certain period of time, the drop in birth rate correlated with the decrease in number of nesting storks!

If it is desired to accommodate *very different values* (e.g. 1 to 100000) on one of the coordinates, either the small values can no longer be plotted individually, or the coordi-

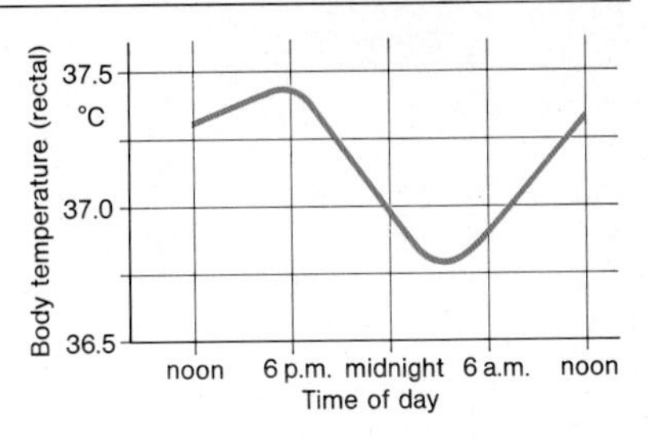

A. Graphic representation of the effect of *time of day* on *body temperature* (measured rectally, at rest)

nate axes have to be too long. In such a case the data can be plotted as their *powers* or as their *logarithms* (see above); 1, 10, 100, 1000, etc., can instead be written as 10^0, 10^1, 10^2, 10^3, and so on, or as the logarithms 0, 1, 2, 3, etc. In this form even the lower numbers can be relatively accurately depicted and the higher numbers still have room on the axis, which is also of a reasonable length (e.g. curves for hearing, → p. 317).

A correlation can be *linear* (→ **B 1**, black line) and in this case obeys the equation

$y = a \cdot x + b$,

whereby a is the *slope* of the line and b its point of intersection with the y axis (= *intercept*).

Many correlations, however, are nonlinear. For simpler functions, a graphic linearization can be achieved by a nonlinear plot of the x- or y-values, or both (e.g. logarithmically). This allows, for example, extrapolation to values beyond the range of measurement (see below) or the establishment of standardization curves from only two points (e.g. → p. 119). Further, the calculation of the "mean" correlation of scattered values of x-y pairs is also possible by this method: *regression lines*.

An *exponential function* (→ **B 1**, red curve)

$y = a \cdot e^{b \cdot x}$

can be linearized by plotting ln y on the y-axis (→ **B 3**):

$\ln y = \ln a + b \cdot x$,
where b = slope and ln a = intercept.

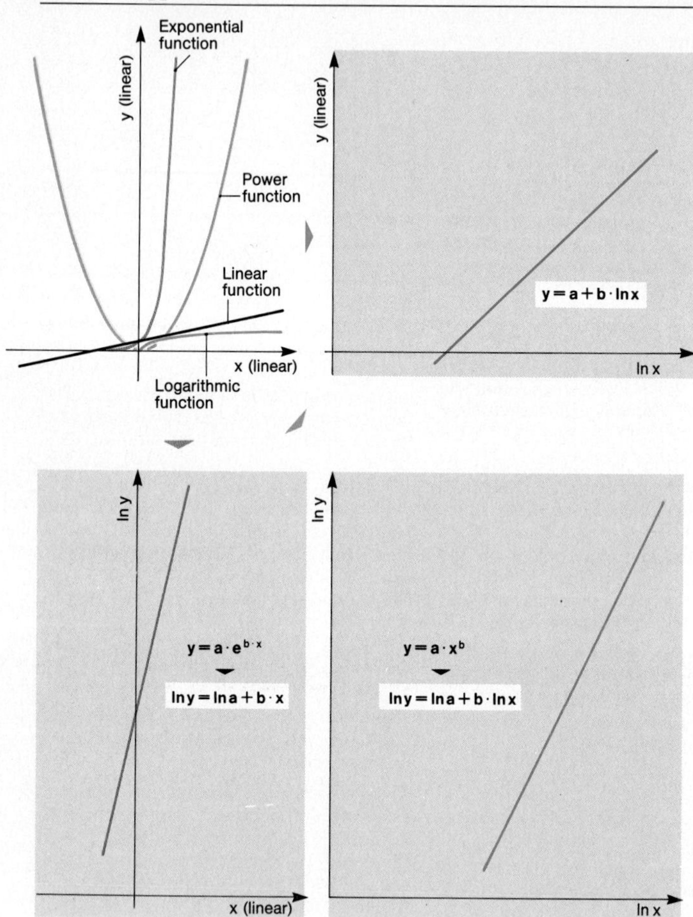

B1. Graphic representation of a linear function (black), an exponential function (red), a logarithmic function (blue), and a power function (green) by *linear* plotting on the x- and y-axes. The 3 curves can be graphically linearized by logarithmic plotting on, respectively, the x-axis (*logarithmic function*: **B2**), on the y-axis (*exponential function*: **B3**), or on both axes (*power function*: **B4**)

A *logarithmic function* (→ **B 1**, blue curve)

$y = a + b \cdot \ln x$

can be graphically linearized by plotting ln x on the x-axis (→ **B 2**), where b = slope and a = intercept.

A *power function* (→ **B 1**, green curve)

$\ln y = a \cdot x^b$

can be graphically linearized by plotting ln x and ln y on the coordinate axes (→ **B 4**), since

$\ln y = \ln a + b \cdot \ln x,$

where b = slope and ln a = intercept.

Note that on logarithmic coordinates x (or y) = 0 does not exist because $\ln 0 = \infty$. Nevertheless, **a** (or ln a) is still called "intercept" if the logarithmic abscissa (→ **B 2, 4**) is crossed by the ordinate at ln x = 0, that is, at x = 1.

Instead of plotting ln x or ln y (or both) on the x- or y-axis (or both), respectively, the **linear** values of x or y (or both) can be plotted on **logarithmic paper** on which the ordinate, the abscissa ("semi-log"), or both ("log-log") are marked in logarithmic divisions. In the latter two cases, a (or ln a) is not called "intercept" anymore because the elevation of **a** depends on where the y axis crosses the x axis (all values of $x > 0$ are possible).

Other nonlinear functions can be graphically linearized by a suitable type of plot on the coordinates, as, for example, the *Michaelis-Menten equation* (→ **C 1**), which governs many enzyme reactions and carrier-mediated transport processes (→ p. 11):

$J = J_{max} \cdot C/(K_m + C),$

where J is the actual rate of transport (e.g. in $mol \cdot m^{-2} \cdot s^{-1}$), J_{max} the maximal rate of transport, C ($mol \cdot m^{-3} = mmol \cdot l^{-1}$) the actual concentration of the substance to be transported, and K_m the (half-saturation-) concentration at 1/2 J_{max}.

One of the three commonly used linearizations of the Michaelis-Menten equation, after *Lineweaver-Burke*, is

$1/J = (K_m/J_{max}) \cdot (1/C) + 1/J_{max},$

so that plotting 1/J on the y axis and 1/C on the x-axis gives a straight line (→ **C 2**). Whereas for a plot of J against C (→ **C 1**),

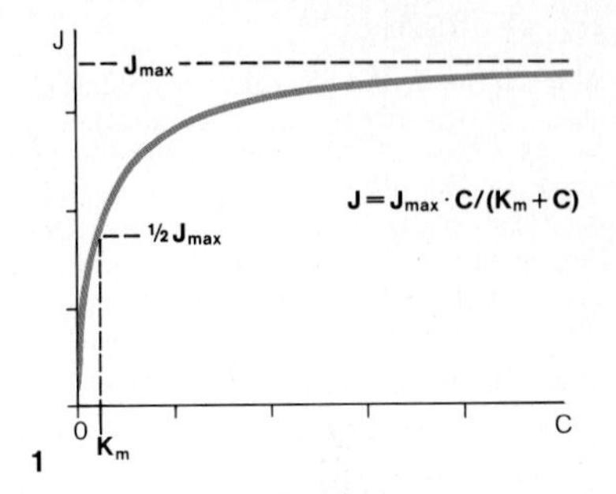

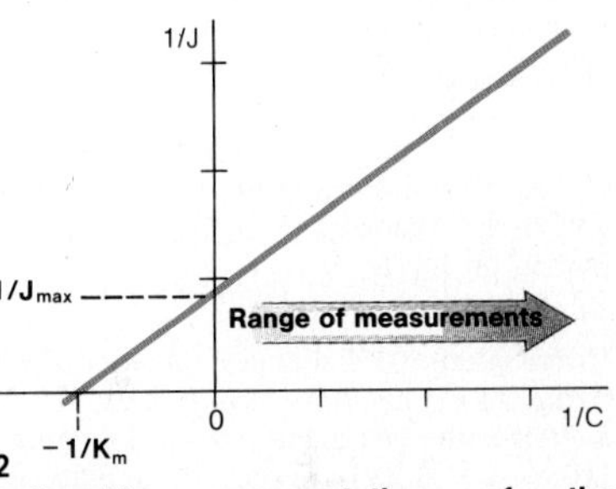

C. Graphic representation of the Michaelis-Menten equation as a curve (J against C; → **C 1**) and in a linearized form (1/J against 1/C; → **C 2**). In the latter form; J_{max} and K_m can be read off by extrapolation beyond the range of measurement (see text)

the accurate experimental determination of J_{max} is impossible (this would require the use of an infinitely high concentration of C), in the linearized form (→ **C 2**) a regression line can be calculated from experimental data, and *extrapolated* to C = infinity. Since, then, $1/C = 1/\infty = 0$, $1/J_{max}$ can be read off the x-axis at its 0-point (→ **C 2**). The reciprocal value of this is J_{max}. If now $1/J = 0$ is substituted in the above Lineweaver-Burke equation, it becomes

$0 = (K_m/J_{max}) \cdot (1/C) + 1/J_{max}$

or $1/K_m = -1/C$, so that K_m can be calculated from the negative reciprocal of the intercept on the x-axis (corresponds to $1/J = 0$) (→ **C 2**).

pH, pK, Buffers

A special unit, the **pH value**, is commonly used for describing the **concentration of the H^+ ions** (= protons). According to *Sörensen*, the pH is the negative common logarithm of the molal H^+ ion concentration in mol/kg H_2O. Currently, however, pH is usually determined by means of glass electrodes that measure the **H^+ activity.**

Thus, the modern definition of pH is:

$$pH = -\log (f_H \cdot [H^+]),$$

where f_H is the activity coefficient for H^+. At the ionic strength of the plasma for instance, $f_H \approx 0.8$.

H^+-activity:

1 or 10^0 mol/kg H_2O : pH 0
0.1 or 10^{-1} mol/kg H_2O : pH 1
0.01 or 10^{-2} mol/kg H_2O : pH 2
and so on, up to
10^{-14} mol/kg H_2O : pH 14

In considering pH changes, the logarithmic nature of pH has to be kept in mind. If, for instance, the pH rises from 7.4 (40 nmol H^+/kg H_2O) to 7.7, the H^+ activity decreases by 20 nmol/kg H_2O. The same pH change in the other direction (i.e. from 7.4 to 7.1) means that the H^+ activity is increased by 40 nmol/kg H_2O.

In principle, the **pK value** is similar to the pH value. It is definied as the *negative common logarithm of the dissociation constant* K_a of an acid, or K_b of a base:

$$pK_a = -\log K_a$$
$$pK_b = -\log K_b$$

(for an acid and its base $pK_a + pK_b = 14$, so that the pK_a can be calculated if the pK_b is known, and vice versa).

When a weak acid (AH) dissociates as follows:

$$AH \rightleftarrows A^- + H^+,$$

according to the Law of Mass Action (→ textbooks of chemistry) the product of the molal concentrations (square brackets) of the reaction products divided by the concentration of the nondissociated substance remains constant:

$$K_a = \frac{[A^-] \cdot [H^+] f_H}{[AH]}$$

Put in logarithmic form the equation becomes

$$\log K_a = \log \frac{[A^-]}{[AH]} + \log [H^+] f_H, \text{ or}$$

$$-\log f_H [H^+] = -\log K_a + \log \frac{[A^-]}{[AH]}, \text{ or}$$

(according to the above definition of the pH and pK values):

$$pH = pK_a + \log \frac{[A^-]}{[AH]}.$$

(Because the concentration and not the activity of A^- and AH is used here, pK_a is concentration-dependent in nonideal solutions.)

This is the general form of the **Henderson-Hasselbalch equation** (→ p. 110ff.), which describes the relationship between the pH value of a solution and the ratio of the (molal) concentrations of the dissociated to undissociated form of a dissolved substance. If $[A^-] = [AH]$ (the ratio is 1/1 = 1), the $pH = pK_a$ since the log of 1 = 0.

A weak acid (AH) and its dissociated form (A^-) can jointly provide a **buffer system** for H^+ ions and OH^- ions:

Addition of H^+: $A^- + H^+ \rightarrow AH$
Addition of OH^-: $AH + OH^- \rightarrow A^- + H_2O$

The *best buffering* action is achieved when $[AH] = [A^-]$, that is, when the pH value of the solution equals the pK value of the buffer. This is illustrated by the following example:

$[A^-]$ and [AH] are each 10 mmol/kgH_2O. The pK value is 7. If 2 mmol/kgH_2O of H^+ are now added, $[A^-]$/[AH] changes from 10/10 to 8/12 since 2 mmol/kgH_2O of A^- are converted by the H^+ into 2 mmol/kgH_2O of AH. The log of $8/12 \approx -0.18$, or, in other words, the pH value changes by 0.18 pH units from 7 to 6.82. If, on the other hand, the ratio $[A^-]$/[AH] had initially been 3/17, the addition of the same quantity of H^+ions would have changed the pH value from (7

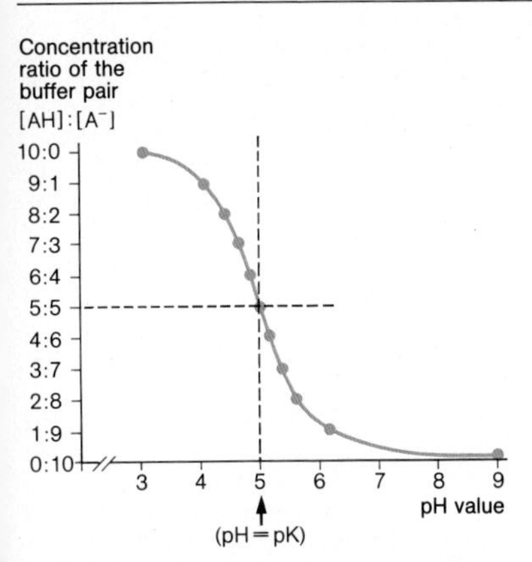

D. Graphic representation of the relationship between *pH value* and the *concentration ratio of buffer acid/buffer base* [AH] / [A⁻]. The values are roughly those of the buffer pair acetic acid/acetate (pK value = 4.7). Optimal buffering is achieved when the pH value of the solution equals the pK value of the buffer, that is, when [AH] = [A⁻] (broken lines)

$+ \log 3/17 =$) 6.25 to ($7 + \log 1/19 =$) 5.7, that is, a change of 0.55 pH units.

The titration of a buffer solution with H^+ (or OH^-) ions can be plotted to give a buffering curve (→ **D**). The steep part of the curve indicates the range over which the buffering action is at its best (i.e. offers the greatest resistance to changes in pH). The pK value lies at the middle of this steep portion (at the point of inflection of the curve). Substances that take up (or lose) more than one H^+ ion per molecule have more than one pK value and can thus exert an optimal buffering action in several regions. Phosphoric acid (H_3PO_4) can give up 3 H^+ ions to form successively, $H_2PO_4^-$, HPO_4^{2-}, and PO_4^{3-}. The buffer pair $HPO_4^{2-}/H_2PO_4^-$ with a pK_a of 6.8 is physiologically important (→ p. 144).

The absolute slope, d(pH)/d[A⁻] of the buffer curve ([A⁻] instead of [AH] : [A⁻] in **D**) is a measure of the **buffer capacity**: $d(pH)/d\,[A^-] = 1/(2.3 \cdot [A^-])$.

Osmolarity, Osmolality, Osmotic and Oncotic Pressure

The **osmolarity** of a solution is the *concentration of osmotically active particles*, irrespective of the substance or substances involved. The unit is the *osmol per liter* (= *osm/l*) (osm is not a SI unit).

Because osmolarity involves volume, and volume is dependent on temperature and on the space occupied by the dissolved substances (plasma roughly 7%, erythrocytes 30%), it is more satisfactory in most cases to employ the term **osmolality**, which involves the mass of solvent: **osm/kg H_2O**. Osmolality (actual or nonideal) is in fact the value measured, and can, as such, be inserted in the *van't Hoff* equation (see below).

The **ideal osmolality** is derived from the mol of the substance in question. If, for example 1 mmol (= 180 mg) glucose is dissolved in 1 kg water (= 1 l at 4 °C), the molality is 1 mmol/kg H_2O and the osmolality is thus also 1 mosm/kg H_2O. The situation is different, however, in the case of electrolytes such as NaCl since they dissociate into ions ($NaCl \rightleftarrows Na^+ + Cl^-$). Each of these ions is osmotically active. Hence, if 1 mmol NaCl is dissolved in 1 kg H_2O, the (ideal) osmolality = molality × number of particles into which one molecule dissociates, and is in this case 2 mosm/kg H_2O. In contrast to NaCl, weak electrolytes are only partially dissociated at physiological pH values, so that the degree of dissociation must also be taken into account.

The above calculation applies only to so-called *ideal*, that is, extremely dilute solutions. The body fluids, however, are **non-ideal solutions** in which the particles are so highly concentrated that they influence one another. In order to convert ideal to non-ideal osmolality, the **osmotic coefficient** (g) has been introduced. This coefficient is dependent on concentration; for NaCl it is 0.926 at (ideal) 300 mosm/kg H_2O. The actual, nonideal osmolality of this solution is thus $300 \times g = 278$ mosm/kg H_2O.

The (real) osmolality is measured with an *osmometer*, which is often based on the principle of the *depression of freezing point*.

Pure water freezes at 0 °C. The greater the number of osmotically active particles dissolved in the water, the greater the depression of the freezing point (ΔT) below 0 °C (this is also the principle underlying the use of salt on the roads in winter to prevent ice formation). Using this (measured) value it is possible to calculate the osmolality of the solution (e.g. plasma: $\Delta T = -0.54\,°C \rightarrow 290$ mosm/kg H_2O).

Solutions with an osmolality equal to that of blood plasma are termed in medicine **iso-osmolal**: those with a higher or lower osmolality are termed **hyperosmolal** and **hypo-osmolal**, respectively.

Tonicity (isotonic, hypotonic, hypertonic), on the other hand, describes the **osmotic pressure** π (in medicine, usually with respect to that of plasma), π is produced by osmotically active particles at **selective membranes**, whereby the selectivity for each of the dissolved substances is described by its **reflexion coefficient** σ (→ p. 10). If the membrane is permeable to the solvent but not to the solute in question ($\sigma = 1$), the membrane is said to be **semipermeable** to the substance concerned in that particular solvent.

If there is a difference in osmolality across the membrane (ΔC_{osm}), the difference in osmotic pressure can be calculated according to *van't Hoff* as:

$$\Delta\pi = \sigma \cdot R \cdot T \cdot \Delta C_{osm};$$

where R is the gas constant ($= 8.314\ J \cdot K^{-1} \cdot osm^{-1}$), T the absolute temperature (in K), and σ the *reflexion coefficient* (→ p. 10). The unit for C_{osm} is mosm/kg H_2O ($\approx$ osm/m^3 solvent water).

An isosmolal solution is hence only isotonic if $\sigma = 1$. Frequently $\sigma < 1$. Cell walls are permeable to urea, for example, and the vessel walls of many organs are (slightly) permeable to proteins. The osmotic (for proteins, oncotic; see below) pressure difference is therefore smaller than would be expected from the concentration difference for urea or protein, respectively.

The osmotic pressure plays an important role at many sites in the organism. As an example, if Na^+ and Cl^- are transported out of the kidney tubule the primary urine becomes slightly hypotonic. An osmotic pressure difference arises; water follows the ions and is also resorbed from the tubules. (The higher the water permeability of the tubule section, the lower the osmotic pressure difference.)

Larger molecules, too (e.g. **proteins** of the blood plasma), exert an osmotic pressure, in this particular case termed **oncotic** or **colloidal osmotic pressure**. It is of such importance in the organism because the capillary walls are (almost) impermeable to the larger protein molecules, whereas water and smaller molecules can readily pass to and fro. Thus the capillaries represent for the proteins the selective membrane that is essential for the osmotic (here, oncotic) pressure to become effective. Here, too, σ (see above) has to be allowed for (→ p. 158).

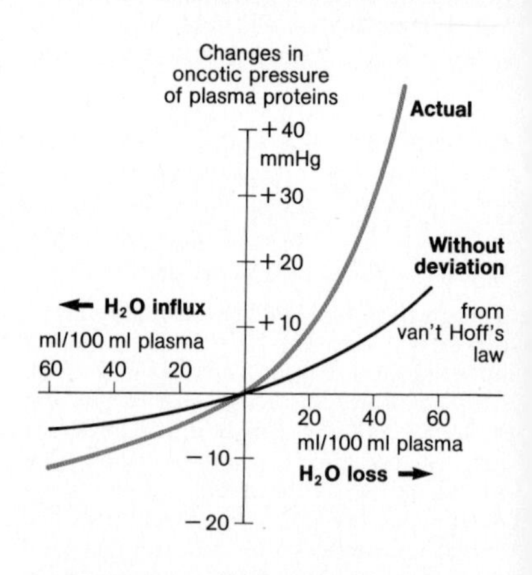

E. Physiological significance of the deviation of the plasma oncotic pressure from the van't Hoff law. Loss of water from the plasma leads to a disproportional rise in oncotic pressure, which counteracts the water loss; conversely, plasma dilution causes only a relatively small reduction in oncotic pressure (From Landis EM, Pappenheimer JR. Handbook of Physiology (Section 2); Circulation, (vol. II). Washington DC: American Physiology Society, 1963: 975.)

If, in spite of this, water escapes from the capillaries (e.g. due to a hydraulic pressure difference; → p. 124 and p. 158), the oncotic pressure rises and increasingly opposes the efflux of water (→ p. 124). The rise in oncotic pressure is appreciably higher than would be expected from van't Hoff's law (→ **E**). This deviation, which plays an important role in maintaining a constant plasma volume, is due to certain biophysical properties of the plasma proteins.

Further Reading

Physiology

Berne RM, Levy MN. Physiology. St Louis: Mosby, 1983.
Berne RM, Levy MN. Principles of Physiology. St. Louis: Mosby, 1990
Guyton AC. Int Rev Physiol 1974–1983: 1–28.
Guyton AC. Textbook of medical physiology. 7th ed. Philadelphia: Saunders, 1986.
McClintic JR. Physiology of the human body. 2nd ed. New York: Wiley, 1978.
Mountcastle VB. Medical physiology. 2nd ed. St Louis: Mosby, 1980. 2 vols.
Rhoades R, Pflanzer R. Human Physiology. Philadelphia: Saunders, 1989.
Ross G. Essentials of human physiology. 2nd ed. Chicago: Year Book, 1982.
Schmidt RF, Thews G. Human physiology 2nd ed. Berlin: Springer, 1989.
Sherwood L. Human physiology. From cells to systems. St. Paul: West Publ., 1989.
Strand FL. Physiology: a regulatory approach. 2nd ed. New York: Macmillan, 1983.
Sturkie PD. Basic physiology. Berlin: Springer, 1984.
Vander AJ, Sherman JH, Luciano DS. Human physiology. 3rd ed. New York: McGraw-Hill, 1980.
Vick RL. Contemporary medical physiology. Reading, MA: Addison-Wesley, 1984.
Weller, H, Wiley RF. Basic human physiology. New York: van Nostrand, 1979.
West JB, ed. Best and Taylor's Physiological basis of medical practice. 11th ed. Baltimore: Williams and Wilkins, 1985.

General and Cell Physiology

Alberts B, Bray D, Lewis J, Raff M, Roberts K, Watson JD. Molecular biology of the cell. New York: Garland, 1983.
Darnell J, Lodish H, Baltimore D. Molecular cell biology. New York: Scientific American Books, 1986.
DeDuve C. A guided tour of the living cell. Vol. 1 and 2. New York: Scientific American Books, 1984
Heinz E. Mechanics and energetics of biological transport. In: Kleinzeller A, Springer GF, Wittmann HG, eds. Molecular biology, biochemistry and biophysics, vol. 29. Berlin: Springer, 1978.
Hopkins CR. Structure and function of cells. Philadelphia: Saunders, 1978.
Schultz SG. Basic principles of membrane transport. Cambridge: Cambridge University Press, 1980. (IUPAB biophysics series, 2.)

Neurophysiology, Muscle

Aidley DJ. The physiology of excitable cells. 2nd ed. Cambridge: Cambridge University Press, 1979.
Blakemore C. Mechanics of the mind. Cambridge: Cambridge University Press, 1977.
Creutzfeld OD. Cortex cerebri. Berlin: Springer, 1983.
Ebashi S, Maruyama K, Endo M. Muscle contraction: its regulatory mechanisms. Berlin: Springer, 1980.
Kandel ER, Schwartz JH. Principles of neural science. 2nd ed. New York: Elsevier, 1985.
Katz B. Nerve, muscle and synapse. New York: McGraw-Hill, 1966.
Kuffler SW, Nicholls JG. From neuron to brain. 2nd ed. Sunderland, MA: Sinauer, 1984.
Ottosen D. Physiology of the nervous system. London: Macmillan, 1983.
Schmidt RF. Fundamentals of neurophysiology. 3rd ed. Berlin: Springer, 1985.

Sensory Physiology

Cagan RH, Kare MR. Biochemistry of taste and olfaction. New York: Academic Press, 1981.
Davson H. The eye. New York: Academic Press, 1962–74. 5 vols.
Gibson JJ. The perception of the visual world. Cambridge, MA: Riverside Press, 1950.
LeGrand Y, El Hage SG. Physiological optics. Berlin: Springer, 1980.
Pickles JO. An introduction to the physiology of hearing. New York: Academic Press, 1982.

Rock J. Perception. New York: Scientific American, 1984.

Schmidt RF. Fundamentals of sensory physiology. 3rd ed. Berlin: Springer, 1986.

Wilson VJ, Melvill-Jones G. Mammalian vestibular physiology. New York: Plenum Press, 1979.

Blood and Immunology

Bellanti JA. Immunology in medicine. 3rd ed. Philadelphia: Saunders, 1985.

Ogston D. The physiology of hemostasis. Cambridge, MA: Harvard University Press, 1983.

Roitt, I. Essential immunology. 6th ed. Chicago: Year Book, 1988.

Respiration

Mines AH. Respiratory physiology. 2nd ed. New York: Raven Press, 1986.

Murray JR. The normal lung: the basis for diagnosis and treatment of pulmonary disease. 2nd ed. Philadelphia: Saunders, 1986.

West JB. Pulmonary physiology: the essentials. 3rd ed. Baltimore: Williams and Wilkins, 1987.

Exercise and Work Physiology

Astrand PO, Rodahl K. Textbook of work physiology: physiological basis of exercise. 2nd ed. New York: McGraw-Hill, 1977.

Lamb DR. Physiology of exercises: responses and adaptations. 2nd ed. New York: Macmillan, 1984.

Shephard RJ. Physiology and biochemistry of exercise. New York: Praeger, 1982.

High Altitude Physiology

Brendel W, Zink RA. High altitude physiology and medicine. Berlin: Springer, 1982.

Sutton JR, Jones NL, Houston CS. Man at high altitude. New York: Thieme-Stratton, 1982.

West JB. High altitude physiology. New York: Van Nostrand Reinhold, 1981.

Heart and Circulation

Berne RM, Levy MN. Cardiovascular physiology. 4th ed. St Louis: Mosby, 1981.

Johnson PC. Peripheral circulation. New York: Wiley, 1978.

Katz AM. Physiology of the heart. New York: Raven Press, 1977.

Noble, D. The initiation of the heartbeat. 2nd ed. Oxford: Clarendon Press, 1979.

Randall WC. Neural regulation of the heart. Oxford: Oxford University Press, 1976.

Shephard JT, Vanhoutte PM. Veins and their control. Philadelphia: Saunders, 1975.

Renal, Fluid, Electrolyte and Acid-Base Physiology

Brenner BM, Rector FC. The kidney. 3rd ed. Philadelphia: Saunders, 1986. 2 vols.

Gamble JL. Acid-base physiology: a direct approach. Baltimore: Johns Hopkins University Press, 1982.

Häusinger, D. pH homeostasis, mechanisms and control. London: Academic Press, 1988

Koushanpour E, Kriz W. Renal physiology. Principles, structure and function. 2nd ed. Berlin: Springer, 1986.

Marsh DJ. Renal physiology. New York: Raven Press, 1983.

Richards P, Trunniger B. Understanding water, electrolyte and acid-base balance. London: Heinemann, 1984.

Seldin, DW, Giebisch G. The kidney: physiology and pathophysiology. New York: Raven Press, 1985. 2 vols.

Sigaard-Andersen O. The acid-base status of the blood. Copenhagen: Munksgaard, 1974.

Smith EKM, Brain EA. Fluids and electrolytes: a conceptual approach. Edinburgh: Churchill-Livingstone, 1980.

Valtin H. Renal function: mechanisms preserving fluid and solute balance in health. 2nd ed. Boston: Little, Brown, 1983.

Willats SM. Lecture notes on fluid and electrolyte balance. Oxford: Blackwell, 1982.

Gastrointestinal Physiology, Nutrition, Digestion, Energy Metabolism

Crane RK. Gastrointestinal physiology. 2nd ed. Baltimore: University Park Press, 1977.

Davenport HW. Physiology of the digestive tract. 5th ed. Chicago: Year Book, 1982.

Johnson LR. Physiology of the gastrointestinal tract. 2nd ed. New York: Raven Press, 1987.

Wieser W. Bioenergetik. Energietransformationen bei Organismen. Stuttgart: Thieme, 1986.

Endocrinology

Degroot LJ. Endocrinology. 2nd ed. New York: Grune and Stratton, 1986. 3 vols.

Hedge GA, Colby HD, Goodman RL. Clinical endocrine physiology. Philadelphia: Saunders, 1987

Williams RH. Textbook of endocrinology. Philadelphia: Saunders, 1974.

Reproduction and Sexual Physiology

Haeberle EJ. The sex atlas. 2nd ed. New York: Crossroad, 1983.

Knobil E, Neill JD. The physiology of reproduction, vol. I and II. New York: Raven, 1988.

Masters WH, Johnson VE. Human sexual response. Boston: Little, Brown, 1966.

Neville MC, Neifert MR. Lactation, Physiology, nutrition and breast-feeding. New York: Plenum, 1983.

Yen SS, Jaffe RB. Reproductive endocrinology: physiology, pathophysiology, and clinical management. Philadelphia: Saunders, 1986.

Animal Physiology

Eckert R, Randall D. Animal physiology: mechanisms and adaptations. 3rd ed. San Francisco: Freeman, 1988.

Schmidt-Nielsen K. Desert animals: physiological problems of heat and water. New York: Dover, 1979.

Schmidt-Nielsen, K. Animal physiology: adaptation and environment. 3rd ed. Cambridge: Cambridge University Press, 1983.

Taylor CR, Johannsen K. Bolis L. A companion to animal physiology. Cambridge: Cambridge University Press, 1982.

Pathophysiology

Altura BM, Saba TM. Pathophysiology of the reticulendothelial system. New York: Raven Press, 1981.

Berlyne GM. A course in clinical disorders of the body fluids and electrolytes. Oxford: Blackwell, 1980.

Brenner B, Coe FL, Rector FC. Renal physiology in health and disease. Philadelphia: Saunders, 1987.

Brooks FP. Gastrointestinal pathophysiology. 2nd ed. Oxford: Oxford University Press, 1978.

Collins RD. Illustrated manual of fluid and electrolyte disorders. 2nd ed. Philadelphia: Lippincott, 1983.

Erslev AJ, Gabuzda TG. Basic pathophysiology of blood. 3rd ed. Philadelphia: Saunders, 1985.

Rose BD. Clinical physiology of acid-base and electrolyte disorders. 2nd ed. New York: McGraw-Hill, 1984.

Rosendorff C. Clinical cardiovascular and pulmonary physiology. New York: Raven Press, 1983.

Ross G. Pathophysiology of the heart. New York: Masson, 1982.

Smith LH, Thier SO. Pathophysiology: the biological principles of disease. 2nd ed. Philadelphia: Saunders, 1985.

Van der Werf T. Cardiovascular pathophysiology. Oxford: Oxford University Press, 1980.

West JB. Pulmonary pathophysiology: the essentials. 3rd ed. Baltimore: Williams and Wilkins, 1987.

Pharmacology

Gilman AG, Goodman LS, Rall TR, Murad F. The pharmacological basis of therapeutics. 7th ed. New York: Macmillan, 1985.

Index

D

H

U